Alert
Ausuittuq
(Grise Fiord)
Resolute
(Qausuittuq)
Ikpiarjuk/Tununirusiq
(Arctic Bay)
Mittimatalik
(Pond Inlet)
Kangiqtugaapik
(Clyde River)
Baffin Bay
Taloyoak
Igloolik
Sanirajak
(Hall Beach)
Qikiqtarjuaq
(Broughton Island)
Uqsuqtuuq
(Gjoa Haven)
Kugaaruk
(Pelly Bay)
Pangnirtung
Naujaat
(Repulse Bay)
Iqaluit
Kinngait
(Cape Dorset)
Salliq
(Coral Harbour)
Qamanittuaq (Baker Lake)
Kimmirut
(Lake Harbour)
Igluligaarjuk (Chesterfield Inlet)
Ivujivik
Salluit
Kangiqsujuaq
Kangiqliniq (Rankin Inlet)
Tikirarjuaq (Whale Cove)
Quaqtaq
Akulivik
Kangirsuk
Arviat
Puvirnituq
Aupaluk
Kangiqsualujjuaq
Tasiujaq
Nain
Kuujjuaq
Hudson Bay
Inukjuak
Hopedale
Makkovik
Postville
Rigolet
Umiujaq
Sanikiluaq
Happy Valley/Goose Bay
Kuujjuarapik

Cover photos: *ITK/Eric Loring; Defence R&D Canada/Janice Lang; Pat Roach*

Published under the authority of the
Minister of Indian Affairs and Northern Development
Ottawa, 2003

www.ainc-inac.gc.ca

1-800-567-9604

QS-8525-002-EE-A1

R72-260/2003-2E

0-662-33468-X

Sources, Occurrence, Trends and Pathways in the PHYSICAL ENVIRONMENT

Editors

T. Bidleman[1], R. Macdonald[2], and J. Stow[3]

Authors and contributors

(in no particular order)

T. Harner; T. Bidleman; L. Jantunen; F. Wania; P. Helm; Y-F. Li; D. Muir; W. Strachan; G. Stern; M. Evans M. Ikonomou; W. Lockhart; D. Rawn; P. Roach; B. Rosenberg; G. Tomy; P. Blanchard; H. Hung; C. Halsall; P. Fellin; L. Barrie; A. Konoplev; B. Rosenberg; S. Friesen; H. Li; R. Bailey; E. Brorström-Lundén; S. Juntto;M. Alaee; C. Cannon; A. Steffen; W. Schroeder; L. Poissant; R. Macdonald; G. Hönninger; G. Lawson; C. Banic; S. Gong; J. Smith; D. Gregor; J. Stow; D. Kennedy; K. Reimer; C. Ollson, Z. Kuzyk; I. Koch; C. Hough; N. Burgess; S. Solomon; M. Diamond; R. Semkin; S. Law; J. Fyfe.

For citation information and individual author and contributor affiliations please see specific chapters in Part II.

ITK/Eric Loring

Acknowledgements

The NCP Phase II studies depended critically on the leadership of Russ Shearer and David Stone who's collaborative approach to designing the blueprints and who's efforts to shepherd the proposed work were monumental; without them, there would have been no program. We also thank the technical review committees for objective assessment of the proposals and for keeping the abiotic work in line with the blueprints. Thanks is also owed to the Northerners who assisted scientists with the collection of samples and data. We are indebted to the administrative personnel at INAC, in particular Jennifer Baizana and Carol Reynolds, for organizing not only the finances but the annual meetings that kept each of us abreast of the other work components. We were grateful throughout the assessment to have a series of accessible, organized Synoptic Reports capably edited by Sarah Kalhok. Finally, we thank three reviewers, Crispin Halsall, Jozef Pacyna, and Don Mackay who read an early draft and provided us with guidance on how to improve the science and its presentation.

[1] Centre for Atmospheric Research Experiments, 6248 Eighth Line, R.R. #1, Essa Township, Egbert, ON, L0L 1N0, Canada

[2] Institute of Ocean Science, Fisheries and Oceans Canada, 9860 West Saanich Road, Sidney, BC, V8L 4B2, Canada

[3] Indian and Northern Affairs Canada, Ottawa, ON, K1A 0H4, Canada.

Executive Summary

The Northern Contaminants Program (NCP) was established in 1991 in response to concerns about human exposure to elevated levels of contaminants in fish and wildlife species that are important to the traditional diets of northern Aboriginal peoples. Early studies indicated that there was a wide spectrum of substances — persistent organic pollutants, heavy metals, and radionuclides — many of which had no Arctic or Canadian sources, but which were, nevertheless, reaching unexpectedly high levels in the Arctic ecosystem.

Under the first phase of the NCP (NCP-I), research was focussed on gathering the data required to determine the levels, geographic extent, and sources of contaminants in the Arctic environment and its people, and the probable duration of the problem. Results generated through NCP-I were synthesized and published in 1997 in the Canadian Arctic Contaminants Assessment Report (CACAR). In 1998, the NCP began its second phase (NCP-II), which will continue until March 2003. NCP-II focussed on questions about the impacts and risks to human health that may result from current levels of contamination in key Arctic food species as well as determining the temporal trends of contaminants of concern in key Arctic indicator species and air. It addressed these issues under a number of subprograms: human health; monitoring the health of Arctic peoples and ecosystems, and the effectiveness of international controls; education and communications; and international policy. The CACAR-II documents NCP-II results in a series of reports that consists of a Highlights report and four technical reports: human health, biological environment, physical environment and knowledge in action.

Under NCP phase-II, studies in the abiotic environment were designed primarily to assist with achieving international controls on contaminants of concern. This was done by advancing research in three priority areas: temporal atmospheric trends, physicochemical processes and modelling, and identification of additional substances of concern. The purpose of this report is to review information that addresses these priority areas and to assess the current state of knowledge on contaminants in the Arctic abiotic environment.

Some key findings which have and will be used by Canada in the context of the POPs Protocol under the Convention on Long-range Transboundary Air Pollution and the Stockholm Convention on POPs include:

Global and regional gridded emissions inventories have been created for some OC pesticides such as HCH and toxaphene. Declines in the global use and emissions of technical HCH are consistent with observed declines of α- and β- HCH in Arctic air, results that are helping to refine global mass balance models. These results also provide strong evidence that international action to reduce global POPs use can have a direct impact on levels in the Arctic. An emissions inventory for toxaphene in the United States suggests that, even though the pesticide was deregistered two decades ago, volatilization of soil residues contributed 360 tonnes of toxaphene to the atmosphere in year 2000.

Decreasing atmospheric trends in most "legacy" persistent organic pollutants (POPs) over the last 5–10 years. The concentration of most POPs which were reviewed in the first CACAR and which have received international attention through the United Nations have been declining in Arctic air. Observations at Arctic monitoring stations, in particular Alert, Nunavut, have confirmed this trend for POPs such as PCBs, DDT, chlordane, and HCH.

The observation of "new chemicals" in the Arctic abiotic environment. A new generation of POPs has been measured in Arctic air, seawater, and freshwater sediments that includes brominated flame retardants (in particular polybrominated diphenyl ethers, PBDEs), perfluorinated alkane compounds (PFA), short chain chlorinated paraffins (SCCPs) and polychlorinated naphthalenes (PCNs). Some pesticides currently used in the circumpolar countries were also identified: endosulfan, trifluralin and methoxychlor. PBDEs have concentrations that are rising. Unlike most other organochlorine pesticides, concentrations of endosulfan in Arctic air have not declined over the last 7 years.

Identification of a dominant oceanic pathway for β-HCH. A comparison of HCH isomer pathways has suggested that most of the β-HCH originating in Asia is deposited in the north Pacific Ocean and then delivered to the Arctic Ocean through the Bering Strait. This hypothesis, if confirmed, implies that other persistent contaminants that partition strongly into water could be delivered to the Arctic by ocean currents.

Some key findings which have and will be used by Canada in the context of the heavy metals Protocol under the Convention on Long-range Transboundary Air Pollution and the assessment for controls on mercury under UNEP include:

New insights into the atmospheric pathways, processes and environmental behaviour of mercury. Under NCP phase-II Canadian scientists discovered a phenomenon we have come to know as mercury depletion events (MDE) wherein atmospheric mercury is converted from the gaseous elemental state to the particulate phase and rapidly deposited on Arctic surfaces during springtime polar sunrise. Enhanced mercury deposition through MDEs could be significantly contributing to the high levels of mercury observed in Arctic biota. Mercury concentrations measured in dated sediment core samples from Arctic lakes suggest that mercury fluxes to the lake sediments has increased by two- to three- fold over the past century.

Relatively stable atmospheric concentrations of anthropogenically derived heavy metals including copper, lead and zinc. Twenty-years of weekly mean concentrations of metals (not including mercury) in Arctic air reveal no significant increasing or decreasing trends for metals derived from anthropogenic sources.

Some key findings which have and will be used by Canada in the context of all of the previously mentioned international agreements include:

Continued development and refinement of contaminant process studies and models for POPs and mercury. Advances in modelling have been enhanced by a growing database of physicochemical properties as a function of temperature. Models of precipitation scavenging of contaminants from air have advanced our understanding of deposition processes. A detailed examination of the Amituk Lake basin has found that small arctic lakes appear not to retain most of the contaminant burden that is delivered from the seasonal snowpack melt, but instead convey the contaminants to other terrestrial and marine systems. Surprisingly rapid rates of microbial decay have been estimated for several POPs in Amituk Lake and for HCHs in the Arctic Ocean. Microbial degradation in freshwater and marine systems is emerging as a dominant removal mechanism for many POPs.

System changes brought on by climate variability or climate change, are having an effect on contaminant pathways. Remarkable changes occurred in the Arctic's climate during the 1990s including wind and weather patterns, ice cover, ice thickness, ice drift patterns, permafrost, hydrology, ocean-currents, precipitation and temperature patterns. These changes have significant consequences for contaminant pathways. These changes have altered the physical pathways that transport contaminants, for example diverting Russian river inflow into the Canada Basin thence to flow out through the Archipelago and altering the drift trajectories of ice within the Arctic. However, even more consequential changes are likely to occur in contaminant magnification pathways including cryoconcentration, attachment to organic-rich particles, and food-web biomagnification. Recent change in the Arctic's ice climate and ecosystem structure require a great deal of caution in interpreting contaminant trend data collected for the past couple of decades.

Résumé

Le Programme de lutte contre les contaminants dans le Nord (PLCN) a été créé en 1991 par suite des inquiétudes que soulevaient les concentrations élevées de contaminants dans les poissons et les autres espèces sauvages qui jouent un rôle important dans le régime alimentaire traditionnel des Autochtones du Nord. Des premières études avaient révélé, en effet, que les concentrations d'un large éventail de substances — polluants organiques persistants, métaux lourds et radionucléides — ne provenant pas, dans bien des cas, de l'Arctique ou du Canada étaient étonnamment élevées dans l'écosystème arctique.

Au cours de la première phase du PLCN (Phase I), les recherches ont été axées sur la collecte des données nécessaires pour déterminer les concentrations, la répartition géographique et les sources des contaminants présents dans le milieu et chez les gens de l'Arctique ainsi que la durée probable du problème. Les résultats de la Phase I ont été présentés dans le *Rapport de l'évaluation des contaminants dans l'Arctique canadien* (RECAC) qui a été publié en 1997. La deuxième phase du PLCN (Phase II) a débuté en 1998 et s'étendra jusqu'en mars 2003. Elle met l'accent sur les conséquences et les risques éventuels pour la santé humaine des concentrations actuelles de contaminants chez les principales espèces consommées, de même que sur les tendances temporelles des contaminants en cause chez des espèces indicatrices et dans l'air de l'Arctique. Divers aspects ont été pris en considération : santé humaine, surveillance de la santé des habitants du Nord, efficacité des mesures internationales de réglementation, éducation et communication, et politique internationale. Les résultats de la Phase II sont présentés dans une série de rapports qui réunissent la Synthèse du RECAC II et quatre rapports techniques portant sur la santé humaine, l'environnement biologique, l'environnement physique et la mise en application des connaissances acquises.

Les études menées en milieu abiotique pendant la phase II étaient essentiellement destinées à aider l'application de mesures internationales de réglementation des contaminants en cause. Pour ce faire, on a privilégié trois domaines sphères de recherche : les tendances temporelles dans l'atmosphère, les mécanismes physicochimiques et la modélisation. D'autres substances préoccupantes ont aussi été décelées. Le présent rapport examine l'information recueillie dans ces domaines et évalue l'état actuel des connaissances relatives aux contaminants présents dans le milieu abiotique arctique.

Parmi les principales constatations qui ont été et seront utilisées par le Canada dans le contexte du protocole sur les POP prévu par la Convention sur la pollution atmosphérique transfrontalière à longue distance et de la Convention de Stockholm sur les POP, citons les résultats suivants :

Création d'inventaires sur des grilles régionales et planétaires des émissions de certains pesticides organochlorés, comme les HCH et le toxaphène. Les baisses de l'utilisation et des émissions mondiales de HCH technique sont compatibles avec les baisses observées α-HCH et de β-HCH dans l'air de l'Arctique, des résultats qui aident à améliorer les modèles de bilan de masse planétaire. Ces résultats donnent également une forte indication que l'action internationale visant à réduire l'utilisation mondiale de POP peut avoir un impact direct sur les concentrations mesurées dans l'Arctique. Un inventaire des émissions de toxaphène aux États-Unis semble indiquer que, même si ce pesticide n'est plus homologué depuis deux décennies, la présence de 360 tonnes de toxaphène dans l'atmosphère en 2000 est attribuable à la volatilisation des résidus au sol.

Régression dans l'atmosphère de la plupart des polluants organiques persistants (POP) « traditionnels » au cours des 5–10 dernières années. Les concentrations de la plupart des POP étudiés dans le premier RECAC, et qui ont fait l'objet d'une attention internationale par l'intermédiaire des Nations Unies, sont en diminution dans l'air de l'Arctique. Les échantillons prélevés aux stations de surveillance arctiques, particulièrement à Alert, au Nunavut, ont confirmé cette tendance en ce qui concerne les POP comme les BPC, le DDT, le chlordane et les HCH.

Observation de « substances chimiques nouvelles » dans l'environnement abiotique de l'Arctique. Des POP de nouvelle génération ont été mesurés dans l'air et les sédiments en eau de mer et en eau douce de l'Arctique, parmi lesquels on trouve des ignifuges bromés (particulièrement des polybromodiphényléthers — PBDE), des alcanes perfluorés, des chloroparaffines à courte chaîne (CPCC) et des polychloronaphthalènes (PCN). La présence d'endosulfan, de trifluraline et de méthoxychlore, pesticides actuellement en usage dans les pays circumpolaires, a également été constatée. Les concentrations de certains PBDE sont en hausse. Contrairement à la plupart des autres pesticides organochlorés, l'endosulfan dans l'air de l'Arctique montre des concentrations qui n'ont pas diminué au cours des 7 dernières années.

Caractérisation d'un mode dominant de circulation océanique du β-HCH. La comparaison des voies de circulation d'un isomère de l'HCH indiquerait que la majeure partie du β-HCH provenant d'Asie est déposée dans le Pacifique Nord, puis est acheminée vers l'océan Arctique par le détroit de Béring. Cette hypothèse, si elle est confirmée, implique que d'autres contaminants persistants qui subissent une forte séparation dans l'eau pourraient être transportés dans l'Arctique par les courants océaniques.

Parmi les principales constatations qui ont été et seront utilisées par le Canada dans le contexte du protocole sur les métaux lourds prévu par la Convention sur la pollution atmosphérique transfrontalière à longue distance et de l'évaluation des mesures de réglementation du mercure sous l'égide du PNUE, citons les résultats suivants :

Nouvelles informations sur les voies atmosphériques, les processus et le comportement environnemental du mercure. Pendant la Phase II du PLCN, les scientifiques canadiens ont découvert un phénomène selon lequel le mercure atmosphérique est converti de la forme élémentaire gazeuse à l'état particulaire et est rapidement déposé sur les surfaces arctiques au moment du lever de soleil polaire au printemps. Cette intensification du dépôt pourrait contribuer de façon appréciable aux concentrations élevées de mercure qui sont observées chez le biote arctique. Les concentrations mesurées dans les carottes datées de sédiments de lacs arctiques semblent indiquer que les flux de mercure dans les sédiments lacustres ont doublé ou triplé au cours du dernier siècle.

Stabilité relative des concentrations atmosphériques de métaux lourds d'origine anthropique, y compris du cuivre, du plomb et du zinc. Les concentrations hebdomadaires moyennes de métaux sur une période de 20 ans (excluant le mercure) dans l'air de l'Arctique ne révèlent aucune augmentation ou diminution significative pour ce qui est des tendances relatives aux métaux d'origine anthropique.

Parmi les principales constatations qui ont été et seront utilisées par le Canada dans le contexte des accords internationaux déjà mentionnés, citons les résultats suivants :

Développement et perfectionnement des études et des modèles des processus des contaminants pour ce qui est des POP et du mercure. La modélisation a tiré profit d'une base croissante de données sur les propriétés physicochimiques en fonction de la température. Les modèles de lessivage par les précipitations des contaminants dans l'air nous ont permis de mieux comprendre les mécanismes de dépôt. Un examen détaillé du bassin du lac Amituk a révélé que les petits lacs de l'Arctique ne semblent pas retenir la majeure partie des contaminants qui y sont déposés par la fonte saisonnière de la neige accumulée; les contaminants sont plutôt acheminés vers d'autres systèmes terrestres ou marins. Étonnamment, des taux rapides de dégradation microbienne ont été prédits pour plusieurs POP dans le lac Amituk et pour les HCH dans l'océan Arctique. La dégradation microbienne se dessine comme le principal mécanisme d'élimination de nombreux POP de l'eau douce et de l'eau de mer.

Les changements systémiques résultant de la variabilité ou des changements du climat ont un effet sur les trajectoires des contaminants. Le climat de l'Arctique a subi de formidables changements au cours des années 1990 qui ont touché notamment les configurations des vents et les conditions atmosphériques, la couverture de glace, l'épaisseur de la glace, les dérives des glaces, le pergélisol, l'hydrologie, les courants océaniques, les précipitations et les températures. Ces changements ont des conséquences importantes pour les voies de transport des contaminants. Les voies physiques ont été modifiées. Ainsi, l'eau des fleuves russes allant dans le bassin Canada a été détournée et s'écoule à travers l'archipel, ce qui modifie les trajectoires de dérive de la glace dans l'Arctique. D'autres répercussions sont également probables sur les trajectoires des contaminants, y compris la cryoconcentration, la fixation à des particules organiques et la bioamplification dans le réseau alimentaire. Compte tenu du changement récent de la structure de l'écosystème et des glaces de l'Arctique, il faut faire preuve d'une grande prudence lorsqu'on analyse les données sur les tendances des contaminants qui ont été recueillies au cours des deux dernières décennies.

Table of Contents

ITK/Eric Loring

D Special Topic 225

E References 281

List of Figures

List of Tables

ITK/Eric Loring

ITK/Eric Loring

PART I

ITK/Eric Loring

ITK/Eric Loring

CHAPTER 1

Introduction

Progress in Understanding Abiotic Pathways and Processes Under the Northern Contaminants Program, Phase II (1997–2002).

The primary goal of the Northern Contaminants Program (NCP) has been to develop a sound scientific basis by which 1) to identify pollutants that pose risks to northern ecosystems and Northerners, and, 2) to bring about international agreements to reduce or eliminate the sources of such chemicals. Under NCP Phase I (1991–1996) five categories of contaminants were studied including chlorinated industrial organic compounds, organic pesticides, polycyclic aromatic hydrocarbons, metals, and radionuclides (Tables 1.1 and 1.2), and the results were summarized in the first Canadian Arctic Contaminants Assessment Report (CACAR; Jensen *et al.*, 1997). The scientific studies conducted under NCP Phase I not only provided considerable understanding about contaminant pathways to and within the Arctic but they also provided the crucial underpinning to convince the international community to institute regional controls for persistent organic pollutants (POPs) and heavy metals, first under the United Nations Economic Commission for Europe (UN-ECE) Convention on Long-Range Transboundary Air Pollution (LRTAP) and subsequently under the global United Nations Environment Programme (UNEP) Stockholm convention on POPs. The abiotic studies in particular were instrumental in defining contaminant levels in abiotic Arctic media, in revealing the pathways connecting these contaminants back to their sources and in defining many of the processes that make the Arctic particularly vulnerable to persistent, semi-volatile contaminants.

TABLE 1.1 Arctic environmental contaminants of concern in 2002

1. Chlorinated industrial organic compounds
Chlorobenzenes (ClBz)
Polychlorinated biphenyls (PCBs)
Dioxins/Furans (TCDD/Fs)
2. Organic pesticides
Chlorobornanes (toxaphene, CHBs)
Hexachlorocyclohexanes (HCHs)
Chlordane
DDT/DDE
Aldrin/Dieldrin
Endosulphan
3. Polycyclic Aromatic Hydrocarbons (PAHs)
e.g., Benzo[a]pyrene (B[a]P)
4. Metals
Cadmium
Lead
Mercury
5. Radionuclides
Cesium (^{137}Cs)
Plutonium ($^{239,240}Pu$)
Iodine (^{129}I)
Strontium (^{90}Sr)
6. "New Chemicals"
Brominated fire retardants (BFRs)
Polybrominated diphenyl ethers (PBDEs)
Short chain chlorinated paraffins (SCCPs)
Polychlorinated naphthalenes (PCNs)
Octachlorostyrene (OCS)
Pentachlorophenol (PCP)
Trifluralin
Mirex
Methoxychlor

TABLE 1.2 Estimates of production and usage of selected persistent organic pollutants (POPs)

	Source/Function	Production/Use/Emissions	Scale/Time period
Industrial organohalogens			
Polychlorinated biphenyls (PCBs)	"Open" usage (plasticizers, inks, paints, sealants, etc.), capacitors and transformers, waste combustion and open burning, soils	Total PCB use 1300 kt. Production of Σ22 congeners = 467 to 703 kt, with 566 kt as mid-value; Emission of Σ22 congeners = 0.4 to 91.7 kt, with 7.7 kt as mid-value (Breivik *et al.*, 2002a, b)	Global, cumulative 1930–2000
Hexachlorobenzene (HCB)	Fungicide, contaminant of other pesticides, industrial byproduct, combustion (Bailey, 2001)	Emissions 0.012 to 0.090 kt/y, with best estimate of 0.023 kt/y (Bailey, 2001)	Global, mid-1990s
Short chain chlorinated paraffins (SCCPs)	Extreme temperature additives in metal working fluids, paints, sealants, leather working industry	Production 45 to 75 kt/y (Muir *et al.*, 2001)	Global, cumulative 1980–99
Polybrominated diphenyl ethers (PBDEs)	Flame retardant additives to polymers and resins	Market demand: penta-BDE 8.5 kt, octa-BDE 3.8 kt, deca-BDE 54.3 kt (DeWit, 2002). Production of total PBDEs: 67 kt (Alaee *et al.*, Section B4.3)	Global, 1999
Polychlorinated naphthalenes (PCNs)	Emissions from the same source categories as PCBs; PCNs are also produced incidentally in combustion processes and have been found in commercial mixtures of PCBs	Global, cumulative production = 150 kt (Falandysz, 1998). Production in USA: 3.2 kt in 1956, 0.32 kt in 1978; Production in Germany: 75 kt (Crookes and Howe, 1993)	Various countries and times
Combustion-related and by-product chemicals			
Octachlorostyrene (OCS)	Industrial by-product of chlorine and magnesium manufacturing		
Polycyclic aromatic hydrocarbons (PAHs)	Products of incomplete combustion, with both anthropogenic and natural contributions		
Polychlorinated dibenzo-*p*-dioxins and dibenzofurans (PCDD/Fs)	Products of incomplete combustion, with both anthropogenic and natural contributions		
Organochlorine pesticides			
Toxaphene (CHBs)	Insecticide	Production 490 kt, emission from soils 0.36 kt (Li *et al.*, 2001b; Section A.2)	USA, cumulative production 1947–86, emissions in 2000
Chlordane	Insecticide	Distribution of 1.4 to 1.6 kt (USHHS, 1993)	USA, 1986
Heptachlor	Insecticide		
Technical hexachlorocyclohexane (HCH)	Insecticide	Usage 9700 kt, emission of α-HCH = 4300 kt. Emissions of α- and β-HCH in 1990 = 44 kt and 0.083 kt; (Li, 1999b; Section A.2)	Global, cumulative 1947–97
Lindane	Insecticide	Usage: Europe, 1970–96 = 81 kt, Canada, 2000 = 0.47 kt, USA, 1990 = 0.11 kt, 1991–93 = 0.028 kt, Mexico 1990 = 0.26 kt. Production in China, 1991–2000 = 11.4 kt. Global usage in 2000 > 3 kt (Breivik *et al.*, 1999; Li *et al.*, 1998a; Section A.2,; Willett *et al.*, 1998)	Various countries and times
DDT	Insecticide	Usage: China, 1951–83 = 260 kt, India, 1948–89 = 330 kt, Japan, 1948–70 = 30 kt (Li *et al.*, 1999; Section A.2)	Various countries and times
Aldrin/Dieldrin	Insecticide	Aldrin usage: 1966 = 8.6 kt, 1970 = 4.8 kt, 1981–85 = 0.45 to 0.68 kt/y. Dieldrin usage: 1966–70 = 0.30 to 0.45 kt/y (USHHS, 1991)	USA, various years
Endrin	Insecticide	Production, 1971–80 = 0.0022 to 0.45 kt, imports 0.021 kt in 1972 (USHHS, 1990)	USA
Endosulfan	Insecticide	Usage in USA, 1992–93 = 0.81 kt; in Ontario 1993 = 0.025 kt (NAFAP, 2000; Harris *et al.*, 2000)	USA and Ontario, 1990s
Pentachlorophenol (PCP)	Wood preservative, industrial by-product	Usage: 400 kt for wood treatment (Lorber *et al.*, 2002)	USA, cumulative 1970–95
Trifluralin	Herbicide		
Mirex	Insecticide, flame retardant	Production in 1972 = 0.019 kt (USHHS, 1993)	USA
Methoxychlor	Insecticide	Production 0.14 to 0.27 kt/y (USHSS, 1993)	USA, 1988–92

In 1998 the UN-ECE LRTAP POPs and heavy metals Protocols were signed by 35 member states and the European Union (for a total of 36). The Protocols will enter into force once 16 countries have ratified it. As of October 1, 2002, 13 countries, including Canada had ratified the Protocols on POPs and heavy metals. The goal of the POPs Protocol is "to control, reduce, or eliminate discharges, emissions, and losses of persistent organic pollutants" (Article 2). To date, the POPs protocol covers 16 organochlorine compound classes; some are subject to outright bans on production and use (aldrin, chlordane, chlordecone, dieldrin, endrin, hexabromobiphenyl, mirex, and toxaphene) while others are scheduled for elimination at a later date (DDT, heptachlor, hexachlorobenzene, and PCBs). Presently, the protocol severely restricts use of DDT, HCHs (including lindane) and PCBs and, furthermore, emissions of PCDD/Fs, PAHs, and hexachlorobenzene are to be reduced to below 1990 levels. The UNEP Stockholm Convention on POPs, which is a legally binding global agreement, had 151 signatories and 22 ratifications as of October 4, 2002. The Stockholm convention covers 12 of the same compounds covered by the UN-ECE protocol and requires ratification by 50 countries before it will enter into force. Canada was the first country to have ratified the Stockholm Convention.

1.1 Findings of the abiotic studies during NCP-Phase I (1991–1996)

The review of progress made during NCP-Phase I studies of the sources, occurrence, and pathways of contaminants entering the Arctic (Barrie *et al.*, 1997) was organized logically around a conceptual model that presented the Arctic as a set of compartments (boxes) which were connected to each other by exchanges (arrows) as shown in Figure 1.1. Under this model, our understanding of any particular contaminant is reflected by how well we can estimate or predict its inventory in the boxes and the rates of its exchange between boxes. Determining inventories requires direct measurements of contaminants in abiotic compartments such as Arctic air, aerosols, water, suspended sediments, soils and sediments, and many such measurements were made during NCP Phase I. However, determining fluxes, which is fundamental to predictive modelling, provides a more severe test of our understanding. It requires studies of chemical processes, both thermodynamic (e.g., partitioning) and kinetic (e.g., exchange or degradation), together with studies of environmental transport rates (e.g., winds, precipitation, ice drift, ocean-currents). These kinds of studies are much more difficult to do, especially in harsh, remote environments like the Arctic and, due to their enormous scope, cannot be conducted in their entirety by the relatively small group of researchers working within the NCP. Accordingly, wherever possible, CACAR I drew on knowledge from a wider Arctic community to conduct its assessment.

ITK/Eric Loring

ITK/Eric Loring

Scientific progress in our knowledge of Arctic contaminant pathways during NCP-Phase I was inferred by Barrie *et al.*, (1997) from progress made in the difficult, but crucial, tasks of quantifying the arrows in Figure 1.1 (cf. Barrie *et al.*, 1997; Table 1.1). Illustrating both the utility and difficulty of constructing even simple models, Barrie *et al.*, (1997) provided global, multi media models and box models for α- and γ-HCH, compounds for which there is the most complete data base.

Based primarily on the this, Barrie *et al.*, (1997) summarized the following critical gaps in knowledge.

1. Temporal trends in air. The lack of temporal trend information in air for most contaminants was identified as the most significant knowledge gap. It was concluded that the most effective medium for temporal trend monitoring is air, where effects of controls will be seen first. Temporal trend data are limited for most persistent organochlorines (OCs) and metals. Weekly air sampling of organochlorines (OCs) and PAHs has been carried out since 1992 at Alert, Nunavut. Sampling for a two-year period (1993–95) was conducted at Tagish, Yukon and Cape Dorset, Nunavut and at Dunai Island in the Russian Arctic. It was determined that further continuous air sampling was needed at Alert, Nunavut and over a two year period at Cape Dorset, Nunavut and at Amderma, Russia.

   Temporal trend data for mercury was judged to be particularly important given evidence of increasing concentrations in the Arctic environment, particularly in marine mammals and seabirds. Continuous measurements of atmospheric mercury were initiated at Alert, Nunavut in 1995. It was determined that further continuous air sampling was needed at Alert, Nunavut, Nunavik and at Amderma, Russia.

2. Predictive Modelling. Modelling has served a number of important purposes, including clarifying the global dimensions to the POPs emission/deposition relationships, and the consequential necessity for emission controls on a hemispheric or global scale. There continues to be a need for modelling in support of international control agreements under the UNEP Stockholm Convention on POPs and the UN-ECE LRTAP Protocols on POPs and heavy metals. These models provide a predictive tool to estimate changes in eventual ambient levels as a result of using particular control strategies.

3. Physical and Chemical Studies. Major knowledge gaps were identified in the availability of reliable physical and chemical properties of contaminants (e.g., partition coefficients, vapour pressures) and these gaps hinder the estimation of a) atmospheric transportation; b) fluxes to improve Arctic mass balance models such as those in the marine environment;

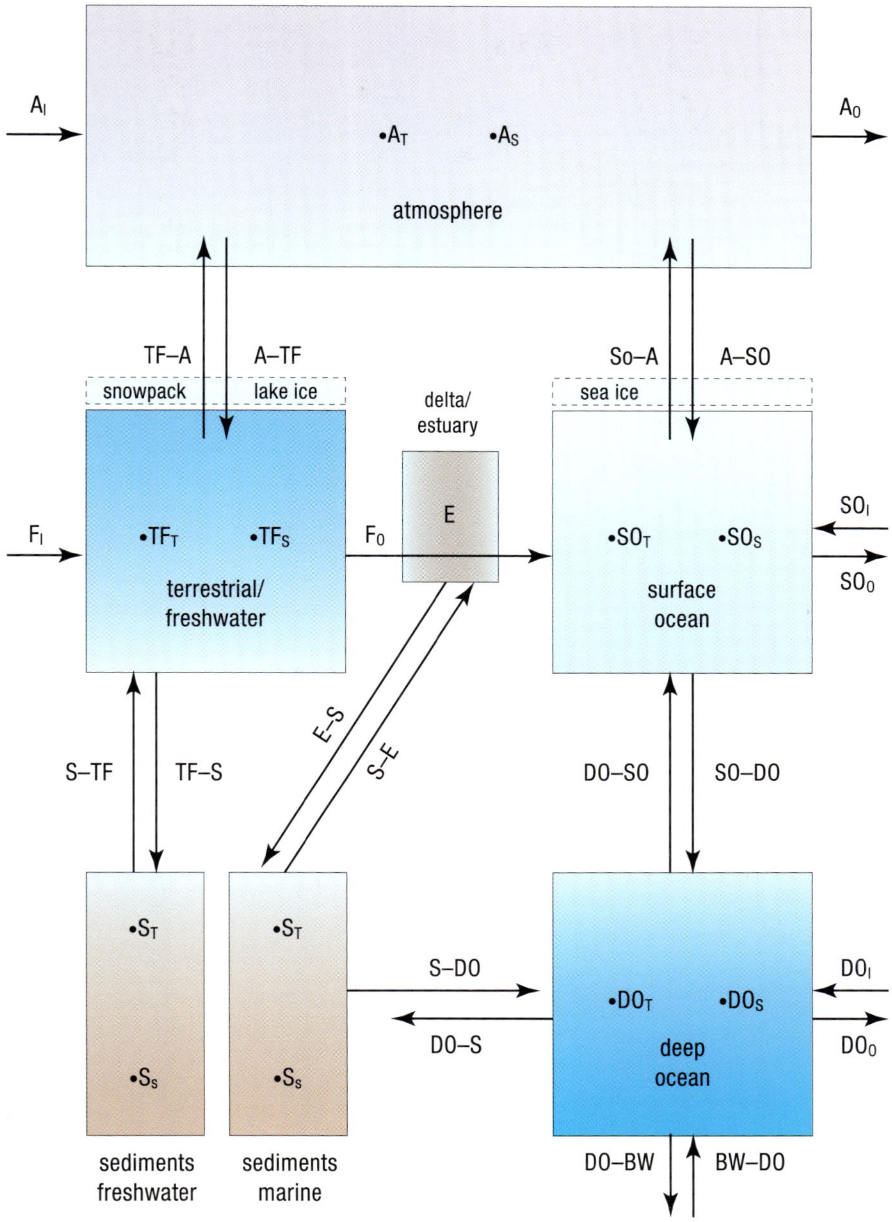

A	atmosphere		SUBSCRIPTS
F	fluvial	I	input
TF	terrestrial–freshwater	O	output
SO	surface ocean	T	transformation
DO	deep ocean	S	internal sources
BW	bottom water		
S	sediments		
E	estuary		

FIGURE 1.1

A box model of the "Arctic System" components and pathways of contaminants.

c) the importance of processes and pathways leading to contaminant availability to biota, particularly aquatic species; and d) the transfer of mercury between the atmosphere and the ocean and its fate in these two environments. It was further recommended that components of HCH and mercury cycles in the Arctic should become priority research topics. Uncertainties in physical and chemical properties of HCH compounds have made it difficult to determine appropriate optimal international control strategies. Humans have augmented the amount of mercury cycling by perhaps a factor of two to three, however, the relative roles of anthropogenic loading and natural processes remains unclear. Of particular concern are the atmospheric mercury depletion events which have the potential to focus globally cycling mercury into Arctic aquatic environments.

4. Other Priority Chemicals. While the list of priority substances monitored in the abiotic environment was substantial there were other chemical contaminant groups for which information was severely limited. These groups included PCDD/Fs, non-*ortho* PCBs, toxaphene, other planar OCs such as polychlorinated naphthalenes (PCNs), polychlorinated and polybrominated diphenyl ethers, current-use pesticides. Setting priorities for which chemical to look for under future studies would require knowledge of the physicochemical properties, toxicity, production, use, and release of these chemicals.

5. Modelling of chemical fate and bioaccumulation. The lack of integration of chemical measurement data with well-established models of chemical fate and food-chain bioaccumulation was identified as a major gap in the program. A future goal is to use the existing database, along with further data collections, to calibrate models and to link chemical fate/distribution model outputs to contaminant bioaccumulation.

1.2 Abiotic studies during NCP–Phase II (1997–2002)

Based on consultation with Northerners and on the findings of NCP-Phase I, blueprints produced for the NCP–Phase II identified three priority areas of study for monitoring and modelling of the physical environment [NCP blueprint 2002–03]:

- Temporal Trend Identification
- Physical and Chemical Studies
- Other Chemicals of Potential Concern

New findings are reviewed under NCP–Phase II, and the current state of knowledge addressing these priority areas is assessed. Accordingly, most of the data considered herein are drawn from the "Abiotic Monitoring and Modelling" subprogram of the Northern Contaminants Program (1997–2002), although relevant published and unpublished results from other contaminant studies conducted in the Canadian Arctic during this period have also been considered. This review takes, as a point of departure, assessments of the contaminants in abiotic media of the Arctic from 1985 to 1996 (Barrie *et al.*, 1992; Barrie *et al.*, 1997). Nevertheless, we include here revisions and additions made to the latter assessment during revision for open literature publication are included (Macdonald *et al.*, 2000). In keeping with the NCP mandate, the geographic scope is restricted mainly to the Yukon, NWT, Nunavut, and coastal Nunavik and Labrador.

ITK/Eric Loring

1.2.1 Organization of the abiotic assessment

Although most of the subchapters of this abiotic assessment can be directly linked to components in the *box and fluxes* scheme (Figure 1.1), the structure of the abiotic assessment is not as closely tied to the scheme in this figure as was the previous assessment. This is partly because the program was considerably re-focussed in the Phase II blueprints and partly because recent findings such as mercury depletion events warranted special treatment. Accordingly, this review is presented in four sections.

1.2.1.1 Section A: Properties and sources

Section A updates our knowledge of chemical properties and emissions and fills an identified gap in our ability to model the transfer of compounds between media. The new data are required especially for the construction of box models and predictive models. Usage patterns for HCH, toxaphene and DDT are also updated. Global emission patterns and trends are crucial forcing components for models and also provide the background against which to evaluate atmospheric time series in the Arctic. The comparison of emissions with atmospheric trends provides a key evaluation of the success of, or compliance to, recent POPs usage protocols. Finally, riverine inputs to the Arctic are summarized. The Russian Rivers in particular were identified as a potentially important, but poorly constrained, component in mass budgets for chemicals like HCHs and PCBs (cf. Macdonald *et al.*, 2000; 2001).

1.2.1.2 Section B: Occurrence and trends

Section B essentially updates our knowledge of the concentrations of contaminants in various boxes in Figure 1.1 and, by directly addressing the first bulleted blueprint item, recent trends in contaminants within those boxes. The section goes far deeper into what drives some of these recent trends, however, especially for mercury. Concentrations and distributions of legacy POPs (HCHs, PCBs, Toxaphene, chlorobenzenes) are reviewed and updated from the last assessment. New chemicals of concern, the third bulleted blueprint item, have specifically been addressed including polybrominated diphenyl ethers (PBDEs) and other brominated fire retardants, polychlorinated naphthalenes (PCNs), short-chain chlorinated paraffins (SCCPs) and other modern pesticides.

1.2.1.3 Section C: Process studies and modelling

Section C provides detailed investigations of transfer processes involving snow and ice. These processes, which are unique to the cryosphere, have enormous potential to transfer and concentrate POPs and metals into Arctic or alpine water or soils. During the previous assessment, snow and ice processes were identified as a key knowledge gap needing proper incorporation into models. Building on the previous assessment, lake sensitivity to mercury and POPs is evaluated by constructing budgets for Amituk Lake, and α- and β-HCH environmental data are used to infer environmental pathway controls on their transport. The global transport and fate of α-HCH and PCBs are evaluated using model simulations. Finally, new model approaches are presented to screen chemicals for environmental persistence based on physicochemical properties.

1.2.1.4 Section D: The interaction of climate change and contaminants

This chapter, which is a considerable departure from the previous assessment, addresses the problem of interpreting contaminant trends and of modelling contaminant distributions in systems subject to change. The box and flux conceptual model (Figure 1.1) used for previous assessments (Barrie *et al.*, 1997; Macdonald *et al.*, 2000) implies that contaminants can be understood as a transient emission released to an environment composed of fixed boxes and arrows. Section D discusses how arrows and boxes are modified under climate variability and global change. In particular, recent changes in the Arctic's physical environment, forced by atmospheric pressure variation (the Arctic Oscillation), are used to demonstrate that transfer functions between physical media have also changed. Although the primary forcing and effects of change are abiotic, large impacts are likely to occur in biological systems and these too affect contaminant transfer, transport, and exposure. For completeness, this section therefore addresses both abiotic and biotic components of change.

ITK/Eric Loring

CHAPTER 2

Summary

2.1 Properties and sources

Contaminants may be either substances unique to human activities whose presence in the environment leaves no doubt as to their artificial origin (e.g., DDT, PCB, toxaphene, ^{239}Pu) or they may be substances produced or released by human activities but which load onto long-standing natural cycles (e.g., Hg, PAHs). "Contamination" does not necessarily imply adverse effects, it merely describes the detection of a substance in the environment that results from human activities. If it can be established that a contaminant is present at a sufficient concentration to cause a deleterious effect(s), then the contamination is, by convention, referred to as pollution (Macdonald and Bewers, 1996). Strictly speaking, the NCP pathways work involves the identification and tracking of contaminants in the Arctic. The judgment of whether these contaminants at the levels found, or predicted by models, cause toxic effects or harm to amenities falls correctly under the biological and human health components of the NCP. Accordingly, the term "contaminant" is used herein, recognizing that some of these chemicals may simply be tracers of human activities whereas others may be shown to have environmental consequences (pollutants). Indeed, we other NCP studies highlight risks associated with particular contaminants found in the abiotic environment, such as those designated as persistent organic pollutants (POPs) and mercury.

2.1.1 Physicochemical properties of contaminants (Harner *et al.*, Section A.1)

The physical and chemical characteristics of a contaminant define how it will interact with the environment. Advances in quantifying partition coefficients and degradation rates in relation to temperature have been incorporated into box models for predicting trends in the global transport and fate of contaminants. This information feeds directly into models which can be used to predict trends. Such models help to compensate for the absence or scarcity of actual field measurements which are often impractical to obtain. Physicochemical data also allow for the evaluation of new (or old, but poorly understood) chemicals and their ability to persist and be transported in the environment, which will lead to improved screening processes.

In the past few years, more physicochemical data have been determined experimentally for several classes of POPs and new compounds of concern. Quantitative Structure Property Relationships (QSPRs) and other estimation techniques are also being developed and used increasingly to provide necessary data where specific measurements have not yet been made.

The transfer of non-polar organic contaminants from one medium to another is described by three key partition coefficients: the octanol-air partition coefficient (K_{OA}); the octanol-water partition coefficient (K_{OW}); and the air-water partition coefficient (Henry's law constant, or K_{AW}). Recent determinations of K_{OA} and K_{AW} for POPs emphasize the development of relationships that cover temperature ranges relevant to transport and deposition in the Arctic. For example, determination of the first temperature-dependent K_{AW} value for β-HCH allows its air-water exchange behaviour in the Arctic to be assessed.

The availability of K_{OA} values has led to semi-empirical K_{OA}-based models of soil-air exchange, of particular interest for OC pesticide residues in soil, that improve the predictive ability of fate and transport models. Once contaminants reach the Arctic, relationships derived for particle-gas partitioning using K_{OA} are useful for estimating rates of deposition.

Also of importance are the rates at which chemicals are transformed or destroyed by hydrolysis, photolysis, and microbial degradation. Because these rates tend to be slower at colder temperatures, chemicals are generally more resistant to chemical degradation in colder climates. However, microbial communities indigenous to the Arctic have proven to be very effective at degrading α- and γ-HCH in Arctic wetlands and lakes, probably contributing the dominant removal mechanism of these HCHs from the Arctic Ocean.

2.1.2 Usage and emissions of organochlorine pesticides (Li *et al.*, Section A.2)

A major focus of the NCP has been to identify global sources of contaminants and quantify emissions for use in global mass budget models. Most of this information, which includes, historic, present, and predicted global use or sales, has been obtained from United Nations and government reports, scientific publications, Battelle Europe, the International Registry of Potentially Toxic Chemicals (IRPTC), and some joint international projects. Under NCP-I, rough estimates of total global use were made for several compounds including PCBs, DDT, toxaphene, lindane, chlordane, aldrin, dieldrin, and endosulfan. Further detail was provided for global use of technical HCH, the most abundant contaminant found in the Arctic. The use of HCH, estimated by country, was gridded globally (1° lat x 1° long) for annual use for α- and γ-HCH for 1980 and 1990.

Under NCP-II the inventory of organochlorine pesticides has been updated and enhanced with the addition of data for β-HCH [globally gridded (1° lat x 1° long)], toxaphene, and DDT. Additionally, estimates of actual emissions to the atmosphere have been made on a global scale for HCHs and in the United States for toxaphene. Estimates of residues left in US agricultural soil were also made for toxaphene and mapped on a grid (1/6° lat x 1/4° long). This new information on use and emissions is critical for predicting trends in contaminant levels in the Arctic.

In most parts of the world, technical HCH was replaced by lindane (γ-HCH) in the late 1970s and 1980s thereby reducing the amount of α- and β-HCH emissions. France was the major user of lindane in Europe and in the world from 1992–1997, with an annual average consumption of 1.6 kt. In 1998, the use of lindane in France was stopped. Until recently, Canada had been the largest known circumpolar user of lindane, mostly for dressing canola seed. An agreement was reached in 2000 between the Canadian Canola Council, the Pest Management Regulatory Agency (PMRA) and pesticide companies who market canola seed treatments containing lindane to cease production of lindane-treated seed. Lindane has also been used in the United States, where it is difficult to estimate exactly how much is being used due to the proprietary nature of commercial seed preparations. Although the use of lindane has declined in recent years, these North American sources as well as sources in Asia and Europe have continued to supply lindane to the Arctic through long-range transport.

Many countries report having banned or severely restricted the use of toxaphene; however, there are 21 countries that appear not to have restricted its use. Whether toxaphene is actually used today in these countries is not known and no updates on global usage are made in NCP-II.

Soil residues of OC pesticides from past and present use in agricultural areas of the world continue to act as sources to the atmosphere and the Arctic through long-range atmospheric transport. The exchange of pesticide residues between soil and air is well described by the K_{OA} or vapour pressure of the chemical. Based on gridded estimates of pesticide distribution and physico-chemical properties, emissions estimates can be derived and used in contaminant transport and fate models. Even though toxaphene was banned in 1986, it is estimated that 29 kt of toxaphene remains in US agricultural soils and that in 2000, an estimated 360 t of toxaphene was emitted from these soils.

2.1.3 Organochlorines in Russian rivers and loadings to the Russian northern seas (Strachan *et al.*, Section A.3)

In the mid-1990s the OC pesticides HCH and DDT were measured in major north flowing Russian rivers and compared to levels for rivers from Canada and Norway. While there has been no substantial study of POPs in rivers in these countries since, some reworking of the Russian data has led to a more complete interpretation.

Comparison of data collected in the 1980s and 1990s shows a decline in loadings of HCH and DDT isomers that coincides with reductions in technical HCH use and a switch to lindane as well as a complete cessation of DDT use for agricultural applications. In general, rivers flowing into the Kara Sea, including the Ob, Nadym, Pur, Taz, and Yenisei, delivered a much greater load of HCHs than rivers flowing into the White and Barents Sea and the Laptev and East Siberian seas. The Ob and, to a lesser extent, Yenisei rivers are the only two that drain agricultural areas in Russia that would have received high pesticide applications. Pesticide use in support of forestry and mining, or other applications, is thought to have contributed to the contaminant load of the other three rivers. Based on observations of HCH concentrations in the Ob, and historic application rates on agricultural land, water concentrations appear to react quickly to changes in use patterns.

For the period of study, the Kara Sea had received 90% of the Russian contaminant discharge to Arctic seas of which the Ob represents 50%, 53% and 72% for each of α-HCH, γ-HCH and *p,p'*-DDT, respectively. The total annual loadings for α and γ-HCH were estimated at 25 tonnes and 44 tonnes, respectively. These values are comparable to estimated annual loadings from flow through Bering Strait (52 and 12 t) and atmospheric deposition (53 and 10 t for α and γ-HCH, respectively).

2.1.4 Local sources of contaminants in the Canadian Arctic (Gregor *et al.*, Section B.9)

Studying local sources of contamination in Canada are not within the mandate of the NCP, but are being assessed by other organizations within INAC as well as other federal and territorial departments. Despite this, there are several reasons to address local sources in this report. The NCP is interested in any information concerning contaminant levels in the North, especially as it relates to traditional/ country foods and spatial trends. Local sources of contamination, in particular contaminated sites, raise serious concerns at a local level. These concerns are often heightened because many of these sites, including Distant Early Warning line (DEW-line) stations, mines, and other industrial sites, are located close to communities. When exchanging contaminants-related information with Northerners the subject of local contaminants concerns is consistently raised and the NCP must be prepared to address these issues.

It is also important to account for local sources of contamination from the perspective of their contribution to global sources. As the direct emissions of many "past use" POPs declines, or is terminated altogether, the remaining global sources of these pollutants are becoming more dominated by soil residues and contaminated sites (Li *et al.*, Section A.3, Blanchard *et al.*, Section B.1, Muir *et al.*, Section B.2). While the contribution of local sources within the Canadian Arctic may be relatively minor to overall Arctic-wide contamination (Barrie *et al*, 1997), their contribution to this remaining global source cannot be ignored. A need to assess these sites with a view to changes that may arise as a result of climate change has also been identified. These changes relate primarily to the destabilization of contaminated materials currently sequestered in permafrost (Macdonald *et al.*, Section D.1).

CACAR-I provided a brief assessment of abandoned DEW-line military radar stations as sources of PCBs. Based on work carried out by the ministry of National Defense, it was concluded that contamination from DEW-line stations generally contribute a 'halo' of elevated PCB levels within a 10 km radius of each site. While the DEW-line stations were considered the most important source of PCB contamination at a local level (accounting for 85% of the source), long-range atmospheric transportation was considered to be the most important source Arctic-wide basis (accounting for 99%).

The Canadian government has undertaken several programs to assess and, if necessary, clean up contaminated sites in the Canadian Arctic, most notably the DEW-line cleanup project, Department of National Defence, and the Arctic Environmental Strategy-Action on Waste program, Department of Indian and Northern Affairs. A summary of the sites that are administered by these two departments and the progress that has been made to date is provided in Section B.9. This section also provides several case studies providing examples of sites that contain relatively large amounts of contamination and that have seen substantial investigation and, in some cases, have been remediated. The extent of contamination at these sites is not representative of all contaminated sites in the North — indeed, in several cases, they represent extreme examples.

The influence of military radar stations on local PCB contamination was examined at a worst-case (exceptionally contaminated) site in Saglek, Labrador. This study concluded that PCB contamination in soil and plants was influenced by site contamination in a halo that extends up to 30 km from the site. Marine sediment PCB contamination displayed a site influence at up to 10 km from the site. Since these studies were initiated, the terrestrial source of PCBs at Saglek has been removed. The source of PCB contamination at similar exceptionally contaminated radar site on Resolution Island, Nunavut has also been removed.

A detailed quantitative ecological risk assessment conducted on PCB-contaminated marine sediments at Saglek identified significant risks to the survival and reproduction of shorthorn sculpin and black guillemots residing close to the site. This study incorporated detailed toxicological effects study on black guillemots and identified significant PCB-associated effects in the virtual absence of other contaminants. Potential options for managing sediment PCB contamination at Saglek are currently under review by the Department of National Defence.

Studies into anthropogenic arsenic contamination resulting from mining activities in Yellowknife, NWT, have revealed the importance of identifying site specific background conditions. The bioavailability of arsenic in tailings and contaminated soils has also been studied and results have been used to assess ecological and human health risks. It was concluded that despite high localized arsenic contamination there were no ecological or human health risks associated with surface soil contamination.

Radionuclide and heavy metal contamination at Port Radium has been delineated and assessed for ecological and human health risks. It was concluded from these studies that, based on site occupancy scenarios of less than three months, there was no increased risk to ecological or human health from the contamination. There remain, however, some risks to ecological receptors. The results of these risk assessments will be used to identify priorities for site remediation.

Sediment core samples collected near a railway tie treatment facility in Carcross, Yukon, reflect past use of PCP as a wood preservative at the site. This study also provided the opportunity to compare contaminant levels influenced by the site to those strictly resulting from long-range atmospheric transport.

Local sources of contamination have provided opportunities for researchers to study contaminants in both abiotic and biotic media and have allowed the examination of intercompartmental processes. Two such cases are presented: a PCB-contaminated radar station at Saglek, Labrador; and, arsenic contamination associated with the Giant mine in Yellowknife, NWT. Both of these sites provided ideal locations for the study of uptake, accumulation, and biological effects over a range of contaminant levels and in the virtual absence of other contaminants. These studies have been applied to ecological and human health risk assessments that have produced conclusions at reduced levels of uncertainty compared to traditional probabilistic methods.

2.2 Occurrence and trends

Contaminant concentrations in the Arctic abiotic environment have been measured since the late 1980s and early 1990s and in some cases even earlier. Most of air monitoring has taken place since 1992 at permanently established stations throughout the circumpolar Arctic that collect air samples year-round. A growing data set for Arctic air is revealing information about global sources, transport pathways, and temporal trends, which together provide much needed validation on the success of international efforts to eliminate sources of pollutants to the Arctic. The air data collection has also helped identify other contaminants that may require international attention. Sampling of seawater has received less attention but has been achieved intermittently through ship-based expeditions organized by circumpolar nations. Spatial trends in seawater are now emerging that elucidate ocean transport pathways of certain contaminants and help explain regional differences in bioaccumulation.

It has been well established that POPs and other contaminants found in the Arctic environment come from distant global sources via long-range atmospheric and oceanic transport. In an effort to monitor the atmospheric influx of these contaminants to the Arctic, several circumpolar air monitoring stations have been established. Atmospheric monitoring is designed to maintain detailed and accurate records of contaminant concentrations year-round and over multiple years.

Contaminant data together with climatic and atmospheric data allow the inference of source regions and conditions that influence contaminant transport and deposition, plus a more robust interpretation of trends. The placement of the monitoring stations at remote locations spanning the entire Arctic Circle also allows for the identification of spatial patterns that help to identify source regions. Air monitoring also provides an early warning for new substances that might be transported to the Arctic. New releases of these substances will show up in air long before they become detectable in biota. All of this information feeds back into global box models along with emissions data and physicochemical data.

Air monitoring data have been collected weekly at Alert, Nunavut (1992–ongoing); Kinngait, Nunavut (1994–1995, 2000–ongoing); Tagish, Yukon (1993–1994); Dunai Island, Russia (1993–1994); and Amderma, Russia (1999–2001). Air is also monitored at various locations throughout the circumpolar region, most notably the Norwegian site at Ny Ålesund (1994–ongoing), and at more temperate regions which provide valuable comparison data.

The collection of weekly data over a long period of time has allowed Canadian scientists to produce sophisticated statistical analyses of temporal trends in levels of atmospheric contaminants. The short record available at the end of NCP-I has now been augmented to build a continuous weekly record at Alert since 1993 allowing a robust interpretation of seasonal variation and longer trends, illustrating the importance of maintaining these kinds of time series (Blanchard *et al.*, Section B.1).

Since NCP-I, contaminants in Arctic seawater have been studied primarily by shipboard scientific cruises organized by circumpolar nations. These cruises have built on the knowledge of previous studies and produced data from various locations of the circumpolar Arctic including the north Barents Sea, the western Canadian Archipelago, northern Baffin Bay, the Beaufort/Chukchi Seas and the east Greenland Sea. These studies used similar methods of sample collection and analysis and have somewhat overcome difficulties of previous years which included analytical sensitivity and sample contamination. Results are beginning to allow the interpretation of spatial trends; however, gaps remain in areas where insufficient water sampling has been carried out (Muir *et al.*, B.2).

2.2.1 Polychlorinated biphenyls (PCBs)

2.2.1.1 Polychlorinated biphenyls in air (Blanchard et al.*, Section B.1)*

In general, ΣPCB concentrations measured in air at Alert, were 5- to 11-fold lower than they were at similar monitoring stations in the vicinity of the Great Lakes. Concentrations of ΣPCB at Alert were also generally lower than at the other Arctic monitoring sites though the differences were relatively minor compared to differences between Arctic and Great Lakes locations.

Cyclic seasonal trends in atmospheric concentrations of PCB-congeners between 1993 and 1997 followed observed seasonal variations in temperature. This seasonal cycle likely reflects the temperature dependence of atmospheric PCB concentrations as well as summer time decreases driven by photochemical and OH-radical reactions, the primary elimination pathway for PCBs in the atmosphere. Analysis of data over the entire collection period revealed a decreasing trend in atmospheric concentrations of lower-chlorinated PCB-congeners (e.g., PCB 28, 31, and 52). This decline, which appears to be a weaker manifestation of the decline observed at temperate latitudes, could result from source depletion in temperate latitudes followed by enhanced atmospheric degradation by OH-radical reactions during the Arctic summer. No long term trend was seen for penta and hexachlorinated congeners at Alert despite observed decreases at temperate latitudes. It is thought that the varying physicochemical properties of different congeners and resulting differences in removal rates and transport mechanisms are responsible for the differing temporal trends that have been observed. Atmospheric half-lives calculated for ten PCB-congeners ranged from 3 to 17 years and were generally longer than those derived for temperate/background rural sites.

2.2.1.2 Polychlorinated biphenyls in seawater (Muir et al.*, Section B.2)*

Data collected for PCBs in seawater have recently improved, overcoming problems associated with detection limits and contamination of samples; however, there still is not enough broadly distributed data of consistent breadth and quality to make interpretations of spatial trends. Despite these shortcomings, some broad conclusions can be drawn from the limited data set. Concentrations of ΣPCBs under the permanent ice cap are much lower than they are in open, nearshore surface waters. The highest ΣPCB concentrations were found in nearshore waters of the Canadian Archipelago, southern Beaufort Sea, and north Baffin Bay. Despite its proximity to urban centres of northern Russia, ΣPCB concentrations in the central basin of the White Sea were lower than in the Canadian Arctic.

PCB-congener profiles measured along the track of a drift station in the Chukchi Sea that passed over the Chukchi Plateau and into the Canada Basin indicated that water entering the Arctic through the Bering Strait was mildly enriched in heavier chlorinated congeners. This enrichment is attributed to Pacific Ocean and Bering Sea water which could be influenced directly by sources of technical PCB mixtures via spills or other terrestrial sources such as contaminated sites. Within the Canada Basin, congener profiles were dominated by mono-, di-, and trichlorobiphenyls consistent with a long-range 'distilled' atmospheric source. Samples from the central Canadian Archipelago displayed a higher proportion of hexachlorobiphenyls than samples from the Chukchi Sea, the central Arctic Ocean, and northern Baffin Bay. The enrichment of higher chlorinated congeners in the central Canadian Archipelago seems to correspond with preferential degradation and evapouration of lower chlorinated congeners in PCB mixtures that are considered to be older and more weathered.

2.2.2 Organochlorine pesticides

2.2.2.1 Organochlorine pesticides in air (Blanchard et al.*, Section B.1)*

Organochlorine pesticides detected in air at Alert include, from most to least abundant: ΣHCHs and HCB > Σchlordanes and endosulfan I > dieldrin >ΣDDT. The relative abundance of the different pesticide groups is likely related to both physical and chemical properties such as volatility as well as the historic and current volumes of global use.

Seasonal cycles similar to those of PCBs, were also observed in atmospheric concentrations of organochlorine pesticides. These cycles were generally characterized by maxima in spring and fall and minima in summer and winter. The seasonal cycles are likely driven partly by temperature and partly by other factors including application patterns in temperate latitudes, photodegradation by OH-radicals in summer, and scavenging by forests during the growing season.

Most of the investigated OC pesticides showed declining trends over the five years of observation. These trends generally mirror those observed in temperate latitudes and likely derive from restrictions and bans placed on certain pesticides. The exceptions were dieldrin and endosulfan I. Endosulfan is currently in use on a global scale.

Isomer ratios of chlordane measured in Arctic air suggest that weathered soil residues are the dominant source. This conclusion is supported by the non-racemic enantiomer ratios of *trans*-chlordane at Alert, Dunai, and Pallas suggesting microbial degradation has occurred.

While these observations are encouraging, spikes in chlordane isomer ratios recently observed at Alert suggest current periodic use of chlordane in the northern hemisphere followed by rapid transport to the Arctic.

Episodic increases in heptachlor concentrations have also been observed at Alert. Heptachlor epoxide (HEPX) in Arctic air and surfaces, however, is enriched in the (+)enantiomer, indicating that microbial degradation has occurred in soils prior to HEPX emission.

The DDT:DDE ratios measured in air at Alert were on the order of 1–1.5 suggesting weathered sources. At Amderma, Russia the measured ratios from a selected number of samples were about 3. Tagish, Yukon, also had elevated DDT:DDE ratios suggesting relatively fresh use of DDT in Asia that is transported across the North Pacific and deposited in the western Canadian Arctic. Further investigation of pesticide use and emissions in Asia is recommended to gain a better understanding of this source and pathway.

2.2.2.2 Organochlorine pesticides in seawater (Muir et al.*, Section B.2)*

Toxaphene has been assessed in relatively few samples of seawater and there are too few results to provide a detailed view of spatial trends. Observations, however, indicate total toxaphene concentrations in north Baffin Bay > south Beaufort Sea > White Sea and north Chukchi Sea. Toxaphene congener profiles measured in the Beaufort signature are dominated by hexa- and hepta-chlorobornanes, in particular B6–923 and B7–1001, terminal residues of toxaphene degradation. In contrast, the White Sea profile is marked by a lack of these terminal residues and a prominence of octa- and nona-chlorobornane congeners signifying a relatively fresh source likely originating in Russia where toxaphene was used into the late 1980s.

DDT and chlordane-related compounds along with dieldrin and HCB were determined in seawater but at concentrations much lower than toxaphene, PCBs, or HCH isomers. Like ΣPCBs, ΣDDT and Σchlordane were higher in the Canadian Archipelago and nearshore Beaufort than they were in the Chukchi Sea or the central Arctic Ocean. Unlike ΣPCBs and ΣDDT, however, Σchlordane and total toxaphene appear to be at maximum concentrations under regions with ice cover.

This is thought to be caused by a summer buildup of these contaminants through seawater-gas exchange and snow melt, important pathways to the Arctic Ocean for these contaminants. The same maximum concentrations are not seen under ice for ΣPCBs and ΣDDT likely because they are hydrophobic, which may cause them to quickly settle out of the water column sorbed to particulate matter. Examination of ΣDDT profiles in different water depths showed that DDT and its isomers were transferred from surface waters to deeper waters in spring when biological activity was at a peak as was the vertical flux of organic particulate matter.

2.2.2.3 Hexachlorocyclohexanes in seawater (Li et al., Section C.3)

A drop in technical HCH use, documented in this report, has been reflected in corresponding reductions in α-HCH levels detected in Arctic air. Two significant drops in α-HCH concentration in Arctic air were identified, between 1982 and 1983, and again between 1990 and 1992. This was noted in NCP-I, but a clear link to changes in global use and emissions was not realized until NCP-II when two drops in global emissions were retrospectively identified, one in 1983 and another in 1990. This discovery provided a clear link between global α-HCH use and emissions and concentrations of α-HCH in Arctic air. Global α-HCH emissions and the concentration of α-HCH in Arctic air are sufficiently correlated that a reasonable estimate of historical air concentrations can be inferred from global emission data.

Hexachlorocyclohexane, primarily the more abundant α- and γ-HCH isomers, has been the focus of many studies that have investigated the mechanisms of contaminant transport and fate in the Arctic Ocean. Under NCP-phase I, HCH investigations allowed for the development of a mass balance for these compounds in the Arctic Ocean. While these more common isomers have been well studied and their environmental behaviour and fate is relatively well understood, β-HCH, which is not as abundant in the Arctic abiotic environment, has not been as well studied. This apparent neglect of β-HCH, due partly to low abundance and partly to analytical challenges, has contrasted growing concern over its potential risks. Beta-HCH is the most bioaccumulating of the HCH isomers and may act as an environmental estrogen (Willett *et al.*, 1998).

Alpha- and β-HCH show strikingly different spatial trends in surface water. Concentrations of α-HCH increase smoothly with latitude from the subtropical seas near China and Japan to the Canada Basin, but β-HCH peaks in the Bering-Chukchi region and then drops off along a transect to the interior Arctic Ocean. The behaviour of α-HCH might be considered a classic example of the "cold condensation effect", where the partitioning from air into water of α-HCH is higher in cold northern waters, although slower degradation at colder temperatures probably plays a role also (see following text).

The unexpected spatial trend for β-HCH is likely due to its lower Henry's law constant, which favours preferential deposition in the North Pacific by rain and air-sea exchange. Ocean-currents then finish the job of delivering β-HCH to the Canada Basin through the Bering Strait.

This hypothesis leads to the conclusion that loading pathways of all HCHs to the North American Arctic Ocean (NAAO) have changed over time. In 1980, direct atmospheric deposition and advection by ocean-currents through the Bering Strait contributed about equal proportions of α-HCH to the NAAO (49% and 51%, respectively). By 1995, atmospheric deposition of α-HCH dwindled due to declining air concentrations, and flow through the Strait took over as the major supply route, delivering 90% of the α-HCH input. Direct atmospheric deposition was always a minor pathway for β-HCH loadings to the NAAO, delivering only 20% of the total input in 1980 and 2% in 1995. For this isomer, the dominant pathway was always through the Bering Strait (Li *et al.*, 2002; in press).

Microbial degradation rates for α- and γ-HCH were estimated from vertical profiles of concentration and (for α-HCH) enantiomer composition in the eastern Arctic Ocean. Microbial degradation appears to be 3–10 times faster than chemical hydrolysis and could account for about a third of the annual loss of α- and γ-HCH from the water column. Still, the combined microbial and hydrolysis rates in the Arctic Ocean are slower than hydrolysis alone in the warm subtropical seas near regions where technical HCH is used.

The removal budget for the NAAO in 1995 indicates that outflow through the Canadian Archipelago and east Greenland Current accounts for most of the HCH output, 65%, 100%, and 81% of total losses for α-, β-, and γ-HCH, respectively. For α- and γ-HCH, microbial degradation is the second largest removal mechanism, accounting for 29% and 16% of total losses. Microbial degradation rates for β-HCH are unknown, but are assumed to be lower than for the other two isomers since it is known to be more recalcitrant towards metabolism.

The lesson learned by comparing the pathways of the HCH isomers is that physicochemical properties can interact with environmental pathways to produce unexpected results. Elucidation of a dominant oceanic pathway for β-HCH has implications for delivery of other long-lived and water-soluble chemicals to the Arctic by ocean-currents — a pathway that has so far been neglected.

2.2.2.4 *Summary — Organochlorine pesticides in air and seawater*

The overall decreasing trends observed for banned OC pesticides in the atmosphere provide encouraging results for international efforts to curb sources of POPs to the Arctic. The declines in atmospheric concentrations of OCs that have been seen over the past 20 years, e.g., α-HCH, have likely been brought about by the collective efforts of individual countries to restrict or ban their use. The evidence suggests that the detrimental accumulation of these contaminants in the Arctic is reversible with appropriate intervention.

Examination of chemical signatures has provided evidence of a degraded state for chlordane and other OC pesticides such as heptachlor and DDT, suggesting that for these largely banned OC pesticides, lingering soil residues, not fresh applications, are becoming the dominant source to the atmosphere. This shift and reduction in atmospheric sources of OCs has been reflected in observations of HCHs in *seawater*. These observations suggest that input of HCHs to the Canada Basin by ocean-currents, primarily through the Bering Strait, is of increasing importance relative to Atmospheric deposition. Evidence of microbial degradation of HCHs in the Arctic Ocean provides a further basis for predicting future reduction of HCHs in the Arctic marine environment. As demonstrated by recent results at Alert; however, fresh applications of current-use OC pesticides (and occasionally banned ones as well) can be detected and are still adding to Arctic contamination. Continuing international effort should be made to eliminate the use of these chemicals.

2.2.3 Polycyclic aromatic hydrocarbons (PAHs)

2.2.3.1 *Polycyclic aromatic hydrocarbons in air (Blanchard* et al.*, Section B.1)*

Polycyclic aromatic hydrocarbons that originate from both natural and anthropogenic combustion sources have been monitored at Alert as a group of 16 compounds (ΣPAH) since 1993. In general, the lighter, more volatile compounds such as fluorene and phenanthrene are the most abundant PAHs found in Arctic air. The heavier compounds, which are more susceptible to atmospheric decomposition and deposition, are likely removed from the atmosphere before they reach the Arctic and are, therefore, not as abundant in Arctic air. Concentrations of PAHs tend to be higher at Spitsbergen, Norway, than at the two Canadian stations of Alert and Kinngait.

Seasonal variations observed in PAH concentrations were associated with the appearance and severity of Arctic haze which is experienced in the late winter and early spring. The extreme seasonality in the PAH data set precluded the quantitative interpretation of temporal trends; therefore, interpretation was based on annual means calculated for each of the six years. These means demonstrated a general decreasing tendency throughout the 1990s, which is consistent with a declining trend in atmospheric sulphate aerosol in the Arctic. This decreasing trend in atmospheric PAHs mirrored trends in industrialized regions where reductions have been attributed to a general decline in the domestic use of fossil fuels, improvement in combustion devices and a decline in industrial activity in the former Soviet Union.

2.2.4 Polychlorinated dibenzo-*p*-dioxins/furans (PCDD/Fs)

2.2.4.1 *Polychlorinated dibenzo-p-dioxins/furans in air (Blanchard* et al.*, Section B.1)*

Polychlorinated dibenzo-*p*-dioxins/Furans, which originate from both natural and anthropogenic combustion sources, were analyzed in a limited number of filter samples collected at Alert during the winter of 2000/1. Results of these analyses showed that ΣPCDD/F concentrations in air at Alert are generally lower than in other locations which are closer to industrial centres and potential sources. The concentrations at Alert were also found to be lower than those observed at Ny Ålesund, which is a Norwegian Arctic site. Five-day Lagrangian back trajectories of air mass movement were examined along with corresponding measures of ΣPCDD/F concentrations to identify potential source areas. Alert was influenced by air masses from the North Atlantic and North America as well as Russia and Eurasia, but peak concentrations corresponded to the air masses originating in Russia and Eurasia.

2.2.5 Persistent organic pollutants in marine sediments (Stern *et al.*, Section B.3)

Very little work has been done to assess contaminant distributions in marine sediments primarily due to low contaminant concentrations and associated analytical limitations. Some sampling efforts have provided limited information on levels in surface sediments for the Canadian Archipelago and Hudson Bay; however, there are very few data upon which to assess larger scale spatial or temporal trends. Limited data are available for the major OCs, PCBs, DDT, chlordane, HCH as well as brominated flame retardants (PBDEs), and SCCP. The detection of PBDEs in Arctic marine sediments is yet another example of the chemical's persistence.

Surface sediment ΣPCB concentrations in samples collected from the Canadian Arctic archipelago and eastern Hudson Bay ranged from 0.28–8.24 ng/g, dw; ΣDDT ranged from 0.01–0.57 ng/g, dw; ΣCHL ranged from 0.01–0.35 ng/g, dw; and, ΣHCH ranged from 0.01–0.35 ng/g, dw. Of the new chemicals, ΣSCCPs ranged from 4.79–77.41 ng/g, dw and ΣPBDEs ranged from 106.7–297.3 pg/g, dw.

2.2.6 Persistent organic pollutants in freshwater sediments (Stern *et al.*, Section B.3)

Compared to ocean sediment sampling, freshwater sediment sampling in the Canadian Arctic has been quite extensive. The records of contaminants in dated sediment cores from Arctic lakes that have provided information about deposition fluxes and trends that are helpful in developing relationships between emissions and deposition in remote Arctic regions have been of particular interest. Vertical contaminant trends in lake cores may be indicative of long-term temporal trends in input of contaminants to the lake from long-range atmospheric sources. However, other factors may also influence the flux of contaminants to lake sediments. For example, temporal shifts in contaminant flux caused by changes in lake characteristics, potentially brought on by climate change, may also play a role in determining sediment core profiles. (Macdonald *et al.*, Chapter D)

In 1997, several lake cores were collected from a remote lake on Devon Is. and, Nunavut. The cores were characterized by distinctive annual layers confirmed by ^{210}Pb dating. Analyses of the core's layers for various POPs revealed a historic record of contaminant deposition in the sediment that corresponded very well with known historic use patterns for the different chemical compounds and provided further evidence and support for theories of long-range atmospheric transport of contaminants. The detailed historic record of contaminant deposition to the lake sediments also provided insight into recent temporal trends in atmospheric contaminant levels. For each of the contaminants DDT, toxaphene, and chlordane, sediment layers near the surface of the core revealed that recent sediment concentrations have been decreasing, implying a potential decrease in atmospheric input.

Analysis of ΣSCCPs in the core revealed a pattern that was similar to historic use patterns though in this case there appeared to have been a 10-year time lag between commercial introduction and arrival at the lake's sediments. This 10-year time lag is also apparent when examining peak core concentrations relative to peak historic use. Consequently maximum ΣSCCPs concentrations

ITK/Eric Loring

possibly corresponding to a peak in industrial application around 1990 were found in surface sediments of the lake.

Polychlorinated dibenzo-*p*-dioxins and PCDFs began to increase in the section of core corresponding to the early 1940s and 1950s, respectively and both peak in 1978. In the earlier half of the record, 1938–1963, the homologue profile is dominated by OCDD, which is consistent with the use of pentachlorophenol wood preservative of which OCDD is a major impurity. OCDD is also a by-product of pentachlorophenol photolytic degradation. Congener profiles in the more recent parts of the core, 1970–1992, appear to be more indicative of wood and coal combustion. Based on the core results, PCDD/F fluxes associated with sedimentation in the Devon Is. lake were calculated at 0.36 $pg \cdot TEQ\ m^{-2} \cdot y^{-1}$ for periods of maximum flux in the late 1970s and 0.1 $pg \cdot TEQ\ m^{-2} \cdot y^{-1}$ for surface sediments corresponding to years prior to sampling. These results imply that fluxes of PCDD/Fs to the Arctic have diminished.

Similar studies have been conducted on cores from other lakes; however, the temporal resolution of the cores has traditionally not been sufficient to identify recent trends. The Devon Island cores thus present an invaluable data set.

Profiles of DDT and toxaphene were examined in sediment cores from numerous lakes in the Yukon. Like the Devon Island cores, several of the Yukon lakes displayed core profiles that corresponded well to historic use patterns. However, several of the lakes also showed evidence of more direct input not related to long-range atmospheric sources. In the case of Hanson lake there was documented use of toxaphene to control the fish population; however, in the other lakes that displayed elevated concentrations of toxaphene, there was no apparent past use within the lake. Toxaphene in these other lakes is suspected to come from its use as an additive to DDT, which was used in the Yukon to control mosquitoes and blackflies.

In the more remote Yukon lakes, PCB profiles in sediment cores correspond well to historic use patterns and long-range atmospheric sources. In select lakes, however, a rise in PCB concentrations in a section of the core that approximately corresponds to 1942 has been attributed to the onset of military activity in the territory. Several lakes appear to have been more directly impacted by military activities as indicated by congener profiles enriched in penta- and hexa-chlorobiphenyls associated with technical PCB mixtures.

Hexachlorocyclohexanes were measured in seven Yukon lakes. A range of maximum core concentrations spanning nearly an order of magnitude is thought to reflect the difference between remote lakes influenced solely by long-range atmospheric transport and lakes that may have been influenced directly by human activity. With the exception of a lake influenced by glacial meltwaters, all of the lake cores exhibited peak concentrations at depth suggesting that inputs of HCHs are consistent with decreases in global use. Glacial melt is thought to potentially release archived contaminant loads previously deposited to the surface of the snow-pack during periods of meltback resulting in enriched sediment concentrations and core profiles that do not correspond to historic atmospheric deposition. This is just one of several ways in which climate change can alter contaminant flux to lake sediments (Macdonald *et al.*, Chapter D).

Chlorophenols were measured in Coal Lake, a remote headwater lake which has not been influenced by local sources of contamination, to provide reference/background data for a contaminated site investigation at an abandoned railway tie treatment plant in Carcross, Yukon. Results showed that chlorophenols appear in sediments corresponding to about 1940, shortly after pentachlorophenol was introduced as a wood preservative. At all depths in the core the chlorophenol profile was dominated by mono-and dichlorophenols which are present as impurities in pentachlorophenol but can also be produced by photochemical reduction of pentachlorophenol in water. Chloroanisoles, the metabolic by-products of chlorophenols, were also identified in the sediment core.

A study was conducted on Great Slave Lake (NWT) to determine sources of contamination to the lake and to assess the effects of resource development in the basins of the Peace and Athabasca rivers. Results showed that the highest concentrations of ΣPCBs, ΣPAHs, and ΣPCDFs were in sediments near the mouth of the Slave river — the Slave carries water from both the Peace and the Athabasca into Great Slave Lake. These results were consistent with elevated sediment fluxes associated with riverine input. ΣPCDDs were present at much lower concentrations than ΣPCDFs but were actually more concentrated in the eastern arm of the lake. Historic trends interpreted from sediment cores indicated that the flux of PCBs and petrogenic PAHs have been increasing over time. ΣPCDDs concentrations, likely associated with pentachlorophenol use, peaked in core sections corresponding to the 1950s and have been declining ever since. The study concluded that Great Slave Lake was a sink for contaminants delivered from the Peace and Athabasca River basins via the Slave River but that these contaminants were not transferred to the Mackenzie River.

2.2.7 New persistent chemicals in the Arctic (Alaee *et al.*, Section B.4)

When found in a remote environment such as the Arctic, a new chemical contaminant may be considered persistent as defined by both the Canadian Toxic Substances Management Policy and the United Nations Economic Commission for Europe (UN-ECE) draft protocols on POPs. If such a contaminant is also found in Arctic food webs, there is the added threat from bioaccumulation/biomagnification ultimately into components of the traditional diet of northerners. Screening the Arctic environment for new chemicals helps to identify those which are persistent, focus studies of bioaccumulation and foster early action to eliminate sources.

Under a "New Chemicals" initiative, a number of chemicals in current-use within the circumpolar region and other industrialized countries have been identified in the abiotic Arctic environment. These include brominated flame retardants (BFRs), short-chain chlorinated paraffins (SCCPs), chlorinated phenolic compounds, and modern agrochemicals.

Other "legacy" industrial chemicals, polychlorinated naphthalenes (PCNs), and coplanar PCBs, which have not been part of the historical NCP monitoring program, were also identified.

2.2.7.1 *Brominated fire retardants*

Brominated fire retardants (BFRs) are a group of chemicals used as additives in textiles, plastics, building materials, and electronic equipment to reduce the risk of combustion. Among the BFR chemicals, polybrominated diphenyl ethers (PBDEs) were the only ones of significance found in Arctic air. Other BFRs such as hexabromocyclododecane and tetrabromo-bisphenol A were below detection limits. ΣPBDE concentrations in air were highest at Tagish, intermediate at Alert, and lowest at Dunai and showed a trend of higher concentrations during the warmer months. The concentrations at Alert and Tagish were higher than air concentrations reported by Strandberg *et al.*, (2001) for the city of Chicago (mean of 52 pg/m^3; 1997–99) and much higher than concentrations at air monitoring sites in the Great Lakes (means of 5–23 pg/m^3 at three sites). In the case of Tagish there is the possibility that incineration of household items in the region could contribute PBDEs to the air at the sampling site. At Alert the source has not been identified.

While there have not been enough data collected to interpret long-term temporal trends in abiotic media there are some biological results that suggest levels of BFRs in the Arctic might be increasing at an alarming rate. The most abundant BFR chemicals found in the Arctic, PBDEs, have been increasing in ringed seal blubber in the Canadian Arctic from the early 1980s to the present (see CACAR-II Contaminant Levels, Trends, and Effects in the Biological Environment)

2.2.7.2 *Short-chain chlorinated paraffins (SCCPs)*

Short-chain chlorinated paraffins (SCCPs), are used in cutting oils and as additives to paints and sealants. They were first used in large volumes in the 1930s as extreme pressure additives in lubricants. The use of SCCPs increased dramatically during the second world war and then again increased to exceed war time production levels during peak use between 1978 and 1985. These chemicals have been found in lake sediments in the Canadian Arctic and in marine sediments from the Archipelago. They were also found in archived air sample extracts from Alert in 1993–94. (See POPs in Marine and Lake Sediments, Section B.3 and New Persistent Chemicals in Arctic Air and Water, Section B.4.)

2.2.7.3 *Modern agrochemicals*

The currently used pesticides endosulphan, trifluralin, and methoxychlor as well as the pentachlorophenol metabolite pentachloroanisole were all identified in Arctic air samples from Alert in 1993–94 and again in 1995–97. Concentrations of endosulfan and pentachloroanisole were exceeded only by HCB, α-, and γ-HCH. Unlike most past-use OC pesticides, endosulfan concentrations showed a slight increase during 1993 and remained stable through 1997.

Despite a relatively short atmospheric half-life of 21–74 minutes, trifluralin was observed at three Arctic monitoring stations, Tagish, Alert and Dunai. These half-life calculations, based solely on photochemical degradation, indicate that trifluralin should not reach to the Arctic at measurable quantities even though relatively large amounts (>5 x 10^6 kg) are applied annually to crops in western Canada and the US Current results indicate that other pathways such as transport on dry particulate or aerosol might be the primary mode of transport for trifluralin. Calculations based in purely gas-phase reactions with OH radicals, present at average tropospheric concentrations, conclude that several conventional OC pesticides also have short atmospheric lifetimes: chlordane = 8 days, DDTs = 2–4 days, and dieldrin = 1 day (Atkinson, 1990). Nevertheless, there is no doubt that these chemicals are present in Arctic air. Such discrepancies need to be resolved, through continued modelling and research.

Endosulphan was also measured in surface water in the eastern Archipelago and Holman Island at concentrations ranging from <0.0005–0.01 ng/L. The average along the Tundra Northwest-99 cruise track in the eastern Canadian Archipelago ranged from 0.004–0.01 ng/L with an average of 0.0063 ng/L similar to concentrations found in the northern Canada Basin in 1994. A 1993 study found that endosulfan concentrations were higher during periods of open water in the central Canadian Archipelago reflecting inputs from gas exchange and freshwater. During this study, endosulfan concentrations increased from 0.002–0.006 ng/L during the late-summer and fall months paralleling increasing air concentrations. This was possibly related to summer use of endosulfan in mid-latitude North America. Current use organophosphates and herbicides were also sought but not detected in the Holman water samples.

2.2.7.4 *Polychlorinated naphthalenes (PCNs)*

Atmospheric ΣPCNs in samples collected in the mid-1990s indicated that the highest levels were found over the Barents Sea followed in abundance by the eastern Arctic Ocean > Dunai > Alert. High levels of ΣPCNs over the Barents Sea were associated with air masses originating in the UK, where PCNs have also been identified in ambient air, and central Europe. Higher concentrations of ΣPCNs at Dunai could be associated with higher concentrations of ΣPCNs in commercial Sovol PCB mixtures found in Russia. ΣPCNs also displayed a seasonality similar to that of PAHs and Arctic haze, although definition of trends are less clear due to the small number of samples. This suggests an influence of combustion-related sources which was corroborated by the enrichment of combustion-related PCN congeners. It is yet unknown whether PCNs are removed from the atmosphere through photochemical reactions.

2.2.7.5 *Coplanar PCBs (CoPCBs)*

Coplanar PCBs (ΣCoPCBs, include mono-*ortho* substituted congeners 105, 114, 118, and 156 and the non-*ortho* substituted congeners 77, 81, 126, and 169) displayed similar spatial trends to the atmospheric ΣPCNs. Similar to ΣPCNs, ΣCoPCBs found over the Barents Sea were associated with air parcels coming from the UK and central Europe. Unlike ΣPCNs, ΣCoPCBs reached a maximum atmospheric concentration in the summer, particularly at Dunai where Russian PCB mixtures enriched in penta-chlorinated PCBs are thought to contribute to the summertime increases.

Assay-derived TEQ values for ΣCoPCBs and ΣPCNs at Alert, Tagish, and Dunai ranged from 0.003 fg/m^3 in the summer to 0.061 fg/m^3 in the winter with the highest values observed at Dunai. In the winter, ΣPCNs accounted for most of the toxicity at Dunai and Alert, whereas most of the toxicity at Tagish was attributed to ΣCoPCBs. In the summer, ΣCoPCBs accounted for most of the toxicity at all stations. Average TEQs for ΣPCDDs and ΣPCDFs were 0.8 fg/m^3, up to 50 times higher than for ΣPCNs and ΣCoPCBs combined.

2.2.7.6 *Octachlorostyrene (OCS)*

Octachlorostyrene (OCS), a by-product of magnesium and chlorine manufacture, was identified in Arctic air at Alert, Tagish, and Dunai during 1993–94 and at Alert during 1995–97. Occasional peak concentrations suggest the influence of as yet unidentified specific long-range sources.

2.2.8 Mercury

2.2.8.1 *Mercury in air (Steffen* et al.*, Section B.5)*

Under NCP-I, weekly measurements of total gaseous mercury (TGM) were collected for a period of one year from August 1992–August 1993 at Alert. The major outcome of this research was the establishment of an important baseline concentration against which future results could be gauged. The mechanisms of mercury behaviour and trends in the Arctic atmosphere, however, remain relatively unknown. Under NCP-II, mercury has been measured continuously at Alert and Kuujjuarapik, Quebec (a sub-Arctic measurement site located on the shore of Hudson Bay) since 1995 and 1999, respectively. The establishment of these two stations enabled the assessment of both spatial and temporal trends in the distribution of atmospheric mercury, as planned, but also revealed an unexpected and highly significant process within the Arctic: the occurrence of mercury depletion events (MDEs) in Arctic air. This phenomenon has proven to be a crucial link in the global Hg cycle, potentially making the Arctic a sink. The transfer of mercury between atmosphere and surface has also been investigated in an effort to improve our understanding of pathways that may connect MDEs to elevated concentrations of mercury in aquatic and terrestrial reservoirs.

Northern hemispheric background concentrations of atmospheric mercury are about 1.5 ng/m^3 as measured in Macehead, Ireland and northern Wisconsin. Gaseous elemental mercury (GEM) concentrations measured at Alert were generally consistent with the northern hemispheric average with annual means of 1.47–1.62 ng/m^3 (1995–2000). Average concentrations of mercury measured at Alert over six years show no statistically significant trend. Mercury concentrations at Alert, however, exhibit a strong seasonal cycle punctuated after polar sunrise by the dramatic loss of gaseous mercury from the atmosphere. These mercury depletion events (MDEs) have recently been observed also at Kuujjuarapik where the annual average GEM concentration was 1.78 ng/m^3. Mercury depletion events take place in the late winter and spring during which concentrations of GEM suddenly plunge to less than 1.0 ng/m^3 and as low as 0.1 ng/m^3 in less than 24 hours.

This is a surprising behaviour for an element with a relatively long atmospheric residence time (6–24 months). The discovery of this phenomenon at Alert in 1995 has spawned major research efforts by scientists from several circumpolar nations that have confirmed its widespread occurrence in the Arctic (Ny-Ålesund, Spitsbergen; Barrow, Alaska; Station Nord, Greenland), in the sub-Arctic (Kuujjuarapik, Quebec), and in Antarctica (Neumayer Station). Mercury depletion during polar sunrise is likely the important process which makes the Arctic particularly vulnerable to the large pool of gaseous mercury circulating in the global atmosphere.

Mercury depletion events at Alert coincide with the appearance of the sun following the long winter period of darkness. Polar sunrise is thought to trigger an oxidative reaction between reactive halogen (and/or halogen oxide) species derived from seawater and Hg (0) resulting in the production of Hg (II) compounds — either reactive gaseous mercury and/or particulate-phase mercury — both of which have much shorter residence times in the atmosphere. These Hg (II) species are much more easily scavenged from the air than the Hg (0) vapour and are readily deposited to Arctic surfaces. The deposition of reactive mercury during and after MDEs has been shown, at times, to result in 3 to 20-fold increases in mercury concentrations in freshly fallen snow at Alert.

Although Hg (II) deposited on the snow surface can be readily dissolved in meltwater, the efficiency of transfer to aquatic ecosystems remains unknown. Once in the aquatic system mercury may be methylated. However, Hg (II) can also be converted back to Hg (0) through photolytic reduction and re-emitted to the atmosphere as Hg (0). Observations in 2000 showed that 50% of the mercury deposited on the snow surface during a MDE at Kuujjuarapik was re-emitted within 12 hours of the event. Research into this phenomenon is ongoing and a complete understanding of these newly discovered reaction cycles of Hg in the Arctic is still pending.

2.2.8.2 *Mercury in sediments (Lockhart et al., Section B.6)*

Mercury concentrations in freshwater sediments were evaluated for spatial trends in the Canadian Arctic, to assess sediment core data as a measure of temporal trends in mercury deposition and to assess the relative importance of anthropogenic and natural sources of mercury.

Mercury concentrations in over 300 surface sediment samples from lakes and rivers of the Canadian North displayed no discernable spatial trend. Of the sites sampled, only about 10% were found to exceed the Canadian sediment quality guideline for mercury of 170 ng/g. This might seem surprising since mercury concentrations in fish exceed human consumption guidelines far more frequently but it is consistent with the general finding that concentrations in fish bear little relation to associated sediment mercury concentrations. The specific biogeochemical characteristics of each lake may play a greater role in the cycling of mercury than simply the basal geological levels and mineralogy.

Examination of lake cores provided some indication of both temporal and spatial trends in mercury deposition. Results generally suggest, at least in some parts of the Arctic, that mercury inputs have increased over the past century. In the eastern Arctic south of 80° N, mercury input appears to have increased by a factor of two, whereas north of 80° N there has been little change. Similarly in the western Arctic, cores from the Mackenzie Delta and the north slope of Alaska indicate little change whereas to the south increased mercury deposition has been inferred from the cores.

Increased mercury deposition may result from increased emissions of mercury from fossil fuel combustion. Other influences, however, such as climate-induced changes in watershed characteristics could also produce increased mercury fluxes to and within lakes (e.g., permafrost destabilization, increased erosion, and loss of ice cover). Emerging analytical techniques may be able to distinguish sources based on isotope composition but this has yet to be done with Arctic samples.

2.2.9 Other metals in air (Gong *et al.*, Section B.7)

Aerosols collected on air filters at Alert were analyzed for metals (Al, Ba, Ca, Fe, Mn, Zn, V, Mg, Cu, Ti, Ni, K, Pb, Na, Hg) over a 20-year period from 1980 to 1999. The origin of these metals varies from anthropogenic to geologic and oceanic and it is often difficult to determine which might be the dominant source. To help identify the origin of the different metal elements found in aerosols at Alert principal components analysis (PCA) as well as lead isotope analysis were used to analyze the data.

Use of PCA revealed that metals such as Pb, V, Mn, and Zn were generally linked to anthropogenic sources which have demonstrated a general decreasing trend over the past 20 years. A smelter factor, dominated by Zn and Cu, showed a slight increase in the 1980s followed by declines since 1991. The metals associated with both of these sources showed seasonal variations with maxima occurring in the winter and minima in spring. Other factors associated with natural sources from soil (Al, Ba, Ca, Fe, Mg, Ti and Mn) and sea-salt (Na, Mg and K) sources were also identified. While soil-related metals did not show significant seasonal variations, sea-salt associated metals showed maxima and minima in the winter and spring, similar to those from anthropogenic sources. Comparison of these results to dominant atmospheric circulation in winter confirms that the primary sources of anthropogenic metals found at Alert are in Eurasia. This is supported by the general spatial observation that average concentrations at

Alert and Barrow were about three to four times lower than concentrations observed in the eastern Russian Arctic.

Lead was assigned to three predominant source categories based on isotope ratios ($^{206}Pb/^{204}Pb$, $^{208}Pb/^{204}Pb$, and $^{206}Pb/^{207}Pb$): i) radiogenic "natural" end-member from Canadian Arctic Islands and the coastal areas of West Greenland, seen dominantly in September; ii) an unradiogenic end-member originating from West and or Northwest Europe that dominates in late fall and winter; and iii) an anthropogenic end-member originating from Asia and Russia that dominates during the late spring and early summer.

Upon examination of the full data sets for individual metals there does not appear to be any obvious increasing trends for anthropogenic components but rather a slight decrease. No long-term trends were detected for Na and Ca, and while Zn and Cu showed a slight increase from 1980 to 1985, they seem to have decreased thereafter. Due to analytical limitations, it was not possible to identify long-term trends in Pb with confidence. Observations based solely on the winter data, however, suggest that over the long-term Pb concentrations have decreased.

2.2.10 Radionuclides in the Arctic Ocean (Smith, Section B.8)

Transport of radioactivity into the Arctic Ocean occurs by long-range atmospheric and ocean-current pathways and by direct injection into shelf seas from local sources. Radioactivity as discussed in this report comes from the decay of radionuclides. Many radionuclides are naturally present in the earth's crust and emit a constant level of background radiation. Anthropogenic radionuclides including: ^{99}Tc, ^{129}I, ^{137}Cs, ^{238}Pu, $^{239+240}Pu$ and ^{241}Am are by-products of nuclear fission. The primary sources of anthropogenic radionuclides to the Arctic include: atmospheric fallout (both local and global) from weapons tests and satellite power unit burnup; discharge from the Ob, Yenisey, and Lena rivers; releases from dumped Soviet and Russian radioactive wastes and accidents; and discharge from nuclear fuel reprocessing plants.

Radionuclides, ^{129}I and ^{137}Cs, discharged into the Irish Sea and English Channel from nuclear fuel reprocessing plants in Sellafield, United Kingdom and La Hague, France have provided scientists with effective tracers of ocean circulation. The documented emission inventories of these point source tracers allow the inference of ocean circulation that is both temporally and spatially resolved. Measurements of ^{129}I in the Arctic Ocean and marginal seas over five years under the submarine Arctic Science Cruise (SCICEX) program has clearly delineated the front between Atlantic-origin and Pacific-origin halocline water that is presently aligned over the Mendeleyev Ridge. These studies also confirmed findings of previous icebreaker studies that indicated a major thermohaline disturbance was underway corresponding with the intrusion of North Atlantic water along the continental margin of the Eurasian Basin into the Canada Basin. This disturbance, also corresponding to strong positive Arctic oscillation, is thought to be indicative of conditions that might arise from climate change and has implications for contaminant transport (Macdonald *et al.*, Section D.2).

Anthropogenic radionuclide contamination from distant sources through long-range atmospheric and ocean-currents generally do not present a major environmental concern in the Arctic (CACAR-II, Human Health report). Local source radionuclide contamination in the Barents Sea and the potential impact on the local fishery, however, is of considerable concern. Chernaya Bay on the south coast of the Russian Island Novaya Zemlya was the site of at least two underwater nuclear tests in the 1950s. Consequently the highest concentrations of $^{239,240}Pu$ ever measured in the marine environment can be found in sediments from Chernaya Bay. Efforts to delineate the extent of discharge into the Barents Sea suggest that influence decreases rapidly in a seaward direction but extends in a northwesterly long-shore direction in excess of 100 km. Within Chernaya Bay biological uptake from sediments to invertebrates and from seawater to brown algae has been measured.

Investigations of sediment contamination in the vicinity of the sunk Russian nuclear submarine Kursk revealed that the submarine itself had not produced additional radionuclide contamination. Studies conducted near the site of the sinking and recovery facilities in Kola Bay suggested that levels of radioactivity were consistent with the rest of the region and ascribed to inputs from global fallout, European nuclear fuel reprocessing plants and the Chernobyl accident. This study did, however, reveal that levels of ^{129}I in the South Barents Sea had increased by 500% and that the $^{129}I/^{137}Cs$ ratio had increased by an order of magnitude between 1992 and 2000 as a result of increased inputs from fuel reprocessing plants at Sellafield and La Hague. These results suggest that regional background levels of radioactivity are varying rapidly on annual time scales and that Europe's nuclear reprocessing facilities are the leading contributor of anthropogenic radioactivity to the region.

2.3 Process studies and modelling

Under NCP-Phase I, investigations of freshwater systems concluded that lakes and rivers received contaminants from long-range atmospheric transport and that discharge from freshwater systems contributed less than 5% of the overall α- and γ-HCH and cadmium loadings to the Arctic Ocean. The fate of contaminants in Arctic lakes was investigated under NCP–Phase II. These investigations had the goal of creating a predictive model that could be applied to new lakes lacking extensive study (Diamond *et al.*, Section C.2). Terrestrial and freshwater systems have been used to investigate the interactions of the atmosphere, Arctic surfaces (snow, ice and land), and freshwater and marine systems, for example, at Amituk Lake, Nunavut.

The studies at Amituk Lake in 1992 and 1994 (Barrie *et al.*, 1997), which included preliminary budget calculations for mercury and selected POPs, have since been updated with models for contaminant run-off and contaminant fate. These studies in combination with investigations into degradation half-lives have significantly enhanced our understanding of contaminant transport and fate in Arctic watersheds.

The investigation of contaminants in ice and snow has been used to assess mechanisms by which they are transferred between the atmosphere and Arctic surfaces and how these processes are influenced by the annual climatic cycle of the Arctic. Studies of Arctic freshwater systems, in particular, have provided a relatively complete picture of the relationships between the atmosphere and terrestrial freshwater systems.

Mass balance studies, modelling, and laboratory studies are providing information on how contaminants are eliminated from the environment. Elimination of contaminants takes place by advective outflow (from lakes and the Arctic Ocean), volatilization, sedimentation, and degradation. Understanding these processes is key to the development of contaminant budgets.

Research into contaminants in the abiotic Arctic environment under the first two phases of the NCP and other programs over the past 10–15 years has provided insight into the global sources, emissions, and exchange of contaminants with the Arctic as well as occurrence, pathways, and fate within the Arctic. This information has underpinned the development of global box models for the prediction of contaminant distribution and trends that are essential for guiding the formulation of international policy. Historical and spatial trends produced by the models have been validated with quantitative measurements in the field, which in turn helps further optimize the models. These modelling efforts are summarized in more detail in the following sections. Understanding the physical and chemical characteristics of contaminants and how these govern transport and fate are key to predicting the effects of climate change on global contaminant cycles. Section 2.4 summarizes how global change may affect contaminant pathways.

2.3.1 Air-snow exchange of POPs (Wania and Halsall, Section C.1)

Snow and ice have a strong impact on how organic pollutants reach Arctic terrestrial and marine ecosystems. In particular, snow is a key component in atmospheric contaminant deposition in Arctic regions where contaminant release to soils, vegetation, and water bodies takes place during snow and ice melt influencing both the location and timing of the delivery.

The efficiency with which snow or rain scavenges contaminants from the atmosphere is quantified using precipitation scavenging ratios. Precipitation scavenging ratios for one contaminant can be highly variable as a result of the different mechanisms operating at temperatures above and below 0°C, and the temperature influence on the partitioning between gas phase, particle phase, and the aqueous phase. The latter influence always leads to higher scavenging ratios at lower temperatures. Whether rain or snow is more efficient in scavenging organic contaminants depends on the properties of the chemicals (Henry's law constant, sorption coefficient on the ice surface) and the characteristics of the precipitation (snow surface area, particle scavenging efficiencies). High snowfall rates and scavenging efficiencies have been identified as causing elevated organic contaminant deposition fluxes at high elevations in mid-latitudes.

Snow's capacity for non-polar organic contaminants is dictated by specific surface area, which in natural snow varies over two orders of magnitude (approximately 0.01 to 1 m^2/g), and typically decreases over the time scale of days as the snow ages. Such decrease may lead to semi-volatile contaminants partially re-volatilizing from aging snowpacks back into the atmosphere.

A simple snowpack model was used to describe the behaviour of selected POPs during snow melt in the Amituk Lake watershed on Cornwallis Island, reproducing the observed preferential elution of HCHs in the first meltwater fractions, and the relative retention of PCBs within the snowpack.

2.3.2 Modelling mercury in freshwater lakes (Diamond *et al.*, Section C.2)

The deposition of mercury from the air onto snow during mercury depletion events has been described in studies at Alert and Kuujuarapik (Steffen *et al.*, Section B.6). Mercury depletion events provide a means to convert gaseous mercury to forms that can be deposited efficiently to surfaces. Mercury so deposited may then enter meltwater and subsequently terrestrial aquatic systems (lakes, rivers, groundwater).

A mass balance for the Amituk Lake watershed revealed that only about 30% of all the mercury that had accumulated in snow in the watershed entered the lake in summer meltwater. While there were uncertainties in the calculations, the remaining 70% of the deposited mercury may have been re-emitted to the atmosphere. Snowmelt accounted for 95% of water flow into the lake, with groundwater and direct precipitation accounting for the other 5%. Of the mercury entering the lake, 59% was estimated to have passed directly across or just beneath the frozen lake surface to the outflow with negligible mixing due to the persisting ice cover. Another 25% of the mercury was deposited on the lake bottom through sedimentation processes leaving only 16% of the mercury in the water column. A large proportion of the mercury that settled on the lake bottom, 39%, was attributed to weathering of rocks within the watershed. Twenty-seven percent (27%) of all mercury entering Amituk Lake in meltwater was estimated to have come from weathered rock. The Amituk Lake mass balance calculations relate watershed characteristics to the efficiency of mercury retention within the lake.

2.3.3 Modelling persistent organic pollutants in freshwater lakes (Diamond *et al.*, Section C.2)

The mass balance for POPs in Arctic lakes was completed to assess previously unknown process rates for water-gas exchange and degradation, and to predict rate of recovery once inputs cease.

Like mercury, most POPs that reached Amituk lake via meltwater passed across or just beneath the frozen surface of the lake to the outflow with very little mixing into the lake. For the more soluble compounds like HCH, 96–99% of the load remains in the water column, whereas only about 32% of the less soluble compounds like PCBs and DDT remain in the water column. Fugacity calculations for air-water exchange of contaminants suggest the lake water to have been at equilibrium with the air and that losses to the atmosphere were negligible.

Degradation half-lives were estimated for α- and γ-HCH as well as endosulfan I using an unsteady-state version of a mass balance model. The half-lives of α-HCH and γ-HCH in Amituk lake were estimated at 0.9 years and 1.25 years, respectively. These half-lives are much shorter than those measured in the Arctic Ocean, which have been estimated at 4.7 and 16.1 years, respectively. Microbial degradation is thought to be the primary degradation pathway with hydrolysis playing a very minor (< 2%) role. Endosulfan-I had the shortest half-life, 0.046–0.147 years, entirely attributed to loss through neutral hydrolysis. The half-life of heptachlor epoxide was difficult to calculate due to highly variable analytical results but was estimated at 3.4–11 years. A model was run with starting concentrations for June 1994 assuming zero input to the lake in subsequent years. Results of the simulation indicated that degradation processes would reduce concentrations of endosulfan by 90% within one year and endosulfan I within 3–4 years. The more persistent POPs would be reduced by 90% within 6–8 years through annual dilution with meltwater and outflow of contaminants.

A major conclusion of this study was that the Arctic Ocean, not Arctic lakes, is more likely to receive contaminants previously deposited on frozen terrestrial surfaces via transportation by meltwaters. Examination of mercury profiles in lake cores has provided evidence that mercury deposition in sediments of Arctic freshwater lakes has increased by a factor of 2.5 since the 1940s in advance of which mercury levels appear to have been stable. The cause of the observed increase is likely due to numerous interrelated factors that could include a combination of increased anthropogenic emissions as well as effects brought on by climate change.

2.3.4 Modelling and mass balance of HCHs in the Arctic Ocean (Li *et al.*, C.3)

A large data base for α-HCH together with multimedia models show this chemical to have exhibited classical 'cold condensation' behaviour. The surface water of the Arctic Ocean became loaded between 1950 and 1990 because atmospheric transport of α-HCH from source regions to the Arctic was rapid and because α-HCH partitioned strongly into cold water. The α-HCH that accumulated under the permanent ice pack of the Canada Basin was not easily dissipated following emission reductions during the 1980s, leading to the situation where the highest oceanic concentrations in the early 1990s were to be found in the Arctic.

A mass balance for the upper 200 m of the Arctic Ocean for conditions typical of the mid-1990s indicated annual input of 186 t of α-HCH and 81 t of γ-HCH (58% and 39% by ocean-currents, 29% and 12% by atmospheric deposition, and 13% and 49% by rivers). Total annual losses were -555 t of α-HCH and –79 t of γ-HCH (49% and 58% by ocean-currents, 43% and 34% by degradation, ~7% by volatilization, and ~1% by ice export for both isomers). Microbial attack was estimated exceed chemical hydrolysis as a degradation mechanism. The relative importance of input processes appears to have changed over time. For example, in 1980 atmospheric deposition and advection through the Bering Strait were estimated to transport equal amounts of α-HCH to the Canada Basin, whereas by 1995 the ocean took over as the main delivery vehicle.

Beta-HCH does not exhibit a strong predisposition to accumulate under the pack ice of the Arctic Ocean as might be inferred from its co-emission with α-HCH and a Henry's law constant twenty times lower. In contrast, β-HCH appears have been less subject to direct atmospheric loadings into the high Arctic. Rather, it was rained out or partitioned into North Pacific surface water and subsequently entered the Arctic by ocean-currents passing through Bering Strait.

The lessons provided by β-HCH are that environmental pathways must be comprehensively understood before attempting to predict the behaviour of one chemical by extrapolation from a seemingly similar chemical. The predominantly oceanic pathway of β-HCH into the Canada Basin suggests that other chemicals, which have high solubility and persistence in water, could also be delivered to the Arctic via ocean-currents.

2.3.5 Modelling in support of international negotiations (Wania, Section C.4)

Zonally averaged global box models have been used to describe the transport and fate of hexachlorocyclohexanes (HCHs) and selected polychlorinated biphenyl (PCBs) congeners over the entire time period of their release to the environment. These simulations reveal the origin of elevated levels of these substances in the Arctic environment, pinpoint the major loss processes on a global scale, and identify the importance of various transport pathways.

Global model calculations reproduce the observed shift to lighter, more volatile congeners of PCBs, which is consistent with fractionation during transport. They further show that these patterns can been established with only minor fractions of the global inventory of PCBs being transferred northward. Atmospheric degradation, particularly of lower chlorinated congeners and transfer to the deep sea through sedimentation have been identified as the major global loss processes for PCBs. As new emissions of PCBs decline, contaminated soils are becoming the dominant environmental storage medium — and potential lingering source — and, therefore, will also control the overall global cycling.

Global models have been used to identify the characteristics that make a chemical susceptible to enrichment in Arctic ecosystems. High potential for contamination in the Arctic is predicted for very hydrophobic, semi-volatile organic substances, if emitted into the atmosphere. Water-soluble, highly persistent substances that are predominantly carried by ocean-currents can experience delayed Arctic contamination as a result of slow oceanic transport to the North.

Models with high spatial resolution have been employed to describe the transport of HCHs and PCDD/Fs to the Arctic region. Using detailed data on global sources of these contaminants, models are able to predict inputs to chosen receptor locations and identify the most important source regions contributing to the contaminant input at those locations.

In direct response to the needs arising from recent UNEP and UN-ECE protocols on POPs, a variety of simple models has been developed that allows the assessment of a chemical's overall persistence and spatial range using information on its physicochemical properties. These models take into account the influence of a chemical's distribution in environmental media on its ability to persist and be transported over long distances. Several models requiring varying amounts of input data have been developed to estimate a chemical's persistence and long-range transport potential. The more accurate models require the largest amount of input data, including physical chemical characteristics, degradation rates in different media, as well as emission characteristics. Simpler models which are useful as preliminary screening tools provide reasonable estimations of persistence and transport potential based solely on physicochemical characteristics.

INAC

2.4 The interaction of climate change and contaminants (Macdonald *et al.*, Section D.1)

The environmental pathways that transport contaminants to the Arctic and redistribute them within the Arctic are subject to alteration as a result of global climate change. Indeed, the Arctic is especially sensitive to change because climate conditions span the phase change between liquid water and ice. Changes in physical pathways can alter the strength of contaminant delivery to the Arctic, the efficiency of transfer between media, and the destination of contaminants within the Arctic. Change in physical conditions (temperature, winds, currents, ice cover) can have dramatic consequences for biological systems (trophic structure and population distribution) which themselves are crucial components of the pathway for contaminants like Hg and the organochlorines.

During the 1990s, the Arctic has shown a surprising capacity for rapid change in wind fields, ice drift, and ice cover, ocean-currents, and in precipitation. These changes, brought on by variation in large-scale atmospheric pressure distribution (tracked by the Arctic Oscillation Index), have altered transport routes and transfer efficiencies for every category of contaminant. The strong positive manifestation of the Arctic Oscillation index in the early 1990s may provide a very good proxy for the sorts of changes likely to occur under global warming and thus provides a valuable warning of future conditions.

Contaminant transport and distribution within the Arctic have undoubtedly changed as a consequence of the observed changes during the 1990s in the circulation of Arctic Ocean water and ice. The ice shifted its trajectories such that the transpolar drift curved over toward the Canadian Basin and the Beaufort Gyre reduced in size, retreating farther into the Canada Basin. Over a longer period of time, ice cover has decreased, especially in the marginal seas. The rivers draining into the Russian Shelves became diverted toward the east (East Siberian Sea) and probably entered the Canada Basin instead of exiting quickly from the Arctic Ocean through the Transpolar Drift.

Further changes affecting contaminants could be that the development of a more ice-friendly climate will encourage commercial transport, industrial activity, and mineral exploration (e.g., offshore oil) each of which bring with them well-known contaminants. Ultimately, regions of the Arctic may begin to exhibit contamination typical of other global transport corridors (oil, metals, antifoulants, sewage, litter).

Trends in contaminant concentrations derived from observational time series and proxies (e.g., sediment cores, ice cores, tree rings) almost never relate directly to emissions. Rather they are affected by changes in the pathway between the emission and the place at which the record is collected, and they are also affected by characteristics of the recorder (e.g., diagenesis in sediments; temperature cycling and sintering in ice cores; lactation in marine mammals). These environmental pathway transfers, which are often exceptionally sensitive to global change, are almost never considered when interpreting trend data. Therefore, time series and spatial data must be interpreted with caution as contaminant patterns may relate to climate change or emission change or, more probably, both.

Under the influence of climate change, enhanced air-sea exchange due to larger areas of open water could lead to greater transfer of oversaturated components from the water to the air (α-HCH) and greater input to water of undersaturated components (PCBs, DDT). Shifts in Arctic Ocean circulation patterns may result in contaminant burdens of the Russian rivers (relatively high concentrations of some POPs, including PCBs), entering the Canada Basin where they will have a longer residence time and where they may then pass into the Canadian Archipelago. Changes to foodweb structure (altering the number of trophic levels) will cause significant contaminant changes in apex feeders, either up or down depending on the direction of change. These changes can be caused from the bottom up (change in nutrient supply, stratification, winds) or top down (change in ice climate or harvesting practices that remove top predators). Change in access to resources through loss of ice cover or loss of permafrost may cause fundamental change to the diets of Northerners who depend on traditional/country foods. Again, this could increase or decrease exposure to POPs depending on how the diet shifts.

Mercury generally enters the aquatic environment either dissolved in meltwater or runoff or by scavenged atmospheric particles settling on the water's surface. It escapes through gaseous evasion of reduced mercury. Decreasing ice cover brought about by climate change will alter the balance of the mercury cycle as it currently functions through a decrease in frozen surface area and an increase in air-water exchange. Deposition through mercury depletion events may change in strength and location. Uptake and accumulation of mercury in the Arctic foodweb may be altered through changes in food-chain structure, mainly an increase in trophic levels. These changes could enhance mercury exposure in apex feeders including Northerners who rely on traditional/country foods. Increased organic carbon and nutrient loadings to lakes (through warming and permafrost degradation) will probably enhance Hg fluxes into lakes and coastal oceans potentially causing increases in concentration.

Climate change induced alterations of oceanic circulation may contribute to changes in metals distribution in the Arctic Ocean. Input of Pb to the Eurasian Basin appears to have occurred via the Atlantic Ocean or the Laptev Sea. Over the past 50 years or so, these pathways have placed little of the contaminant Pb into the Canada Basin, a situation that may have changed with the high Arctic oscillation index of the early 1990s. The natural geochemical cycle of Cd in the upper water column is the single most important control on Cd distribution in the Arctic Ocean. Recent changes in climate have the potential to alter Cd pathways by altering the inflow the Pacific water through Bering Strait (Cd rich waters) or upwelling along shelf margins or in the Archipelago.

Diversion of Russian rivers to the east and alteration of ice-drift trajectories may change where artificial radionuclides discharged to the Laptev and Kara Seas go. In particular, as was seen under strong AO+ conditions of the early 1990s, there is a greater chance for these contaminants to enter the Canadian Arctic and the Archipelago. Probably the most significant change for radioactivity will come from the natural radiodecay series of Uranium; ^{222}Rn will escape more easily from non-frozen soil to enhance environmental fluxes of ^{210}Pb and ^{210}Po.

ITK/Eric Loring

CHAPTER 3

Conclusions and Recommendations

3.1 Properties and sources

3.1.1 Physicochemical properties of persistent organic pollutants

Conclusions:

- The physicochemical properties and degradation rates of a chemical control its transport and distribution among environmental compartments. These properties and rates depend strongly on temperature, which must be taken into account when modelling chemical behaviour in the Arctic.
- Research has begun to quantify physicochemical properties, such as octanol-air partition coefficients (K_{OA}) and Henry's law constants (HLC), and microbial degradation rates in seawater under Arctic conditions, and to apply the results to descriptions of chemical pathways.

Recommendations:

- Physicochemical properties for more substances should be measured as a function of temperature. Similarly, degradation rates of chemicals in air, water and soil in cold climates are often lacking and should be determined. This information is required to assess chemical partitioning and fate in the Arctic.
- Quantitative Structure Property Relationships (QSPRs) and other estimation techniques should continue to be investigated and developed. These techniques are particularly useful in the assessment of new and existing substances for which measurements are lacking.

3.1.2 Usage and emissions of organochlorine pesticides

Conclusions:

- Global emissions data are important as a key input for global POPs models and for the study of sources and pathways of contaminants to the Arctic.
- Globally gridded (1° lat x 1° long) estimates for chemical use and emissions were derived for β-HCH and updated for α- and γ-HCH.
- Emissions inventories combined with results of long-term atmospheric monitoring in the Arctic suggest that a reasonable estimate of air concentrations of α-HCH in the western Arctic can be inferred from global emissions data.
- Gridded (1/6° lat x 1/4° long) inventories for toxaphene present in US agricultural soils and associated emissions inventories have also been derived.

Recommendation:

- International cooperation is required for continuous updates of global emissions datasets and to expand these inventories to more and new compounds.
- Further research should expand current emission inventories, especially in eastern Asia.
- Linkages should be improved between atmospheric monitoring results and emission estimates to facilitate the determination of transport pathways and identification of major source regions. This research should also expand to include new substances.

3.1.3 Organochlorines in Russian rivers and loadings to Russian northern seas

Conclusions:

- Previous estimates of loadings of α-HCH and γ-HCH (15 and 6 t/y, respectively) from Russian rivers for the early 1990s have been revised upwards to 20 and 35 t/y, respectively. These higher loadings imply that Russian rivers accounted for about 40% of total inputs of α-HCH and 80% of γ-HCH (via precipitation, gas exchange and riverine flow and excluding ocean-currents) to the Arctic Ocean at that time.
- Fluxes of DDT related compounds to the Arctic Ocean via Russian rivers are large (14.5 t/y). In the early 1990s, DDT fluxes estimated for the Ob River were about 240-times higher than those for the Mackenzie River (~0.03 t/y).
- POPs concentrations in Russian rivers appear to be declining in response to the cessation of select organochlorine pesticide use. Whereas concentrations of 9–69 ng/L for α-HCH and 16–114 ng/L for γ-HCH were estimated in the previous assessment for rivers entering the Kara Sea, measurements in 2000–01 suggest concentrations now to be <0.5 ng/L.
- The input of PCB to the Arctic Ocean from Russian rivers remains a major uncertainty. PCBs are still used in electrical systems in Russia and thus urban areas along north flowing rivers are likely significant sources. Preliminary measurements suggest low ng/L concentrations similar to estimates made for the AMAP Assessment. However, due to uncertainties about contamination of water samples, reliable PCB data are not yet available for any Arctic rivers.

Recommendation:

- Due to the importance of riverine inputs of POPs to the Arctic Ocean and nearshore waters, circumpolar measurements of POPs in river water should be undertaken. Results for PCBs in river water need to be validated to assure that they do not reflect contamination during collection or analysis.
- A much wider array of chemicals including the more water soluble chemicals (e.g., current-use pesticides, surfactants, personal care products and pharmaceuticals) should be measured in the future.
- New technologies, such as semi-permeable membrane devices should be evaluated for obtaining long-term concentrations in river waters, especially for PCBs. Sites for long-term monitoring of temporal trends should be selected.

3.2 Occurrence and trends

3.2.1 Persistent organic pollutants in the Arctic atmosphere

Conclusions:

- Atmospheric concentrations of the lower-chlorinated PCBs in Arctic air declined from 1993 to 1997, probably reflecting atmospheric declines in temperate regions. No clear trend was apparent for the penta- and hexachlorinated congeners, likely due to slower degradation rates in the Arctic coupled with global fractionation or high-temperature emission sources maintaining Arctic air concentrations.
- Most of the OC pesticides showed declining trends at Alert over the five years, except dieldrin and endosulfan I. Several OCs were still being used within the last 10 years and endosulfan is a current-use pesticide. Concentrations of OCs showed a strong seasonal cycle with maxima before and after the warmest time of the year.
- Episodes of elevated OC pesticide air concentrations at a high altitude site in the Yukon were associated with air originating from Asia.
- Significant decreasing trends of chlordanes along with their chemical signatures has provided evidence that emission of old soil residues is replacing new usage as an important source to the atmosphere.
- PAH concentrations at Spitsbergen were generally 2–3 times higher than those found at Alert and Kinngait (eastern Arctic). The PAH levels in the Arctic mimic those at mid-latitudes consistent with long-range transport to the Arctic, particularly for lighter PAHs. A decline in particulate PAH was observed, similar to SO_4^{2-} and attributed to the collapse of industrial activity in the former Soviet Union between 1991 and 1995.
- PCDD/Fs were found in air during a wintertime short study at Alert. Although levels were low, PCDD/Fs could accumulate in other Arctic environmental media due to the low temperatures and slow degradation processes.

ITK/Eric Loring

Recommendations:

- Continuous atmospheric measurements are still required for "legacy" POPs such as PCBs and OC pesticides. The 5 to10 years of atmospheric data currently available barely provide enough information to develop statistically significant trends for forecasting purposes.
- OC pesticides are still currently used in some countries and re-emitted from contaminated soils in others. Further research is required on atmospheric concentrations, agricultural usage, and soil concentrations in eastern Asia to gain knowledge on the role this region plays as a source of pesticides to the Arctic.
- Correlation of emission inventories and OC levels in Arctic air, coupled with source-sink fingerprinting of OC compositions, should be investigated to facilitate the determination of transport pathways and identification of major sources.
- Detailed fingerprinting of PAH emission profiles and their relationships with combustion indicators, e.g., black carbon and SO_4^{2-}, in different source regions is required. Source fingerprinting and trends in PAH concentrations/profiles in Arctic air may reflect changes in global energy consumption resulting from changing economic conditions. These include fluctuation of petroleum prices, movement of industrial activities to Asia following China joining the World Trade Organization, and government decisions on ratification of the Kyoto Protocol. These factors would also affect the atmospheric levels of PCDD/Fs.
- There is a need to include new chemicals in the atmospheric monitoring network to serve as an early warning for their transport to the Arctic and to address growing concerns of their environmental impacts.
- The PCB class of compounds is a good POPs surrogate to be used for the development and verification of regional/global transport models, and research in this area is recommended.

3.2.2 Persistent organic pollutants in seawater

Conclusions:

- New measurements of POPs have filled gaps in knowledge of spatial trends especially for HCH isomers. Measurements of other POPs like DDT and chlordane in seawater remain limited especially in the European Arctic. For PCBs there are major differences in concentrations reported for the same geographical area and depth.
- All ΣPCB values greater than about 0.04 ng/L appear to be invalid due to contamination. The most reliable results suggest that Σ10PCB ranges from about 0.001–0.010 ng/L in open waters of the Bering, Chukchi, Barents and Greenland Seas and from 0.0005–0.001 ng/L under the permanent ice pack (Barents Sea area). Total (dissolved plus particulate) ΣPCB concentrations estimated during the previous AMAP assessment can be considered five times too high. Low water concentrations imply that air-to-water exchange probably accounts for the major input pathway to the Arctic Ocean.

Recommendations:

- High-quality data on the inventories and concentrations of PCBs in Arctic Ocean seawater are urgently required to validate models, to interpret trends in marine biota and to understand spatial variation. Further work is urgently needed to compare different sampling methods and to evaluate the extent of contamination on ships and in the laboratory.
- A program to measure legacy and emerging POPs in seawater in the Canadian Arctic should be built on the work conducted in the 1990s. Parallel time-series data for POPs in seawater are crucial for understanding the fate and trends of contaminants, especially in Arctic food webs.
- Retrospective studies of β-HCH illustrate that ocean transport of some POPs may have been widely underestimated. There is a need to conduct similar studies for other persistent and less volatile chemicals.
- New technologies should be investigated for obtaining long-term concentrations in ocean waters and seasonal variation, especially of PCBs. Sites for long-term monitoring of temporal trends should be selected.

3.2.3 Persistent organic pollutants in marine and lake sediments

Conclusions:

- Major OC group concentrations were the highest in marine surface sediments from Hudson Bay.
- Total PBDE concentrations in surface sediment samples from Barrow Strait, Penny Strait, and Nanisivik ranged from 100 to 300 pg/g, and ΣSCCP ranged from 4.8 to 77.4 ng/g.
- Concentrations and congener profiles for ΣCHB (toxaphene) suggest only atmospheric sources in Laberge and Kusawa Lakes, and contamination by non-atmospheric sources in Watson, Fox, and Hanson Lake sediments.
- Clorophenols appear to have entered the sediments of a small Yukon sub-alpine headwater lake (Coal Lake) via long-range atmospheric transport.
- An undisturbed, laminated lake-sediment core collected from Devon Island contained vertical profiles for DDT, toxaphene, and chlordane reflecting their known usage in North America. Maximum SCCP concentrations occurred in the surface sediment and at a sediment depth corresponding to 1956. PCDFs levels start to increase in the early 1950s, and peak in 1978. Levels of PCDDs start to increase about 10 years earlier but also peak in 1978. Older samples (1938–1963) have homologue profiles consistent with impurities in pentachlorophenol whereas later samples (1970 to 1992) imply combustion of coal and wood as sources.
- In Great Slave Lake, PCB, PAH, and PCDD concentrations were highest in surface sediments offshore of the Slave River mouth and decreased with increasing distance from the mouth. The West Basin of Great Slave Lake seems to be depositional sink for contaminants entering the lake with Slave River inflow (and from the atmosphere).

Recommendations:

- Laminated cores like the ones collected from Lake DV-09 on Devon Island contain exceptionally valuable historical records of contaminants undisturbed by sediment mixing. Other Canadian Arctic lakes with similarly laminated sediments should be sampled and analyzed for HOCs, especially current-use contaminants such as PBDEs, SCCPs and endosulfan to provide a comparative, spatially large data set.
- Further water column (dissolved and particulate), surface sediment and plankton HOC measurements should be made in Arctic marine waters to assess the importance of biogeochemical processes (primary production, vertical flux) relative to other processes such as air-water exchange.

3.2.4 New persistent chemicals in the Arctic environment

Conclusions:

- Chemicals that are currently used within the circumpolar region and other industrialized countries have been identified in the abiotic Arctic environment. These include brominated flame retardants (BFRs), short-chain chlorinated paraffins (SCCPs), chlorinated phenolic compounds and modern agrochemicals. Other "legacy" industrial chemicals were identified which have not been part of the NCP monitoring program in the past: polychlorinated naphthalenes (PCNs), coplanar PCBs and octachlorostyrene.
- Among the BFR chemicals, polybrominated diphenyl ethers (PBDEs) were the only ones of significance found in Arctic air. PBDE levels in air were highest at Tagish, intermediate at Alert and lowest at Dunai and were higher during the warmer months. Hexabromocyclododecane and tetrabromobisphenol A were below detection limits.
- SCCPs have been found in lake sediments in the Canadian Arctic and in marine sediments from the Archipelago but have not been measured in Arctic air.
- The currently used pesticides endosulfan, trifluralin and methoxychlor and the pentachlorophenol metabolite pentachloroanisole were identified in Arctic air. Endosulfan was found in surface water of the eastern Archipelago and Holman Island at concentrations similar to those of the banned OC pesticide chlordane.
- Highest levels of atmospheric PCNs were found over the Barents Sea, followed in abundance by the eastern Arctic Ocean, Dunai, and Alert. Episodes of PCNs and coplanar PCBs over the Barents Sea were associated with air masses originating in the U.K. — central Europe region.
- PCNs in air at Dunai and Alert were generally higher in winter-spring than in summer-fall, suggesting an association with combustion and combustion marker PCN congeners were identified. Coplanar PCBs were highest in spring-summer, particularly at Dunai. Russian PCB mixtures enriched in pentachlorinated PCBs may be responsible.

Recommendations:

- Surveillance of these substances should be continued and searches made for additional new contaminants in the Arctic. Presence in the Arctic and other remote regions is recognized internationally as a key point in evaluating chemicals for persistence.
- The ability of currently used chemicals to undergo long-range transport should be assessed through modelling.
- Photochemical products of contaminants should be identified and evaluated with regard to their potential for atmospheric formation and transport.

3.2.5 Mercury in the Arctic atmosphere

Conclusions:

- Arctic springtime mercury vapour depletion events (MDEs) occur each year after polar sunrise in the Arctic at Alert, Kuujjuarapik and elsewhere (Alaska, Svalbard, Greenland, Russia).
- MDEs provide a mechanism by which a large pulse of bio-available mercury is introduced to Arctic ecosystems at a time when the Arctic emerging from a long polar night.
- During MDEs, gaseous elemental mercury is converted through reactions involving sea salts to more reactive species either in reactive gaseous form or associated with particles. The mercury concentration in snow during and after MDEs increases significantly, demonstrating that the converted mercury species are deposited onto the snow surface. A proportion of the deposited mercury, presently not well quantified, is subsequently lost from snow.
- Depleted levels of mercury are observed mainly in the lower 1 km of the atmosphere.

Recommendations:

- Interdisciplinary studies should be initiated to determine the significance of MDEs to Arctic ecosystems by observing not only the MDEs themselves, but the pathways taken by the mercury after deposition including re-evasion to the atmosphere, entry into water and entry into biota. Such studies should be designed to produce a mercury multi-media mass balance including atmosphere, snow, ice, water and biota.

- Further work is required to understand the chemical reactions involved in MDEs and their sensitivity to climate variables such as ice cover, snow formation and transport, and the injection of salt onto ice and snow surfaces. To determine the spatial scale of MDEs, comparative studies over inland locations remote from the sea should be undertaken.
- Present methodology, which yields operationally-defined mercury components (e.g., RGM, TGM), needs to be enhanced through the development of analytically rigorous and species-specific methods.
- Instrumental time series for atmospheric mercury (reactive gaseous, particle-phase, Hg (0)) should be maintained to establish seasonal variation and whether concentrations are increasing or decreasing with time.

3.2.6 Mercury in abiotic samples

Conclusions:

- Sediments, primarily from freshwaters, contain mercury, mostly less than 0.17 µg/g, the guideline indicated by Canadian Sediment Quality Guidelines.
- Amounts and forms of mercury in water and ice are unknown because samples, freshwater or marine, have not been analyzed. Consequently, the importance of potential transport pathways like rivers, ocean-currents, and pack ice is unknown.
- Sources of the mercury in abiotic samples remains difficult to determine. Sediment core profiles suggest that approximately half the mercury retained by lakes in the eastern Canadian Arctic may be of anthropogenic origin with the other half supplied by local geology. In contrast, lakes in the northwestern Arctic and the extreme High Arctic may have received less anthropogenic input.
- Experience elsewhere with effects of temperature and flooding of northern soils by hydroelectric developments suggests that climate warming may accelerate the movement of mercury into biological food chains.

Recommendations:

- The amounts and forms of mercury in water, both freshwater and ocean water, should be measured together with stream flow to estimate mercury fluxes and budgets. A long-term water-monitoring program should be considered.
- Improved estimates are required of the relative contributions of natural and/or anthropogenic sources of mercury.
- The absolute and relative biological availability of anthropogenic and natural forms of mercury should be measured using recently developed methodology (genetically modified strains of microorganisms)
- Research should be conducted on the bio-availability of mercury in sediments including methylmercury and other mineralogical forms and their relationships to, for example, organic carbon and oxides of iron and manganese.
- The sensitivity of mercury cycling to climate change (temperature, hydrology) should be investigated, especially processes like methylation, demethylation and release of geological forms of mercury from soils.

3.2.7 Other heavy metals in Arctic air

Conclusions:

- Based on observations between 1980 and 2000 there does not appear to be increasing trends for anthropogenically derived metals in the atmosphere but rather a slight decrease. While Zn and Cu showed a slight increase from 1980 to 1985, they seem to have decreased thereafter.
- Due to analytical limitations of the Pb data, it was not possible to identify long-term trends with confidence. Observations based solely on the winter data, however, suggest that over the long term Pb concentrations have decreased.
- Anthropogenic-derived metals such as Pb, Zn, and Cu and sea-salt components Na and Mg exhibited a maximum in winter and minimum in summer. Similar variations were observed for non-soil fractions of V and Mn. No significant seasonal variations were found for the soil-related metals Al, Ba, Ca, Fe, and Mn. No long-term trends were detected for Na and Ca.

Recommendations:

- More detailed analyses should be done to identify the source and transport patterns of various metals. A chemical transport model with anthropogenic source emissions, deposition, and transport should be developed to more quantitatively characterize the metals at the Arctic region.

3.2.8 Radionuclides — Local and long-range sources

Conclusions:

- The artificial radionuclides, ^{129}I and ^{137}Cs, discharged from European nuclear fuel reprocessing plants can be measured at very low levels throughout the entire Arctic Ocean. Radioactivity from these sources does not presently present a radiological threat in the Arctic to human health or to the environment.
- Local sources of radioactivity in the Russian Arctic Ocean, such as Chernaya Bay and other fjords on the island of Novaya Zemlya in the Kara and Barents Seas represent potential future threats to the environment, but are not currently producing significant contamination of the Arctic Ocean.
- Large inventories of radioactivity remain in floating storage on ships and submarines near Murmansk on the Kola Peninsula in Russia. The secure storage and future disposal of these materials remains a formidable challenge for Russian authorities.

Recommendations:

- Measuring sources of radionuclides to the Arctic and the assessment of temporal trends should continue in order to monitor future global emissions and assist with the detection of input from Russian shelves and rivers.
- Monitoring of contaminated sites in the Kara and Barents Seas is necessary to ensure that the inventories of radioactivity at these locations remain isolated and secure.
- Canada should endorse activities to secure stored Russian radioactive waste.

3.2.9 Local sources of contamination in the Canadian Arctic

Conclusions:

- Local sources of contamination in the Canadian Arctic continue to be a concern for Aboriginal people whose communities and traditional hunting grounds tend to be close to these sites.
- At a military radar station (Saglek, Labrador), chosen as a worst-case site, PCB contamination was found up to 30 km away in soil and plants and up to 10 km away in marine sediments. Terrestrial PCB contamination here and at Resolution Island has since been cleaned up.
- PCB-contaminated marine sediments at Saglek were assessed to be significant risks to the survival and reproduction of local shorthorn sculpin and black guillemots.
- Mining activities near Yellowknife, NWT, produced high, localized arsenic contamination. Based on bioavailability, the arsenic-contaminated surface soil does not present ecological or human health risks.
- Radionuclide and heavy metal contamination at Port Radium has been delineated but remediation awaits the results of ongoing ecological and human health risk assessments.
- Sediments in the vicinity of a railway tie treatment facility in Carcross, Yukon, reflect past use of PCP as a wood preservative at the site.

Recommendations:

- Integrated abiotic/biotic studies at locally-contaminated sites should be used to evaluate contaminant uptake by — and effects on — local ecosystems. The wide range of contaminant concentration provides a solid basis to reduce uncertainty in estimating the ecological and human health risks.

3.3 Process studies and modelling

3.3.1 Air/snow exchange of persistent organic pollutants

- Snow is a key component in atmospheric contaminant deposition in Arctic regions. Contaminant release to soils, vegetation, and water bodies takes place during snow and ice melt, influencing both location and timing of the delivery.
- Precipitation scavenging ratios for one contaminant can be highly variable as a result of both the different mechanisms operating at temperatures above and below 0°C, and the temperature influence on the partitioning between gas phase, particle phase, and the aqueous phase. The latter influence always leads to higher scavenging ratios at lower temperatures.
- High snowfall rates and scavenging efficiencies have been identified as causing elevated organic contaminant deposition fluxes at high elevations in mid-latitudes.
- The specific surface area of natural snow varies over two orders of magnitude and typically decreases over the time scale of days as the snow ages. Such a decrease may lead to semivolatile contaminants partially revolatilizing from aging snowpacks.
- A simple snowpack model was used to describe the behaviour of selected POPs during snow melt in the Amituk Lake watershed on Cornwallis Island, reproducing the observed preferential elution of HCHs in the first meltwater fractions, and the relative retention of PCBs within the snowpack.

Recommendations:

- Additional research is needed to gain an improved quantitative and mechanistic understanding of the snow scavenging of organic contaminants, behaviour of the organic contaminants in snowpacks, and to understand the Arctic snowpack as a transient sink for atmospherically deposited contaminants and source to aquatic systems.
- Models of snow scavenging and snowpack behaviour of organic contaminants should be further developed to a point where they can be incorporated in general fate and transport models.

3.3.2 Mass balance and modelling of contaminants in lakes

Conclusions:

- Studies in the Amituk Lake watershed on Cornwallis Island showed that atmospheric deposition to the snowpack is the main source of contaminants to Arctic freshwater systems and that the snowpack is a transient sink for atmospherically deposited compounds such as Hg and POPs.
- Bedrock weathering was estimated to contribute about 27% of total runoff loadings of Hg to Amituk Lake with the remainder originating from atmospheric deposition. Sediment core data suggest that recent activities have increased Hg deposition and retention in Amituk Lake by about 2–2.5 times compared to before the 1950's.
- Atmospherically deposited Hg in early fall snowfalls may be subject to volatilization as the snow is scoured during winter. During polar sunrise, Hg concentrations in the snowpack can increase by up to two orders-of-magnitude coincident with atmospheric ozone depletion. Much of this Hg volatilizes from the snowpack and/or is retained by watershed substrates during snowmelt. Thus, measured the Hg burden in the snowpack is not a reliable estimate of loadings to surface waters.
- About 25% of Hg loadings to Amituk Lake was retained in the sediments, with almost 60% leaving the lake through the outflow. Similarly, most of the POPs that were released from the snowpack to the lake during melt were discharged from the lake at the outflow. The percent retention in the sediments was low in comparison to temperate systems. Arctic lakes convey most loadings from the watershed to a downstream sink such as the ocean.
- Amituk Lake is near equilibrium with air, suggesting that the lake is not a source of POPs to the atmosphere.
- Microbial degradation rates, calculated by means of a mass balance model for Amituk Lake, were surprisingly fast for α- and γ-HCH ($t_{1/2}$ = 0.6–1.7 years) and endosulfan I ($t_{1/2}$ = 0.05–0.14 years).
- Degradation of HCHs in Arctic wetlands, small streams and lakes with long water residence times is promoted by contact time with microbially rich substrates. For α-HCH, enantioselective degradation is more favoured in nutrient-poor waters of the Arctic than in nutrient-rich waters of temperate regions. This is hypothesized to be due to the opportunistic nature of oligotrophic microbial consortia in these systems.

- Lake waters are predicted to respond rapidly (<1–2 years) to loading changes; however, sediments will take decades to respond.

Recommendations:

- Arctic lake waters should be used to monitor temporal changes in loadings to aquatic systems because of their relatively rapid response time.
- Further work is required to gain a better understanding of the role of the Arctic snowpack as a transient sink for atmospherically deposited contaminants and source of contaminants to aquatic systems.
- Studies of chemical degradation rates in Arctic lakes and watersheds should be carried out to reduce the uncertainty in estimating the response time of the Arctic to loading changes and to discriminate between processes whereby contaminants are transported from one system to another (e.g., from lakes to ocean or air) versus elimination.

3.3.3 Transport and fate of hexachlorocyclohexanes in the North American Arctic Ocean

Conclusions:

- Loadings of α-HCH to the high Arctic via direct atmospheric deposition and transport by ocean currents are estimated to have been about equally important in 1980, shifting to a dominance of the ocean pathway by the mid-1990s. This is reflective of reduced global use and atmospheric transport of technical HCH.
- β-HCH is thought to be deposited in the North Pacific via precipitation scavenging and sea-air exchange processes that are promoted by its extremely low Henry's law constant. This process only delays transport of β-HCH to the Arctic Ocean which eventually occurs through the Bering Strait. This process may result in higher food chain exposures to β-HCH in the Bering and Chukchi seas than in the rest of the Arctic.
- Delivery of other long-lived and water-soluble chemicals to the Arctic by also take place by ocean currents.
- The distinctly different behaviour of the HCH isomers provides a caution that time series need to be constructed for individual congeners rather than for the sum of congeners.

Recommendations:

- The hypothesis concerning selective removal of β-HCH from the atmosphere into the North Pacific followed by ocean current advection into the Canada Basin should be tested by studies in the North Pacific, Bering, and Chukchi regions.
- Additional studies of β-HCH in the food chain of the Bering, Chukchi, and western Canada Basin are needed to improve the linkage between the physical and biological components of these ecosystems.
- The possibility of ocean transport should be examined for substances with properties similar to β-HCH (low HLC, high persistence in water). Chemicals can be screened to find those that are amenable to this pathway, followed up by appropriate field investigations.
- Surveys are needed to provide a better understanding of spatial and vertical distribution of other POPs in Arctic and subArctic waters. Formulating a mass budget for HCHs was possible because of the relatively large database for HCHs in water and air, coupled with a complete knowledge of their physicochemical properties.

3.3.4 Modelling in support of international negotiations

Conclusions:

- Zonally averaged global box models reveal the origin of elevated PCB and HCH levels in the Arctic environment, pinpoint the major loss processes and identify the importance of various transport pathways. Spatially highly resolved modelling approaches have been employed to describe the transport of HCHs and PCDD/Fs to the Arctic region.
- Simple models have been developed that allow the assessment of a chemical's overall persistence and spatial range from information on its physicochemical properties.
- Atmospheric degradation and transfer to the deep sea have been identified as the major global loss processes for PCBs. The relative importance of these processes depends on the degree of chlorination. As new PCB emissions decline, soils become the dominant environmental storage medium, controlling the overall global loss rate.

- Global model calculations reproduce observed shifts in the relative composition of PCBs that are consistent with the "global fractionation" hypothesis. They further show that these patterns can been established with only minor fractions of the global inventory of PCBs being transferred northward.
- Elevated levels of α-HCH in the Arctic Ocean are a result of efficient northward transport in both atmosphere and oceans during periods of high usage in Asia, followed by relative enrichment caused by degradation that is much slower in the Arctic than in source regions.
- Global models have been used to identify the chemical characteristics that make a chemical susceptible to enrichment in Arctic ecosystems. A high Arctic contamination potential (ACP) is predicted for very hydrophobic, semi-volatile organic substances, if emitted into the atmosphere. Arctic contamination by water-soluble, highly persistent substances can be delayed as a result of slow oceanic transport to the North.

Recommendations:

- Results of assessment models for long-range transport potential (LRTP) should be verified by field observations; e.g., by developing means of deriving empirical transport distances from spatial concentration gradients measured in the environment.
- POP candidate substances should be evaluated in terms of their potential for long-range transport and Arctic contamination using existing assessment models. Candidate substances are those which have a high production volume and a high potential for release to the environment as well as partitioning and persistence characteristics that lead to unusually high LRTP and ACP.
- Compilation of global scale usage, emission estimates and temperature dependent physicochemical property measurements should be continued, with a shift in focus from "legacy" chemicals to "new" and candidate POPs.
- Emphasis in development of global transport models should be placed on improving the quantitative description of key environmental fate processes (e.g., air-surface exchange, gas-particle partitioning, vertical and horizontal transport in the oceans) rather than increasing spatial resolution.

3.4 Special topic

3.4.1 The interaction of climate change on contaminant pathways to and within the Canadian Arctic

Conclusions:

- Remarkable changes occurred in the Arctic's climate during the 1990s including wind and weather patterns, ice cover, ice thickness, ice drift patterns, permafrost, hydrology, ocean-currents, precipitation and temperature patterns. These changes have significant consequences for contaminant pathways.
- Change will continue to occur globally, especially within the Arctic, and inevitable surprises will occur because our understanding of couplings and transfers of contaminants in environmental systems remains incomplete.
- To understand how change will affect contaminants, physical and biological pathways must be studied as systems that interact with one another.
- The pathway of a contaminant from its emission to its accumulation in an Arctic ecosystem includes many steps each of which can be altered by global change. Risks of contaminants to Arctic ecosystems cannot be understood and assessed without considering the entire transport pathway.
- Almost all contaminant emissions have exhibited an initial rapid rise followed by a dramatic decrease when bans or controls are instituted. The atmosphere often predominates initially as a rapid transporting medium but with time other slower-moving (ocean) or stationary media (lake water, soil, vegetation) become the dominant reservoirs or transporting media.
- Models provide guidance to our understanding of contaminant transport, but many crucial processes remain insufficiently understood to be properly incorporated in such models.
- There is a lack of adequate time series for contaminants in many media (e.g., seawater). The dramatic change that occurred in the Arctic during the early 1990s makes it difficult to interpret time series spanning that change.
- Changes associated with ice and precipitation (the zero-degree isotherm), and with ecosystems (trophic structure, pathway bifurcation) have a great potential to alter pathways in ways that are particularly difficult to predict.

ITK/Eric Loring

- Alteration of the coastal margin's ice climate will lead to increased risk of contamination due to shipping, oil production and tourism. Local contaminants presently depending on permafrost for containment (sewage lagoons, dumps, tailings ponds, pipelines) are likely to become risks.
- Mercury continues to provide the greatest concern because the Arctic possesses unique, climate-sensitive processes that capture globally cycling atmospheric mercury. The mercury pathway between its deposition to ground and its concentration in apex aquatic feeders is very poorly known.
- Sea ice provides a 'short-cut' pathway for radionuclides. Change in ice-drift patterns has likely increased the potential for ice formed over the Russian shelves to reach Canadian Arctic coasts.
- Known effects organochlorines (immune function, reproduction, and endocrine disruption) can interact with nutritional stress and pathogen vectors, both altered by climate change, to produce epidemics like those observed in European seal populations.

Recommendations:

- In the future it is crucial that climate change and contaminant scientists work together, not only to better understand pathways and their variability, but also to maximize benefit and reduce costs of observation networks.
- To advance the skill of predictive models requires more research on the physical and biogeochemical linkages between systems and the ways in which such systems may change.
- Atmospheric and biological time series collections need to be accompanied by similar data for ocean water, rivers and lakes to provide a sound basis for interpretation. Such time series (type and frequency of data collection) must be interpreted in the context of natural cycles such as the Arctic Oscillation (5–7 years).
- Future emphasis should be placed on Canada's marginal seas where change will occur first and most dramatically. Hudson Bay appears to be an especially important location that has already undergone significant change in physical and biological systems.
- Disposal sites within the Arctic should be designed (or improved) given the likely loss of permafrost as a containment mechanism.
- For Hg, research is needed on the mechanisms and efficiency of transfer of Hg into water and aquatic ecosystems.
- Further surveys of radioactivity in ice, especially in the Canadian Archipelago, should be made.
- For many organochlorine compounds, fundamental physical and chemical properties still need to be measured. Partitioning onto snow or organic-rich surfaces are particularly sensitive to climate change.

PART II

Properties and Sources

A.1 Physicochemical properties of persistent organic pollutants

Authors:

T. Harner (1); T. Bidleman (2); L. Jantunen (1); F. Wania (3); and P. Helm (4).

Affiliations:

1. Meteorological Service of Canada, Environment Canada, Toronto, ON, M3H 5T4, Canada;
2. Centre for Atmospheric Research Experiments, 6248 Eighth Line, R.R. #1, Essa Township, Egbert, ON, L0L 1N0, Canada;
3. Division of Physical Sciences, University of Toronto at Scarborough, 1265 Military Trail, Scarborough, ON, M1C 1A4, Canada;
4. Freshwater Institute, Fisheries and Oceans Canada, 501 University Crescent, Winnipeg, MB, R3T 2N6, Canada.

A.1.1 Introduction

Phase partitioning of pollutant chemicals in the environment is often approximated with the help of empirical relationships involving partitioning equilibria with pure phases (pure chemical phase, gas phase, aqueous phase, n-octanol). The advantage of the latter physicochemical properties is that they are inherent to a chemical and not dependent on environmental characteristics. Key properties include: vapour pressure (p°, Pa) and the octanol-air partition coefficient (K_{OA}) which are measures of a chemical's volatility; the Henry's Law Constant (H, Pa m^3/mol) or air-water partition coefficient (K_{AW}) which describe partitioning between air and water; and the octanol-water partition coefficient (K_{OW}) and aqueous solubility (S_W, g/m^3) which are useful for describing partitioning in aquatic systems.

The partitioning behaviour of a non-polar organic chemical in the environment can be estimated using the three key partition coefficients — K_{OA}, K_{OW}, and K_{AW} (Gouin *et al.*, 2000). If two of these parameters are known, the third can be estimated (e.g., $K_{AW} = K_{OW} / K_{OA}$). Similarly, the Henry's Law Constant "H" can be estimated from the ratio of the vapour pressure and the water solubility, if both apply to the liquid or solid state. These relationships linking the physicochemical properties, however, are only approximations (Beyer *et al.*, in press). It is, therefore, preferable to measure each property independently to avoid measurement errors that are propagated in the calculation.

Of particular importance for quantitatively addressing issues related to the contamination of the Arctic with POPs, is information on the temperature dependence of these physicochemical properties. A partition coefficient at 25°C is of little relevance under Arctic conditions, if even a temperature difference as small as 5°C has been shown to have a notable impact on key physicochemical properties. (see Chapter D, Figure D.1.23, where the effect of climate change on the fate of POPs is considered). The integrated Clausius-Clapeyron equation is often used to express the temperature dependence of a physicochemical partitioning property. For a partition coefficient K_{XY}, the relationship takes the form:

$$\ln K_{XY} = m / T + \text{constant}$$

If a partition coefficient is measured at several temperatures "T", the internal energy of phase change, ΔU_{XY} (J/mol) can be calculated from the slope "m" using:

$$\Delta U_{XY} = -m \cdot R.$$

where: R is the ideal gas constant.

Internal energies of phase change (or enthalpies of phase change ΔH_{XY} if the gas phase concentration in the partition coefficient is expressed in units of pressure) are presented in Table A.1.1. It is important to realize that the ΔU (or ΔH) values for various properties are related in a similar fashion as the properties at 25°C. These relationships are explained in the literature (Paasivirta *et al.*, 1999; Mackay *et al.* (1991-1997); Cole and Mackay, 2000).

TABLE A.1.1 Recent measurements of key physicochemical properties and their temperature dependence

Chemical	Reference	T (°C)	ΔH or ΔU (kJ/mol)
Henry's Law Constant (H, Pa m³/mol)			
Toxaphene (technical mixture)	Jantunen and Bidleman, 2000	+10 to +40	61.3
HCHs	Jantunen and Bidleman, 2000 Sahsuvar *et al.*, 2002 Altschuh *et al.*, 1999	+10 to +40 +25	α-HCH = 62.6 to 63.1 β-HCH = 67.1 γ-HCH = 57.5 to 63.8
Endosulfan	Rice *et al.*, 1997	+20	not measured
OC Pesticides, PAHs, nitro PAHs	Altschuh *et al.*, 1999	–25	not measured
PCBs (n = 26, 1-Cl to 8-Cl)	Bamford *et al.*, 2000	–4 to +31	14.5 to 167
PAHs (n = 13)	Bamford *et al.*, 1999	+4 to +31	34.2 to 100.9
Chlorinated alkanes	Drouillard *et al.*, 1998a	+25	not measured
Aqueous Solubility (s, g/m³)			
Chlorobenzenes (n = 7, 2-Cl to 6-Cl), PCBs (n = 5, 1-Cl to 6-Cl) Dibenzofuran	Shiu *et al.*, 1997	+5 to +45	18.2 to 33.7 24.5 to 41.3 33.3
Chlorinated alkanes	Drouillard *et al.*, 1998b	+25	not measured
Super-cooled Liquid Vapour Pressure (p^o_L, Pa)			
Polychlorinated Naphthalenes (PCNs) (n = 17 + 58 est., 1–8 Cl)	Lei *et al.*, 2000	50 to 150	58 to 96
Polybrominated Diphenyl Ethers (PBDEs) (n = 6, 3–7 Br) (n = 23 + 7 est., 1–7 Br)	Tittlemier and Tomy, 2001 Wong *et al.*, 2001	130 to 175 90 to 200	94 to 121 63.7 to 115.8
PAHs (n = 35)	Lei *et al.* submitted	50 to 200	56.1 to 104.2
Toxaphene congeners	Bidleman *et al.*, 2002b	70 to 130	78.2 to 92.0
β-HCH	Bidleman, MSC, 2002, pers. comm.	70 to 110	66.6
Chlorinated alkanes	Drouillard *et al.*, 1998a	+25	not measured
Octanol-Air Partition Coefficient (K_{OA})			
Chlorobenzenes (n = 6, 2-Cl to 6-Cl) PCBs (n = 5, 1-Cl to 6-Cl) *p,p'*-DDT	Harner and Mackay, 1995	-10 to +20	-49 to -75 -64 to -76 -75.7
PCBs (n = 15, 1-Cl to 7-Cl)	Harner and Bidleman, 1996	-10 to +30	-66 to -93
PCBs (n = 16, 2-Cl to 7-Cl)	Kömp and McLachlan, 1997	+10 to +43	-71 to -93
PAHs (n = 3)	Harner and Bidleman, 1998a	0 to +40	-75.5 to -82.9
PCNs (n = 24, 2-Cl to 6-Cl)	Harner and Bidleman, 1998a		-62 to -96
PCDDs (n = 9, 1 to 7 Cl) and PCDFs (n = 1)	Harner *et al.*, 2000a	0 to +50	-61 to -106
PBDEs (n = 13, 3-Cl to 7-Cl)	Harner and Shoeib, 2002	0 to +50	-72.8 to -107
OC Pesticides (n = 19)	Shoeib and Harner, 2002	+15 to +45	-58 to -137
PCBs	Zhang *et al.*, 1999	0 and +20	not reported
Chlorobenzenes (n = 12) PCNs	Su *et al.*, submitted	+10 to +50 +10 to +50	-43.9 to -68.6 -58.3 to -106.6
Octanol-Water Partition Coefficient (K_{OW})			
Chlorobenzenes (n = 7)	Bahadur *et al.*, 1997	+5 to +45	-17.1 to -24.4
Toxaphene congeners	Fisk *et al.*, 1999		
Chlordanes	Simpson *et al.*, 1995	+25	
Chlorobenzenes (n = 10) PCBs (n = 3) PAHs (n = 9)	Lei *et al.* 2000	+5 to +55	-15 to -24.3 -20.9 to -24.0 -15.7 to -27.7

A.1.2 Incentives to measure physicochemical properties

In recent years chemical assessment activities have focused on risk associated with exposure to substances that are persistent, toxic, and bioaccumulative. Key examples include the Canadian Environmental Protection Act (CEPA), the POPs Protocol under the UN-ECE LRTAP Convention, and the development of a global POPs treaty under the United Nations Environment Programme (UNEP). Persistent organic pollutants (POPs) have received special attention because of their ability to be transported in the atmosphere to regions such as the Arctic that are far removed from sources.

The use of simple box models that consider chemical partitioning in a multi-media environment has become increasingly popular (see Section C.4). These models are used to screen, assess, and set priorities for chemicals based on risk. Ultimately, these findings will be used to formulate government policy and to regulate chemicals. Physicochemical property data and dependence on temperature are key data required for the successful application of these approaches. It is not possible to measure all of the necessary input data, however, because of the large number of chemicals that need to be evaluated. Quantitative Structure Property Relationships (QSPRs) and other estimation techniques are being developed and used increasingly to fill this gap (Lee *et al.*, 2000; Chen *et al.*, 2001b; Boethling and Mackay, 2000)

A.1.3 New physicochemical property data

At the time of writing the first CACAR report, temperature-dependent physicochemical properties were available for only a few POPs. In CACAR-I, vapour pressures at 20°C and ΔH values at -2°C were presented for several relevant POPs. In the past few years more and more temperature-dependent partitioning data for several classes of POPs and new compounds of concern have been determined experimentally. A comprehensive compilation of temperature-dependent properties and physical constants for several classes of POPs can be found in two recent reviews by Shiu and Ma (2000a; b). Staudinger and Roberts (2001) and Paasivirta *et al.* (1999) also presented compilations of temperature-dependent partitioning properties and estimation methods of physicochemical properties for several classes of POPs. Table A.1.1 summarizes data that are most relevant to the NCP.

A.1.4 Advances in describing chemical partitioning

A.1.4.1 Octanol-air partition coefficient

In the past few years K_{OA} has become a preferred descriptor of chemical partitioning between the atmosphere and environmental organic phases such as aerosols, soil and vegetation. Previously, partitioning to these phases was described using the sub-cooled liquid phase vapour pressure (p^o_L). K_{OA} has the advantage that it can be measured directly, unlike p^o_L which is a hypothetical state for most POPs at ambient temperatures. K_{OA} has been used extensively to describe the absorptive partitioning of chemical between the atmosphere and environmental organic phases (Finizio *et al.*, 1997a; b; Harner and Bidleman, 1998b; Pankow, 1998; Falconer and Harner, 2000)

A.1.4.2 Partitioning to aerosols

Helm *et al.* (2002a) have recently determined gas-(C_A) and particle-phase (C_P) concentrations of polychlorinated naphthalenes (PCNs) and coplanar (mono- and non-ortho-substituted) PCBs in Arctic air at Dunai, Russia, and Alert, Nunavut Territory. K_{OA}- and p^o_L-based models were applied to the particle-gas partitioning of PCNs and coplanar PCBs at ambient Arctic temperatures during the winter months. The resulting relationships are depicted in Figure A.1.1. The slopes of log C_P/C_A (normalized to total suspended particulate) versus log K_{OA} and log p^o_L were similar indicating that either K_{OA} or p^o_L -based models will adequately describe the partitioning of these compounds in Arctic air. These relationships may be incorporated into models estimating the role of deposition to the Arctic environment.

A.1.4.3 Partitioning to soil

Partitioning to soil has also been successfully modelled using K_{OA}. The soil acts as a reservoir of previously emitted POPs that are hydrophobic and partition strongly to organic media. Thus soil may buffer air concentrations of POPs. Revolatilization of POPs from soil is also important in the global distillation ("grasshopping") of chemicals from source regions to the Arctic. Hippelein and McLachlan (1998; 2000) and Cousins *et al.* (1997) have developed K_{OA}-based empirical relationships for describing the soil-air partition coefficient (K_{SA}) for various classes of POPs. Harner *et al.* (2001b) used K_{OA}-based relationships for particle-gas partitioning and soil-air exchange in the context of a multimedia fate model to describe the emission of pesticide residues from agricultural soils in the US. Results of this exercise show that several tonnes of past-usage POPs such as toxaphene, DDE, and dieldrin continue to be emitted to the atmosphere each year from previously contaminated agricultural soils. These chemicals become available for long-range transport to ecologically sensitive regions such as the Arctic (see Sections A.2.3.6, A.2.3.8, and B.1.2.2). Recently, Meijer *et al.* (2002a; in press) measured soil-air partition coefficients for OC pesticides in field experiments for various soil types, further refining the K_{OA}-based models. Results from controlled laboratory measurements on the same soils (Meijer *et al.*, 2002b; in press) indicate that the enthalpy of phase change for K_{SA} is on the order of 75–110 (kJ/mol), similar to what is reported for K_{OA} (Shoeib and Harner, 2002).

A.1.4.4 Air-water exchange

Transfer of contaminants across the air-water interface is an important pathway that controls the delivery of POPs to the marine environment and its ecosystems. This pathway has been considered in CACAR-I and recently reviewed by Wania *et al.* (1998). Further improvements to air-water exchange estimates rely on temperature-dependent measurements of the Henry's Law Constant, H.

A.1.4.5 Partitioning to snow and ice

For some contaminants, deposition onto snow may be important in their delivery to the Arctic ecosystem. This pathway is discussed separately in Section C.1.4.

A.1.5 Reaction rates

To investigate the fate of chemicals in the Arctic ecosystem, degradation rate data associated with photolysis, hydrolysis, and microbial degradation must be known at Arctic conditions. Gas-phase OH radical reaction rates have been measured as a function of temperature for several classes of POPs including PCBs (Anderson and Hites, 1996; Brubaker and Hites, 1998a), PCDDs/Fs and PAHs (Brubaker and Hites, 1997; 1998b) and α-HCH, γ-HCH, and HCB (Brubaker and Hites, 1998c). When measured data are not available it is possible to estimate reaction rates using group contribution methods such as that proposed by Kwok and Atkinson, 1995.

In a modelling scenario of PAH transport from the UK to the Russian Arctic over a 5-day transect, Halsall *et al.* (2001) found that reaction with OH was a major removal mechanism for gas-phase PAHs especially in the temperate atmosphere. Much slower rates of removal were demonstrated during the winter and at higher latitudes. This was attributed to greatly reduced OH concentrations under these scenarios. Temperature has a very minor direct effect on OH reaction rates (Brubaker and Hites, 1998b).

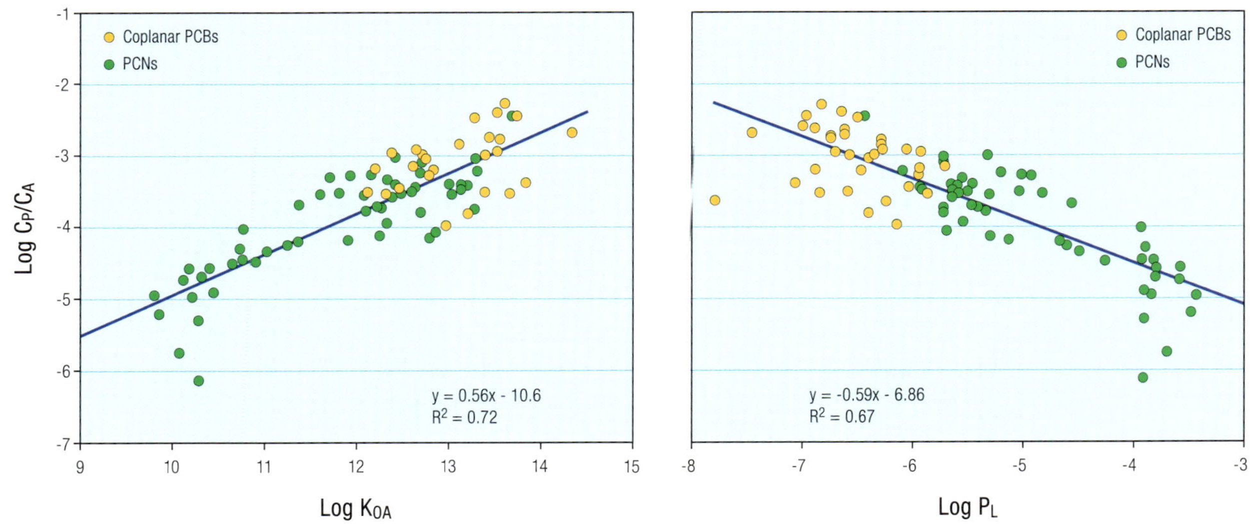

FIGURE A.1.1

Particle-gas partitioning relationships (log C_p/C_a) to the octanol-air partition coefficient (K_{oa}) for coplanar PCBs and PCNs at Alert, NU and Dunai, Russia during January–March, 1994 and November 1994–January 1995. C_p represents particle phase concentration normalized to total suspended particles and C_a is the gas phase concentration.

Microbial degradation is an important removal process for some compounds. Harner *et al.* (1999) have approximated in-situ microbial degradation rates for α- and γ-HCH in the eastern Arctic Ocean. In a revised budget of HCHs in the Arctic Ocean, microbial degradation is shown to be a dominant removal pathway — eliminating approximately 204 t/y of α-HCH and 23 t/y of γ-HCH (Harner *et al.*, 2000a). Other studies have observed high rates of microbial degradation of HCHs in Arctic lakes (Law *et al.*, 2001) and watersheds (Helm *et al.*, 2000b).

Chemical removal by hydrolysis and microbial degradation is sensitive to temperature, with higher rates being observed at warmer temperatures. The influence of temperature on the rate constant, k, is usually described using the Arrhenius expression and is discussed in greater detail in Sections C.3.7 and D.1.6.3.4. Ngabe *et al.* (1993) investigated the effect of temperature and pH on the rates of hydrolysis for α-HCH and γ-HCH, the most abundant pesticides in Arctic waters. The authors showed a strong temperature dependence on reaction rates. This was attributed to an increased OH- rate constant and also greater water dissociation at higher temperatures (greater activity of OH-).

A.2 Usage and emissions of organochlorine pesticides

Authors:

Y-F. Li (1); and T. Bidleman (2).

Affiliation:

1. Meteorological Service of Canada, Environment Canada, 4905 Dufferin Street, Downsview, ON, M3H 5T4, Canada;
2. Centre for Atmospheric Research Experiments, 6248 Eighth Line, R.R. #1, Essa Township, Egbert, ON, L0L 1N0, Canada.

A.2.1 Introduction

In the Phase-1 Assessment Report (Jensen *et al.*, 1997), usage of several organochlorine pesticides was summarized from information given in Barrie *et al.* (1992); Li *et al.* (1996; 1997), and Voldner and Li (1995). Here the global usage is updated for technical hexachlorocyclohexane (HCH), lindane, toxaphene, and DDT, and details are presented for China (HCHs, DDT), Canada (lindane), and the US (toxaphene). Improvements have been made by introducing the crop surrogate method of estimating pesticide applications in areas where no primary data exist (Li, 1999a; 2001; 2002; Li *et al.*, 1998b; 1999; 2001a; b), and for the first time estimates of actual emissions to the atmosphere are made for HCHs (Li *et al.*, 2000; Li, 2002 to be submitted) and

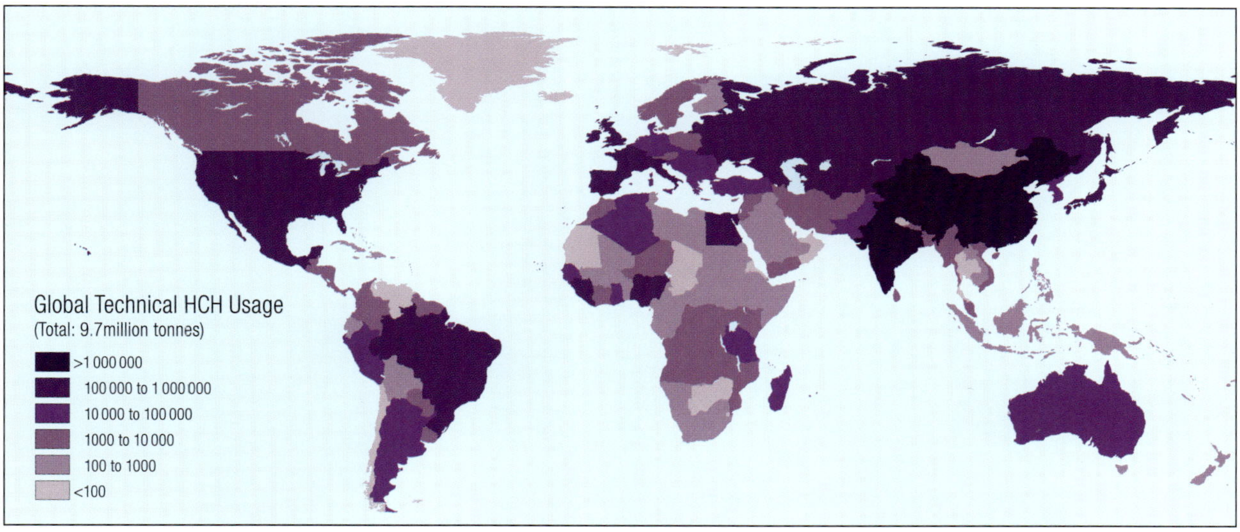

FIGURE A.2.1
Estimated technical HCH usage in each country from 1948–1997 after both temporal and spatial interpolation. The total global usage is estimated as 9.7 Mt (Li, 1999a).

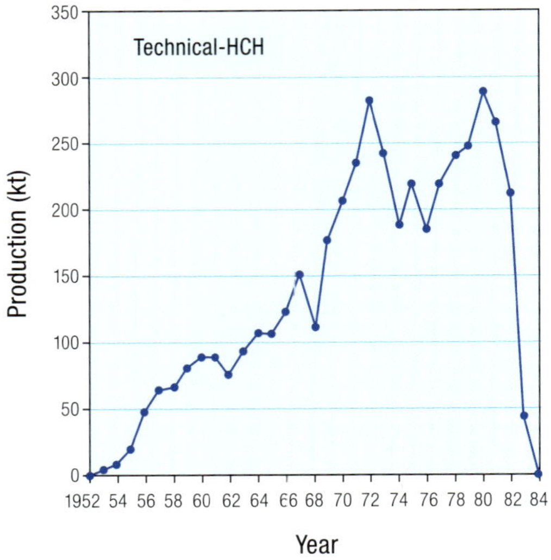

FIGURE A.2.2
Annual technical HCH production in China from 1952–1984 (Li *et al.*, 1998a).

for toxaphene (Li *et al.*, 2001a), and estimates of actual residues left in the agricultural soil are also made for toxaphene (Li *et al.*, 2001a).

A.2.2 Hexachlorocyclohexane (HCH)

1,2,3,4,5,6-Hexachlorocyclohexane (HCH), also called benzene hexachloride (BHC), is an organochlorine insecticide, and available in two formulations: technical HCH (containing 60–70% α-HCH, 5–12% β-HCH, 10–15% γ-HCH, and other isomers) and lindane (almost pure γ-HCH). As a broad-spectrum insecticide, technical HCH has been used for both agricultural and non-agricultural purposes. Due to its effectiveness and low price, technical HCH was one of the most widely used insecticides in the world.

A.2.2.1 Global technical HCH usage

Estimates of global technical HCH usage on a 1° latitude x 1° longitude grid were done using a global cropland data set with 1 km^2 resolution, apportioning HCH within the grids where usage was known and interpolating in regions where usage was unknown (Li, 1999a; b). Estimated technical HCH usage in each country from 1948 to 1997 is depicted in Figure A.2.1 by shading, and shown for the whole country rather than attributed to a specific region. The cumulative global technical HCH usage without spatial interpolation is 9.7 Mt, of which 8.4 Mt was used by the ten countries listed in Table A.2.1 and led by China (4.46 Mt), India (1.06 Mt) and the former Soviet Union (693 kt) (Li *et al.*, 1999a). The next section presents details for China, because of the overwhelming usage there.

A.2.2.2 Technical HCH usage in China

China started to produce and use technical HCH in 1952, and banned its application in 1983. The total amount of technical HCH produced in China was 4.46 Mt (Table A.2.1) (Li *et al.*, 1998a; b). The annual production of technical HCH (Figure A.2.2) (Li *et al.*, 1998b), increased almost every year, reaching a peak in 1972. After decreasing for a few years, production increased again and reached a second peak in 1980, decreasing thereafter to zero in 1984.

TABLE A.2.1 Top 10 countries with highest technical HCH use (Li,1999a)

Country	Usage (kt)	Year of ban
China	4464	1983
India	1057	Unknown
Former Soviet Union	693	1990
France	520	1988
Egypt	479	1981
Japan	400	1972
United States	343	1976
Germany (East)	142	1982
Spain	133	1992
Mexico	132	1993

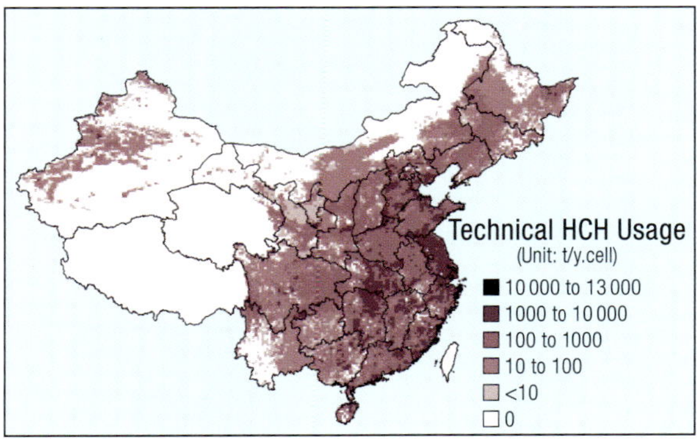

FIGURE A.2.3
Distribution of technical HCH usage in agricultural fields in China between 1952–1984 with 1/4° longitude x 1/6° latitude resolution (each cell has a size of about 20 km x 20 km) (Li *et al.*, 2001a).

Technical HCH in China was mainly used in agriculture, although a small portion was also used in forestry and public health (Li *et al.*, 1998a). Crops on which HCH was widely used were rice, wheat, soybean, maize, and sorghum. The total quantity of technical HCH applied in agriculture was about 4.40 Mt and of this total, about 50% was used on rice paddies, 25% on wheat and <10% on each of soybean/sorghum, and maize. The distribution of total technical HCH usage in agricultural fields in China with 1/4° longitude by 1/6° latitude resolution is shown in Figure A.2.3 for 33 years from 1952 to 1984 (Li *et al.*, 2001b). Heaviest use of technical HCH in China took place in Jiangsu, Hunan, and Henan provinces, reaching 950 t per grid cell in some areas in 1980 and 13 kt per cell over 33 years. Use in Jiangsu province reached as high as 500 kt between 1952 and 1984, followed by Hunan, 400 kt. The total in 10 provinces was 3.2 Mt, 72% of total usage in China. A detailed comparison of estimated versus actual usage was carried out for the Taihu Lake Basin, one of the highest HCH usage areas in China (Li *et al.*, 2001b). Cumulative HCH usage between 1952-84 in two grid cells was estimated by the cropland surrogate method at 13.5 kt, agreeing excellently with 13 kt based on reported HCH applications.

TABLE A.2.2 Lindane use regulation for some countries (AMAP, 1998; PAN, 1995; Breivik *et al.*, 1999; CIEL, 1998; Li *et al.*, 2001a)

Country	Action	Year	Restrictions/Remaining uses
Argentina	severely restricted	1995	
Austria	banned	1992	n/a
Bangladesh	banned	1991	
Belgium	severely restricted	1989	veterinary use and wood protection
Belize	severely restricted	1991	
Brazil	banned	1992	
Bulgaria	banned	1993	
Canada	restricted		
China	severely restricted	1991	use started in 1991
Cyprus	severely restricted	1987	paints and wood protection; agricultural use banned
Czechoslovakia	severely restricted	n/a	n/a
Denmark	banned	1995	
E.C.	severely restricted	1978/1988	EEC Directive 79/117 on HCH containing less than 99% γ-HCH
Finland	banned	1987	
France	banned	1998	
Germany	banned	1988	
Hungary	severely restricted	1968	registered as grain treatment for winter wheat and nurseries
Iceland	restricted use		
Italy	severely restricted	n/a	n/a
Netherlands	severely restricted	1991	use for agricultural applications prohibited
Norway	banned	1991	
Poland	phaseout	n/a	
Portugal	banned	1974	ban on products using mixed isomers
Sweden	banned	1989	
Switzerland	severely restricted	1986	prohibition of sale of mixed isomers; sale of γ-HCH for some applications
U.K.	banned	1979	sale of γ-HCH allowed
United States, Alaska	banned	1983	cancelled for most uses 1983
United States	restricted		
Former U.S.S.R	severely restricted	1988	prohibited for use as pesticide
Russia			technical HCH probably still in use; lindane still in use

n/a = no information available

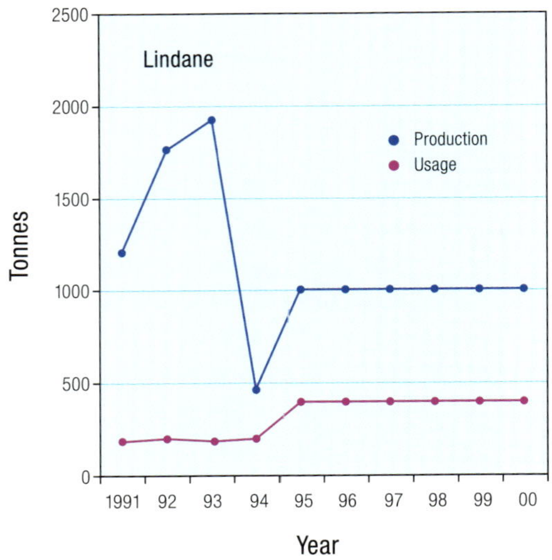

FIGURE A.2.4

Annual lindane production and usage in China between 1991–2000 (Li *et al.*, 2001b).

TABLE A.2.3 Lindane use for some countries (t)

Country	Volume (t)	Time period	Reference
Canada[1]	9100	1970–2000	this study
Canada[1]	500	2000	this study
USA[2]	1600	1980–1995	this study
USA[2]	28	1995	this study
Europe	16200	1992–1997	CIEL, 1998
Europe	81000	1970–1996	Breivik *et al.*, 1999
Europe	2700	1970	Breivik *et al.*, 1999
Europe	2200	1996	Breivik *et al.*, 1999
France	9600	1992–1997	CIEL, 1998
France	1600	1996	CIEL, 1998
China	3200	1991–2000	Li *et al.* 2001
China	500	2000	Li *et al.* 2001
Brazil	3.3	1990	Alho and Vieira, 1997
Brazil	3.9	1991	Alho and Vieira, 1997
Brazil	12.5	1992	Alho and Vieira, 1997

(1) Use on canola and corn seed treatment only.
(2) Use on seed treatment not included.

A.2.2.3 Global lindane usage

Lindane, containing almost pure γ-HCH (the only insecticidally active isomer), replaced technical HCH in Canada, the United States and western Europe during the late 1970s–early 1980s and in China in 1991. Worldwide lindane usage between 1948 and 1993 was estimated to be 720 kt (Voldner and Li, 1995). Although lindane is still important on a global scale, many countries have restricted or eliminated its usage. Tables A.2.2 and A.2.3. give the registration status and usage for lindane in some countries.

A.2.2.4 Lindane usage in Europe

Breivik *et al.* (1999) give the total lindane application in Europe as 81 kt between 1970 and 1996, and 2.2 kt in 1996. According to Centre International d'Etudes du Lindane (CIEL, 1998), the average lindane consumption in Europe was 2.1 kt/y from 1992–1997. France was the major user of lindane in Europe and in the world during this period, with an annual average consumption of 1.6 kt (CIEL, 1998), more than 76% of total lindane usage in Europe. In July 1998, lindane usage was stopped in France (CIEL, 1998).

A.2.2.5 Lindane usage in China

China banned the use of technical HCH in 1983 and lindane has been produced and used since 1991 (Li *et al.*, 2001b). Figure A.2.4 shows the annual production and usage of lindane. The total production was estimated to be 11.4 kt between 1991 and 2000. While the total technical HCH produced in China was considered to have been used there, only 3.2 kt of lindane was applied in China between 1991 and 2000. The rest was exported or stockpiled. The major uses of lindane in China are to kill locusts, wheat mole crickets, midges, and wood moths in forestry. The wheat land data set was used to distribute application of lindane in China, and the result for 1995 is shown in Figure A.2.5 (Li *et al.*, 2001b).

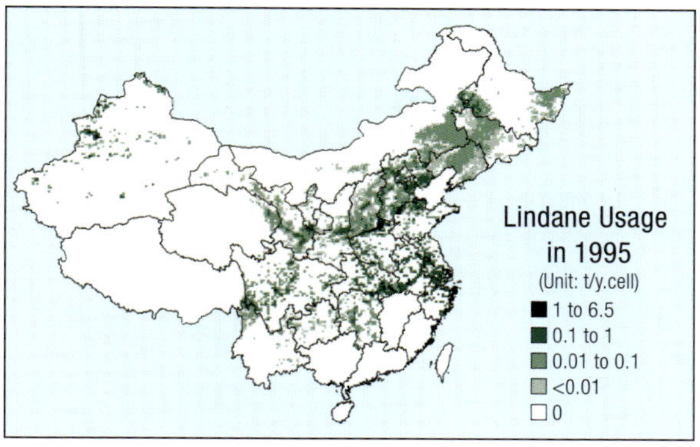

FIGURE A.2.5

Distribution of lindane usage in China for 1995 with 1/4° longitude x 1/6° latitude resolution (Li *et al.*, 2001b).

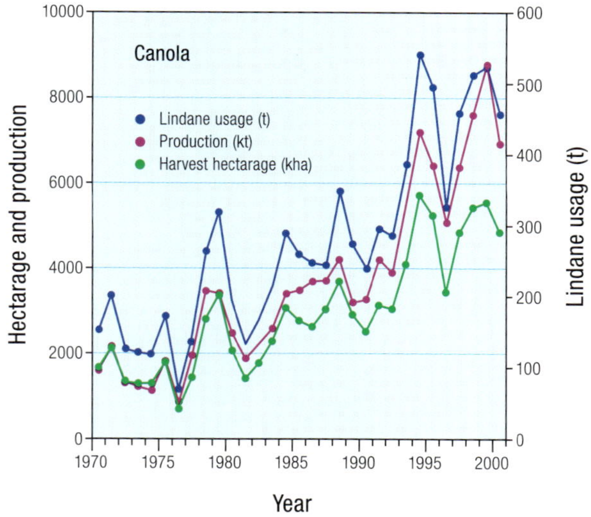

FIGURE A.2.6

Annual harvest hectarage and production of canola (Market Analysis Division, Agriculture and Agri-Food Canada) and estimated lindane usage between 1970–2000 in Canada.

A.2.2.6 Lindane usage in Canada

In Canada, lindane has been used since the 1960s, and technical HCH was banned in 1971 (Voldner and Smith, 1986; Barrie *et al.*, 1992). In 1968, usage in Canada was reported to be 16 t, decreasing to 3 t in 1969 and 1970, and 2.8 t in 1971 (Voldner and Smith, 1986); however, usage may have been underestimated. During the 1990s lindane imports of 100–150 t/y were reported (Barrie *et al.*, 1992). Lindane became one of the top ten insecticides used in Canada during the 1990s (Agriculture Canada-Environment Canada, 1992). Pesticide information from Canadian companies is proprietary, and although Agriculture and Agri-Food Canada has information on company and product sales statistics as well as sales summaries broken down by province, product market type, use pattern and active ingredient, only composite information can be divulged to ensure that the competitive position of a producer is not jeopardized (Voldner and Li, 1995). Surrogate cropland information was therefore used to estimate lindane usage for Canada, as outlined below. Most lindane applications have been on canola seed in the prairie region of Canada, and on corn seed in central and eastern Canada.

Canola is primarily grown in the prairie region of Canada, the provinces of Alberta, Saskatchewan, and Manitoba. Annual harvest hectarage and production of canola between 1980 and 2000 are shown in Figure A.2.6. The area farmed with canola increased from around 2 Mha in 1980 to 5.6 Mha in 1999, and production increased from 2.5 Mt in 1980 to 8.8 Mt in 1999.

Lindane is frequently applied to canola seed in combination with some fungicides by the seed companies before selling to the farmers. Waite *et al.* (2000) calculated usage of lindane on Canadian canola fields by assuming: 1) average seed treatment is 15.3 g/kg of seed; 2) 95% of the canola seed is treated; and 3) average seeding rate is 6.5 kg/ha. For example, 455 t and 510 t of lindane were estimated to have been used in the prairie region of Canada in 1997 and 1998, since the canola hectarages were 4.82 Mha and 5.40 Mha in the same region for these years. The annual use of lindane in Canada for canola seed treatment between 1970 and 2000 was estimated using Waite *et al.* (2000) and the results are shown in Figure A.2.6. Annual lindane applications to canola seed increased from about 156 t in 1970 to 530 t in 1999, and 460 t in 2000, with a cumulative usage of 8.66 kt.

A voluntary agreement on canola seed treatment has been reached between the Canadian Canola Council, the Pest Management Regulatory Agency (PMRA), and the pesticide companies who market canola seed treatments containing lindane (Interprovincial CO-OP, Rhône Poulanc, Uniroyal, and Zeneca Agro) (PMRA, 2000). According to the agreement, lindane seed treatments for canola are no longer being produced, which means that the use of lindane in Canada will decrease substantially.

Corn is Canada's third largest grain crop (after wheat and barley) and the most important one in central and eastern Canada where livestock and feed grains are key components of the agricultural sector. In 1999, almost 9.2 Mt of corn was produced on about 1.2 Mha of land in Canada, and about 97% of this was produced in Ontario and Quebec (Market Analysis Division, Agriculture and Agri-Food Canada). Figure A.2.7 gives the annual harvest hectarage and production of corn in Canada between 1970 and 2000. Corn hectarage increased from about 0.5 Mha in 1970 to 1.2 Mha in 1999, and the production has increased from 2.6 Mt in 1970 to 9.2 Mt in 1999.

Lindane has been used in Canada as a dressing for corn seed (as a mixture of 10-15% diazinon, 10-15% Captan, and 15–20% lindane). Corn seed dressing is done twice, during industrial treatment, and in the field directly by the farmer. About 2–4 t of lindane was used in Quebec around 1994 in the field application (Poissant *et al.*, 1994; Poissant and Koprinivik, 1996). Assuming that 4 t of lindane was used in 1994 for corn seed (by industrial treatment and in the field) on 285 kha of Quebec land, the rate of application was 0.014 kg/ha harvested land. The amount of lindane used was estimated by using this application rate in other provinces and for other years (see Figure A.2.7). Annual lindane usage for corn seed dressing in Canada increased from 7 t in 1970 to 16.6 t in 2000 and the cumulative usage of lindane on corn was about 410 t.

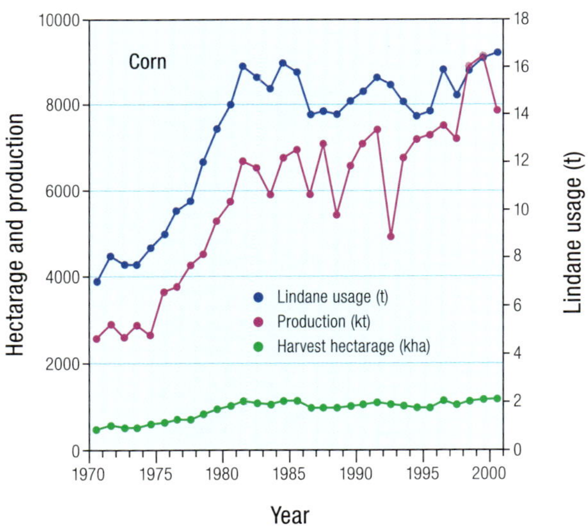

FIGURE A.2.7

Annual harvest hectarage and production of corn (Market Analysis Division, Agriculture and Agri-Food Canada) and estimated lindane usage in Canada between 1970–2000 (total usage is 410 t).

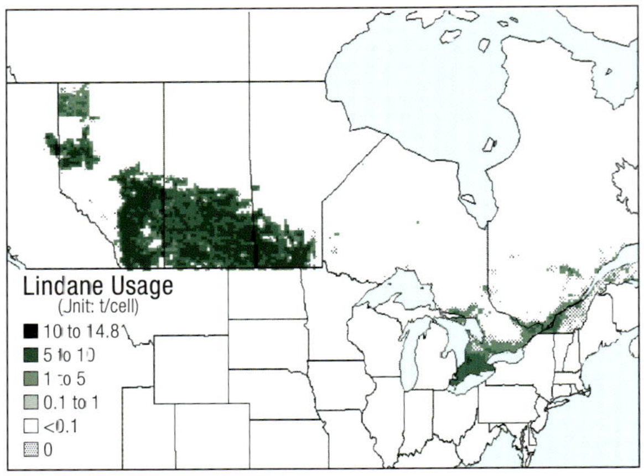

FIGURE A.2.8

Total lindane usage on Canadian canola and corn lands between 1970–2000 with 1/4° longitude x 1/6° latitude resolution.

The purpose of usage gridding is to allocate the use of pesticides on areas where pesticides are actually used, instead of distributing equally to all of the country or to all provinces. For example, lindane applications were primarily concentrated in the southern part of Canada, and very little or none was used in Yukon and the Northwest Territories. Even in a province, lindane could be used in a large amount in some areas, but none at all in others. Accurate allocation of the use of pesticides is very important for both policy makers and modelling scientists, although it is impossible to know exactly where and when this pesticide was used.

Gridded lindane usage with 1/6° by 1/4° latitude/longitude resolution on both canola and corn is shown in Figure A.2.8 between 1970 and 2000. The total usage of lindane on these two crops during these 30 years is 9 kt, and the intensive use of lindane on croplands was concentrated in the prairie region of Canada, the provinces of Alberta, Saskatchewan, and Manitoba. Besides canola and corn, lindane could be used on other crops and livestock. The PMRA indicated recently, however, that all other Canadian uses of lindane are minor compared with canola seed treatment (Don Waite, Environment Canada, personal communication). Use of lindane on canola and corn in Canada between 1970 and 2000 gives a good picture of temporal and spatial use trends of this pesticide over the last 30 years.

A.2.2.7 Lindane usage in the USA

As in Canada, pesticide information from US companies is proprietary. Information on agricultural pesticide use is available from several national pesticide use surveys and reports, along with various research reports on specific crops or states. The US Department of Agriculture's (USDA) National Agricultural and Statistics Service (NASS) provides state summaries of pesticide use annually for major field crops, and conducts a second survey that targets specialty crops. The NASS data are based on a statistical sample of farms in states that account for at least 80% of United States production for a specific crop. The United States Census of Agriculture conducts a survey every 5 years of all farms within the United States that have an annual farm income of $1000 or more. This enumeration includes information on the type, quantity, and cost of agricultural chemicals used on each farm. The data, reported by county, include the number of farms and the number of acres on which broad classes of chemicals are used (for example, herbicides and insecticides).

In February 1995, the National Center for Food and Agricultural Policy (NCFAP), a nonprofit, nongovernmental organization based in Washington, D.C., released the National Pesticide Use Database, containing over 15 000 individual records that quantify the use of specific active ingredients by crop and state (Web site: http://NCFAP.org/pesticid2.htm). The NCFAP database contains use estimates

TABLE A.2.4 Lindane use (t) in the USA

Year	SRI[1]	USDA[2] (Agr only)	Eckerman, 1984	Gianessi, 1986[3]	NCFAP[4]
1964		617			
1966		319.3			
1971		294.8			
1976		79.8[5]	400		
1980	348.5				
1981				39	
1982	174.2		268		
1984	261.4				
1986	304.9				
1989	217.8				
1989–1990					28
1991	43.6				
1991–93					27.8
1995					

Source: [1]SRI: Pesticides, the Chemical Economics Handbook — SRI International, February 1994;
[2]USDA: Data for 1964 and 1966: Eichers *et al.*, 1970 Rp136; Data for 1971: Andrilinas, 1974 Rp137; Data for 1976: Eichers *et al.*, 1978 Rp135;
[3]Usage on crop only, and seed treatment not included;
[4]NCFAP: http://NCFAP.org/pesticid2.htm, Usage on crop only, and seed treatment not included.

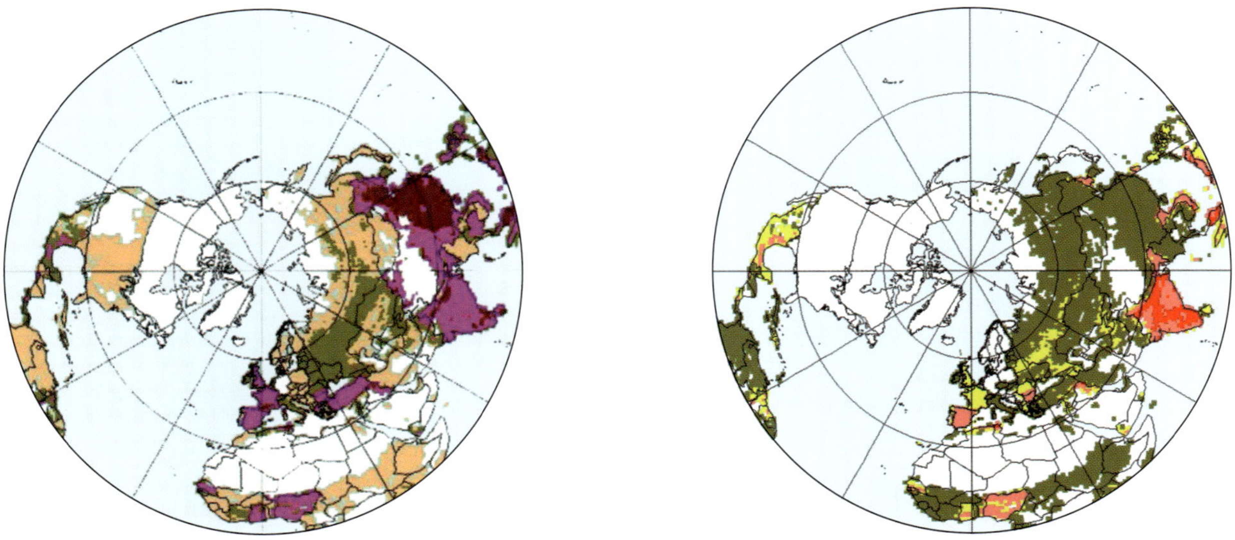

FIGURE A.2.9
Gridded global annual emissions for α-HCH in 1980 (a) and 1990 (b) (Li *et al.*, 2000).

TABLE A.2.5 Quantities of lindane used by farmers on crops, livestock, and for other purposes in the United States (active ingredients, t)

Year	Total	Crops	Livestock	Others	Sources
1964	617	433	177	6	Voldner and Smith, 1982
1964	458. 6		163.3		Eichers *et al.*, 1970
1966	319.3	179. 2	132.9	7.3	Eichers *et al.*, 1970
1971	294.8	79.8	188.7	26.3	Andrilinas, 1974
1976			79.8		Eichers *et al.*, 1978
1976	402	224	64	114	Eckerman, 1984
1982	263	140	56	67	Eckerman, 1984
1989–90		28			NCFAP
1991–93		27.8			NCFAP

for 200 active ingredients, including lindane, used on 87 crops in the 48 contiguous states for a typical crop year during 1991-1993 (D:\databk2\pest\source\country\usa\IntNet\Source\Sources.doc).

The state-level information on pesticide use rates available from the NCFAP for 1991–1993 and 1995 were used by the US Geological Survey's National Water Quality Assessment (NAWQA) Program to estimate county-based pesticide use for the contiguous United States with the help of the county-level information on harvested crop acreage based on the 1992 Census of Agriculture (http://water.wr.usgs.gov/pnsp/use92/index.html). The average annual pesticide use, the total amount of pesticides applied, and the corresponding areas treated were compiled for the 208 pesticide compounds that are applied to crops in the conterminous United States. Pesticide use was ranked by compound and crop based on the amount of each compound applied to 86 selected crops.

Although the NCFAP database offers a unique source of pesticide use data for the United States, the data on lindane are for crop spray only and seed dressing data are not included. In the following estimates, the use of lindane for seed dressing is not discussed because of lack of information.

Technical HCH use in the US began in the late 1940s and the use of lindane followed. In 1978, technical HCH was banned while lindane usage continued (Barrie *et al.*, 1992). As previously noted, US information is limited regarding use patterns of lindane. The import of lindane to the United States was 1.6 kt in 1974, 2.9 kt in 1975, 13.1 kt in 1976, and 152 kt in 1977 (Eichers *et al.*, 1978; Voldner and Smith, 1986), an increase of almost a factor of 100 over four years. The great increase in lindane imported in 1977 could be due to the ban of technical HCH in following year. Lindane consumption in the United States for several years from different sources are is shown in Table A.2.4. Usage of lindane by type of treatment for different years is given in Table A.2.5. In recent years this has amounted to ~28 t/y on crops, but as previously noted, seed treatment is not included.

A.2.2.8 Alpha-HCH emissions and their impact on the Arctic atmosphere

A.2.2.8.1 Global emissions

The gridded global annual emissions for α-HCH are given in Figure A.2.9 for 1980 and 1990. Total global α-HCH emissions in 1980 are estimated to be 184 kt, 92.5% of which is contributed by the 1980 current year use, and 7.5% by the residues from the preceding 15 years. In 1990, total global α-HCH emissions are estimated to be 44 kt, 93% contributed by the 1990 current year use, and 7% by the residues from the preceding 15 years (Li *et al.*, 2000). About 4.3 Mt of α-HCH was emitted to the atmospheric between 1945 and 1997 (Figure A.2.10) (Li, 1999b; Li *et al.*, 2000).

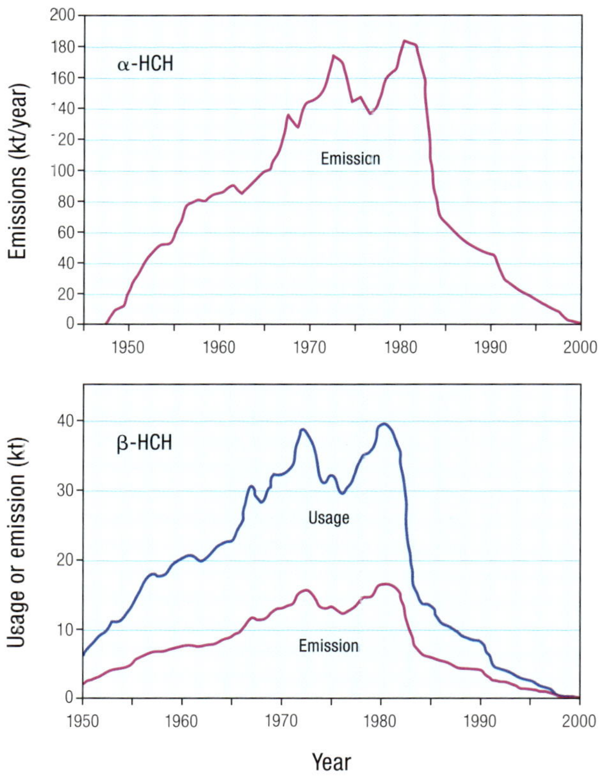

FIGURE A.2.10
Global usage and emissions of α-HCH (a) and β-HCH (b) from 1950–2000 (Li *et al.*, 2000).

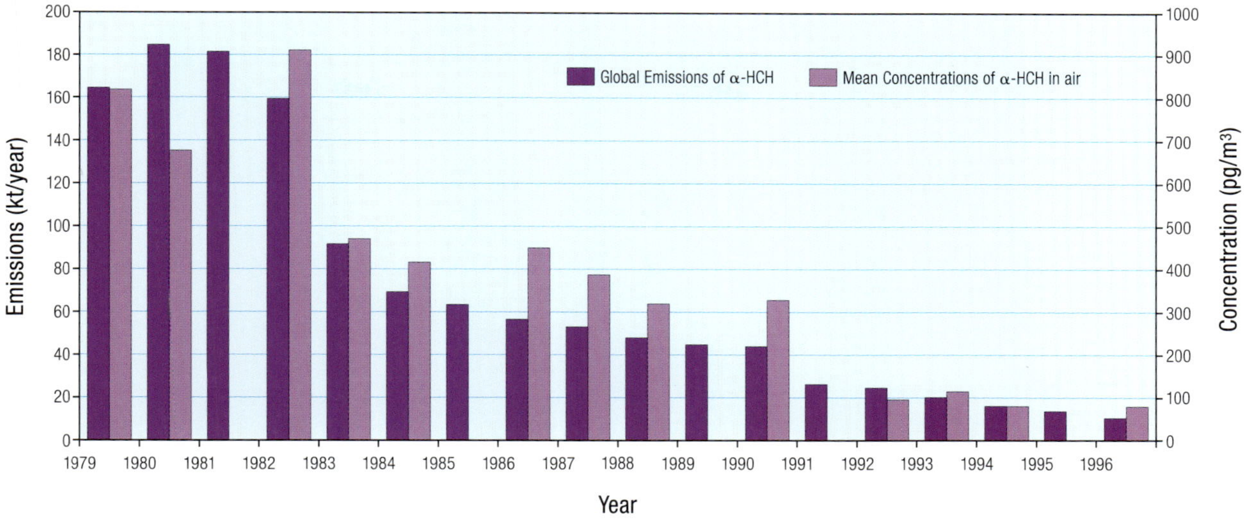

FIGURE A.2.11

Global emissions of α-HCH (left bars, dark purple) and mean concentrations of α-HCH in Arctic air (right bars, light purple) from 1979–1996. Air concentration data for α-HCH in the Arctic have been measured at different stations by several research groups referenced in Bidleman *et al.*, 1995a; and by Falconer *et al.*, 1995; Harner *et al.*, 1999; Halsall *et al.*, 1998; Hung *et al.*, 2002; and Jantunen and Bidleman, 1995, 1996.

A.2.2.8.2 Concentrations in Arctic air

The relationship between the global technical HCH use trends and their effect on the Arctic atmospheric environment has been studied. Two significant drops in technical HCH usage were identified (Li *et al.*, 1998a), one in 1983 and another one in 1990. These correspond to significant drops in α-HCH concentration in Arctic air between 1982 and 1983, and again between 1990 and 1992 (Bidleman *et al.*, 1995b; Li *et al.*, 1998a). Figure A.2.11 shows the long-term trends of global emissions of α-HCH and its mean air concentrations in the Arctic regions from 1979 to 1996. The trends shown here are similar to those shown in Li *et al.* (1998a), which compared air concentrations with global technical HCH usage. The relationship between α-HCH emissions and concentrations in Arctic air (Figure A.2.11) are even better than the same relationship shown for technical HCH in Li *et al.* (1998a).

A plot of mean air concentrations of α-HCH in the Arctic (C, pg/m³) versus global α-HCH emissions (E, kt/y) (Figure A.2.12) shows strong correlation with $r^2 = 0.90$ and the least-squares fit:

$$C = 4.31E + 83.5 \quad (1)$$

It is interesting to see that the data in Figure A.2.12 cluster into three groups. There are two gaps between these three groups, representing the two drops of both global emissions and air concentrations of α-HCH, one after 1983 and another after 1990.

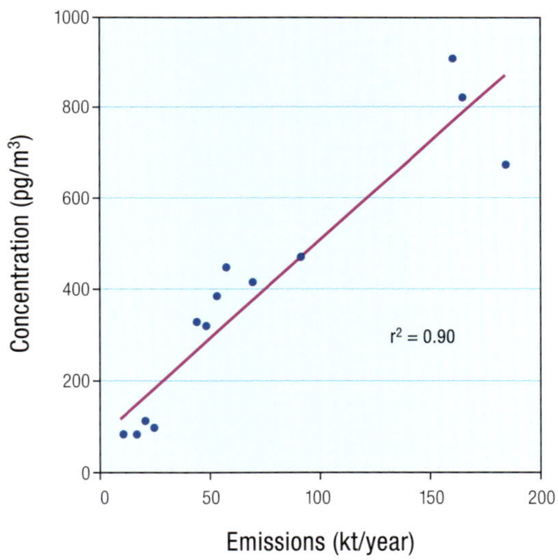

FIGURE A.2.12

Correlation between global emissions of α-HCH and its concentration in Arctic air, $r^2 = 0.90$. The least-squares line is given by Equation 1.

The intercept (83.5 pg/m^3) of Equation 1 can be interpreted as the background air concentration in the Arctic. Some of this background HCH might be supported by air-surface exchange with α-HCH reservoirs in the ocean and soils (Koziol and Pudykiewicz, 2001; Wania *et al.*, 1999c). Even after global α-HCH emissions approach zero, a background concentration of α-HCH will remain in Arctic air and decline slowly as levels in seawater and soils dissipate. Concentrations of α-HCH at Alert varied from 60-61 pg/m^3 in 1993-94 and decreased to 47 pg/m^3 in 1997 (Hung *et al.*, 2002a) Analysis of this trend yields an estimated time for 50% decrease of 10-16.9 y (Hung *et al.*, 2002a).

Equation 1 is derived by using the data between 1979 and 1996. This equation is not very accurate for calculating the air concentration of α-HCH in the Arctic from global emissions before 1980. By assuming the background value decreases linearly from 1990 to zero in 1945, a modified equation is proposed for the calculation of air concentrations of α-HCH in the Arctic prior to 1980:

$$C = 4.31E + 2.46(Y-1945) \quad (2)$$

Where: Y is the year.

The concentrations of α-HCH in Arctic air calculated from the annual global emissions using Equation 1 for 1980–96 and Equation 2 for 1945-80 are presented in Figure A.2.13, and the measurement data in the Arctic atmosphere are also presented for comparison. The calculated emissions fit the measurement data very well between 1980 and 1990, but deviate somewhat after 1990 where a yearly comparison between emissions and measurements shows closer agreement (Figure A.2.11).

A.2.2.9 Beta-HCH emissions and their impact on the Arctic atmosphere

A.2.2.9.1 Global emissions

The gridded global annual emissions for β-HCH are given in Figure A.2.14 for 1980 and 2000 (Li *et al.*, 2002). Total global β-HCH emissions in 1980 were 16.6 kt, 85.7% contributed by the current-use of β-HCH in 1980, and 14.3% by the residues due to use of technical HCH since 1966. Total global β-HCH emissions in 2000 were 83 t, all these from residues, since it is assumed that there was no technical HCH use in that year. Figure A.2.10 shows the long-term trends of global usage and emissions of β-HCH from 1950 to 2000. As for β-HCH, the two significant drops of global β-HCH emission since 1980 can also be identified in Figure A.2.10.

A.2.2.9.2 Concentrations in Arctic air

Monitoring data from Alert show β-HCH concentrations of 0.9 pg/m^3 in 1988, 0.75–0.39 pg/m^3 in 1993–94 and 0.24–0.09 pg/m^3 in 1996–1997 (Hung *et al.*, 2002a; Li *et al.*, 2002; Patton *et al.*, 1991). A relationship similar to Equation 2 for β-HCH was derived for estimating air concentrations of β-HCH in the Arctic before the beginning of 1990; however this gave unrealistically high values for β-HCH in Arctic air, in the range of 57–72 pg/m^3 from 1970 to 1980, and 70–20 pg/m3 from 1981 to 1990. The problem is that emissions of both isomers are not transported to the Arctic with the same efficiency. In fact, α- and β-HCH appear to behave quite differently

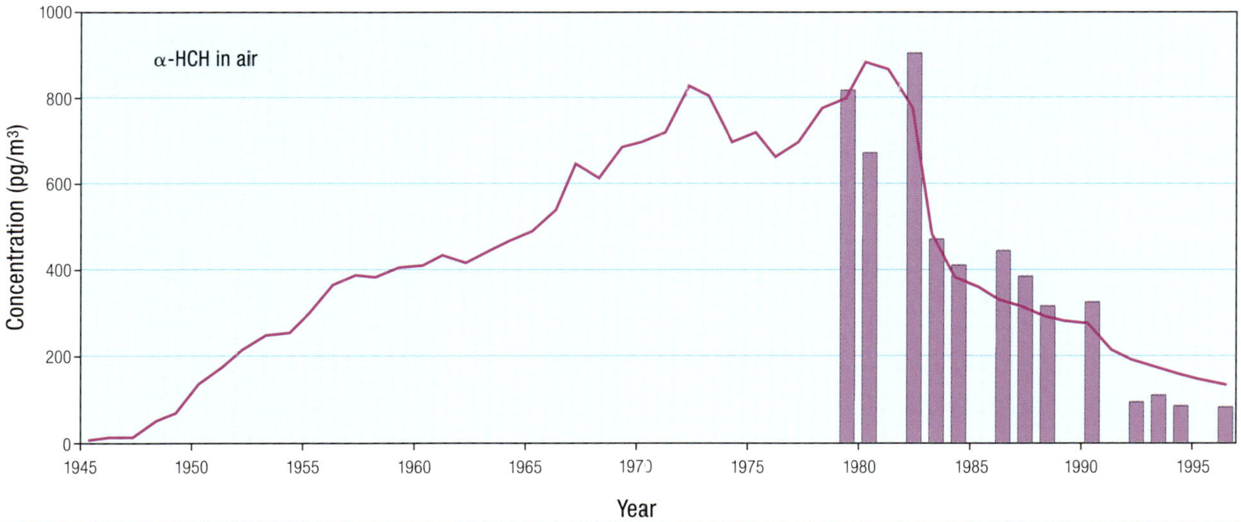

FIGURE A.2.13

Concentrations of α-HCH in Arctic air calculated from the annual global emissions using Equation 1 for the years between 1980–1996 and Equation 2 (see text) for the years between 1945–80 (curve). Measurement data in the Arctic (bars) are also presented for comparison.

during long-range atmospheric transport, with β-HCH being much more subject to deposition into the ocean by precipitation and gas exchange. The diverging pathways of these two HCH isomers can be explained by differences in their Henry's Law Constant, which is about a factor of 20 lower for β-HCH. Further discussion is provided in Section C.3 and in Li *et al.* (2002).

A.2.2.10 Gamma-HCH emissions and their impact on the Arctic atmosphere

A.2.2.10.1 Northern Hemisphere sources

Recent sources of the γ-HCH in Arctic air are unclear. Among circumpolar countries, four have banned the use of lindane: Finland in 1987, Sweden in 1989, Norway in 1991, and Denmark in 1995. Although lindane is still registered in the United States, there has been very little usage in Alaska. Applications in Russia, according to Breivik *et al.* (1999) are not significant. The use of lindane in Iceland is also very limited. Canada, with an annual lindane consumption of ~500 t in 2000, had become the biggest lindane user among the circumpolar countries. Other potential sources include France, which until lindane was banned in 1998, was the highest user in Europe. Use of lindane in China was 400 t in 2000 (Figure A.2.4). Transport of lindane from all of these regions, plus emissions of γ-HCH from the few countries where technical HCH is still used could contribute to levels of γ-HCH currently seen in Arctic air.

A.2.2.10.2 Concentrations in Arctic air

Oehme *et al.* (1995) found lower mean concentrations for γ-HCH in spring 1992 at Ny-Ålesund compared to the same months in 1984, and also declines in γ-HCH at two sites in mainland Norway between 1984 and 1992. However, γ-HCH levels were higher in 1992 at all three Norwegian air monitoring sites than in 1984. This probably reflects the increased use of lindane and declining release of γ-HCH as use of technical HCH products ceased in Europe during the 1980s. Concentrations of γ-HCH at Alert varied from 10–11 pg/m^3 in 1993-94 and decreased to 6.4 pg/m^3 in 1997 (Hung *et al.*, 2002a). Analysis of the five-year record yields an estimated time for 50% decrease of 4.3–4.9 years.

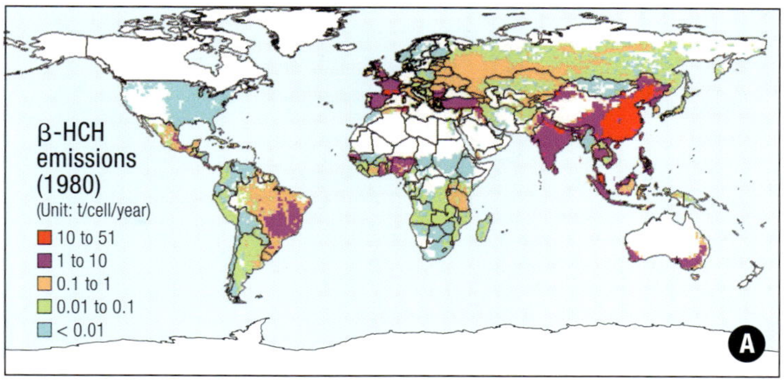

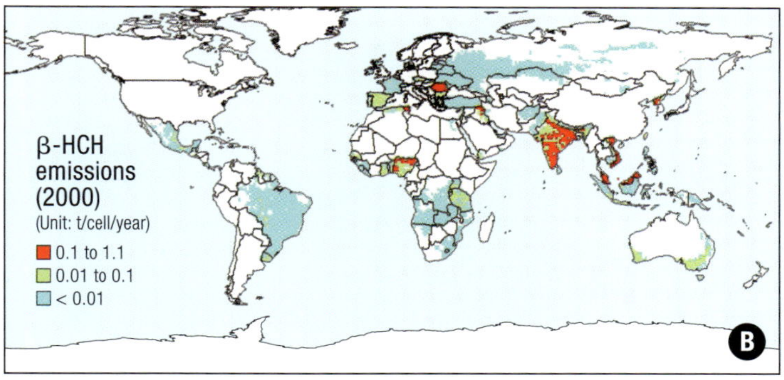

FIGURE A.2.14

a) Gridded global annual emissions for β-HCH in 1980; the total emission was 16.6 t (Li, 2001a). b) Annual emissions in 2000 due to residues resulting from use in the preceding 15 years. It was assumed that there was no technical HCH used in 2000. The total emission was 83 t (Li, 2001a).

TABLE A.2.6 Current status of registration of toxaphene in some countries for which information is available

Country	Status	Year	Reference	Country	Status	Year	Reference
Argentina	SR	1993	Avendano *et al.*, 1995	Iceland	NR	B, 1995	AMAP, 1998
Armenia	SR	1991	PAN, 1991	India	B	1991	PAN, 1991
Australia	WD	1993	PAN, 1993	Indonesia	B	1995	PAN, 1995
Austria	B	1992	Avendano *et al.*, 1995	Ireland	B	1985	Avendano *et al.*, 1995
Azerbaijan	SR	1991	PAN, 1991	Israel	NB	1995	PAN, 1995
Bangladesh	NB	1995	PAN, 1995	Italy	B	1989	PAN, 1989
Belgium	B	1974	Avendano *et al.*, 1995	Ivory Coast	NB	1995	PAN, 1995
Belize	B	1985	IRPTC/PAN	Jamaica	B	1995	PAN, 1995
Bolivia	B	1991	IRPTC/PAN	Japan	UR	1991	PAN, 1991
Brazil	B	1985	Avendano *et al.*, 1995	Jordan	B	1995	PAN, 1995
Bulgaria	B	1991	IRPTC/PAN	Kazakhstan	SR	1991	PAN, 1991
Burkina Faso	B	1991	IRPTC/PAN	Kenya	B	1987	UN, 1994
Bylorussia	SR	1991	PAN, 1991	Kyrgyzstan	SR	1991	PAN, 1991
Cameroon	NB	1995	PAN, 1995	Latvia	SR	1991	PAN, 1991
Canada	WD	1982	UN, 1994	Lebanon	NB	1995	PAN, 1995
Chad	NB	1995	PAN, 1995	Liechtenstein	B	1986	UN, 1994
Chile	NB	1995	PAN, 1995	Lithuania	B	1985	UNEP/IRPTC, 1993
China	B	1975	National EPA of China, 1992	Luxembourg	NB	1995	PAN, 1995
Colombia	SR	1988	UN, 1994	Madagascar	B	1995	PAN, 1995
Congo	U	B & A 1961	El-Sebae *et al.*, 1993	Malaysia	NB	1995	PAN, 1995
Costa Rica	B	1988	Avendano *et al.*, 1995	Mauritania	NB	1995	PAN, 1995
Cuba	B	< 1989	IRPTC/PAN	Mauritius	NB	1995	PAN, 1995
Cyprus	B	1989	PAN, 1989	Mexico	B	1993	PAN, 1993
Denmark	B	1987	AMAP, 1998	Moldova	B	1995	PAN, 1995
Dominica	SR	1986	UN, 1994	Mozambique	B	1995	PAN, 1995
Dominican Republic	B	1991	Avendano *et al.*, 1995	Netherlands	B	1984	UN, 1994
EC	B	1988	Avendano *et al.*, 1995	New Zealand	WD	< 1990	UN, 1994
Ecuador	B	1985	UN, 1994	Nicaragua	B	1995	PAN, 1995
Egypt	B	1962	El-Sebae *et al.*, 1993	Norway	NU		AMAP, 1998
El Salvador	B	1989	PAN, 1989	Pakistan	B	1989	PAN, 1989
Estonia	B	1984	UNEP/IRPTC, 1993	Panama	B	1987	UN, 1994
Ethiopia	U	B & A 1961	El-Sebae *et al.*, 1993	Papua New Guinea	NB	1995	PAN, 1995
Fiji	NB	1995	PAN, 1995	Paraguay	B	1993	Avendano *et al.*, 1995
Finland	B	1969	AMAP, 1998	Peru	B	1992	Avendano *et al.*, 1995
France	B	1991	PAN, 1991	Philippines	B	1989	PAN, 1989
Georgia	SR	1991	PAN, 1991	Poland	UR	1995	PAN, 1995
Germany East	B	1990	H. Bleitz, 1993	Portugal	B	1988	Avendano *et al.*, 1995
Germany West	B	1986	UN, 1994	Russia	SR	1991	PAN, 1991
Greece	NB	1995	PAN, 1995	Singapore	B	1984	UN, 1994
Guatemala	B	1989	Avendano *et al.*, 1995	South Africa	SR	1985	Avendano *et al.*, 1995
Honduras	B	1995	PAN, 1995	South Korea	B	1986	UN, 1994
Hong Kong	NR	1991	PAN, 1991	Soviet Union	SR	1992	AMAP, 1998
Hungary	NB	1995	PAN, 1995	Soviet Union	SR	1986	UN, 1994

A.2.6 continued

Country	Status	Year	Reference	Country	Status	Year	Reference
Spain	NB	1995	PAN, 1995	Turkey	SR	< 1990	UN, 1994
Sri Lanka	WD	1991	IRPTC/PAN	Turkmenistan	SR	1991	PAN, 1991
St. Lucia	B	1989	PAN, 1989	Uganda	U	B & A 1961	El-Sebae *et al.*, 1993
Sudan	UR	1993	PAN, 1993	Ukraine	SR	1991	PAN, 1991
Suriname	NB	1995	PAN, 1995	United Kingdom	B	1991	PAN, 1991
Sweden	B	1960	AMAP, 1998	United States	B	1982	Chemical and Engineering News, 1982
Switzerland	B	1986	UN, 1994	Uruguay	NB	1995	PAN, 1995
Tadzhikistan	SR	1991	PAN, 1991	Uzbekistan	SR	1991	PAN, 1991
Taiwan	B	1995	PAN, 1995	Venezuela	SR	1983	UN, 1994
Tanjania	U	B & A 1961	El-Sebae *et al.*, 1993	Yemen	NB	1995	PAN, 1995
Thailand	B	1983	UN, 1994	Yugoslavia	SR	1972	UN, 1994
Togo	NB	1995	PAN, 1995	Zambia	NB	1995	PAN, 1995
Tonga	B	1993	PAN, 1993	Zimbabwe	UR	1993	PAN, 1993

A.2.6 continued

SR = severely restricted; WD = deregistered and voluntarily withdrawn by manufacturer; B = banned; NB = not banned; U = in use; NR = never registered; UR = unregistered; NU = never used

A.2.3 Toxaphene

Toxaphene is a complex mixture of polychlorinated bornanes (CHBs) and camphenes. It is also referred to as camphechlor or polychlorinated camphenes (PCCs), although CHBs is preferred as a generic designation (Muir and deBoer, 1993). The theoretical number of different chlorinated compounds is ~16000 (Vetter, 1993), but only a few hundred have environmental significance (Hainzl *et al.*, 1994).

A.2.3.1 Global toxaphene usage

The registration status of toxaphene in 108 countries, for which information is available, is listed in Table A.2.6, summarized by number of countries: banned 55, never used 1, severely restricted 20, never registered 2, deregistered and voluntarily withdrawn by manufacturer 4, and not banned 21. This information should be used only as a general indication of status in the country. In cases where the product has not been specifically noted as

TABLE A.2.7 Toxaphene production and use for the world and some countries

Country	Mode	Volume (kt)	Time period	Reference
World	P	> 500	1947–77	Zell and Ballschmiter, 1980
World	U	1330	1950–93	Voldner and Li, 1995
USA	U	490	to 1985	Li, 2001
Mexico	U	22	1978	Burton and Philogene, 1985; Maltby, 1980
Mexico	U	20	1988	Burton and Philogene, 1985; Maltby, 1980
Latin America	U	17.338	1978	Burton and Philogene, 1985; Maltby, 1980
Latin America	U	11	1988	Burton and Philogene, 1985; Maltby, 1980
Soviet Union	U	< 100	1950s–1980s	Voldner and Li, 1995
China	P	3.6	1967–1972	Cai, 1997
Egypt	U	54	1956–61	El-Sebae *et al.*, 1993
Nicaragua	U	65	1974–90	Carvalho *et al.*, 1999

P = production; U = usage

banned or prohibited, it has been assumed that the product is in use. According to the survey done by UNEP for the POPs convention negotiations, toxaphene manufacture has ceased in recent years. Canada and most of the European countries had banned the use of toxaphene before 1982 due to its hazards. According to the information in Table A.2.6, toxaphene may be still used in some countries.

Globally, toxaphene has been one of the most heavily used pesticides, with an estimated cumulative usage of 1.33 Mt (Voldner and Li, 1995). Table A.2.7 shows toxaphene production and use for the world and some countries. The United States consumed the highest amount of toxaphene in the world, and a fairly accurate estimate of cumulative USA usage is ~490 kt (see following).

A.2.3.2 Toxaphene usage in Asian countries — The former Soviet Union and China

Toxaphene was introduced in the former Soviet Union in the 1950s and was still widely used in Russian agriculture at the end of the 1980s. Total toxaphene usage within the former Soviet Union was less than 100 kt (Voldner and Li, 1993; 1995), and was applied mostly in the Ukraine (Kundiev and Kagan, 1993). The application of strobane (mixtures of toxaphene and chlorinated α-pinene) was banned in 1981 (Fedorov, 1999); however, toxaphene was still used to control sugar beet pests with application rates of 1.25–1.5 kg/ha and to perennial herbs with application rates 0.8–1.25 kg/ha by the year of 1993 (Kundiev and Kagan, 1993). High production rates of sugar beets were sustained by the intensive use of organochlorines, including toxaphene, and manual labour (Fedorov, 1999). As of 1992, toxaphene was categorized as a "severely restricted" pesticide (Voldner and Li, 1993).

China produced toxaphene between 1967 and 1972 with the highest annual production of about 1 kt in 1970, and total production of about 3.6 kt. Toxaphene was sprayed on cotton to kill boll worms between July and September.

A.2.3.3 Toxaphene usage in Mexico, Central America, and Colombia

Toxaphene was used on cotton crops and rice paddies in some Central American countries. Nicaragua produced this pesticide to serve the needs of the cotton market in Central America. The Nicaraguan plant, which was the last world producer of toxaphene was closed in 1991 (Avendano *et al.*, 1995). During 1974 to 1979 toxaphene production averaged about 10 kt/y, and dropped to approximately 3.5 kt/y from 1980 to 1985, and to 0.6 kt in 1988. In 1990 the production of toxaphene ceased, and its use was discontinued in 1993. During the last years, toxaphene was exclusively produced for Nicaraguan use (Castillo *et al.*, 1997). During 1974 to 1989, therefore, the total toxaphene production was 86 kt, of which 20 kt were used in Guatemala, El Salvador, Costa Rica, and Honduras.

Toxaphene was used in Mexico in the early 1950s and was banned in 1993 (Pesticide Action Network, 1993). The total use of toxaphene in Mexico was about 68 kt and the highest use occurred in 1974 when about 3.0 kt was used. Usage in 1981 was 1.8 kt, and gradually reduced to zero in 1995.

Colombia was a big user of toxaphene in northern South America in the 1970s, but the use of toxaphene in this country was severely restricted in 1988 (Avendano *et al.*, 1995). The highest application in Colombia (1.3 kt) occurred during the 1970s (Food and Agriculture Organization (FAO) Production Yearbooks, 1970–1980). In 1990, 335 t of toxaphene was reported to be used (Battelle Europe, World Agrochemical Data Bank, Geneva). The total consumption of toxaphene in Colombia between 1967 and 1990 was about 19 kt.

A.2.3.4 Toxaphene usage in Canada

Toxaphene has never been licensed for use in Canada (INAC, 2000). A small amount of toxaphene was used in Canada for scabies treatment on livestock (HWC, 1977) and as fish poison (Stringer and McMynn, 1960; Donald *et al.*, 1998).

A.2.3.5 Toxaphene production and usage in the United States

Toxaphene was originally produced in the US in 1946 by Hercules Inc. (now Nor-Am Chemical Co.), which was the principal manufacturer from 1946 to 1981, ceasing production in 1981. After the end of 1960s, there were three additional major producers, Tenneco Chemicals, in New Jersey, Riverside Chemicals in Texas (previously operated by the Sonford Corp.), and Vicksburg Chemical Co. in Mississippi. Tenneco Chemicals reported production from 1970 to 1976 (USEPA, 1986). Vertac Inc., initially a subsidiary of Vicksburg Chemical began to produce toxaphene in 1975 and by 1984 was the last plant in operation in the United States. The production of toxaphene was 30 kt in 1966, peaked at 55 kt in 1974, gradually decreased to 10 kt in 1981 and was finally discontinued in 1984 with the ban on the use of toxaphene in agriculture (Rice and Evans, 1984). Production had slowed down around 1976 in anticipation of the ban (Howdeshell and Hites, 1996), but the total estimated production between 1946 and 1982 was ~720 kt.

The use of toxaphene in the United States can be divided into four different periods between 1947 and 1986, with total usage of about 490 kt, more than 1/3 of the total global use (Li, 2001b).

1. 1947–1971. Application of toxaphene was mainly on cotton with annual consumption of 11.5 kt. Total usage during this period was ~290 kt.
2. 1972–1975. The ban of DDT in 1972 stimulated an increase in toxaphene use, especially on cotton, soybeans, corns, and other field crops. Toxaphene became the most heavily used general purpose insecticide in this period with 30 kt used annually, and a total usage of ~150 kt.
3. 1976–1982. Toxaphene applications to cotton decreased significantly after 1976 due to the introduction of synthetic pyrethroids (Voldner and Ellenton, 1987). The average rate was ~7.6 kt/y, with a total of ~45 kt.
4. 1983–1986. Toxaphene registrations were cancelled in October 1982 (Chemical and Engineering News, 1982; Rapaport and Eisenreich, 1986). However, toxaphene in stock at that time was still allowed to be used through 1986 for specific purposes (Chemical and Engineering News, 1982). Total usage during this period was 1.75 kt.

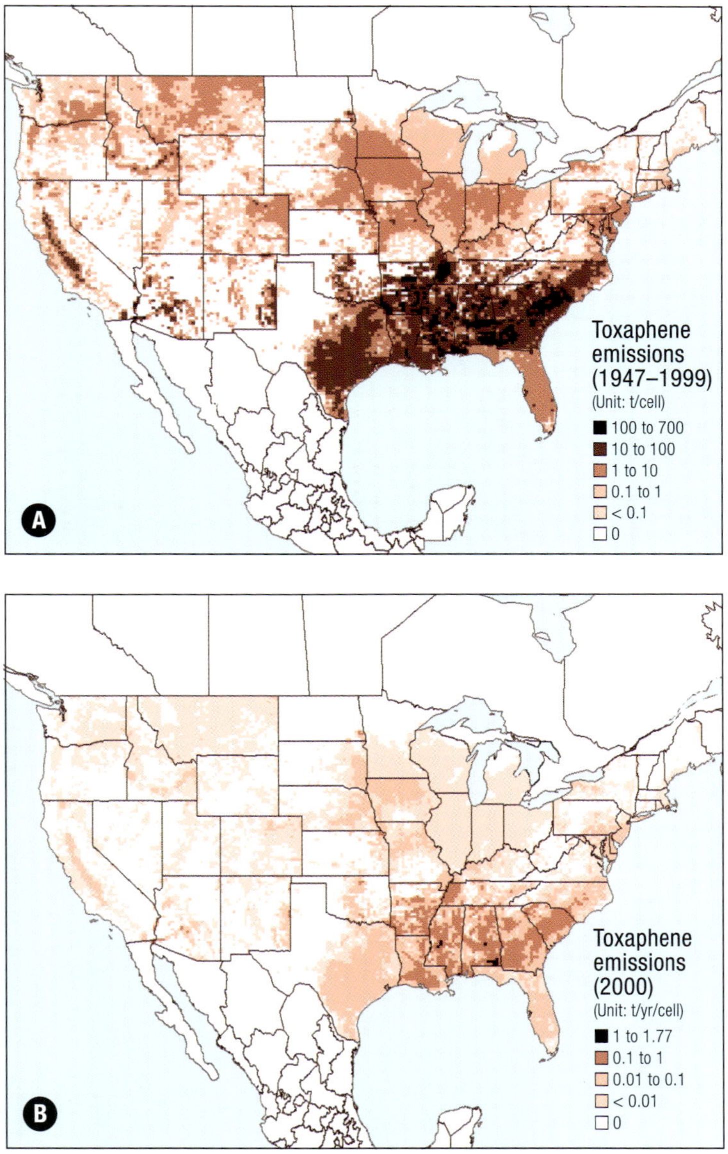

FIGURE A.2.15

Total toxaphene emissions in the United States on a 1/4° longitude x 1/6° latitude resolution grid system. a) 1947–99, b) 2000. Around 90% of total emissions in 2000 are from the southeastern part of the United States (Li *et al.*, 2001a).

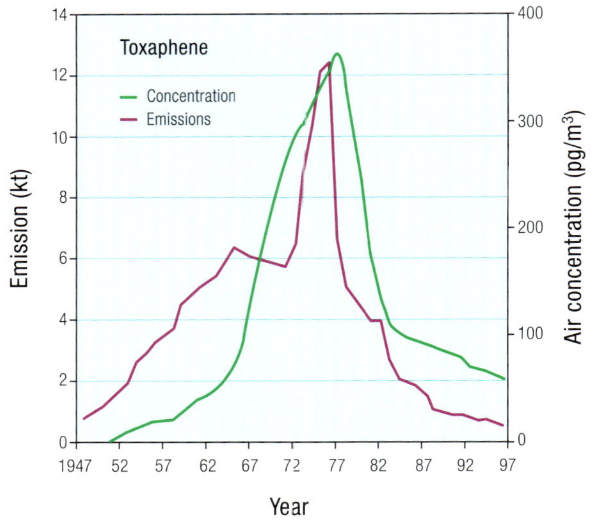

FIGURE A.2.16

Total emissions of toxaphene from use on crops and livestock and from manufacture in the United States (red curve) and predicted air concentrations of toxaphene from 1948–1996 over the Great Lakes (green curve). The prediction for air concentration was based on air measurements and the historical production curve of toxaphene (Swackhamer *et al.*, 1999).

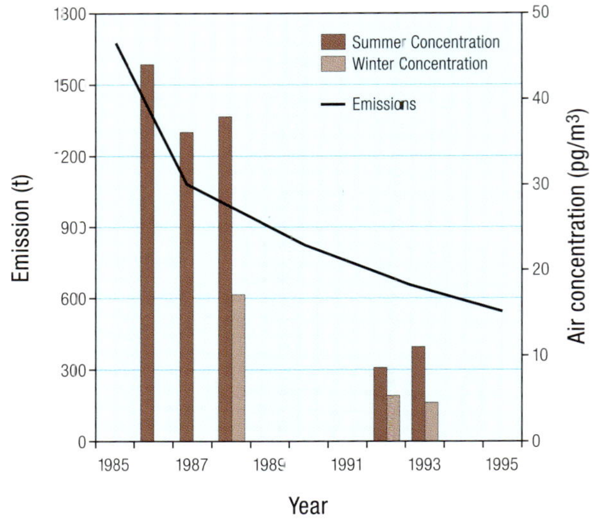

FIGURE A.2.17

Temporal trends in concentration of toxaphene in Canadian Arctic air (medium brown bars for the air concentration in summer, light brown bars for winter) and the calculated emissions of toxaphene from the United States (curve) from 1985–1995. Sources of air concentration data: 1996–97, Patton *et al.*, 1989; 1988, Patton *et al.*, 1991; Hinckley *et al.*, 1991; 1992, Bidleman *et al.*, 1995; Fellin *et al.*, 1996; 1993, AMAP, 1998; Macdonald *et al.*, 2000.

A.2.3.6 Toxaphene emissions and residues in the United States

The total gridded emissions for toxaphene between 1947 and 1999 are given in Figure A.2.15a and for the year 2000 in Figure A.2.15b (Li *et al.*, 2001a). Total toxaphene emissions were 194 kt between 1947 and 1999, 80% of which was contributed by the use of toxaphene in agriculture, and 20% of which was contributed by the use on livestock. Residues in the cropland of the United States have been modelled between 1947 and 2004 (Li *et al.*, 2001a). Agricultural lands most contaminated by toxaphene are in the southeastern part of the United States, with residues as high as 150 t per cell in some places. The total toxaphene residues left in the agricultural soil in the United States at the beginning of 2000 were around 29 kt. In 2000, almost 20 years after banning the use of this pesticide ~360 t of toxaphene were estimated to be emitted from the agricultural soil in the United States (Figure A.2.15) (Li *et al.*, 2001a). These estimated emissions agree well with the observed atmospheric input function of toxaphene to dated peat cores from the Great Lakes region and eastern Canada (Li *et al.*, 2001a; Rapaport and Eisenreich, 1986) and predicted air concentrations over the Great Lakes (Li *et al.*, 2001a; Swackhamer *et al.*, 1999) (Figure A.2.16).

A.2.3.7 Toxaphene in Arctic air

There is a large discrepancy between various measurements of toxaphene in Arctic air. Mean concentrations of total toxaphene in the 1990s were 11–19 pg/m^3 at Alert, Tagish, and Dunai, compared to 5–7 pg/m^3 at Resolute Bay and the Bering-Chukchi seas (AMAP, 1998). The former values were determined by gas chromatography with electron capture detection (GC-ECD), and later adjusted to a scale of measurements by GC — negative ion mass spectrometry (GC-NIMS). The lower results at Resolute and Bering-Chukchi were directly measured by GC-NIMS, using the same techniques employed in 1986–88 studies at the Canadian Ice Island (Patton *et al.*, 1989) and Alert (Patton *et al.*, 1991).

Figure A.2.17 shows the temporal trends of toxaphene in Canadian Arctic air (the bars) and the estimated emissions of toxaphene from the United States (the curve) from 1985 to 1995. Despite discrepancies in the toxaphene air data from the 1990s, levels have apparently dropped since the mid-late 1980s. The estimated emission trend of toxaphene from the US is consistent with the declining levels of toxaphene in Arctic air during summer.

A.2.3.8 Sources of toxaphene to the Arctic

What are the sources of toxaphene found in Arctic air today? Are they atmospherically transported from countries where they are still applied (new sources), or are they volatilized of residues left in soils and water bodies after application (old sources)? Clues can be found in the levels and congener profiles of toxaphene in air of suspected source regions. All of the measurements described in the following paragraphs were done by GC-NIMS and should be comparable.

As discussed in Section A.2.3.6, the large reservoir of toxaphene in soils of the southern US is likely a major source of today's emissions. Concentrations of total toxaphene in South Carolina and Alabama air during 1994–97 averaged 160–180 pg/m^3 (Bidleman *et al.*, 1998a; Jantunen *et al.*, 2000) — an order of magnitude higher than concentrations near and over the Great Lakes (James and Hites, 2002; Jantunen and Bidleman, 2002). A soil-air exchange model predicted that soil emissions could account for a large fraction of the observed toxaphene in air (Harner *et al.*, 2001b). Donald *et al.*, (1998) suggested that atmospheric transport from the United States could be an important source of toxaphene to alpine lakes in western Canada. Although the southern US appears to be the source of highest toxaphene emissions, modelling of soil residues also shows contamination in some of the northwestern states (Li *et al.*, 2001a). These could be the closest sources of toxaphene to western Canada and the Canadian Arctic.

Total toxaphene in air over Lake Baikal and in Irkutsk, Russia ranged from 11–21 pg/m^3 during June 1991 (McConnell *et al.*, 1996), similar to levels measured near or over the Great Lakes in the 1990s (Glassmeyer *et al.*, 1999; James and Hites, 2002; Jantunen and Bidleman, 2002; McConnell *et al.*, 1998; Shoeib *et al.*, 1999). Total toxaphene in Belize, Central America during December 1995 to January 1996 and July-August 1996 averaged 30 pg/m^3, again similar to Great Lakes values (Alegria *et al.*, 2000). Concentrations in Chiapas, Mexico during 2000–01 were much higher, averaging 490 pg/m^3 (Alegria *et al.*, 2001).

Chromatographic profiles of toxaphene in ambient air of the southern US and Chiapas, and in southern US soils, show depletions of two labile octachlorobornanes, designated as Parlar 39 and Parlar 42 (Alegria *et al.*, 2001; Bidleman and Leone, unpublished; Jantunen and Bidleman, 2000). Similar depletions are seen in air samples collected over Lake Superior (Jantunen and Bidleman, 2002). This suggests that soil-derived toxaphene is contributing to residues observed in ambient air.

A.2.4 Dichlorodiphenyltrichloroethane (DDT)

A.2.4.1 Global DDT use

The pesticide DDT was one of the first synthetic chemicals to be produced in large quantities and one of the most widely used chemicals for controlling insects on agricultural crops and controlling insects that carry such diseases as malaria and typhus. From 1946–1972 DDT was the most widely used agricultural insecticide in the world. The registration status of DDT in 95 countries, for which information is available, is listed in Table A.2.8. As of 1995, DDT has been banned in 59 countries, severely restricted in 20 countries, never registered in 2 countries, unregistered and voluntarily withdrawn by the manufacturer in 1 country, and not banned in 13 countries. The Table A.2.8 registration status is most likely for DDT in agriculture only, since many countries still allow DDT to be used for vector control and public health.

Between 1945 and 1972 the total cumulative production in the United States was 1.34 Mt, domestic sales accounted for 645 kt, and exports were estimated at 825 kt. In 1972, some producers ceased operations. The largest known current producer is India (~7 t/y). DDT usage in agriculture has been compiled from different sources, and temporal and spatial interpolations have been conducted. The total estimated DDT usage in agriculture from 1950 to 1993 is ~2.6 Mt (Voldner and Li, 1995). The top ten countries with highest usage in agriculture are listed in Table A.2.9. The gridded global annual usage of DDT from farmland is given in Figure A.2.18 for 1980 and 2000. Global DDT usage in agriculture was ~40 kt in 1980 and ~0.1 kt in 2000.

A.2.4.2 DDT use in China, India, and Japan

In 1951, China started to produce DDT; it was used in agriculture and a small amount was used in forestry and for public health(Cai et. al. 1992) . Annual DDT production from 1951 to 1983 is depicted in Figure A.2.19. The data were acquired from Chinese government agencies (Chinese Botanical Protection Bureau, Chinese Ministry of Agriculture and others), research institutes, and pesticides manufacturers. The total estimated amount of DDT produced in China during this period was 270 kt.

TABLE A.2.8 Current status of registration of DDT in some countries for which information is available

Country	Status	Year	Regulation	Source
Argentina	B	1992		PAN, 1992
Armenia	B	1991		PAN, 1991
Australia	B	1993		PAN, 1993
Azerbaijan	B	1991		PAN, 1991
Bangladesh	SR	n/a		UNEP/IRPTC, 1993
Belgium	B	1989		PAN, 1989
Belize	SR	1985	prohibited except for health	UNEP/IRPTC, 1993
Benin	B	1995		PAN, 1995
Bolivia	SR	n/a		UNEP/IRPTC, 1993
Brazil	SR	1984	not authorized for agriculture	UNEP/IRPTC, 1993
Bulgaria	B	1971	maximum air level	UNEP/IRPTC, 1993
Burkina Faso	B	< 1991	banned for agriculture	UNEP/IRPTC, 1993
Cameroon	NB	1995		PAN, 1995
Canada	WD	1969	major reductions	UNEP/IRPTC, 1993
Chad	NB	1995		PAN, 1995
Chile	SR	Jan. 1985	prohibited	UNEP/IRPTC, 1993
China	B	1983		National EPA of China, 1992
Colombia	B	1986	used for health emergency only	UNEP/IRPTC, 1993
Costa Rica	B	< 1991		PAN, 1991
Cuba	B	1970	prohibited	UNEP/IRPTC, 1993
Cyprus	B	1976	banned	UNEP/IRPTC, 1993
Czechoslovakia		Oct. 1978	maximum food level	UNEP/IRPTC, 1993
Denmark	B	Oct. 1984	prohibited	UNEP/IRPTC, 1993
Dominica	SR	1986	severely restricted	UNEP/IRPTC, 1993
Dominican Republic	B	1992		PAN, 1992
EC	B	July 1984	maximum food level	UNEP/IRPTC, 1993
Ecuador	SR	July 1985	prohibited	UNEP/IRPTC, 1993
Egypt	B	< 1989		PAN, 1989
El Salvador	B	< 1989		PAN, 1989
Estonia	B	1977		UNEP/IRPTC, 1993
Finland	B	1976	banned	UNEP/IRPTC, 1993
Germany East	B	1988		H. Ble tz, 1993
Germany West	B	Nov. 1972	prohibited	UNEP/IRPTC, 1993
Greece	SR	1993		PAN, 1993
Guatemala	B	1980	import for cotton only	UNEP/IRPTC, 1993
Honduras	NB	1995		PAN, 1995
Hong Kong	UR	n/a		UNEP/IRPTC, 1993
Hong Kong	B	1995		PAN, 1995
Hungary	B	Jan. 1968	banned	UNEP/IRPTC, 1993
Iceland	B	1996	never registered	AMAP, 1998
India	SR	1995		PAN, 1995
Indonesia	B	1991		PAN, 1991
Ireland	B	< 1991		PAN, 1991

A.2.8 continued

Country	Status	Year	Regulation	Source
Israel	SR	1958	use on cotton only	UNEP/IRPTC, 1993
Italy	B	n/a	maximum air level	UNEP/IRPTC, 1993
Ivory Coast	B	< 1989		PAN, 1989
Jamaica	B	< 1995		PAN, 1995
Japan	B	Oct. 1981	severely restricted	UNEP/IRPTC, 1993
Jordan	B	< 1995		PAN, 1995
Kenya	SR	Feb. 1987	restricted to health reasons only	UNEP/IRPTC, 1993
Latvia	B	< 1991		PAN, 1991
Liechtenstein	B	1986	banned	UNEP/IRPTC, 1993
Lithuania	B	1968		UNEP/IRPTC, 1993
Luxembourg	SR	1993		PAN, 1993
Madagascar	B	< 1995		PAN, 1995
Malaysia	NB	1995		PAN, 1995
Mauritania	NB	1995		PAN, 1995
Mauritius	SR	1970	restricted	UNEP/IRPTC, 1993
Mexico	SR	June 1973	maximum aquatic level	UNEP/IRPTC, 1993
Mozambique	B	< 1995		PAN, 1995
Netherlands	B	Jan. 1981	prohibited	UNEP/IRPTC, 1993
New Zealand	NB	1995		PAN, 1995
Nicaragua	B	< 1989		PAN, 1989
Norway	B	1988		AMAP, 1998
Pakistan	NB	1995		PAN, 1995
Panama	B	Sept. 1987	banned	UNEP/IRPTC, 1993
Papua New Guinea	NB	1995		PAN, 1995
Paraguay	B	< 1995		PAN, 1995
Peru	NB	1995		PAN, 1995
Philippines	SR	< 1989		PAN, 1989
Poland	B	1976	banned	UNEP/IRPTC, 1993
Portugal	B	Jan. 1974	banned for household product	UNEP/IRPTC, 1993
Romania		1975	maximum air level	UNEP/IRPTC, 1993
Singapore	B	Apr. 1984	banned	UNEP/IRPTC, 1993
South Africa	SR	< 1991		PAN, 1991
South Korea	B	June 1986	banned	UNEP/IRPTC, 1993
Soviet Union	B	1971		Kundiev and Kagen, 1993
Spain	B	Dec. 1975	prohibited	UNEP/IRPTC, 1993
Sri Lanka	B		banned	UNEP/IRPTC, 1993
St. Lucia	B	n/a		UNEP/IRPTC, 1993
Sudan	NB	1995		PAN, 1995
Suriname	NB	1995		PAN, 1995
Sweden	B	Feb. 1975	banned	UNEP/IRPTC, 1993
Switzerland	B	1986	prohibited	UNEP/IRPTC, 1993
Taiwan	B	< 1995		PAN, 1995
Thailand	SR	1983	banned except for health emergency	UNEP/IRPTC, 1993
Togo	B	1981	banned	UNEP/IRPTC, 1993
Tonga	NB	1995		PAN, 1995

A.2.8 continued

Country	Status	Year	Regulation	Source
Turkey	SR		severely restricted	UNEP/IRPTC, 1993
United Kingdom	B	1984	banned	UNEP/IRPTC, 1993
United States	B	July 1972	banned except for health reasons	UNEP/IRPTC, 1993
Uruguay	NB	1995		PAN, 1995
Venezuela	SR	1983	severely restricted	UNEP/IRPTC, 1993
Yemen	B	< 1995		PAN, 1995
Yugoslavia	SR	1972	severely restricted	UNEP/IRPTC, 1993
Zambia	UR	< 1995		PAN, 1995
Zimbabwe	B	< 1993		PAN, 1993

A.2.8 continued

n/a = no information available; SR = severely restricted; WD = deregistered and voluntarily withdrawn by manufacturer; B = banned; NB = not banned; U = in use; NR = never registered; UR = unregistered; NU = never used

Amounts of DDT used in China, India, and Japan are summarized in Table A.2.10, along with information on the years of ban (for China and Japan), the arable land in 1977, and the annual average and total amounts of DDT used on each hectare of arable land (Li *et al.*, 1999). It was assumed that China used all the DDT produced, and the DDT used in these three countries was applied to the total arable lands (even though a large part of this insecticide was used for vector control in India), and the amount of arable land in any application year is the same as that in 1977. The results show that China, with 101 Mha of arable land, used 270 kt of DDT in 33 years (from 1951 to 1983). In 42 years (from 1948 to 1989), India, with 168 Mha of arable land, applied 330 kt. Japan, with 5 Mha of arable land, used 30 kt of DDT in 23 years (from 1948 to 1970).

TABLE A.2.9 Top 10 countries with highest DDT use in agriculture between 1948 and 2000

Country	Usage (kt)	Year of ban
United States	590	1972
Soviet Union	320	1971
China	260	1983
Mexico	180	n/a
Brazil	106	1998
India	75	n/a
Egypt	66	1972
Guatemala	60	1985
Italy	46	n/a
Hungary	43	n/a

n/a = no information available

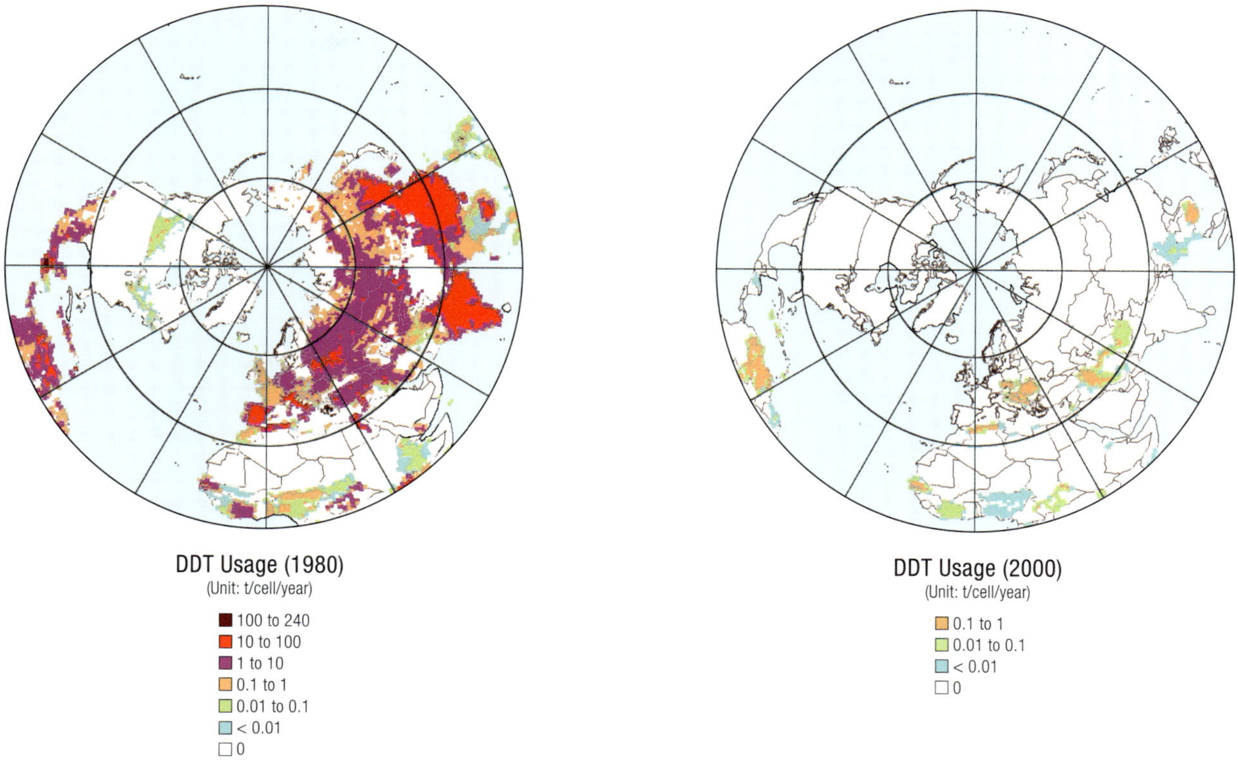

FIGURE A.2.18

Global gridded usage for DDT in (a) 1980 and (b) 2000.

The total amount of DDT used in these three countries (for arable land and other uses) was 630 kt. Although India was the biggest consumer of DDT, Japan applied the largest amount per ha of arable land (i.e., highest total usage density).

Besides the use in agriculture, DDT has been widely used for vector control. India began importing DDT for public health purposes following WW II, and agricultural usage began in the late 1940s. In India, consumption of DDT was ~14 kt/y in the late 1980s. Out of this, ~85% of DDT (12 kt) use is for indoor spraying of rural dwellings for mosquito control (Singh *et al.*, 1988). Gupta (1986) gave cumulative DDT usage to 1965 for public health as 126 kt, while an informal report citing the trade publication Europa Chemie, gave India's cumulative DDT usage to 1989 as 280 kt for malaria control and 50 kt in agriculture, this gives the total DDT usage in India to 1989 as 330 kt (PANNA, 1990).

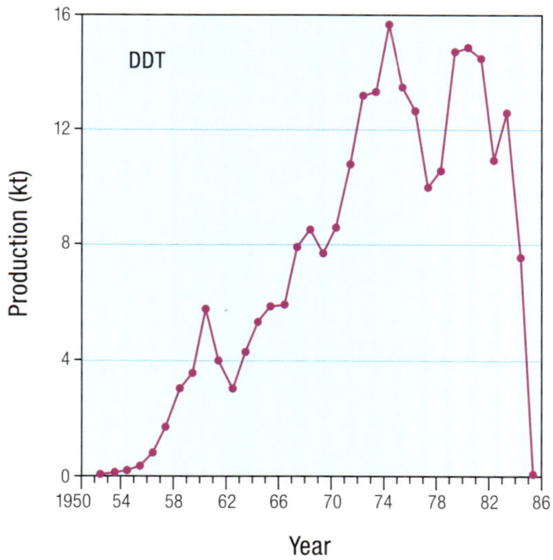

FIGURE A.2.19

Annual DDT production in China from 1952–1984.

TABLE A.2.10 DDT usage and usage density for China, India, and Japan[1]

Country	Year	Year of banning use	Total usage (kt)	Arable land (kha; 1977)[2]	Average usage density (kg/ha•year^{-1})	Total usage density (kg/ha)
China	1951–1983	1983	260	100 523	0.08	2.7
India	1948–1989	n/a	330	168 260	0.005	2.0
Japan	1948–1970	1972	30	4985	0.26	6.00

(1)Total usage for India are from PANNA Outlook 1990, and that for Japan from Loganathan *et al.* 1989.
(2)From FAO, 1993.
n/a = no information available.

A.2.5 Summarized usage and emission estimates

Table A.2.11 summarizes the amounts of HCHs, toxaphene, and DDT used globally. These figures were derived from new information or from an earlier report by Voldner and Li (1995). Estimates of global emissions for the HCHs are also shown. Although global emission estimates could not be made for toxaphene and DDT, toxaphene emissions in the US were estimated as 194 kt between 1947–99 and 360 t for the year 2000 (Section A.2.3.6).

TABLE A.2.11 Global cumulative usage and emissions of selected pesticides for various periods

Pesticide	Usage	Emission	Source
Tech. HCH	9.7 million t (1948–1997) 446 kt (1980) 29 kt (1990)		Li, 1999a Li, 1999b
Lindane	> 720 kt (1950–1993) > 3 kt (2000)		Voldner and Li, 1995 Li, present study
α-HCH	6.3 million t (1948–1997) 290 kt (1980) 59 kt (1990)	4.3 million t (1948–1997) 184 kt (1980) 44 kt (1990)	Li, 1999a, b; Li *et al.*, 2000
β-HCH	39.7 kt (1980) 8.4 kt (1990)	3.3 kt (1980) 0.9 kt (1990)	Li, 2001a
γ-HCH	> 3 kt (2000)	> 2 kt (2000)	Li, present study
Toxaphene	1.3 million t (1950–1993)		Voldner and Li, 1995
DDT	2.6 million t (1950–1993)		Voldner and Li, 1995

A.3 Organochlorines in Russian rivers and loadings to Russian northern seas

Authors:

W. Strachan (1); and D. Muir (1).

Affiliations:

1. Environment Canada, National Water Research Institute, 867 Lakeshore Rd., Burlington ON, L7R 4A6, Canada.

A.3.1 Introduction

The previous AMAP POPs assessment reported concentrations of the OC pesticides HCH and DDT in major north-flowing rivers in Russia and compared them to results from Canada and Norway from sampling and analysis campaigns conducted the mid-1990s (de March *et al.*, 1998). Macdonald *et al.*, (2000) also summarized results for selected POPs in Canadian and Russian Arctic rivers. In the past five years there have been no new studies of POPs in river water in the Canadian Arctic and northern Norway. A project underway in Russia in 2001–02 is studying contaminants in selected Arctic rivers (RAIPON/AMAP/GEF Project, 2001) and preliminary results for concentrations of persistent organochlorines in the Pechora, Yenisey, Katanga and Kanchalan rivers are included here for comparison with previous studies.

A.3.2 Verification of older Russian river water data

The data from Russian rivers presented in the AMAP Report (de March *et al.*, 1998) and CACAR-I (Jensen *et al.*, 1997) were received late in the assessment process and were not thoroughly evaluated. Since then, several papers have been published giving a more comprehensive description of the quality assurance, as well as the levels and trends of OC pesticides, in major Russian rivers (Alexeeva *et al.*, 2001; Zhulidov *et al.*, 1998). The available results are, therefore, re-examined here. Zhulidov *et al.* (2002) have also reported temporal trends of HCH and DDT isomers in major rivers from the late 1980s to early 1990s.

All sampling and analyses of POPs in Russian rivers during the 1990-96 period were carried out by regional laboratories (UGMSs) of ROSHYDROMET (Federal Service of Russia on Hydrometeorology and Environment) with method development and quality assurance roles by a central co-ordinator (the Hydrochemical Institute, located in Rostov-on-Don). Methods for measuring the OC pesticides used solvent extraction (hexane) of 1 L unfiltered water and gas chromatographic determination. An improved cleanup, using H_2SO_4 became available in 1995 and was in use in 1996 in all UGMS labs. The internal UGMS results for both precision and accuracy for all contaminants were very good (< 30 % deviation) for the six OC compounds determined (α-HCH, β-HCH, γ-HCH, *p,p'*-DDE, *p,p'*-DDT, and dihydroheptachlor (DHH; a heptachlor analog not determined in European or North American studies of OC pesticides).

A.3.2.1 Sampling program

In Russia sampling took place at stations along major rivers which flow into the Arctic Ocean; however, most attention has focused on the stations furthest downstream in each river to estimate loadings to the Arctic Ocean. Some of the stations were several hundred kilometres upstream of the geographical river mouth, as defined by the headlands on the mainland. In some cases, this is because the river is tidal/saline many kilometres upstream; in others, it is a matter of logistical convenience for sample collection. In some cases, there are cities with significant populations located downstream of the river mouth monitoring stations. Such cities and rivers include: Onega on the river Onega; Novodvinsk and Archangelsk on the river Severnaya Dvina; Narian-Mar on the river Pechora; and, Dudinka on the river Yenisey.

TABLE A.3.1 Seven-year weighted mean concentrations (μg/L) of contaminants in Russian northern rivers, 1990–96[(1)]

River	α-HCH	γ-HCH	*p,p'*-DDT	*p,p'*-DDE
Kola	0.003	0.003	< 0.001	< 0.001
Onega	0.003	0.003	< 0.001	< 0.001
S. Dvina	0.001	0.001	< 0.001	< 0.001
Mezen	0.004	0.004	0.004	< 0.001
Pechora	0.002	0.012	0.001	0.001
Ob	0.030	0.053	0.020	0.002
Nadym	0.029	0.065	0.020	0.007
Pur	0.069	0.107	0.027	0.006
Taz	0.058	0.114	0.038	< 0.001
Yenisey	0.009	0.016	< 0.001	< 0.001
Anabar	< 0.001	< 0.001	< 0.001	< 0.001
Olenek	0.001	< 0.001	< 0.001	< 0.001
Lena	< 0.001	0.001	< 0.001	0.008
Kolyma	< 0.001	0.003	< 0.001	< 0.001

[(1)]Weighted average calculated for all data for the study period — weighted for the number of contributing samples for each annual mean.

Estimation of loading of OC pesticides was difficult because of many "non-detected" (nd) values. This does not mean that the chemical was not present, only that if present, its concentration was less than the detection limit. In the data sets reported herein, a zero was substituted for "nd" for spreadsheet purposes whenever the mean for an analyte was calculated or other aspects of reporting were needed. This biases the mean result towards a low value; substitution of the detection limit, however, would bias the results towards a high value. The true mean lies between these values and the means reported here are, therefore, conservative. During examination of the data, it was seen that an occasional maximum value represented the determining value in the mean, usually when there were only a small number of data points. Where a reason to reject such points was discovered by the analysts, the mean was recalculated; where there was no reason for rejection the data were included. All data presented have been "vetted" in this manner. The extent of the nd values in the data reported here is unknown but is believed to be extensive and the means and loadings presented must therefore be conservative. Better reporting of means will require much more improved detection limits for all contaminants mentioned in this report.

Comparability of OC pesticide concentrations for the rivers Ob, Yenisey, Lena and Kolyma was assessed by Alexeeva *et al.* (2001) using independent data reported by Zhulidov *et al.* (1998). The mid-range of the data from the "independent specialists" part of that report was used for comparison since means were not presented. The mid-range values from the independent specialists and the concentration means here (Table A.3.1) gave inconsistent ratios. In the Ob, values from the specialists were 3–6x higher than those reported by UGMS labs for both α-HCH and γ-HCH; for the Yenisey, the concentrations given in both reports were similar (within 20%) while levels for the Lena and Kolyma were below those of the specialists.

A.3.2.2 DDT, α-HCH, and γ-HCH levels and spatial/temporal trends

Monitoring data on OC pesticides in river water in Russia from the 1980s and early to mid-1990s has recently been published. Providing a historical perspective of pesticide loadings to Russian northern seas (Petrosyan *et al.*, 1998; Gordeev and Tsirkunov, 1998; Alexeeva *et al.*, 2001; Zhulidov *et al.*, 2002). Gordeev and Tsirkunov (1998) summarized the estimated fluxes of ΣHCH and ΣDDT for 32 rivers in Russia including 11 flowing to the Arctic Ocean. To obtain fluxes, they used annual arithmetic mean concentrations, summarized by Petrosyan *et al.* (1998), and river discharges taking into account seasonal variability in flows. Alexeeva *et al.* (2001) using a similar approach to estimate fluxes for the same pesticides for 1990–1996, used monitoring data from regional laboratories (UGMSs) of ROSHYDROMET and data from the same pesticides monitoring program as Gordeev and Tsirkunov (1998). The combined results (Figure A.3.1) show a general decline in loadings of HCH and DDT isomers from 1981–1996 coinciding with the reduction in use of HCH in Russia following the conversion to lindane, and the complete cessation of agricultural use of DDT in the 1970s. DDT use in Russia continued until 1990 when 0.084 kt was reported to be used, down from 0.463 kt in 1980 (HELCOM, 2001).

Zhulidov *et al.* (2002) reported temporal trends of ΣDDT (*p,p'*-DDT, DDE, and DDD) and ΣHCH (and isomers) in water and sediments from eight Russian Arctic rivers from 1988 to 1994. DDT was not detected in any sediments and DDE and DDD were only detected in the sediments of three rivers. DDD was not detected in river water and DDT and DDE were only present in North Dvina and Pechora river water. The detection limits, however, are high: 2.5 ng/g in sediment and 5 ng/L in water. Both HCH isomers α and γ were detected in all river sediments, except in the Kolyma River, and were detected in water from all rivers. Concentrations of both DDT and HCH declined significantly from 1988-1992 and were near detection limits for all chemicals in most rivers by 1992 (Figure A.3.2).

The available concentration data was limited to annual arithmetic means, ranges,and the number of samples (log-normal distributions were reported for the larger of these data sets). This data set, consisting of ~15 rivers and six regularly monitored OC contaminants over seven years, was examined for each river to assess whether temporal or geographic pattern(s) existed. A significant trend of concentration with time was apparent only for γ-HCH in the Pechora River (1990–1996) but not for any other pesticide or river. In the Pechora River, the origin of γ-HCH must be relatively local because similar patterns were not observed in other nearby rivers. There has been no reported agricultural use of HCH in this basin, however, during the years reported. It may be that the source of HCH is urban wastewaters from domestic use in towns and cities within the watershed, or possibly from its use for biting fly control in urban and other non-agricultural settings.

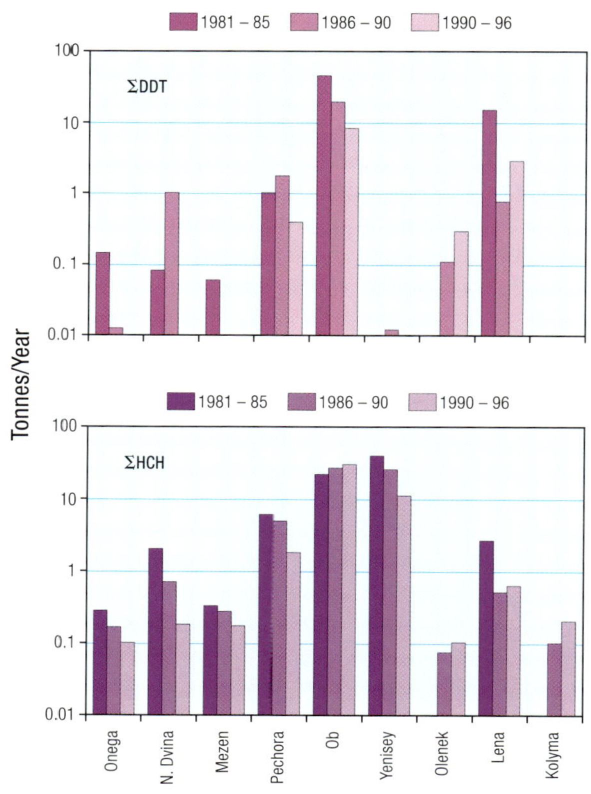

FIGURE A.3.1

Temporal trends in fluxes of ΣDDT and ΣHCH fluxes in major rivers flowing into Russian northern seas for the periods 1981–85, 1986–90 and 1990–1996 from Gordeev and Tsirkunov (1998) and Alexeeva *et al.* (2001).

Consequently, with no consistent temporal patterns, the concentration data for the rivers were averaged for the study period, and weighted for the number of contributing samples for each annual mean (Table A.3.1).

The α-HCH and γ-HCH concentrations in the rivers draining to the White/Barents Seas (Table A.3.1) were in the low µg/L range (means of 0.0023 and 0.0041 µg/L, respectively). DDT and DDE were non-detectable except for the Pechora River in 1993 and 1996. Even the maximum contaminant levels for the rivers do not appear to change in any consistent fashion over the reporting period. Gamma-HCH in the Pechora River seemed to display higher mean and maximum levels in the 1990-1994 time period, but these were only 2-3x those found elsewhere in the same region. DDT and DDE detection limits were high, and it is not surprising that almost no residues were found to be above the low µg/L range.

Four of the five monitored rivers flowing to the Kara Sea (the Ob, Nadym, Pur, and Taz) were characterized by much higher levels of the reported OC pesticides than rivers flowing to the White and Barents Seas; the fifth river, the Yenisey River, is significant for loadings but less so for concentrations. The averages of the annual means for α-HCH, γ-HCH, *p,p'*-DDT and *p,p'*-DDE in these four rivers were: 0.028, 0.050, 0.015 and 0.0022 µg/L, respectively, although the contributing means had a high level of variability and the Ob had more samples contributing to its annual means than the other rivers. Babkina (1999) reported that agricultural usage and soil concentrations of γ-HCH in the river basins and sub-basins studied here were significant only in the upper Ob-Irtysh basin and to a lesser extent in the upper Yenisey basin. The Ob River (and its tributary the Irtysh River) drain agricultural areas in the south of Russia and in Kazakhstan, where they have received high pesticide applications, including γ-HCH. The observation that the shorter, non-agricultural Nadym, Pur, and Taz Rivers exhibited γ-HCH levels as high or higher than those of the Ob, lends weight to the premise that forestry, mining or other usage was significant in those basins.

There appears to be a transition for the HCHs in the Ob river around 1993-94; values before this period are much higher than those in the mid-1990s. Alexeeva *et al.* (1997) reported usage of total-HCH in Russia at roughly 5000 tonnes/annum for 1970 through to the mid-1980s falling to 600-700 t/y by the late 1980s. HELCOM (2001) reports lindane use in Russia of 13.7 t in 1990 and 5.9 t in 1996. Unpublished data (Babkina, 1999) show that there was a further sharp reduction in use of HCH in the upper Ob-Irtysh basin around 1992 and that collectively, the usage in this basin

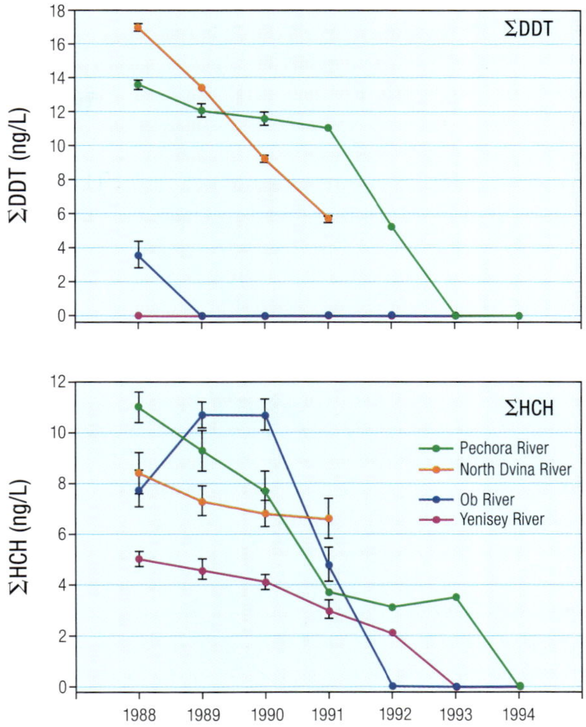

FIGURE A.3.2

Temporal trends in concentrations of ΣDDT and ΣHCH in water of major rivers flowing into Russian northern seas for the periods 1988–1994 (Zhulidov *et al.* 2002; in press).

before this time was a significant fraction (ca. 1/3) of the amounts reported for all of Russia and the USSR in the late 1980s. The reductions in HCH usage in these upper reaches of the river are roughly reflected in the water concentrations at the river mouth and indicate a fairly rapid system response time.

There were few observations of OC pesticides above the detection limits for any of the rivers flowing into the Laptev and E. Siberian Seas (only for HCHs and that mainly in 1990). The region is characterized by tundra over much of the basins and there is less reason to expect the use of γ-HCH in the region. It is noted that the 1990 observations were during the period when HCH usage for agriculture was still high. While there are no significant agricultural activities in the region, mining is extensive and the pesticide use in support of such activities, as with forestry, is unknown.

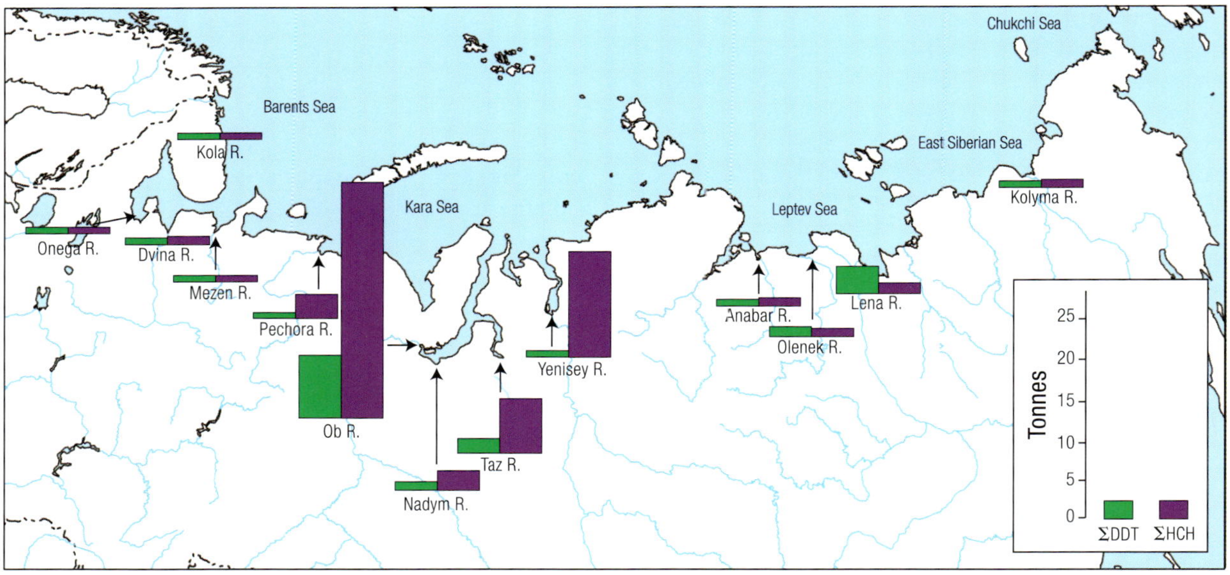

FIGURE A.3.3
Average fluxes of ΣDDT and ΣHCH from Russian rivers to the Arctic Ocean — 1990–96 (Alexeeva *et al.*, 2001).

A.3.2.3 Other chlorinated pesticides in northern Russian rivers

Beta-HCH was not found at any site in the northern rivers of Russia. This was expected, given that the α- and γ-HCH isomer levels are near their limits of detection. In Canadian Arctic rivers, levels of β-HCH were also less than detection limits (ca. 0.00002 µg/L) (Jeffries, D.S., Backus Consulting for Environment Canada, pers. comm. 1999); in Norway, they were not reported (Holtan *et al.*, 1994).

The rivers flowing to the Kara Sea were the only ones in which the insecticide degradation product DHH (dihydroheptachlor) was investigated. This compound may have been used in forests against biting insects, as well as in the agricultural industry. The overall mean of the reported years/rivers was 0.0042 µg/L and, at least for the Ob, the sample numbers were considerable. In all of the former Russia/USSR in 1990 and 1991, 580 and 520 tonnes/annum of DHH were used (Alexeeva *et al.*, 1997) and the Ob basin accounted for 12% of these amounts in both years. Data for later years are not available. This pesticide has not been investigated in Canadian or in Norwegian rivers.

A.3.2.4 Loadings of OC pesticides into the Russian northern seas

The loadings of ΣHCH and ΣDDT to Russian northern seas are illustrated in Figure A.3.3. The Kara Sea received by far the largest fraction of the OC pesticide discharges going to the northern seas, well over 90% of the totals for three of the four measured organics. The two HCHs and *p,p'*-DDT flowed to the Arctic mostly via the Ob River (α-HCH, 50%; γ-HCH, 53%; *p,p'*-DDT, 72%); in 1990, DDE largely (76%) entered the Laptev Sea via the Lena River. Since DDT has been banned for more than 25 years, this material may be partly from soil and deposited material applied earlier (Harner *et al.*, 1998), although a higher proportion of DDE would be expected than what was observed. The loadings of HCHs were greater than those of DDT and DDE, except for DDE in the Lena River in 1990, for which there is no known explanation.

The Ob River carried the major loadings of the two HCHs which, in the case of α-HCH, was 50% of the total for all of the rivers studied; the Yenisey was responsible for a further 24% of the total. The Ob carried 53% of the γ-HCH to the Kara Sea, while the Yenisey carried 18% of this isomer. A major fraction (⅓ based on data for pre-1990 use) of Russia's agricultural use of HCH occurs in the upper reaches of these two rivers (and these are the only reported uses in the northern river basins). The observation that the Nadym, Pur, and Taz Rivers also have significant loadings (and high concentrations) of both of these isomers relative even to the Ob, indicates other anthropogenic sources in these non-agricultural watersheds. Forestry and mining are prominent industries in the area and both use significant quantities of pesticides to provide more acceptable working conditions in the region.

These rivers represent the major but incomplete loadings from the northern flows. Using the AMAP data for total flows to several seas (Gregor *et al.*, 1998), the basin loadings from the monitored rivers in each northern sea were scaled up to account for the rivers not monitored (those measured represented 58–80 % of several basin flows). The total annual loadings of the OC pesticides were estimated to be 25 t for α-HCH, 44 t for γ-HCH, 13 t for *p,p'*-DDT and 6 t for *p,p'*-DDE. The HCH values can be compared with several fluxes to the Arctic previously reported (Macdonald *et al.*, 2000). Delivered via the Bering Strait, the loadings of the two HCH isomers were 52 and 12 t/y (α-HCH and γ-HCH, respectively) and via the atmosphere 53 and 10 t/y.

A.3.3 Recent studies of organochlorines in Russian freshwater

Samples of freshwater (15 to 45 kg) were collected from lakes and rivers in four regions of the Russian Arctic [Kola Peninsula, Pechora Basin, and Taimyr Peninsula (Dudinka and Khatanga) and Chukotka] in 2000–2001 for analysis of OCs (RAIPON/AMAP/GEF Project, 2001). Samples from Lovozero are of lake water and were collected at various depths (0–30 m). There did not appear to be trends in OC concentrations with depth and these data have been combined. Samples from all other sites were collected from rivers near the surface (0–1m) and at depth (up to 7 m) and results for various depths were combined to obtain an average concentration for each location. Water (unfiltered) was extracted by solid phase extraction on C18-SPE disks. PCBs were the predominant OC followed by DDT and chlorobenzenes (Figure A.3.4). Toxaphene (sum of Parlar 26, 52 and 60, detection limits ~ 0.1 ng/L) and polybrominated diphenyl ethers (PBDEs, detection limits ~ 0.5 ng/L) were not detected in any water sample but the detection limits of this study are above the levels that would normally be anticipated for these compounds in the water column. Blank results from the water analysis showed non-detect (< 0.05 ng/L) concentrations of all PCB congeners. The extent of contamination caused during sampling, however, which is a problem in the collection of seawater for PCB analysis (see section B2.1), was not investigated.

No geographical trends were obvious for any OC group, although levels were slightly higher in the lake from the Lovozero region (Figure A.3.4). Whether this is due to these being lake rather than river samples cannot be assessed with the information available. Organochlorine concentrations are in the same range as levels reported previously for Russian Rivers.

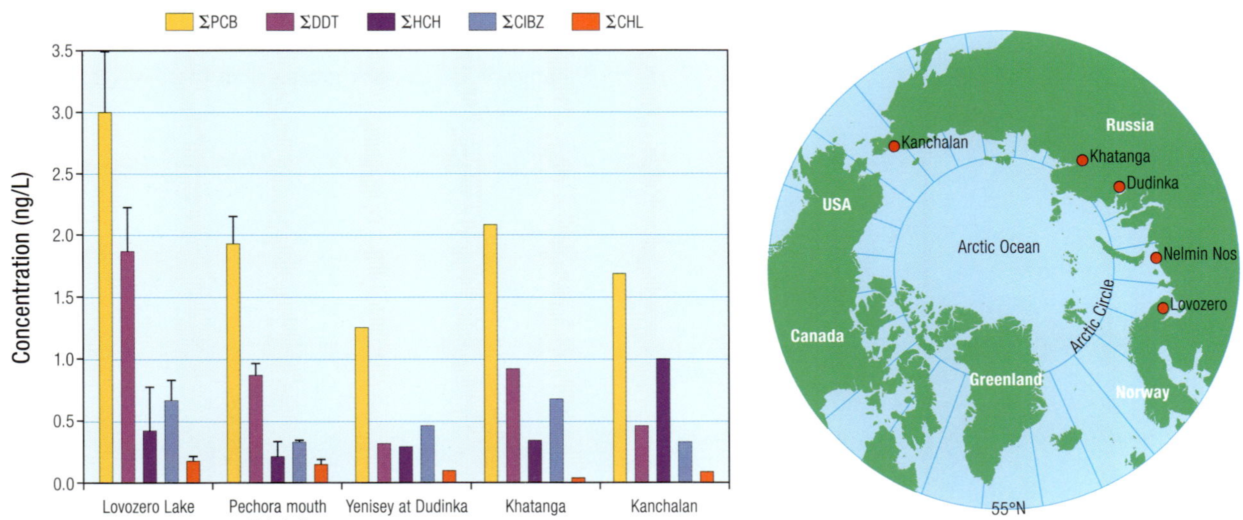

FIGURE A.3.4

Concentrations (ng/L) of OCs in lake (Lovozero only) and river water from regions within Russia (RAIPON/AMAP/GEF Project 2001). Locations are arranged from west to east. Bars are means or means ± 1 SE when more than 2 samples are available. All samples are pools of numerous samples collected in 2000–2001. Samples from Lovozero included surface and those at depth; OC concentrations varied little with depth.

Defence R&D Canada/Janice Lang

Occurrence and Trends

B.1 POPs in the atmosphere

Authors:

P. Blanchard (1); H. Hung (1); C. Halsall (2); T. Bidleman (3); G. Stern (4); P. Fellin (5); D. Muir (6); L. Barrie (7); L. Jantunen (1); P. Helm (4); and A. Konoplev (8).

Contributors:

B. Rosenberg (4); S. Friesen (4); H. Li (5); R. Bailey (9); E. Brorström-Lundén (10); and S. Juntto (11).

Affiliations:

1. Meteorological Service of Canada, 4905 Dufferin Street, Downsview, ON, M3H 5T4, Canada;
2. Environmental Science Dept., Institute of Environmental and Natural Sciences, Lancaster University, Lancaster, LA1 4YQ, UK;
3. Centre for Atmospheric Research Experiments, 6248 Eighth Line, R.R. #1, Essa Township, Egbert, ON, L0L 1N0, Canada;
4. Freshwater Institute, Fisheries and Oceans, 501 University Crescent, Winnipeg, MB, R3T 2N6, Canada;
5. AirZOne, Inc., 2240 Speakman Drive, Mississauga, ON, L5K 1A9, Canada;
6. National Water Research Institute, Canada Centre for Inland Waters, 867 Lakeshore Road, Burlington, ON, L7R 4A6, Canada;
7. Atmospheric Research and Environment Programme, World Meteorological Organization, 7 bis avenue de la Paix, CH-1211 Geneva 2, Switzerland;
8. SPA Typhoon, Lenin Avenue 82, Obninsk, 249038, Russia;
9. Department of Chemistry & Biochemistry, University of Regina, Regina, SK, S4S 0A2 Canada;
10. Swedish Environmental Research Institute, Box 47086, Gothenburg, SE-402 58, Sweden;
11. Finnish Meteorological Institute, Sahaajankayu 20E, Helsinki, FIN-00810, Finland.

B.1.1 Introduction

Weekly air samples (particulate and gaseous), for the measurement of atmospheric pollutants, including PCBs, PAHs, and organochlorine (OC) pesticides, and synthetic industrial compounds, have been collected from the Canadian Arctic sampling sites of Alert, Nunavut (1992–ongoing); Tagish, Yukon (1993–1994); Dunai Island, Russia (1993–1994); Amderma, Russia (1999–2001); and Kinngait, Nunavut (1994–1995, 2000–ongoing). In CACAR-I, Arctic atmospheric data for PCBs, PAHs, and OC pesticides through 1994 were presented for the sites of Alert, Tagish, and Dunai Island in Russia. Results included a variation in the PCB homologue pattern according to the origin of air masses affecting the receptor sites. For the OCs, the lack of spatial differences indicated widespread atmospheric contamination while levels at Tagish reflected relative proximity to North American and Asian source regions. The particle/gas partitioning of PAHs during the haze season at Alert was successfully represented using the Junge-Pankow model. The PAH partitioning was also well correlated with vapour pressure for Alert and Tagish, similar to temperate sites around the Great Lakes. This demonstrated that temperature-dependent vapour pressure was a good descriptor of gas/particle distribution of PAHs even at extremely cold Arctic temperatures.

B.1.2 Data and time series analysis

Trend analysis was conducted using data up to the end of 1997 for Alert, Nunavut, to investigate whether atmospheric concentrations are declining in the Arctic. Atmospheric concentrations for semi-volatile pollutants, including PCBs and OCs, often exhibit seasonality. If a temporal trend is to be determined, it is important that the intra-annual variability be reduced. As part of the data analysis, a statistical time series analysis technique (digital filtration, DF) was used to smooth any seasonality in the air data (Nakazawa *et al.*, 1997). The digital filters extracted variabilities that were longer than four months and shorter than two years to obtain the overall seasonal cycle, and variabilities that were longer than two years to obtain the long-term trends (Hung *et al.*, 2001; 2002a). Results of the trend analyses are presented in Sections B.1.2.1 and B.1.2.2 for PCBs and OCs. Results of a study on the correlation between chlordane isomers and enantiomers, and the time trends of chlordane isomers in air based on compiled data from 1984 to 1998 observed at four Arctic stations, including Alert, are also summarized in Section B.1.2.2 (Bidleman *et al.*, 2001).

Recent PAH results at Alert are reported in Section B.1.2.3 where annual means are compared to other Arctic sites. As a special study, polychlorinated dibenzo-p-dioxins and dibenzofurans (PCDD/Fs) were measured in filter samples from Alert in the winter of 2000–2001. Results from this short study are presented in Section B.1.2.4. Trans-Pacific transport is examined in detail using Tagish data in Section B.1.2.5. Atmospheric concentrations at Alert, Dunai, and Tagish for 1994–95 were also determined for a number of non-traditional chemicals, including polychlorinated naphthalenes (PCNs), short-chain chlorinated paraffins (SCCPs), coplanar PCBs, and brominated flame retardants. These results are presented and discussed in Section B.4.

B.1.2.1 Polychlorinated biphenyls (PCBs)

The mean and range of ΣPCB concentrations in both particulate and vapour phases for 102 congeners, as well as a subset of the ten Arctic Monitoring Assessment Programme (AMAP) congeners are listed in Table B.1.1. As observed in CACAR-I for 1993, the PCB levels for 1995–1996 were generally lower at Alert than those reported for other AMAP Arctic sites. For example, annual means (1995 and 1996) for the ten AMAP congeners were 16.9 and 19.4 pg/m^3 at Storhofdi, Iceland and 68.2 and 165 pg/m^3 at Spitsbergen, Norway (AMAP, 1999). The mean vapour phase concentrations observed more recently (1999–2001) at Amderma, Russia and Kinngait, Nunavut were similar to that observed at Alert. A selected set of 25 samples collected at Amderma between 1999 and 2000 showed a mean vapour phase concentration of 4.05 pg/m^3 (AMAP subset). At Kinngait, results of 28 samples collected between August 1, 1994 and April 16, 1995 and 12 samples between November 20, 2000 and March 12, 2001 showed mean vapour phase concentrations of 2.74 and 3.28 pg/m^3, respectively (AMAP subset). In comparison to the temperate IADN sites of Eagle Harbour (ΣPCB = 128 ± 12 pg/m^3), Sleeping Bear Dunes (160 ± 10 pg/m^3) and Sturgeon Point (315 ± 20 pg/m^3) over 1991 to 1995 (Hillery *et al.*, 1997), levels observed at Alert were 5- to 11-fold lower.

Figure B.1.1 shows the seasonal cycles and trends of PCB 31, 101, 138, and 180 at Alert determined using the DF technique. The weekly average temperature recorded at the sampling site is also included in Figure B.1.1. The seasonal cycles likely result from the temperature dependency of atmospheric PCB concentrations as well as photochemical and radical (OH) reactions during the Arctic summer. This is in contrast to what was reported for two years of data (1993–94) in CACAR-I and is simply a result of the longer time series now available. By comparing the trends, it was noted that the lower chlorinated congeners, PCB 28, 31, and 52, declined from 1993 to 1997, as exemplified by PCB 31 in Figure B.1.1. In each case, both the seasonal minimum and maximum declined over the five years; however, no clear trend was apparent for the penta- and hexa-chlorinated congeners.

First order half-lives (dividing ln 2 with the regression slopes) were estimated for forecasting purposes through linear regression of the DF trend lines. The regression results for the ten AMAP congeners are presented in Table B.1.2. The strongest correlations between concentration and time occurred for the lighter congeners (r^2 ranged from 0.58 to 0.87), with evidence of declining trends for these compounds over the five years of monitoring.

TABLE B.1.1 Annual mean and range of ΣPCBs concentrations (pg/m^3) of total PCBs (102 congeners) and the AMAP subset (10 congeners) in vapour (PUF) and particulate (GFF) phases at Alert (Hung *et al.*, 2001)

	Total PCBs[1]						AMAP subset[2]					
	PUF			GFF			PUF			GFF		
Alert	Mean	(SD)	Range	Mean	(SD)	Range	Mean	(SD)	Range	Mean	(SD)	Range
1993	29	(18)	3.2–98	3.7	(1.9)	1.5–15	5.6	(4.6)	0.18–26	0.33	(0.70)	0.089–5.0
1994	30	(20)	8.0–150	3.7	(2.5)	2.6–21	5.8	(3.8)	1.0–25	0.27	(0.23)	0.15–1.6
1995	30	(14)	12–94	3.3	(0.8)	2.6–8.3	4.9	(3.1)	2.0–22	0.25	(0.18)	0.16–0.88
1996	26	(13)	7.5–100	3.4	(0.9)	1.9–8.2	4.2	(2.4)	0.69–17	0.30	(0.17)	0.11–0.97
1997	23	(6.7	10–42	3.9	(2.2)	2.7–16	3.4	(1.2)	0.99–6.4	0.39	(0.64)	0.15–4.2

(1)Sum of 102 PCB congeners. Refer to Stern *et al.* (1997) for congener numbers.
(2)Sum of PCB 28, 31, 52, 101, 105, 118, 138, 153, 156 and 180.

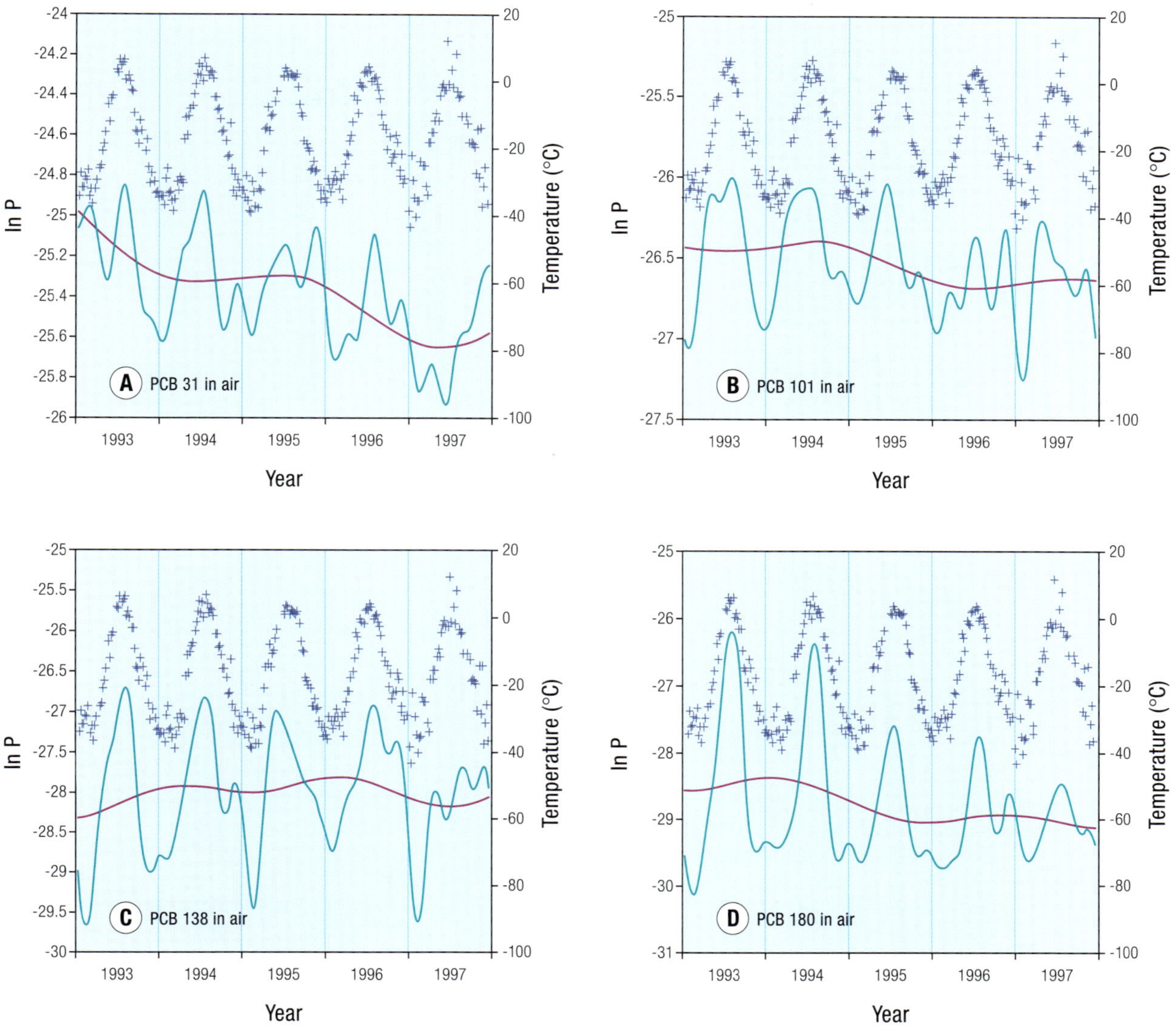

FIGURE B.1.1

Seasonal cycles and trends of PCB 31, 101, 138 and 180 at Alert.

The r^2 values for the heavier congeners were generally lower (< 0.2), except for PCB 180, which showed a highly correlated decreasing trend. However, no pattern was apparent between half-lives and chlorine numbers, with half-lives ranging from 3 to 17 years. The half-lives at Alert were typically longer than those derived from temperate background/rural sites from the early 1990s (Table B.1.2).

The lack of a declining PCB temporal trend for the more chlorinated congeners may be due to several factors. The lower temperatures encountered in the Arctic, coupled to winter darkness, will slow both biotic and abiotic degradation relative to temperate regions. Importantly, global fractionation (Wania and Mackay, 1996) — a process where chemicals may be latitudinally "fractionated" according to the ambient temperature and their physicochemical properties — may have the effect of maintaining Arctic air concentrations, while levels in temperate regions are showing apparent reductions. Nevertheless, levels of some of the lower chlorinated congeners do appear to be dropping in the Arctic atmosphere, albeit at a slower rate than shown in the temperate studies. This trend for the lighter congeners is probably due to the dominating effect of long-range transport, whereby levels in the Arctic will simply reflect the falling levels encountered in the source regions. This, in combination with their relatively fast OH depletion rates during the Arctic summer, may explain the temporal decline of the lighter congeners. For example, at an average temperature of -17.6°C at Alert, the reaction rate constants of PCB 31 (2,4',5-trichlorobiphenyl) and 110 (2,3,3',4',6-pentachlorobiphenyl)

TABLE B.1.2 Half-lives of various POPs at Alert compared to those at temperate locations

PCBs	Alert, this study		PCBs	Sleeping Bear Dunes[1]		Sturgeon Point[1]	
	$t_{1/2}$ (yr)[2]	r^2		$t_{1/2}$ (yr)[2]	rel SE (%)[3]	$t_{1/2}$ (yr)[2]	rel SE (%)[3]
28	12	0.58	28+31	2.7	22	3.0	27
31	6	0.87	–	–	–	–	–
52	3	0.86	52	4.3	30	7.5	56
101	11	0.73	101	2.9	24	4.0	32
105	INC	0.021	118	2.1	19	2.1	24
118	INC	0.093	105+132+153	1.8	22	1.8	17
138	INC	< 0.01	138+163	2.2	23	2.8	28
153	17	0.18	–	–	–	–	–
180	4	0.78	180	1.7	18	3.7	46

OC	Alert, this study		Eagle Harbor[4]		Sleeping Bear Dunes[4]		Sturgeon Point[4]	
	$t_{1/2}$ (yr)[2]	r^2	$t_{1/2}$ (yr)[2]	rel SE (%)[3]	$t_{1/2}$ (yr)[2]	rel SE (%)[3]	$t_{1/2}$ (yr)[2]	rel SE (%)[3]
α-HCH	16.9	0.79	3.0	6.2	3.2	11	3.0	10
γ-HCH	4.9	0.88	4.4	14	3.4	19	4.4	21
T-CHLOR	8.3	0.48	6.9	42	5.2	33	3.2	18
C-CHLOR	4.1	0.71	*23*	*110*	9.7	50	5.9	25
T-NONA	6.2	0.71	*33*	*160*	6.0	35	5.0	30
DIELD	INC	< 0.10	3.7	18	2.4	17	2.9	21

[1] Simcik *et al.* (1999) Supporting Information. $t_{1/2}$ calculated using multiple linear regression method. Only $t_{1/2}$ calculated with airmass originating over land is presented.

[2] $t_{1/2}$ = half-life, calculated as ln 2/slope. Data with increasing time trend are indicated as INC = increasing trend.

[3] rel SE (%) = Relative Standard Error (%).

[4] Cortes and Hites (2000). The study periods were Nov. 90 to Dec. 97 for Eagle Harbor, Dec. 91 to Dec. 97 for Sleeping Bear Dunes and Sturgeon Point. Half-lives that are statistically significant with 95% confidence are indicated in normal fonts, those that are not significant are in italics.

would be 8.0 x 10^{-13} and 3.4 x 10^{-13} cm^3/s, respectively [estimated using Arrhenius parameters given by Anderson and Hites (1996)]; combining with an estimated OH radical concentration of 1.7 x 10^5 molecules/cm^3 in the Arctic atmosphere in April (Barrie *et al.*, 1988), the lifetime of PCB 31 would be 85 days compared to 202 days for PCB 110. However, the picture is complicated by notable exceptions; for instance, PCB 180 (heptachlorinated) shows a clear downward trend and has a calculated half-life of only four years at Alert (compared to 17 years for PCB 153). It is possible that heavier congeners might be subjected to different removal processes than lighter ones. Precipitation or scavenging processes, e.g., particulate deposition and snow scavenging, could be more important for the heavier congeners, while degradation via reaction with OH radicals is the dominant process for the removal of lighter congeners. It is believed that the relative removal rates pertaining to the different processes and the physicochemical properties of the congeners may be responsible for the differing trends observed.

B.1.2.2 Organochlorine pesticides

The annual average total concentrations (particle and gaseous) at Alert are shown in Figure B.1.2. Beta-HCH, heptachlor, *c*-nonachlor, photomirex, endrin, trifluralin and the DDT group (aside from *p,p'*-DDE) were detected at very low levels (45% to 95% of all samples were below the method detection limit in gas phase). The relative levels were generally: (ΣHCHs and HCB) > (Σchlordanes and endosulfan I) >dieldrin> ΣDDT.

Alpha and γ-Hexachlorocyclohexanes (α-, γ-HCHs), and hexachlorobenzene (HCB), which are the lightest of all the OCs, achieved the highest levels in all five years. Annual average air concentrations observed at Alert were similar to those found at Spitsbergen, e.g., in 1996, ΣHCHs and HCB were 63.7 and 63.4 pg/m^3, respectively at Alert and 73.1 and 92.5 pg/m^3, respectively at Spitsbergen (AMAP, 1999). For Alert, HCB shows the highest percentage of breakthrough (25.9%); thus, the concentrations might be underestimated. Preliminary data on α-HCH vapour phase concentrations from

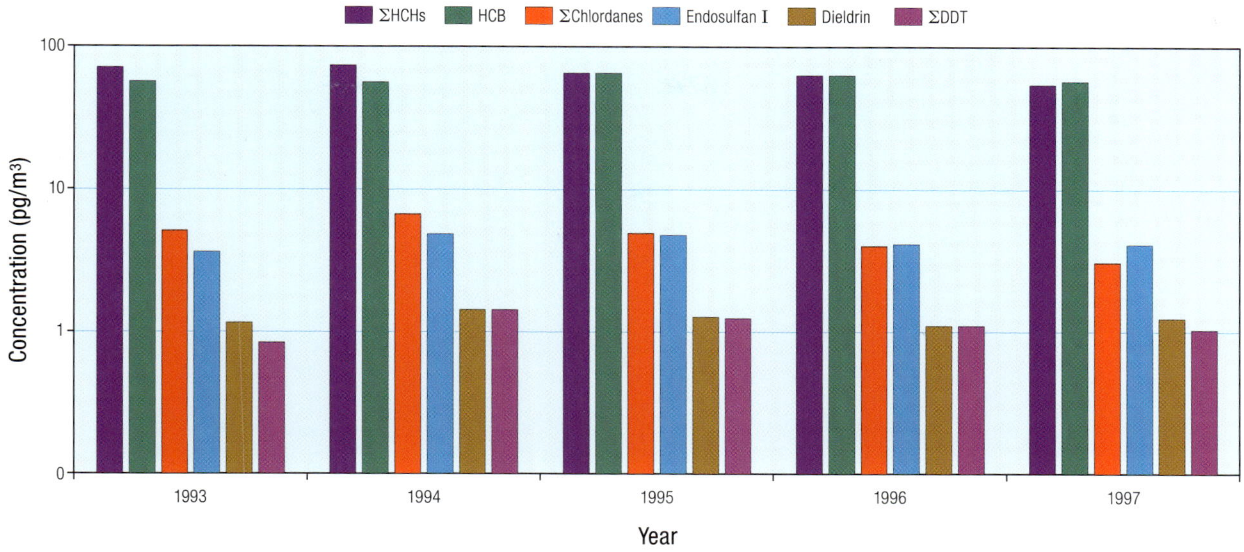

FIGURE B.1.2

Relative concentrations of selected OCs at Alert (1993–1997).

Kinngait, Nunavut, have shown lower levels in 2000–2001 than 1994–1995 (Figure B.1.3). This decrease is expected, since the use of technical HCH [containing 55–80% α-HCH (Li, 1999b)] has been banned by its two major users, China and the former USSR, in 1983 and 1990, respectively (Li *et al.*, 1998a).

After 1996, the endosulfan I concentration at Alert exceeded Σchlordanes (Figure B.1.2). Endosulfan is a current-use pesticide while the chlordanes have been deregistered in most European and North American countries since the 1980s (Barrie *et al.*, 1992) and Velsicol, its sole producer, permanently ceased production in May 1997 (Pesticide Action Network North America, 1997). Summertime DDT/DDE ratios at Alert were on the order of 1–1.5, whereas for the Russian Arctic site of Amderma, the DDT/DDE ratios were about 3. Technical DDT is made up of 80% *p,p'*-DDT and 20% *o,p'*-DDE (Ramesh *et al.*, 1989). A larger DDT/DDE ratio has also been observed at Tagish in Western Canada and is linked to trans-Pacific transport from Asia (see Section B.1.2.5.).

Figure B.1.4 shows the seasonal cycles and trends of selected OCs. It can be clearly seen that elevated OC

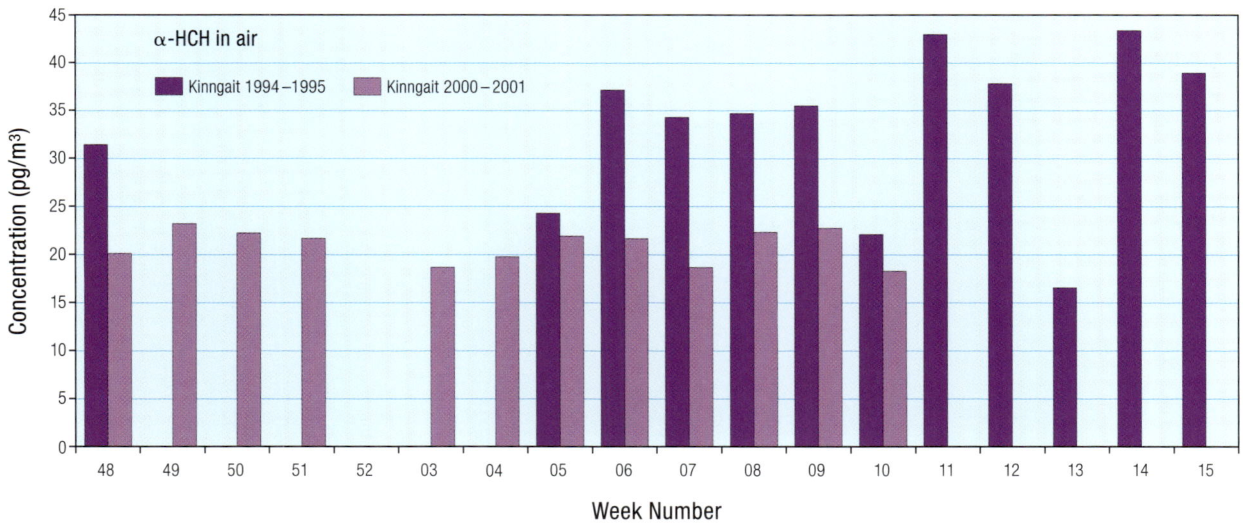

FIGURE B.1.3

Alpha-HCH vapour concentrations at Kinngait, Nunavut, (Hung *et al.*, 2002b).

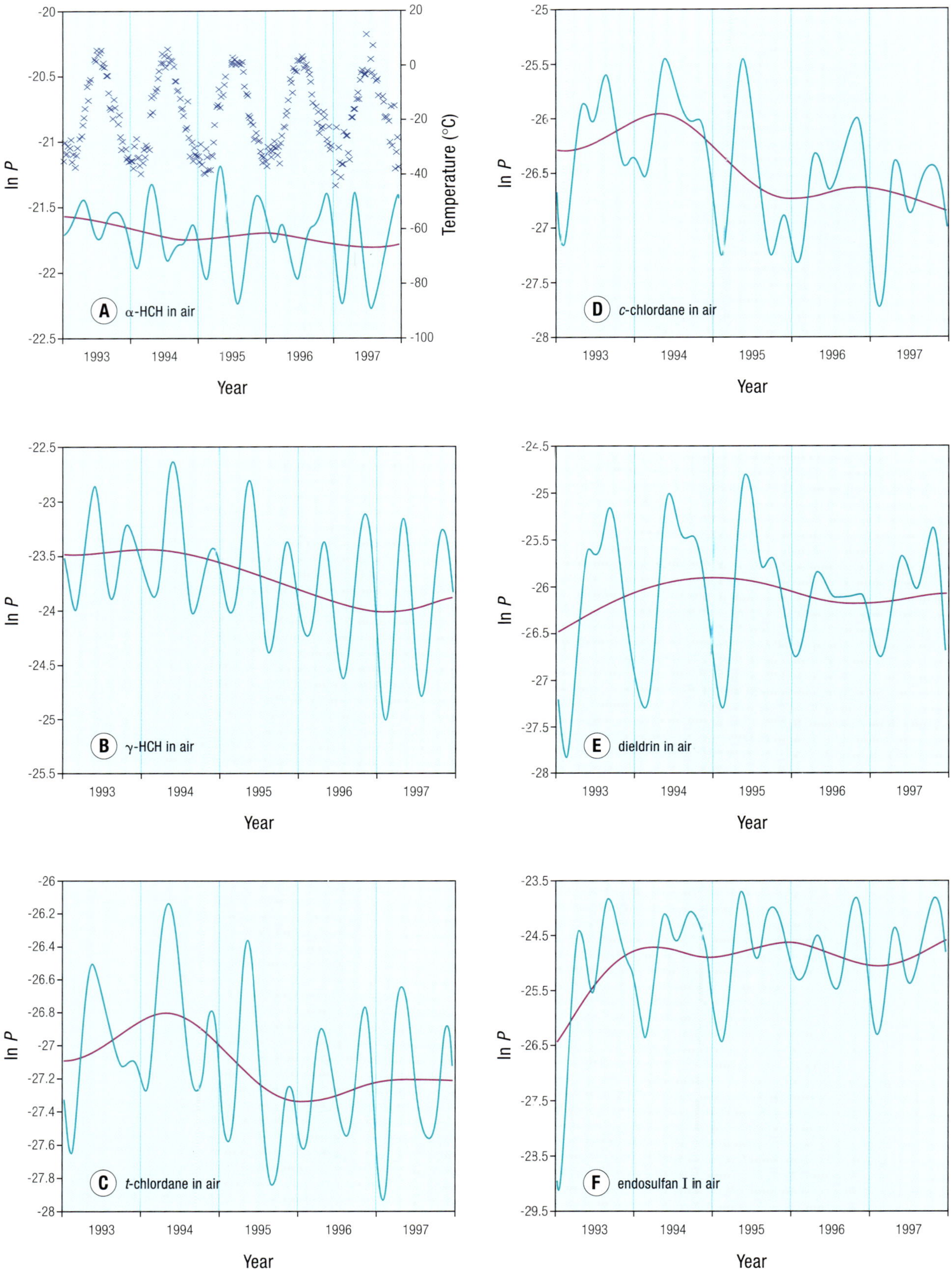

FIGURE B.1.4

Trends and seasonal cycles of a) α-HCH, b) γ-HCH, c) *trans*-chlordane, d) *cis*-chlordane, e) dieldrin, and f) endosulfan I in air at Alert ("x" in a) gives temperature profile) (Hung *et al.,* 2002a).

concentrations were generally observed before and after the warmest time of the year. The seasonal cycles were probably the result of application patterns at temperate locations suggesting that the atmosphere at Alert was mainly influenced by long-range transport (LRT). Other possible explanations include a summertime increase in photodegradation or depletion reactions via OH radicals and the scavenging of these compounds by forests along the transport pathway to the Arctic during the growing season. Most of the selected OCs showed declining trends over the five years, except dieldrin and endosulfan I. Unlike the half-lives observed for most of the heavier PCBs, which were typically longer than those derived from temperate background/rural sites, those of the OCs measured at Alert were comparable to the three Great Lakes sites with the exception of α-HCH (Table B.1.2). While PCBs have been banned in most countries for about three decades, most OCs were still being used within the last 10 years and some are, in fact, current-use pesticides in the northern hemisphere. Changes in atmospheric levels of the OCs in the Arctic appear, therefore, to mirror those in the temperate source regions.

The effect of long-range transport at Alert can be further illustrated by examination of the *trans-/cis*-chlordane ratio (TC/CC) using DF (Figure B.1.5). The ratio of TC/CC in technical chlordane mixtures used in the US is about 1:1 (Jantunen *et al.*, 2000; Mattina *et al.*, 1999). The average ratios in the Arctic atmosphere have been reported as < 1 in recent years (Oehme *et al.*, 1996; AMAP, 1998). This indicates that a more weathered chlordane profile is affecting the Arctic compared to previous decades since the *trans*-isomer is relatively less stable and more susceptible to degradation by microorganisms in soil (Beeman and Matsumura, 1981). A slightly increasing trend in TC/CC is revealed in Figure B.1.5, however, and is marked in particular by episodes when the ratio was > 1. The higher TC/CC ratios from 1995 to 1997 may indicate fresh use of chlordane-based pesticides. In an attempt to trace the sources of high TC/CC episodes, Lagrangian 5-day back trajectories (Olson *et al.*, 1978) were computed every six hours (925 hPa) for those weeks when TC/CC ratios were > 1. The trajectories were then assigned to geographical sectors of the Arctic defined by Stern *et al.*, (1997). A trajectory was assigned to a certain sector when ~70% of the points fell within that sector. The frequency distribution of trajectories in each sector for the periods during which TC/CC ratios were > 1 is summarized in Figure B.1.6. It shows that Alert was evenly influenced by the air

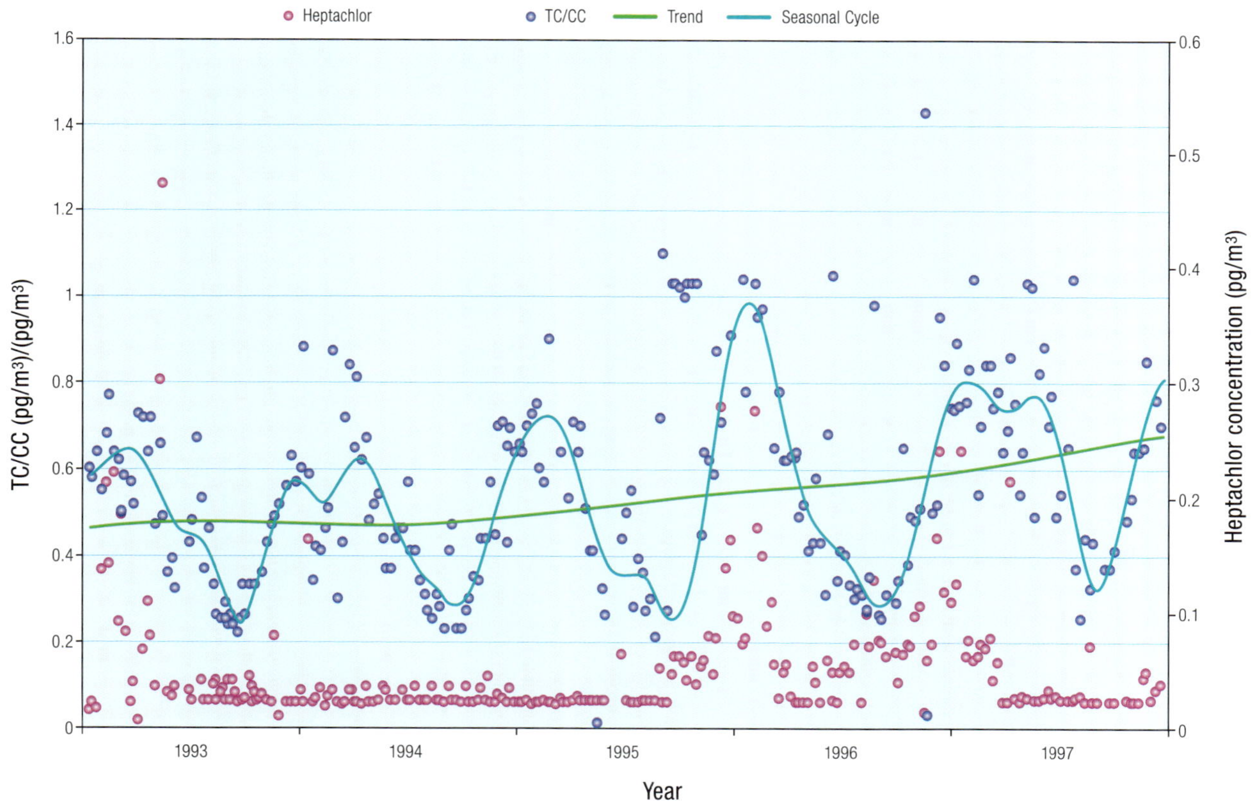

FIGURE B.1.5

Trans-/cis-chlordane (TC/CC) ratios and heptachlor concentrations in air at Alert (Hung *et al.*, 2002a).

mass originating from East/West Russia (Sector 3+4), N. America (Sector 1) and North Atlantic/Greenland (Sector 5+6) with about 30% in each sector. A summary of back-trajectories from January 1988 to December 1997, calculated at four different endpoint heights, has shown that Alert was almost equally affected by air from N. America (Sector 1, 25.9%), the Pacific region (Sector 2, 23.7%), Eastern Russia (Sector 3, 21.0%), and North Atlantic/Greenland (Sector 6, 18.5%) (Worthy *et al.*, 1999). In Figure B.1.6, there was a definite shift of the dominant sectors eastwards into Eurasia but the actual source regions of the fresh inputs were not apparent.

Another indication that chlordane-based pesticides might still be used sporadically was shown by the levels of heptachlor. These were usually very low and mostly below detection limits prior to 1995, but increased to relatively higher levels (0.2–0.3 pg/m^3) during the winter months of 1995/96 and 1996/97 (Figure B.1.5). If heptachlor was truly a "past-use" pesticide, the residue should mainly be heptachlor epoxide and there should not be fresh peaks of heptachlor. Karlsson *et al.* (2000) have observed cyclic temporal patterns of heptachlor at Senga Bay, S. Africa between 1997 and 1998 which indicate that there has been repeated use of heptachlor rather than technical chlordane. According to a survey on sources of persistent organic pollutants (POPs) performed by the United Nations Environment Programme (UNEP, 1996), the global import figures of heptachlor in 1993 and 1994, based on the response of 61 countries, were 389 and 435 t, respectively. Therefore, fresh inputs into the Arctic atmosphere are not surprising. Nonetheless, technical chlordane mixtures usually contain some heptachlor (~ 7%) (Gooch *et al.*, 1990) and technical heptachlor is contaminated with chlordane isomers, especially *trans*-chlordane (Bidleman *et al.*, 2001). For instance, the TC/CC ratio peaking together with heptachlor from February 5 to 12, 1996 may suggest fresh usage of either heptachlor or chlordane mixtures.

Bidleman *et al.* (2002a) have compiled *trans*-chlordane, *cis*-chlordane, and *trans*-nonachlor (TC, CC and TN) measurements in Arctic air from 1984–1998 at the monitoring stations of Alert (1993–97), Tagish (1993–94), Dunai (1993–94), and the AMAP station of Pallas, Finland (68°58'N, 24°07'E) (1996–98). [Data from Pallas was kindly supplied by Eva Brorström-Lundén (Swedish Environmental Research Institute) and Sirkka Juntto

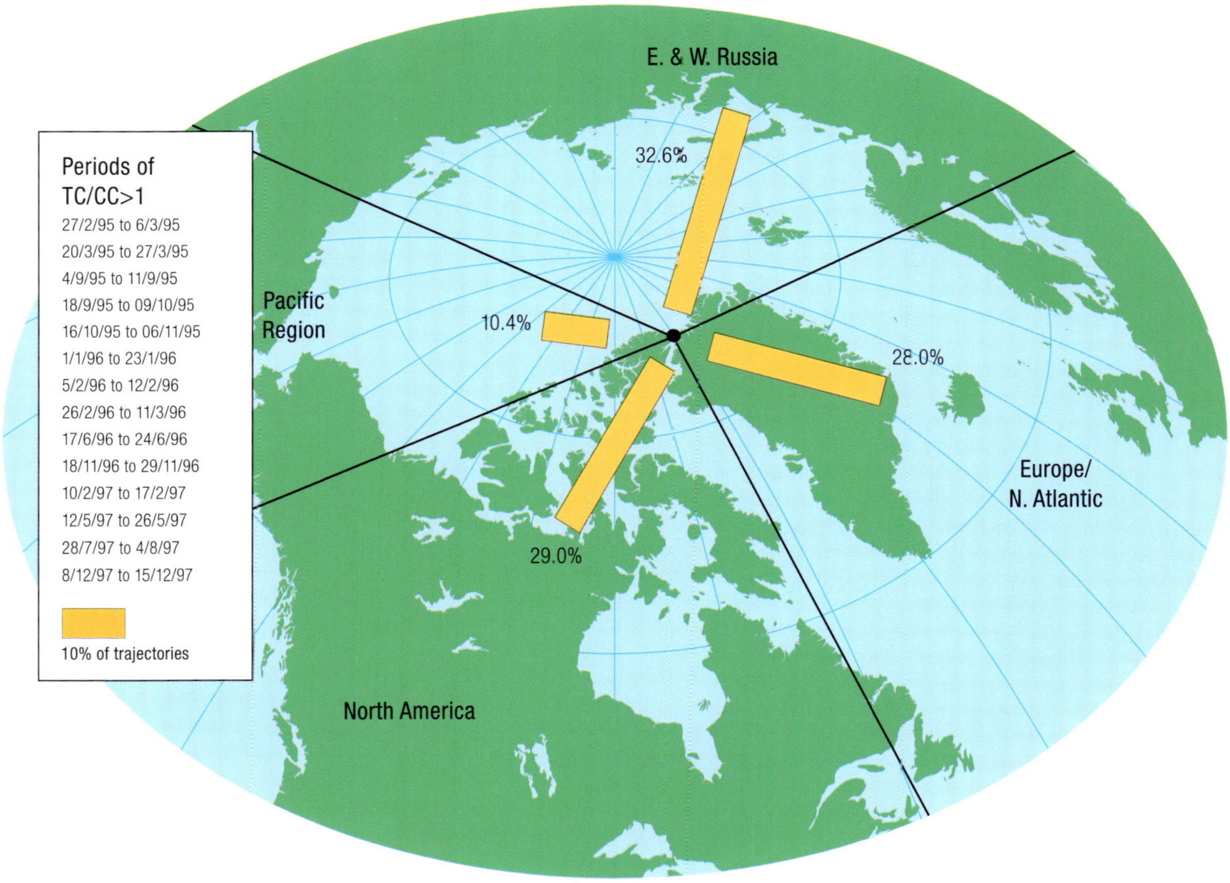

FIGURE B.1.6

Frequency distribution of trajectories to each sector when TC/CC > 1. (Hung *et al.*, 2002a)

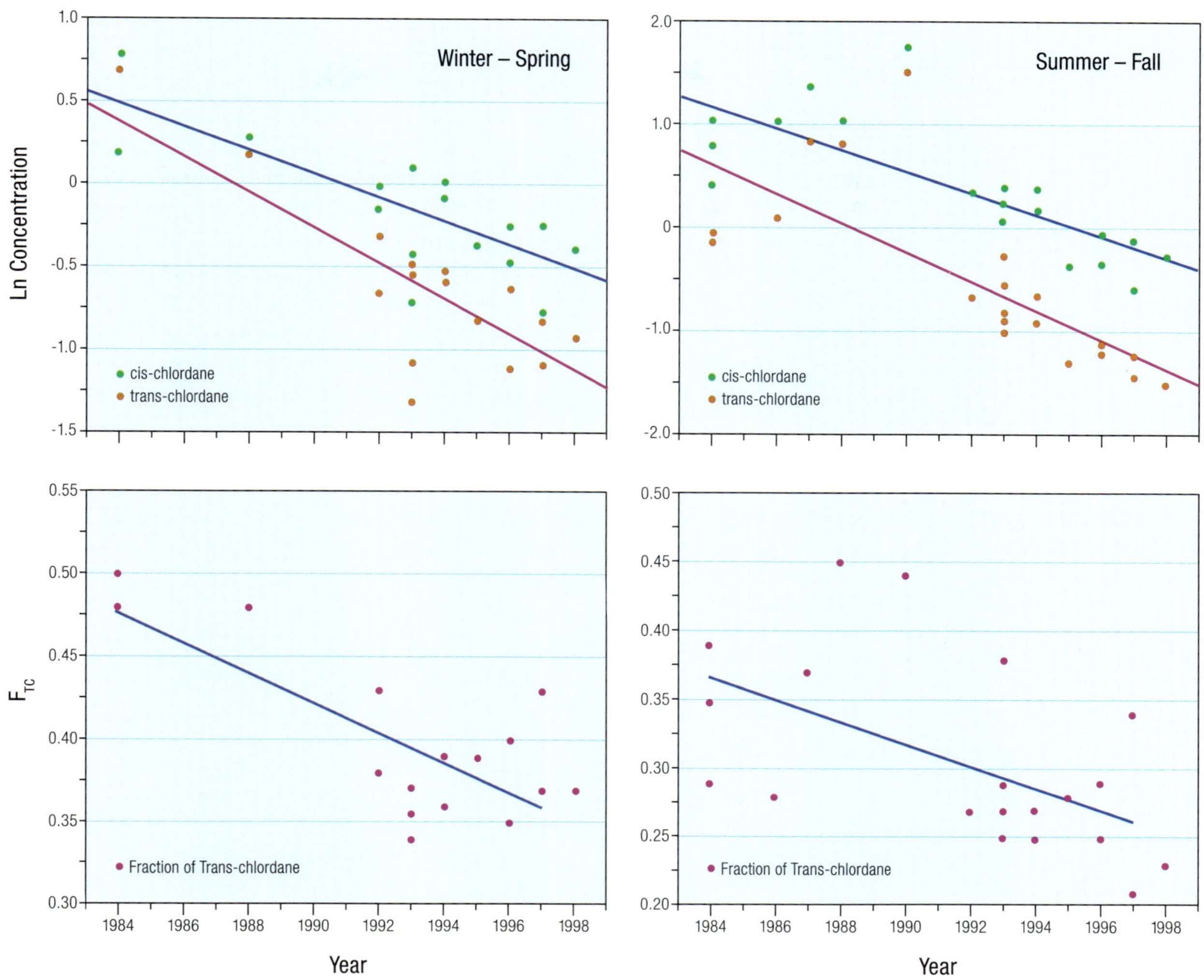

FIGURE B.1.7

Long-term trends of *trans*- and *cis*-chlordane concentrations and fraction of *trans*-chlordane, F_{TC} = TC/(TC+CC), in Arctic air (Bidleman *et al.,* 2002a).

(Finnish Meteorological Institute)]. The data set contains over 500 measurements, 70% from 1993–98 and 30% from earlier. Significant declines in TC, CC, and TN concentrations over the 14-y period in both summer-fall (May-October) and winter-spring (November-April) were observed ($p < 0.001$) (Figure B.1.7). Apparent values of half-lives, $t_{1/2}$, obtained by regressing the natural logarithm of concentration versus year (first order plot) were (summer-fall, winter-spring) 4.9, 6.4 y (TC), 6.7, 9.7 y (CC) and 7.4, 6.3 y (TN). They are similar to the time trends found by Hung *et al.* (2002a) as described previously for the 1993–97 period at Alert only, which were 8.3 y for TC, 4.1 y for CC, and 6.2 y for TN (Table B.1.2).

As seen in Figure B.1.5 and in other studies over many years (Bidleman, 1999; Halsall *et al.*, 1998; Oehme *et al.*, 1996), the ratio of TC/CC undergoes seasonal cycles in Arctic air, being higher in winter-spring and lower in summer-fall. Between 1984–97, the fraction of TC = FTC = TC/(TC+CC) declined significantly in both winter-spring ($p < 0.001$) and summer-fall ($p = 0.014$) seasons (Figure B.1.7). This suggests that sources of chlordane in Arctic air have changed over the last 14 years and are now more depleted in TC. The FTC in technical chlordane is 0.54 and, when weighed for the slightly higher vapour pressure of TC, the expected atmospheric FTC in equilibrium with technical chlordane is 0.62 (Jantunen *et al.*, 2000). The FTC in soils is about the same or lower than in technical chlordane (Aigner *et al.*, 1998; Jantunen *et al.*, 2000; Mattina *et al.*, 1999). Confounding the interpretation of FTC values is the fact that technical heptachlor was also used in agriculture and as a termiticide. As mentioned previously, technical heptachlor is contaminated with chlordane isomers, especially TC, so a higher FTC in soils or air could be associated with this source.

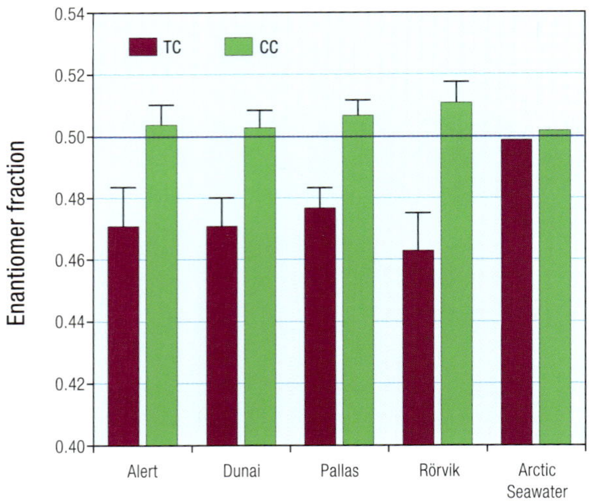

FIGURE B.1.8

Enantiomer fractions of *trans*-chlordane (TC) and *cis*-chlordane (CC) in air, means ± SD, at three Arctic stations (Alert, Dunai, Pallas), the temperate EMEP station Rörvik on the Swedish west coast, and in Arctic seawater, displaying racemic EFs, sampled on AOS-94 (Jantunen and Bidleman, 1998).

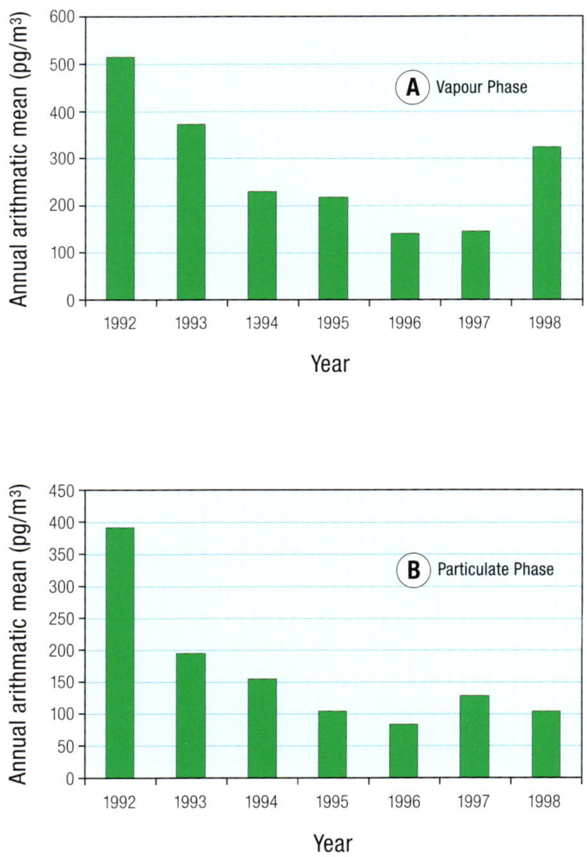

FIGURE B.1.9

Annual arithmetic means of ΣPAHs (16 compounds) in a) vapour and b) particulate phases at Alert.

Chlordane enantiomers in air samples from the Arctic stations at Alert and Dunai (mostly 1994–95, with a few from 1993 and 1996), Pallas (1998), and a European Monitoring and Assessment Programme (EMEP) station at Rörvik on the Swedish west coast (57°25'N, 11°56'E) were determined by Bidleman *et al.* (2002a) as an indicator of historical residue emissions, since non-racemic chlordanes have been found in soils (Aigner *et al.*, 1998; Jantunen *et al.*, 2000; Meijer *et al.*, 2001; Wiberg *et al.*, 2001) and in ambient air from the Great Lakes region (Bidleman *et al.*, 1998c; Ulrich and Hites, 1998). The samples from Alert and Dunai were mostly monthly composites of archived weekly extracts with a few additional individual weekly samples. Those from Pallas and Rörvik were seven-day samples collected once per month. Enantiomers were determined using published methods (Aigner *et al.*, 1998; Bidleman *et al.*, 1998c). Depletion of the (+)TC and (-)CC enantiomers was found for most of the samples. The average enantiomer fraction, EF = (+)/[(+) + (-)], of TC was different from racemic (racemic EF = 0.500) with a high level of significance ($p < 0.001$) at all three Arctic stations and the temperate coastal station Rörvik (Figure B.1.8). The average EFs of CC were also significantly non-racemic at Rörvik ($p < 0.001$), Pallas ($p < 0.01$), and Alert ($p < 0.05$), but only weakly at Dunai ($p < 0.1$). At Alert, the EF of TC was positively correlated with FTC ($r^2 = 0.41$, $p = 0.0032$), but not at Dunai ($r^2 = 0.0024$, $p > 0.9$) nor Pallas ($r^2 = 0.029$, $p > 0.6$).

Have chlordanes in Arctic air always been non-racemic, even in the mid-1980s and earlier? This is difficult to answer because no air samples from the 1980s are available for examination. However, chlordanes in surface water collected from the northern Canada Basin in 1994 were racemic (EFs = 0.499 for TC and 0.501 for CC). HEPX (heptachlor exo-epoxide) in seawater was non-racemic with an average EF = 0.639. Most of these samples were collected between 72–90° latitude and under ice cover (Jantunen and Bidleman, 1998). Chlordane and HEPX EFs in Arctic cod (*Boreogadus saida*), which feed on zooplankton, are similar to those in water (Wiberg *et al.*, 2000). It is likely that these enantiomer signatures represent chlordane and HEPX loadings to the Arctic Ocean sometime in the past. The EFs in water imply that chlordanes in Arctic air have not always been non-racemic, but HEPX may have been. Sources of HEPX include the metabolism of heptachlor in soils and the photochemical oxidation of heptachlor. The former process is thought to yield HEPX enriched in the (+) enantiomer, as occurs when heptachlor is metabolized by rat liver microsomes, while the second yields racemic HEPX (Buser and Müller, 1994). HEPX in ambient air samples from temperate regions is also non-racemic, suggesting that emission from soils is the primary source (Bidleman *et al.*, 1998b).

In summary, most of the OCs monitored have shown declining trends over the five years of study. Similar to the PCBs, there was a lack of relationship between temperature and OC air concentrations at Alert, and the slopes of ln P versus 1/T plots of most of the OCs were much lower than those observed at temperate locations. This again confirms that the Arctic atmosphere is strongly influenced by long-range transport and is acting as a sink for both OC pesticides and PCBs. Some of the supposedly "past-use" pesticides, such as chlordane and heptachlor, might still be in use sporadically, resulting in occasional spikes of high TC/CC. From the study of enantiomer fractions, however, it appears that general sources of chlordane have shifted over time, from atmospheric transport of the freshly applied pesticide to a greater proportion of chlordane "recycled" from soils. The non-racemic HEPX in Arctic air and seawater exemplifies the volatilization of soil residues on a large scale and their influence on atmospheric levels in remote regions.

B.1.2.3 Polycyclic aromatic hydrocarbons (PAHs)

Figure B.1.9 presents the annual arithmetic means of ΣPAH air concentrations (vapour + particle phase) at Alert from 1993 to 1998, which ranged from 91 to 205 pg/m^3 (ΣPAH = 16 compounds; acenaphthylene, acenaphthene, fluorene, phenanthrene, anthracene, fluoranthene, pyrene, benzo[*a*]anthracene, chrysene, benzo[*b*]fluoranthene, benzo[*k*]fluoranthene, benzo[*e*]pyrene, benzo[*a*]pyrene, indeno[1,2,3-*c,d*]perylene, dibenzo[*a,h*]anthracene, benzo[*g,h,i*]perylene). Attempts were made to extract trends using the DF technique from the entire PAH dataset and from segretated winter/spring data, but the data were extremely scattered and reliable trends could not be derived. This is mainly due to the extreme seasonality of ΣPAH concentrations at Alert (Halsall *et al.*, 1997); more years of data are required to obtain reliable temporal trends. Nonetheless, annual means can be used to discuss general tendencies and for comparing the Alert measurements to observations from other locations.

The decrease observed at Alert for gas phase ΣPAH (Figure B.1.9) between 1993 and 1996 follows measurement trends around the Great Lakes (Cortes and Hites, 2000) and at urban sites in the UK (Halsall *et al.*, 1994; Coleman *et al.*, 1997). Decreasing trends in the UK have been attributed to a substantial decline in domestic consumption of fossil fuels for space heating and improvement in domestic and industrial combustion devices. Given the importance of long-range transport of PAHs to the Arctic (Halsall *et al.*, 2001), it is not surprising that the levels observed at Alert mimic the declining trends at mid-latitudes. The decline in particulate phase PAHs from 1993 to 1996 (Figure B.1.9) is consistent with the general declining trend of atmospheric sulphate aerosol, SO_4^{2-}, at Alert after 1991 reported by Sirois and Barrie (1999). Particulate-sulphate accounts for most of the aerosol loading in the Arctic (Halsall *et al.*, 1997) and the size distribution of SO_4^{2-} (≤ 1 µm) corresponds to the range where most PAHs have been observed (Allen *et al.*, 1996). Sirois and Barrie (1999) have shown that between 1985 and 1990, there was a lack of a decline in concentration at Alert and Barrow, reflecting emission data from the former Soviet Union. Between 1991 and 1995, they

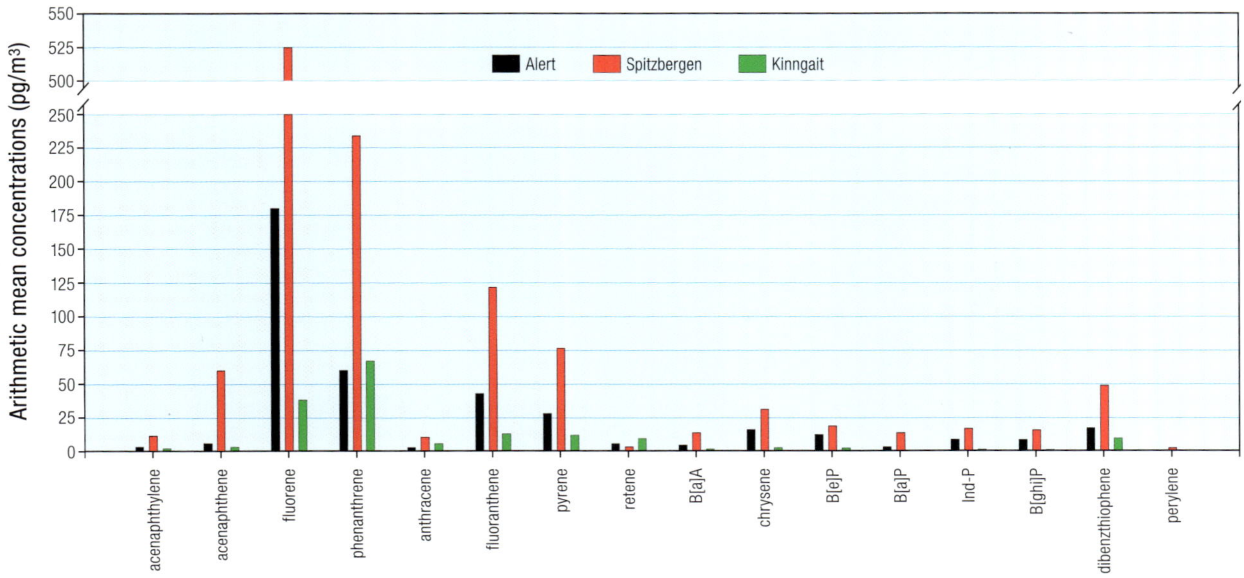

FIGURE B.1.10

PAH profiles at Alert, Spitsbergen and Kinngait in 1994. Kinngait data are arithmetic mean concentrations from week 16, 1994 to week 15, 1995.

TABLE B.1.3 Comparison of air concentrations of PCDD/F at Alert with those of other locations[(1)]

Location (Time/Type)	Total PCDD (fg/m³)	Total PCDF (fg/m³)
Alert filter only (week 49: 27/11–4/12/2000)	2.1	2.4
Alert filter only (week 3: 15–22/1/2001)	13	46
Alert filter only (week 7: 12–19/2/2001)	3.7	9.9
Ny-Ålesund (21/4–17/5/95)[(2)]	28	76
Ny-Ålesund (21/7–23/8/95)[(2)]	16	51
Rörvik, Sweden (background air only)[(3)]	220–470	84–240
Central McMurdo, Antarctica 1992–1993[(4)]	610	560
Central McMurdo, Antarctica 1993–1994[(4)]	440	50
Trout Lake, WI (rural)[(4)]	240	180
Research Triangle Park, NC (suburban)[(4)]	710	1100
Bloomington, IN (suburban)[(4)]	1200	630
Hamburg, Germany (suburban)[(4)]	1800	740
Indianapolis, IN (urban)[(4)]	2500	2600
Nordhein–Westfalen, Germany (urban and industrial)[(4)]	3200	5500
Barbados (18/3/1996–13/8/1997)[(5)]	6.8	12
Bermuda (22/8/1993–30/11/1994)[(5)]	31	50
Bermuda (4/6/1996–20/8/1997)[(5)]	25	16

[(1)]All concentrations, other than those from Alert, are total concentrations (vapour + particle phases).

[(2)]Schlabach *et al.*, 1996. For sample taken during 21/4–17/5/95, recoveries of H6CDD/F were low (20%); and those of H7CDD/F and OCDD/F were unacceptable (0–4%).

[(3)]Tysklind *et al.*, 1993.

[(4)]Adopted from Table 3 of Lugar *et al.*, 1996.

[(5)]Calculated from supplementary material of Baker and Hites, 1999. Remote marine atmospheric data. Barbados: average of 22 samples. Bermuda (22/8/1993–30/11/1994): average of 21 samples. Bermuda (4/6/1996–20/8/1997): average of 16 samples.

attributed the decline in concentrations to the collapse of industry during the early years of the new Eurasian republics.

Figure B.1.10 shows the profile of the PAH compounds in 1994 at Alert, Spitsbergen, and Kinngait. The concentrations observed at Spitsbergen were generally much higher than those found at Alert and Kinngait. For the dominant compounds, including fluorene, phenanthrene, fluoranthene and pyrene, the annual mean concentrations were higher at Spitsbergen by a factor of two to three. The observed profiles at all locations confirm the modelling result presented by Halsall *et al.* (2001). They have found that lighter compounds, such as fluorene and phenanthene would be able to reach the Arctic while most of the heavier compounds, such as benzo[*a*]pyrene (B[*a*]P), would have been removed from the atmosphere before reaching the Arctic, either by reaction with OH radicals or deposition along the transport pathway.

B.1.2.4 Polychlorinated dibenzo-p-dioxin and dibenzofurans (PCDD/Fs)

During the winter of 2000/2001, 15 weekly filter samples (particulate phase) from Alert were analyzed for polychlorinated dibenzo-*p*-dioxins and dibenzofurans (PCDD/Fs) (Hung *et al.*, 2002b). The sampling period coincided with the occurrence of the Arctic haze when airborne particulate levels were high. Since PCDD/Fs have very similar properties to PAHs, e.g., both originate from combustion sources and have low volatilities with high tendencies to partition to aerosols, they are expected to follow similar seasonal cycles. Halsall *et al.* (1997) previously showed that atmospheric levels of PAHs were the highest in winter/early spring of 1992/93 and 1993/94 at Alert.

Table B.1.3 compares the atmospheric concentrations observed at Alert (weeks 49 of 2000, weeks 3 and 7 of 2001) with those found at other locations. The concentrations of both ΣPCDDs and ΣPCDFs observed were lower than those found at Ny-Ålesund and other locations worldwide. This is attributable to the remoteness of Alert and the absence of anthropogenic activities nearby. Figure B.1.11 shows the average five-day Lagrangian back trajectories (Olson *et al.*, 1978) for the first few weeks of 2001 along with the corresponding sum of PCDD/Fs. The trajectories were computed every six hours at an elevation of 925 hPa. The five concentric circles show the spread of the trajectories for each day back. Four geographical source regions (or sectors) as defined by Stern *et al.* (1997) are also shown in Figure B.1.11. Alert was mainly affected by air originating from the North Atlantic Ocean and North America during the first week of 2001. The origin of the air mass shifted eastwards further into Russia and Eurasia during the second and third weeks when the air concentrations of ΣPCDD/Fs peaked at Alert. Then, the influence from Eurasia decreased when the concentrations tapered off during the fourth and fifth weeks while the North Atlantic sector regained dominance.

Although levels of PCDD/Fs found in air were low, there are still possibilities of significant accumulation of these compounds in other Arctic environmental media due to the low temperatures in the Arctic and slower degradation processes. In order to gain more insight into the fate of these compounds in the Arctic and the risk that they pose to the local population, further research is required involving the simultaneous sampling of various environmental media.

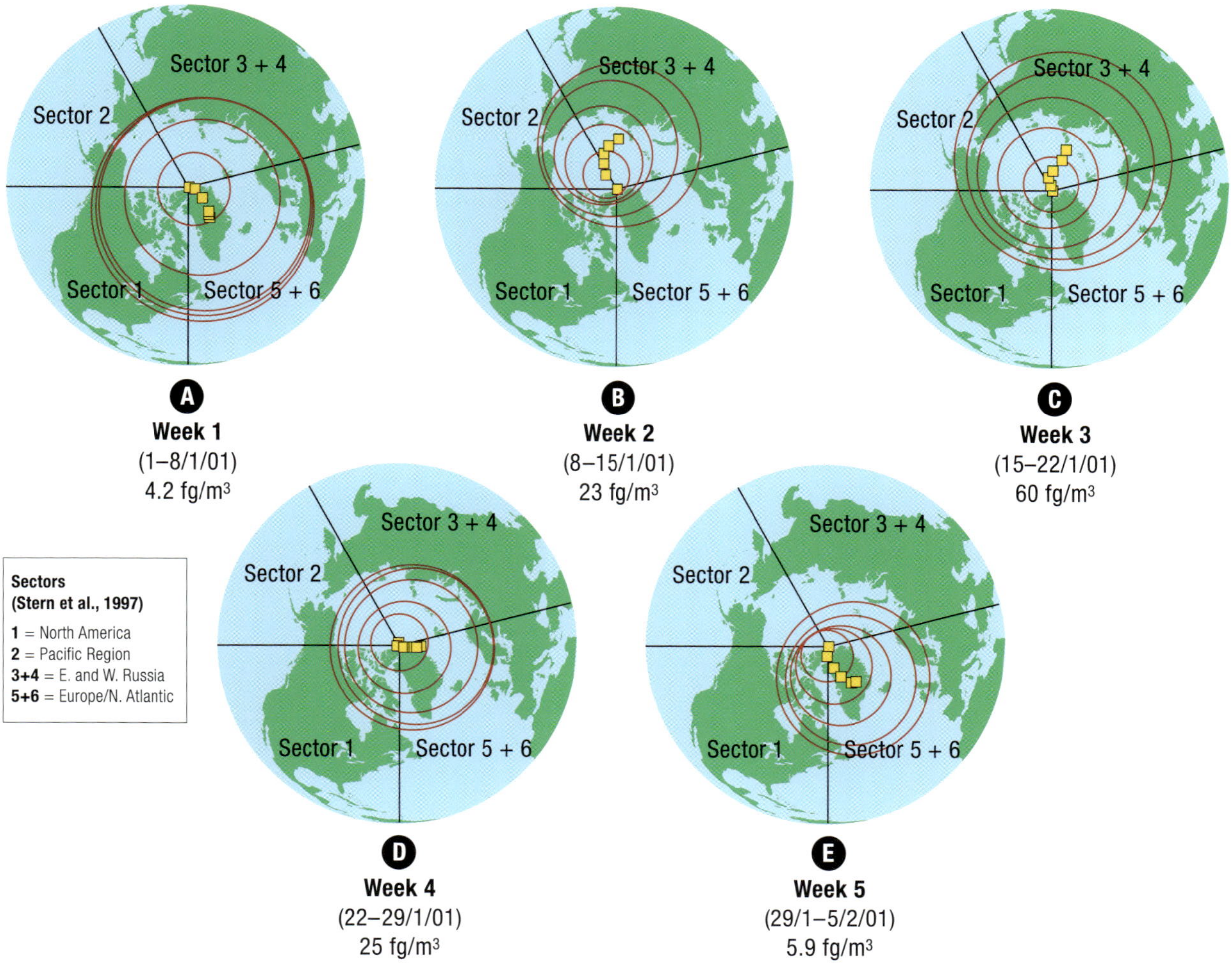

FIGURE B.1.11

Average 5-day Lagrangian back trajectories at Alert (925 hPa) for the first six weeks of 2001. Squares show six-hourly intervals. The five concentric circles indicate the spread of trajectories for each day back (Hung *et al.*, 2002b).

B.1.2.5 Evidence of trans-Pacific transport

Air monitoring at Tagish in the western Yukon has provided an interesting database consisting of over two years of weekly measurements (December 1992 to March 1995). Tagish represents the western Arctic, typically receiving air from a westerly direction over the Pacific Ocean and, unlike the other monitoring stations, is unique in that air is sampled at a higher altitude and away from the margins of the Arctic Ocean. In this respect, Tagish is less influenced from known source regions such as Europe/Russia or North America. For example, atmospheric levels of PAHs often associated with Arctic haze during the winter months have been observed to be much lower at Tagish in contrast to the high Arctic sites of Alert and Dunai (Siberia) (Halsall *et al.*, 1997). Importantly, analysis of the data has revealed "episodes" of elevated OC air concentrations, which occur largely during the winter months, notably when airflow is predominantly from the Pacific Ocean (Bailey *et al.*, 2000).

Lagrangian air mass back trajectories were calculated throughout each season to show the general air mass movement. These were calculated for 700 hPa as Tagish tends to receive air from aloft due to its altitude. Figure B.1.12 shows trajectories arriving at Tagish, with groupings based on either "strong" or "moderately strong" *trans*-Pacific transport, with the air originating over the Asian landmass in most cases. Five-day trajectories corresponding to strong transport out of eastern Asia and with moderately strong transport reaching almost to the coast of east Asia occurred mostly during the winter months. During this period, levels of OCs such as *p,p'*-DDE and *trans*-chlordane reached a maximum, and high levels of *cis*-chlordane and heptachlor epoxide had also

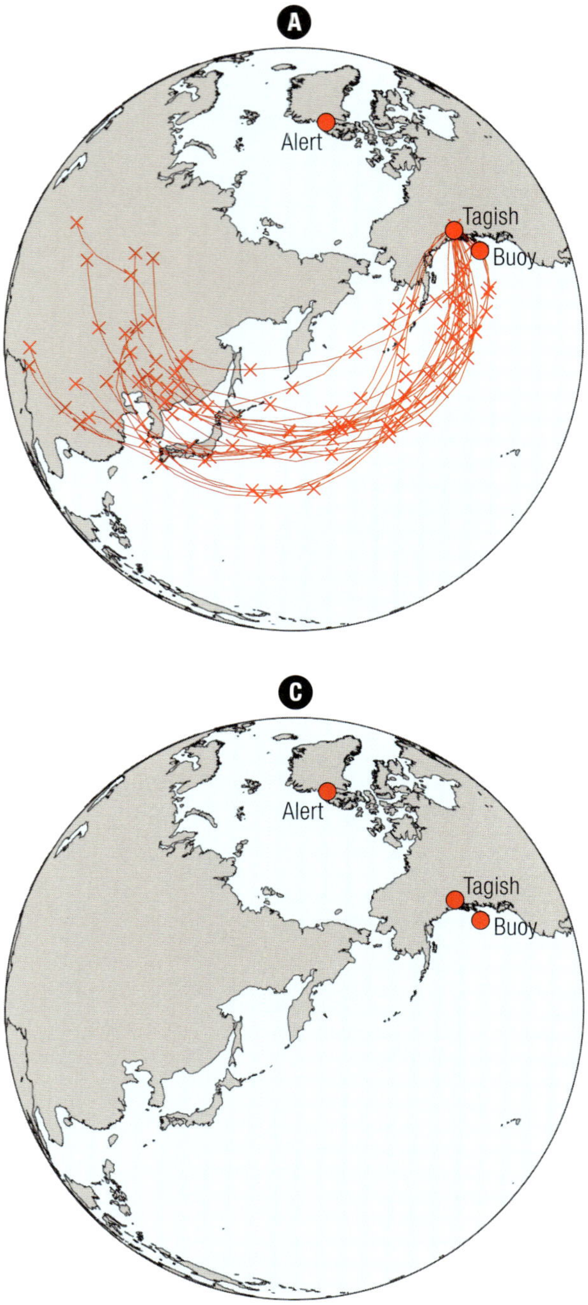

FIGURE B.1.12

Trajectories to Tagish arriving at 700 hPa level (~1500 m) during (A-D) strong and (E-H) moderately strong trans-Pacific transport out of Asia by season: (A) and (E), winter (December–February); (B) and (F), spring (March–May); (C) and (G), summer (June–August); (D) and (H), fall (September–November). No strong trans-Pacific transport out of Asia occurred during the summer (Bailey *et al.,* 2000).

been observed. Trans-Pacific airflow from Asia is not as frequent or as strong during the summer months. Air masses formed over southerly regions of Asia, such as India, can also be transported around the eastern side of the Tibetan Plateau to the western North Pacific or the more northerly latitudes of Asia. Examples of this type of transport are illustrated in Figure B.1.12. Trans-Pacific airflow originating over eastern Russia near the Bering Strait also occurred, but was not included in the Asian category due to the substantial differences in usage patterns of pesticides between the China-India region of Asia and eastern Russia. In eastern Russia, use of

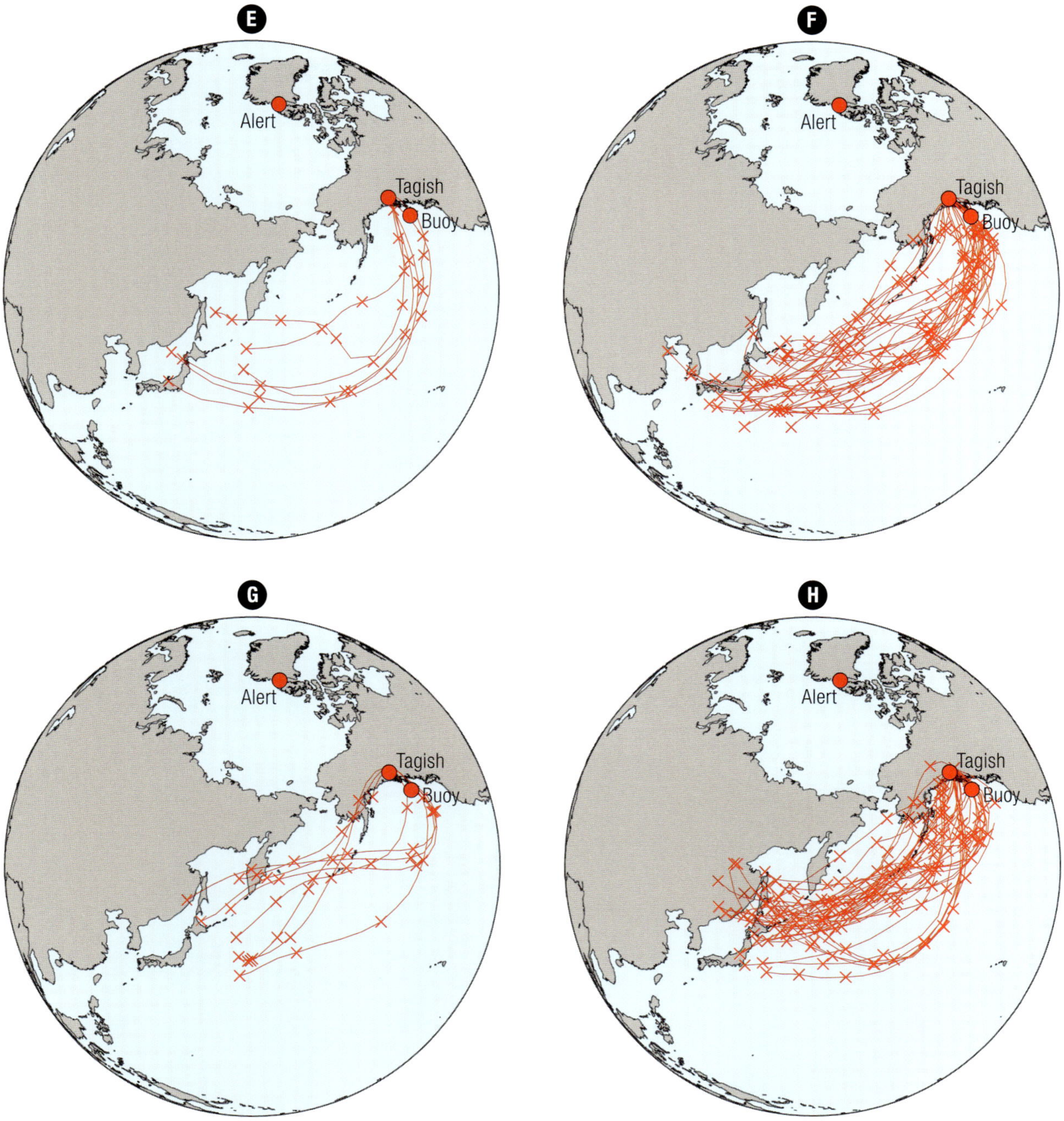

FIGURE B.1.12 (CONT.)

pesticides is expected to be minimal because of the mountainous terrain and colder climate (Li, 1999a). No significant airflow from Europe or western Russia occurred for this period. Clearly Asia is an important source region for these compounds, similar to North America, and air mass transport from Asia can influence contaminant levels in the western Arctic atmosphere. Further research is required on atmospheric concentrations, agricultural usage, and soil concentrations in eastern Asia in order to gain knowledge on the role this region plays as a source of pesticides.

B.2 Persistent organic pollutants in seawater

Authors:

D. Muir (1); and W. Strachan (1).

Affiliations:

1. Environment Canada, National Water Research Institute, 867 Lakeshore Rd., Burlington ON, L7R 4A6, Canada.

B.2.1 Introduction

The marine environment has historically received the most attention with regard to organochlorine and PAH contaminants in the Arctic environment. It was also the first Arctic system to be examined for the presence of OC contaminants (Holden, 1970). The level of research on OC contamination of the marine environment has continued to out distance efforts in the freshwater and terrestrial Arctic. This is clearly due to the higher levels observed in the marine system and the role of marine organisms in northern societies.

A large number of measurements of organochlorine pesticides and PCBs have been made in Arctic Ocean waters from 1996 to 2001 (Table B.2.1) mainly as a result of scientific cruises organized by circumpolar countries. During this period, there were also several peer-reviewed publications on results of measurements of organochlorines in seawater pre-1996 (Rice and Shigaev, 1997; Strachan *et al.*, 2001; Hargrave *et al.*, 1997; Jantunen and Bidleman, 1998) however, those results were included in the previous assessment (de March *et al.*, 1998) and reviews (Macdonald *et al.*, 2000) and are included only for discussion here. This section discusses the "legacy" POPs: PCBs, chlorinated benzenes, and organochlorine pesticides. The occurrence of modern agrochemicals in Arctic seawater is discussed in Section B.4.5.2.

The measurements were, for the most part, conducted from oceanographic ships in a series of cruises (Figure B.2.1). The Swedish "Oden" cruise in 1996 conducted sampling from the ice edge to the North Pole in the northern Barents Sea (Harner *et al.*, 1999). In 1997, a Canadian supply trip (JOIS-I) for the Surface Heat Budget of the Arctic (SHEBA) study (Perovich *et al.*, 1999) was used to obtain seawater from the western Canadian Archipelago in September and in May–June, 1998 the Northwater Polyna study in Northern Baffin Bay (NOW '98) provided another platform for sampling. In 1998–99 the SHEBA study permitted continuous sampling of seawater over the Beaufort/

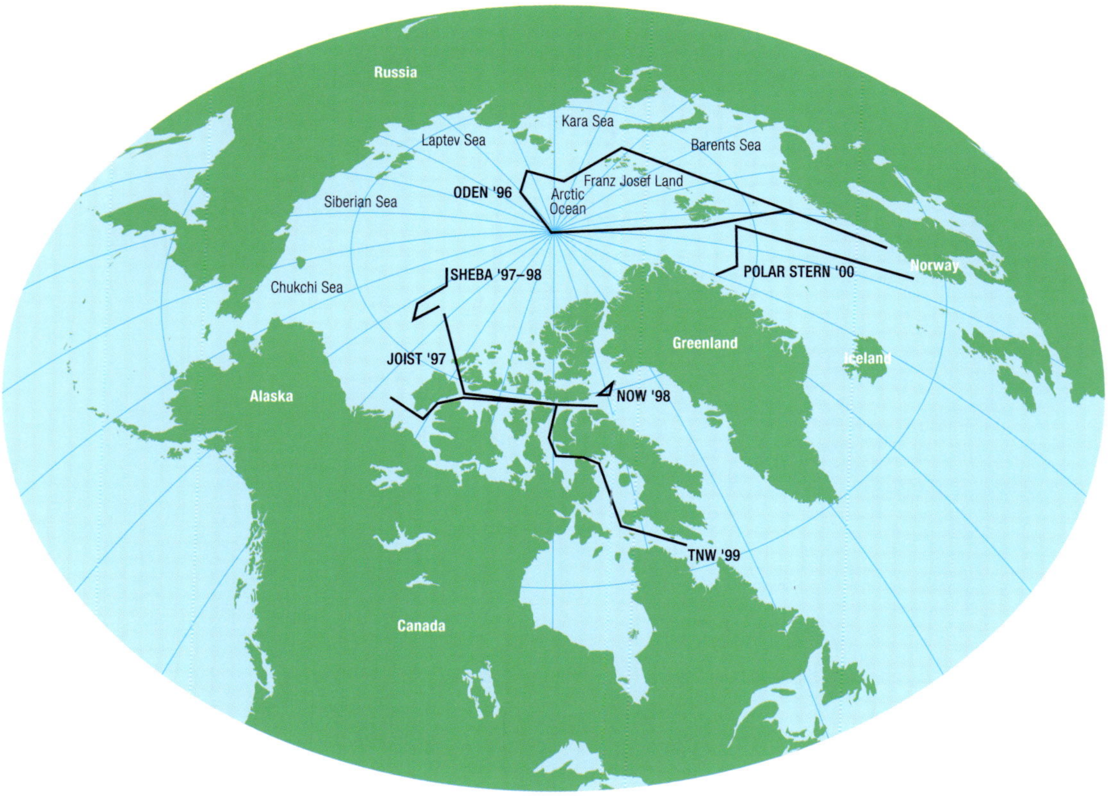

FIGURE B.2.1
Cruise tracks for seawater collection, 1996–2000.

Chukchi Seas [The Swedish Tundra Northwest (TNW '99) traversed the Canadian archipelago during July–August 1999]. An Arctic to Antarctic cruise by the RV Polar Stern provided seawater samples from the East Greenland sea in 2000 (Lakaschus *et al.*, 2002). Schultz-Bull *et al.* (1998) reported PCBs in large volume samples collected in the North Atlantic Ocean around Iceland which are also relevant to this discussion and Sobek and Gustaffson (2002) presented preliminary results from the Oden cruise to the North Pole in 2001.

The new measurements have filled in some of the gaps in information on spatial trends of OCs in the Arctic Ocean. Sampling and extraction techniques were more uniform than in the 1980s and early 1990s with large volume (20 to >100 L) samples collected and extracted using resin columns to minimize airborne contamination. Some investigators used filtration (Harner *et al.*, 1999) or continuous centrifugation (Strachan *et al.*, 2001) to remove particles while samples for HCH analysis were often unfiltered (Lakaschus *et al.*, 2002).

B.2.2 Hexachlorocyclohexanes (HCHs)

Hexachlorocyclohexane isomers were the most commonly measured chemicals in the Arctic Ocean and adjacent seas. These compounds are present at ng/L concentrations and there are few problems with contamination on ships or in the laboratory. Recent measurements are summarized in Table B.2.1 and also in Table C.3.2. In general, surface concentrations of HCHs are highest in the central Canadian Archipelago, intermediate in the Beaufort and Chukchi seas and at the North Pole, and lowest in the Greenland Sea and northern Barents Sea (Figure B.2.2). The large number of measurements of HCH during the 1980s and 1990s in the Bering/Chukchi/Beaufort seas, as well as in the Greenland Sea, permits an examination of mass balances over time and temporal trends. These are discussed in Section C.3.

Waters of the southern Beaufort Sea (Hoekstra *et al.*, 2002a; Fisk, unpublished) and in the central and western Canadian Archipelago (Jantunen *et al.*, unpublished; Macdonald *et al.*, 1999b; 2000; Strachan *et al.*, 2000) show higher ΣHCH levels than those in the Northwater Polynya in northern Baffin Bay (Fisk, unpublished; Moisey *et al.*, 2001; Strachan *et al.*, 2000) (Table B.2.1). Enantiomer fractions (EFs) of α-HCH in the central and western Archipelago (Jantunen *et al.*, unpublished) and in the Northwater Polynya (Moisey *et al.*, 2001) were in the range 0.441–0.459, indicating preferential degradation of (+)α-HCH.

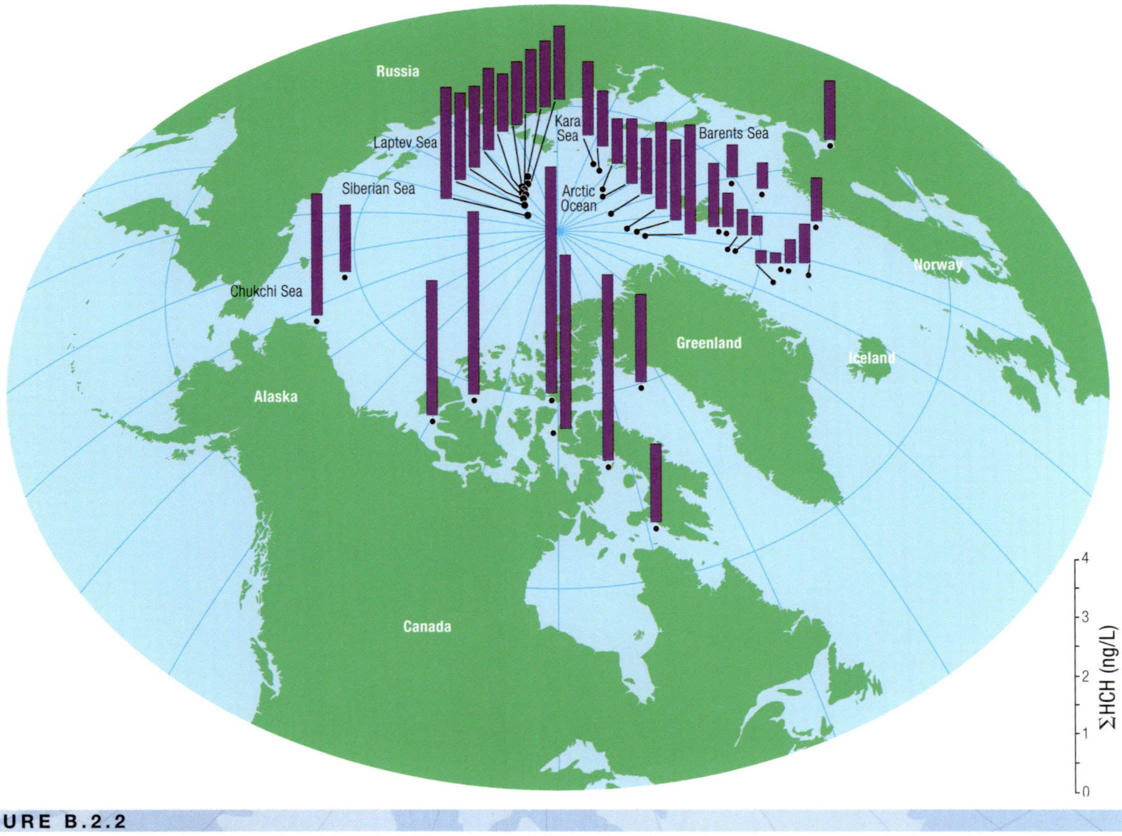

FIGURE B.2.2

ΣHCHs in seawater, 1996–2000.

TABLE B.2.1 Concentrations (Mean ± SD) of major OC groups in surface sea water samples (ng/L) from the Arctic (1994–2001)

Region	Location/year/ depth[1]	N	ΣClBz[2]	ΣHCH[3]	ΣCHLOR[4]	ΣDDT[5]	Endosulfan	Toxaphene[6]	ΣPCB[7]	Σ10PCB[8]	Reference Cruise[9]	Method
Central Archipelago	Resolute, 1999, surface	1	-	3.9	-	-	-	-	-	-	1	SPE
Central Archipelago	Hudson Strait, 1999, surface	9	-	3.1	0.013	-	0.0091	-	-	-	1	SPE
Central Archipelago	Foxe Basin/Channel, 1999, surface	4	-	3.0	0.005	-	0.0049	-	-	-	1	SPE
Central Archipelago	Melville Sound/McClure Strait, 1997, surface	10	0.04 ± 0.03	3.09 ± 0.136	0.007 ± 0.001	0.024 ± 0.032	0.002 ± 0.002	-	0.740 ± 0.731	0.208 ± 0.209	2	"Go-Flo" & XAD
Central Archipelago	Hudson Strait, Foxe Basin, Pr. Regent Inlet, 1999 surface	6	0.017 ± 0.003	1.33 ± 0.081	0.011 ± 0.007	0.017 ± 0.014	< 0.001	-	0.277 ± 0.126	0.065 ± 0.004	3	"Go-Flo" & XAD
Chukchi Plateau	SHEBA 1998 (July and Sept., 10–30m)	4 4	0.013 ± 0.007	1.10 ± 0.80	0.005 ± 0.003	0.001 ± 0.001	0.0007 ± 0.0004	0.040 ± 0.016	0.042 ± 0.022	-	4	"Infiltrex", *in situ*
North Barents Sea	Oden cruise, 1996, surface	20	0.004 ± 0.002	0.967 ± 0.499	0.009 ± 0.014	0.003 ± 0.003	0.001 ± 0.002	-	0.161 ± 0.074	0.057 ± 0.027	5	"Go-Flo" & XAD
North Barents Sea	Oden cruise, 1996, > 100m	21	0.086	0.549	0.007	0.008	0.002	-	0.053		5	"Go-Flo" & XAD
North Barents Sea	Oden cruise, 1996 50–300m	17	-	0.971 ± 0.45	-	-	-	-	-		6	"Go-Flo" & XAD
North Barents Sea	Oden cruise, 1996, 500–1000m	10	-	0.465 ± 0.23	-	-	-	-	-		6	"Go-Flo" & XAD
Southwest Beaufort Sea	Barrow, AK, 1999–2000, surface	7	0.120 ± 0.006	2.11 ± 0.46	0.010 ± 0.004	0.019 ± 0.006	0.001 ± 0.001	0.188 ± 0.046	0.231 ± 0.090	0.058 ± 0.058	7	"Infiltrex", *in situ*
Southeast Beaufort Sea	Holman, NWT, 1999, surface	3	0.043 ± 0.004	2.28 ± 0.15	0.008 ± 0.002	0.005 ± 0.001	0.002 ± 0.001	0.150 ± 0.040	0.135 ± 0.018	0.039 ± 0.013	7	"Infiltrex", *in situ*
North Baffin Bay	North water Polynya, 1998, > 100m	9	0.094	0.959	0.004	0.015	0.002	-	0.405	-	8	"Go-Flo" & XAD
North Baffin Bay	North water Polynya, 1998, surface	8	0.018 ± 0.005	1.42 ± 0.273	0.010 ± 0.004	0.008 ± 0.006	0.003 ± 0.002	-	0.187 ± 0.098	0.058 ± 0.046	8	"Go-Flo" & XAD
NW Russia White Sea	West Onega Bay; Central Basin, 1999–00, surface	10	0.009 ± 0.004	0.84 ± 0.273	0.007 ± 0.004	0.003 ± 0.001	0.003 ± 0.002	0.041 ± 0.008	0.105 ± 0.057	0.028 ± 0.014	9	"Infiltrex", *in situ*
Greenland Sea, East	74.6N,12.0E, surface	5	-	0.572 ± 0.203	-	-	-	-	-		10	
Greenland Sea, East	74.9N,4.2E, surface	2	-	0.371	-	-	-	-	-		10	
Greenland Sea, East	76.0N,1.2E, surface	1	-	0.487	-	-	-	-	-		10	
Greenland Sea, East	62.1N,5.9W, surface	2	-	0.698	-	-	-	-	-		10	
Greenland Sea, West	74.6N,6.0W, surface	1	-	0.131	-	-	-	-	-		10	
Greenland Sea, West	68.2N,10.1W, surface	1	-	0.438	-	-	-	-	-		10	
Greenland Sea, West	77.2N,11.1W, surface	1	-	0.26	-	-	-	-	-		10	
Northwest Atlantic	63.67N,33.0W, 50m	1	-	-	-	-	-	-	-	0.00195	11	"Kiel" *in situ*
Northwest Atlantic	63.67N,33.0W, 100m	1	-	-	-	-	-	-	-	0.00066	11	"Kiel" *in situ*
Greenland Sea	68.2N,22.67W, 300m	1	-	-	-	-	-	-	-	0.00209	11	"Kiel" *in situ*
Greenland Sea	68.2N,22.67W, 870m	1	-	-	-	-	-	-	-	0.00054	11	"Kiel" *in situ*
Norwegian Sea	62.11N,4.57E 66.43N,10.32E	2	-	-	-	-	-	-	-	0.0014 ± 0.00023	12	On-line GFF/XAD
Barents Sea	77.50N,29.54E 88.57N,1.13W	4	-	-	-	-	-	-	-	0.00055 ± 0.00043	12	On-line GFF/XAD

[1]surface = ≤ ~10 m depth; other depths are indicated.

[2]ΣClBz (chlorobenzenes) = sum of 1,2-diClBz, 1,4-diClBz, 1,2,3-triClBz, 1,2,4-triClBz, 1,3,5-triClBz, 1,2,3,4-tetraClBz, 1,2,3,5-tetraClBz, pentaClBz, and hexaClBz.

[3]ΣHCH (hexachlorocyclohexanes) = sum of β-HCH, γ-HCH, except for Harner *et al.*, sum of α-HCH, and γ-HCH.

[4]ΣCHLOR (chlordanes) = sum of *cis*-chlordane, *trans*-chlordane, oxychlordane, *cis*-nonachlor, *trans*-nonachlor, heptachlor, and *cis*-heptachlor epoxide.

[5]ΣDDT (dichlorodiphenyltrichloroethane isomers) = sum of *o,p'*-DDD, *p,p'*-DDD, *o,p'*-DDE, *p,p'*-DDE, *o,p'*-DDT, and *p,p'*-DDT.

[6]TOX (toxaphene) = total toxaphene determined by GC-ECNIMS.

[7]ΣPCB (polychlorinated biphenyls) = sum of congeners 4/10, 7/9, 6, 8/5, 19, 12/13, 18, 15/17, 24/27, 16, 32, 54/29, 26, 25, 50, 31/28, 33/21/53, 51, 22, 45, 46, 52/49, 43, 47/48, 44, 59, 42, 64, 41/71, 40, 100, 63, 74, 76/98 70, 95, 66, 91, 55, 56/60, 92/84, 101, 99, 119, 83, 97, 87, 81, 85, 136, 110, 82, 151, 135, 144, 107/147, 149/133, 118, 114, 143, 141, 145, 153, 132, 105, 141/179, 137, 176/130, 163, 138, 158, 129/178, 175, 187 182, 183, 128, 167, 185, 174, 177, 171, 156, 202/173, 172, 197, 180/193, 191, 199, 170/190, 198, 201, 176/203, 189, 206, 195, 207, 194, 205, 208, and 209.

[8]Sum of PCB congeners 28,31, 52, 101,105,118,138,153,156 and 180 except for results of Schultz-Bull *et al.* (1998) which are for 23 congeners including the 10 listed here.

[9]References: 1 – Jantunen and Bidleman, unpublished – Tundra 99; 2 – Strachan *et al.* 2000 – JOIST cruise – 1997; 3 – Strachan *et al.* 2000, Tundra 99; 4 – Macdonald *et al.* unpublished; 5 – Strachan *et al.* 2000 – ODEN cruise; 6 – Harner *et al.* 1999; 7 – Hoekstra *et al.* 2001; 8 – Strachan *et al.* 2000 – Northwater; 9 – Muir *et al.* 2002; 10 – Lakaschus *et al.* 2002; 11 – Schultz-Bull *et al.* 1998; 12 – Sobek and Gustafsson (2002).

Harner *et al.* (1999) determined HCH isomers and enantiomers ratios of α-HCH in a large series of surface and deep water samples from the northern Barents Sea and central Arctic Ocean (80–88°N, 11–143°E). The mean concentrations for surface water were 1.18 ± 0.408 ng/L for ΣHCH (Table B.2.1). These agreed well with measurements by Strachan *et al*, (2001) on the same cruise and also with values reported by Gaul (1992) for a 1985 survey of the northeast Atlantic (ΣHCH, 1.250 ng/L). Both α- and γ-HCH increased with latitude between 74–88°N (r^2 = 0.58 and 0.69 for α- and γ-HCH, respectively). Values similar to those on the Oden cruise were found during 1994 in the eastern Arctic Ocean north of Spitsbergen (1.170 ng/L for ΣHCH) and in the Greenland Sea (0.830 ng/L) (Jantunen and Bidleman, 1997; 1998).

Vertical profiles at 11 Oden stations indicated that concentrations at 250–1000 m were ~40% of surface values for α-HCH and ~60% for γ-HCH (summarized as ΣHCH in Table B.2.1). Tracer studies in this region (Nansen Basin) indicate that the water masses at 250–1000 m have ages on the order of 12–20 years which coincide with the time of heaviest usage of HCH (Li *et al.*, 1998a). Enantiomer fractions of α-HCH averaged 0.465 in surface water and decreased with depth to values as low as 0.152, similar to vertical profiles seen in the northern Canada Basin (Jantunen and Bidleman, 1997). The vertical profiles of concentration for α- and γ-HCH and the EFs of α-HCH allowed estimates to be made of *in situ* microbial degradation rates (Harner *et al.*, 1999), which suggested half-lives of 5.9 and 23.1 y for (+) and (–) enantiomers of α-HCH, respectively, and 18.8 y for γ-HCH. These are 3–10 times faster than half-lives estimated for hydrolytic breakdown of the HCHs. Microbial degradation may be responsible for up to ⅓ of the total annual HCH loss budget for the Arctic Ocean (Harner *et al.*, 2001a; Macdonald *et al.*, 2000).

Lakaschus *et al.* (2002) determined α-and γ-HCH in surface (11 m) seawater along an east-west transect in the Greenland Sea. Concentrations in this region (ΣHCH = 0.57 +/- 0.20 ng/L east; 0.13–0.698 ng/L west) were similar to those in the surface and at depth in the Barents Sea (Harner *et al.*, 1999) but two to four times lower than in the Beaufort Sea (Table B.2.1). Concentrations of ΣHCH in the Greenland Sea (68–75°N), however, were three to five times higher than in the eastern Atlantic, near the Bay of Biscay, because of much higher levels of α-HCH in Arctic waters (Lakaschus *et al.*, 2002).

Concentrations of ΣHCH in the White Sea in northwestern Russia (0.84 ng/L, Kandalasksha Bay and Central basin) (Muir *et al.*, 2002; unpublished data) were similar to those in the north Barents Sea (0.90 ng/L) (Harner *et al.*, 1999).

B.2.3 Polychlorinated biphenyls (PCBs)

Polychlorinated biphenyls are the next most prominent contaminants in seawater (Table B.2.1); however, there are far fewer measurements for PCBs than for HCHs. Measurements of PCBs in seawater are challenging because of the low levels and the potential for shipboard as well as laboratory contamination. Comparisons among different studies and with previous work on PCBs in the Arctic Ocean and nearby northern seas (Iwata *et al.*, 1993a; Schultz-Bull *et al.*, 1998; Sobek and Gustafsson, 2000) raise important questions about contamination and effect of different sampling techniques. These include possible differences between whole (unfiltered) water versus filtered samples, and use of *in situ* collection with remotely deployed samplers versus submersible pumping onto the ship. Schultz-Bull *et al.* (1998) and Sobek and Gustafsson, (2002) have reported concentrations of PCBs which are about 10 times lower than other measurements summarized in Table B.2.1. These latter groups have used ultra-clean techniques. Schultz-Bull *et al.* (1998) employed the "Kiel" *in situ* sampler, a remotely deployed filtration-extraction system (Petrick *et al.*, 1996) to sample large volumes of seawater (230–1070 L) through glass fibre filters (GFF) and XAD-2 resin. Sobek and Gustafsson (2002) employed a stainless steel seawater intake system directly into the Oden followed by online collection using GFF and polyurethane foam (PUF) in a "clean" room on board. Seawater sampling on the earlier Oden, TNW, and JOIST cruises (Table B.2.1) for PCB analysis involved use of a stainless steel-lined "Go-Flo" bottle followed by GFF and XAD-2 resin but did not employ clean room techniques during the sampling phase. Sampling on the SHEBA study and nearshore samples at Barrow, Holman and in the White Sea used *in situ* samplers (Axys Instruments, Sidney BC) and GFF/XAD-2 resin deployed from ships or through sea ice.

The results in Table B.2.1 summarize the current state of the art of PCB measurements in the Arctic Ocean. Up to 100 PCB congeners were determined by some investigators, however, for this assessment results for the sum of 10 major PCB congeners were also included, where possible, to make comparisons more equitable. Although all studies employed solid phase extraction (mainly XAD-2 resin) for the collection of PCBs, and followed strict protocols to avoid contamination, there is nevertheless major disagreement between measurements, especially between those made using samples collected with "Go-Flo" bottles and those with *in situ* or direct online sampling waters. There are also major differences between the results for PCBs determined by *in situ* sampling with "Infiltrex" systems and those using the "Kiel" *in situ* sampler.

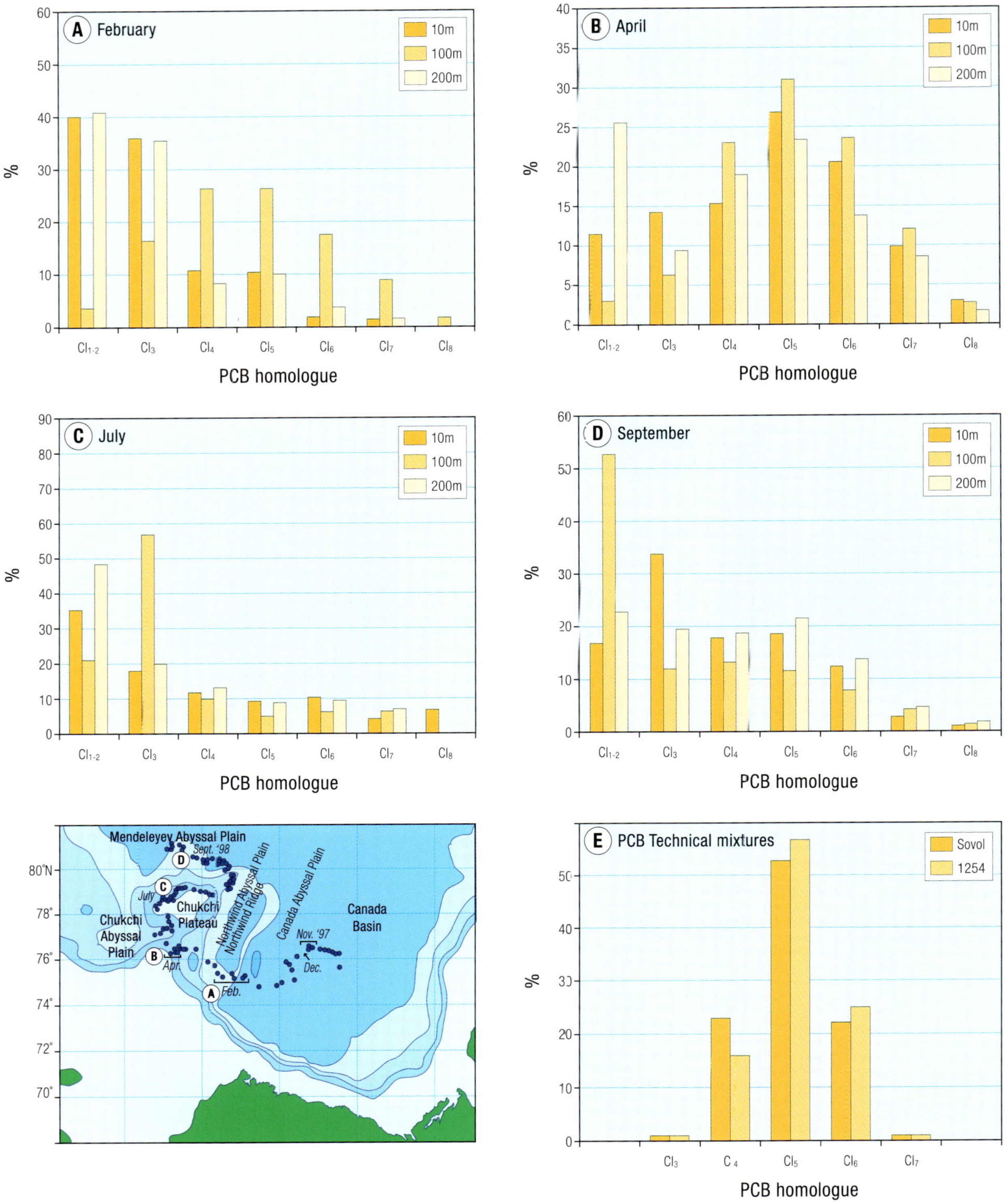

FIGURE B.2.3

Seawater PCB homologue distribution for the vertical profiles collected in a) February, b) April, c) July and d) September during the SHEBA study. The drift of the ship is indicated on the map by the blue circles which identify where conductivity-temperature-depth (CTD) profiles were collected. Also shown (e) is a representative PCB distribution for the technical mixtures Aroclor 1254 and the Russian formulation Sovol.

There are similar irreconcilable differences between the results from Strachan *et al.* (2001) for PCBs in the Bering Sea and results from Iwata *et al.* (1993a) from the Bering Sea (previously reported in the 1998 AMAP and CACAR Assessments). The latter study used GFF/PUF sampling from on-line water system while the former used a submersible pump, a continuous centrifuge to separate particles and solvent extraction with dichloromethane. Iwata *et al.* (1993a) reported mean concentrations of 0.0084 ng/L at three locations in the Chukchi Sea and 0.012 ng/L for four locations in the Bering Sea from their 1990 cruise. Strachan *et al.* (2001) reported dissolved phase concentrations averaging 1.0 ng/L in the Bering Sea (seven sites) and means of 0.6–0.9 ng/L in the Chukchi Sea. Given that both cruises covered the same area and sampled approximately the same depth it is difficult to attribute the differences except to contamination introduced from the ship. The results of Iwata *et al.* (1993a) for the Chukchi Sea are generally consistent with those of Sobek and Gustafsson (2002) and Schultz-Bull *et al.* (1998) for the range of concentrations but these studies cannot be compared directly because different numbers of congeners were quantified.

With these caveats in mind we have limited the discussion of results to data from Schultz-Bull *et al.* (1998) and Sobek and Gustafsson (2000) and to the "Axys" *in situ* sampler work conducted on the SHEBA study and in nearshore waters recognizing that even the latter results may be confounded by contamination introduced on handling and storage or in the laboratory (Table B.2.1).

There are too few PCB measurements to assess spatial trends of PCBs; nevertheless, some general conclusions can be drawn from the data of Schultz-Bull *et al.* (1998) and Sobek and Gustafsson (2000). ΣPCB concentrations in the upper 50 m water column under the permanent ice cap are much lower (Mean Σ10PCB = 0.00055 ± 0.00043 ng/L) than in the northern North Atlantic (range — 0.0014–0.0021 ng/L). *in situ* sampling during the SHEBA study (Macdonald *et al.*, 2001b) and in nearshore waters (Hoekstra *et al.*, 2002a) also found higher ΣPCB concentrations in nearshore waters of the Canadian archipelago and southern Beaufort Sea (Σ10PCB means of 0.040–0.060 ng/L). Σ10PCB levels in the Central Basin of the White Sea (0.028 (± 0.014 ng/L) were lower than in the Canadian Arctic archipelago, using the same sampling equipment and analytical techniques, despite the proximity of the urban areas of Dvina Bay (eastern White Sea) and Kandalaksha Bay (western arm).

Although the PCB measurements in the northern Chukchi Sea (SHEBA study Figure B.2.3) may be confounded by contamination they are internally consistent in that they show a remarkable transition from the Canada Basin to the Chukchi Plateau. East of the plateau (about 77°N, 160°W) the PCB profile consisted mainly of mono-, di- and trichloro CB congeners (Figure B.2.3a). Over the southern edge of the Chukchi Plateau (April), the profile looks far more strongly influenced by the technical mixture component (Figure B.2.3b). As SHEBA drifted back toward the interior ocean, the profiles in July (Figure B.2.3c) and September (Figure B.2.3d) revealed a strengthened lower chlorinated PCB congener profile. The PCB profile in waters west of the Chukchi Plateau bore a greater resemblance to technical PCB mixtures such as the Russian "Sovol" (Schulz *et al.*, 1989; Ivanov and Sandel, 1992; Figure B.2.3e) and Aroclor 1254. There were also major differences in proportions of the homologue groups between surface waters (10–30 m) and deep samples (150–250m) especially in the Canada Basin. Given the stratification of the water column and the ice cover, these changes were not due to ice melt or biological process. It seems much more plausible that the PCB compositions along the track reflect the various water masses through which SHEBA passed. The dominant contribution of the technical mixture, seen in April (Figure B.2.3b), suggests strongly that this profile has been imported from the Pacific Ocean/Bering Sea. The profiles collected in February, July and September, on the other hand, probably reflect a stronger influence of Canada Basin interior water which has accumulated an important component of its PCB inventory through the atmosphere. The inference to be drawn here is that PCBs have been delivered via the Bering Sea in water directly contaminated through, for example, runoff or spills, whereas the interior Arctic Ocean owes a significant portion of its PCB content to condensation and/or air-sea exchange which favour the lighter components as one progresses northward (Wania and Mackay 1993b; Muir *et al.*, 1996a).

B.2.4 Toxaphene

Only a limited number of new measurements of toxaphene in seawater have been reported since the last AMAP POPs assessment and they are too few to fully assess spatial trends. Total toxaphene levels measured in surface seawater during recent campaigns ranged from 0.04–0.25 ng/L (Table B.2.1). By comparison, the range was 0.014–0.096 ng/L in the Chukchi Sea, northern Canada Basin and central Archipelago in 1992–94 (Bidleman *et al.*, 1995a; Hargrave *et al.*, 1997; Jantunen and Bidleman, 1998). Recently measured toxaphene levels in the North American Arctic Ocean were higher in northern Baffin Bay than in the southern Beaufort Sea (Hoekstra *et al.*, 2002a). Lower toxaphene levels were found in seawater in the White Sea and in the northern Chukchi Sea than in the southern Beaufort Sea or north Baffin Bay (Table B.2.1). The recent measurements have included chlorobornane congeners which have allowed insights into the degradation of toxaphene. Levels of individual chlorobornanes were in the low pg/L range (Figure B.2.4). The congener profile of toxaphene in the Beaufort Sea was dominated by hexa- and heptachlorobornanes especially B6–923 and B7–1001 (Figure B.2.4). These are terminal residues from dechlorination of many toxaphene congeners (Fingerling *et al.*, 1996) and are prominent in lake water in the Great Lakes (Muir *et al.*, 1999b; 2001). The White Sea had a different congener pattern with undetectable levels of B6–923 and more prominent octa- and nona-chloro congeners. This pattern suggests a fresher, less degraded, source of the technical product. Polychlorocamphene or toxaphene was widely used in Russian agriculture at least until the late 1980s (McConnell *et al.*, 1996). Toxaphene was also present in sediments of a harbour area in Kola Bay (near Murmansk) indicating possible other uses such as for insect control on ships.

B.2.5 Other "legacy" organochlorine pesticides

DDT and chlordane-related compounds, as well as dieldrin and HCB were determined in several studies (Table B.2.1) but were present at much lower concentrations than toxaphene, PCBs, or HCH isomers. Similar to the pattern for PCBs, DDT residues were higher (by 5 to 20 times) in the Canadian archipelago and nearshore waters of the Beaufort Sea than over the Chukchi Plateau or in the central Arctic Ocean. Total chlordanes had a similar pattern although differences were not as pronounced suggesting that gas exchange and melting are important sources for DDT and CHL. Hargrave *et al.*, (1997) found generally higher levels of *cis*- and *trans*-chlordane and toxaphene under ice-covered conditions in the central Canadian archipelago compared to the open water season. These authors concluded that net deposition by seawater-air exchange of toxaphene and chlordanes during the open water period was equivalent to 50% to greater than 100% of the surface layer inventory. Thus lower concentrations of the more hydrophobic OCs such as PCBs and DDT under the ice pack are probably due to slow removal by sedimentation and lack of resupply from the atmosphere.

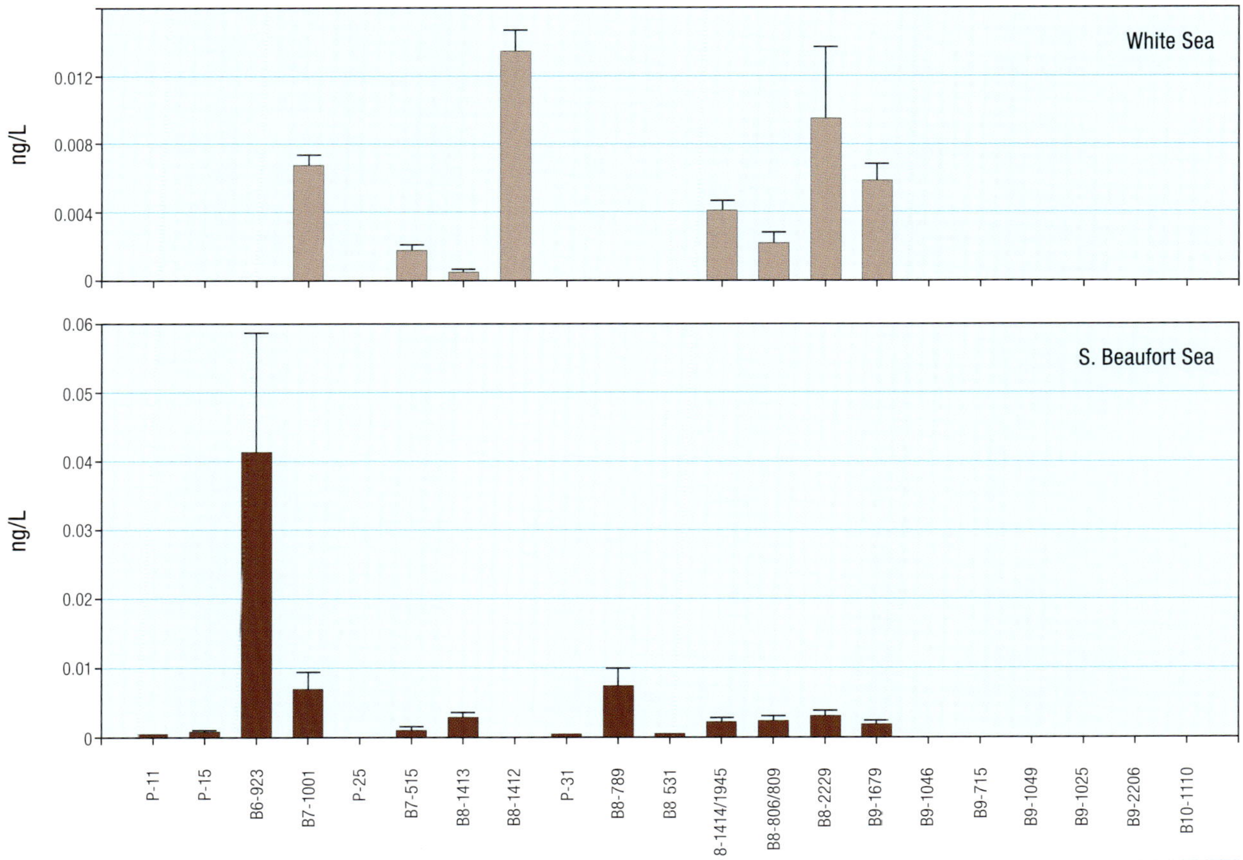

FIGURE B.2.4

Comparison of mean (± SE.) concentrations of individual toxaphene congeners (ng/L) in surface waters from the White Sea (n = 5) and the southern Beaufort Sea.

Levels of ΣDDT in the White Sea were lower than in the southern Beaufort Sea or in the northern Barents Sea (Table B.2.1). Despite possible ongoing use in urban areas in the White Sea region there was relatively little spatial variation of ΣDDT among ten sampling sites (0.002–0.005 ng/L) ranging from Dvina Bay (eastern White Sea) to the western end of Kandalaksha Bay (western arm).

Macdonald *et al.* (2001) reported on the vertical distribution of DDT, dieldrin, and chlordane over the Canada Basin and Chukchi Plateau (Figure B.2.5, a to c) as part of the SHEBA study. The vertical distributions of these OC pesticides were very different from those observed for the HCHs, where concentrations were found to decrease from their highest values near the surface to very low values below 200 to 300 m. This may have been due to changes in water mass caused by watermass change. However, taking DDT as perhaps the best example, it seems likely that there is also a seasonal signal related to the biological cycle. In April and February of 1998, the DDT profiles were somewhat noisy, but reasonably uniform at about 0.002 ng/L through the water column down to 250 m (Figure B.2.5a). By July, however, this uniformity had been destroyed with much lower DDT concentrations evident in the top 100–150 m and higher concentrations at about 200–50 m. In September, the surface concentrations of DDT were again centred at about 0.002 ng/L, but higher concentrations were evident at or below about 150 m. Although water mass changes may have contributed partly to the observed behaviour, there appears also to have been a transfer of DDT from surface to deeper water in spring when biological production produced a vertical flux of organic-rich particles.

B.2.6 Chlorinated benzenes

Chlorobenzenes (ClBz) are prominent contaminants in Arctic seawater. Dichloro- to hexachlorobenzenes have been measured (Strachan *et al.*, unpublished data) (Table B.2.1) and ΣClBz ranged from 0.009 ng/L in the White Sea to 0.14 ng/L in northern Baffin Bay. Concentrations of HCB appear to be relatively uniform compared to other compounds. Concentrations ranged from 0.005–0.006 ng/L in the southern Beaufort Sea and the White Sea. The number of chlorobenzene measurements is, however, limited. For example, there are no data for the European Arctic except for the White Sea.

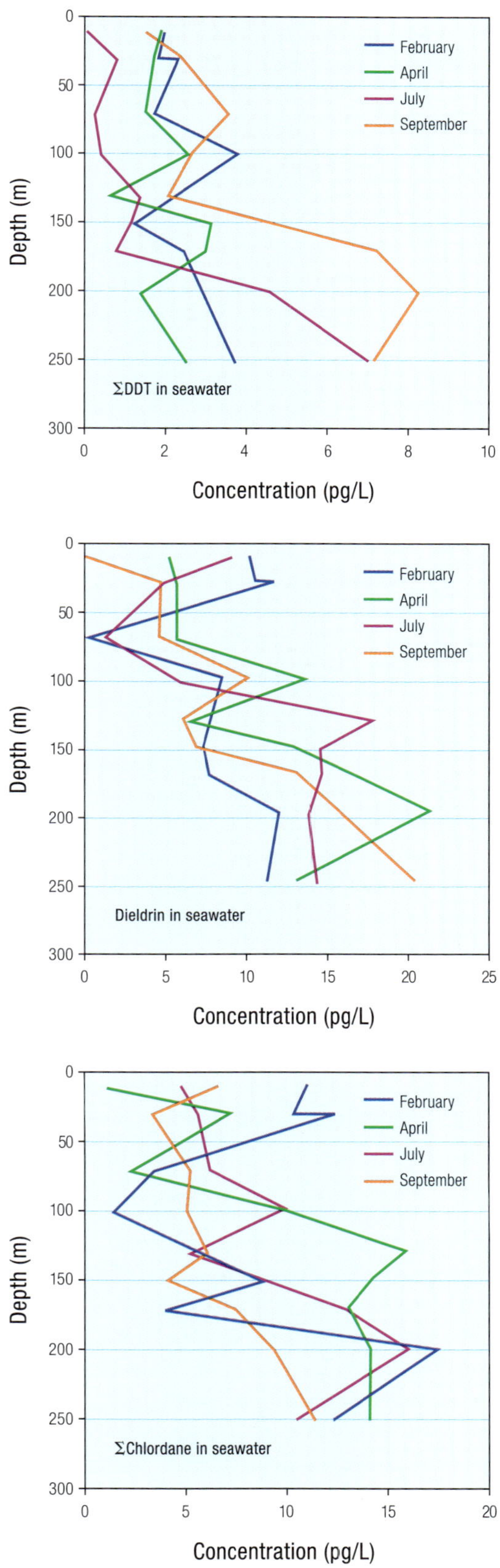

FIGURE B.2.5

Vertical profiles of "legacy" organochlorine pesticides during four sampling intervals (February, April, July and September, 1998) for ΣDDT, dieldrin, and Σchlordane (Macdonald *et al.*, 2001)

B.3 Persistent organic pollutants in marine and lake sediments

Authors:

G. Stern (1); and M. Evans (2).

Contributors:

M. Ikonomou (3); W. Lockhart (1); D. Muir (4); D. Rawn (5); P. Roach (6); B. Rosenberg (1); and G. Tomy (1).

Affiliations:

1. Freshwater Institute, Department of Fisheries and Oceans, 501 University Crescent, Winnipeg, Manitoba, R3T 2N6, Canada;
2. Aquatic Ecosystem Protection Research Branch, National Water Research Institute, Environment Canada, 11 Innovation Boulevard, Saskatoon, Saskatchewan, S7N 3H5, Canada;
3. Institute of Ocean Sciences, Fisheries and Oceans Canada, 9860, West Saanich Road, Sidney, British Columbia, V8L 4B2, Canada;
4. National Water Research Institute, Environment Canada, 867 Lakeshore Road, Burlington, Ontario, L7R 4A6, Canada;
5. Food Research Division, Health Canada, Tunney's pasture, Ottawa, Ontario, K1A 0L2, Canada;
6. Department of Indian Affairs and Northern Development, 345–300 Main Street, Whitehorse, Yukon, Canada

B.3.1 Introduction

In CACAR-I, latitudinal and temporal variations of organochlorine contaminant deposition in sediment cores for eight lakes along a mid-continental transect from 49 to 82°N and the Yukon were reported. Sub-Arctic lakes showed an onset of PCBs in the 1940s in reasonable agreement with historical usage patterns. In high Arctic lakes, however, no significant PCB inputs were observed until the 1960s, lagging the initiation of production by approximately 20 years. This difference was attributed to the lag time in diffusion of PCBs released at mid-latitudes to the North. In the Arctic lakes, HCB and ΣPCB levels also appeared to peak 5 to 10 years after the period of maximum production.

As part of NCP-Phase II, additional compounds such as toxaphene and chlorophenols have been analyzed in some of the same cores studied in NCP-Phase I. In addition, sediment cores from other Arctic lakes such as Great Slave Lake, Coal Lake in the Yukon Territories, and a very rare laminated sediment core from Lake DV09 on Devon Island have been collected, dated, and analyzed for a wide range of OC contaminants. Up until recently, few contaminant measurements in Canadian Arctic marine sediments have been reported. Sediment grab samples, from various regions of the Canadian Arctic archipelago, have been analyzed for organochlorine pesticides, PCBs, and current-use contaminants such as short-chain (C_{10}-C_{13}) chlorinated paraffins (SCCPs) and PBDEs.

B.3.2 Marine sediments

Until recently, few contaminant measurements in Canadian Arctic marine sediments have been reported. Where organochlorine concentration measurements have been made, with the exception of the highly productive regions, levels were often found to be below detection limits (Macdonald *et al.*, 2000).

In 1997 and 1998 as part of the Joint Ocean Ice Studies (JOIS), 35 sediment cores and grab samples were collected from various regions of the Canadian Arctic archipelago. Sediment samples from northern Baffin Bay were collected as part of the 1998 Northwater (NOW) Polynya study and the Hudson Bay samples during sampling cruises held in 1992 and 1993. To date, a selected number of the grab samples have been analyzed for organochlorine pesticides, PCBs (Stern *et al.* 2002; unpublished data) and current-use contaminants such as short-chain (C_{10}-C_{13}) chlorinated paraffins (SCCPs) (Bidleman *et al.*, 2000) and PBDEs (Stern *et al.* 2002 (unpublished data)). Highest concentrations of the major OC groups occur in the Hudson Bay surface sediments (Table B.3.1) with general decreasing trends from south to north and from east to west (Figure B.3.1 and B.3.2). Hudson Bay is a relatively shallow body of water (average depth ~130 m) and has a high sedimentation rate (736–1180 $g/m^2 \cdot y^{-1}$, Lockhart *et al.*, 1998). The combination of the high K_{OC} (i.e., tendency for a chemical to sorb to the organic portion of sediment) of DDT, PCB, CHL, and CHB congeners, and the shallow depth of Hudson Bay, makes the burial or sedimentation of these particle-bound compounds very rapid (i.e., the sediments act as a sink for the hydrophobic OCs) and reduces the probability of the OC-bound organic carbon being remineralized while still in the water column. Because of the lower K_{OC} values associated with HCH isomers, their tendency would be to reside in the water and not in the sediments.

Sediment ΣSCCP concentrations ranged from 4.8–77.4 ng/g and, as with the OC pesticides and PCBs, a clear decreasing trend in concentrations was observed from south (Barrow Strait) to north (Peary Channel). Western Arctic (M'Clure Strait/Viscount Melville) sediment concentrations were lower than in the eastern Arctic (Strathcona Sound/Lancaster Sound) sediments (Table B.3.1; Figure B.3.3).

	ΣPCB	Site
1	**80** ng/g OC	Beaufort
2	**15** ng/g OC	M'Clure St.
3	**40** ng/g OC	V. Melville Sd
4	**88** ng/g OC	Barrow St.
5	**33** ng/g OC	Peary Ch.
6	**9.6** ng/g OC	Penny St.
7	**170** ng/g OC	Barrow St.
8	**37** ng/g OC	Massey Sd.
9	**98** ng/g OC	Wellington Ch.
10	**150** ng/g OC	Norwegian Bay
11	**54** ng/g OC	Lancaster Sd.
12	**350** ng/g OC	Strathcona Sd.
13	**190** ng/g OC	Jones Sd.
14	**120** ng/g OC	N. Baffin Bay
15	**920** ng/g OC	E. Hudson Bay

FIGURE B.3.1

ΣPCB (ng/g, organic carbon) concentrations in Arctic marine surface sediments.

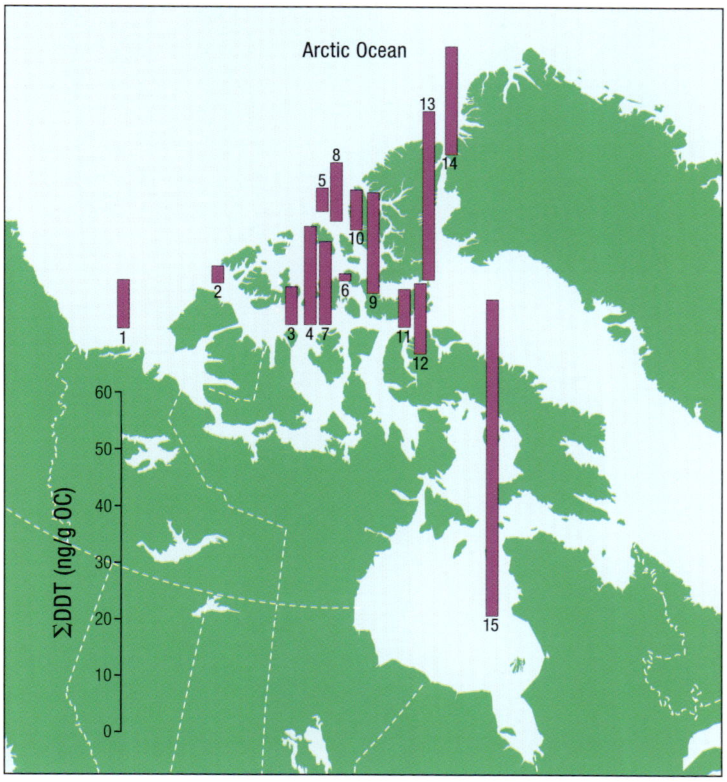

	ΣDDT	Site
1	**8.6** ng/g OC	Beaufort
2	**2.5** ng/g OC	M'Clure St.
3	**6.0** ng/g OC	V. Melville Sd
4	**17** ng/g OC	Barrow St.
5	**3.7** ng/g OC	Peary Ch.
6	**1.2** ng/g OC	Penny St.
7	**14** ng/g OC	Barrow St.
8	**10** ng/g OC	Massey Sd.
9	**18** ng/g OC	Wellington Ch.
10	**6.8** ng/g OC	Norwegian Bay
11	**6.2** ng/g OC	Lancaster Sd.
12	**12** ng/g OC	Strathcona Sd.
13	**30** ng/g OC	Jones Sd.
14	**19** ng/g OC	N. Baffin Bay
15	**56** ng/g OC	E. Hudson Bay

FIGURE B.3.2

ΣDDT (ng/g, organic carbon) concentrations in Arctic marine surface sediments.

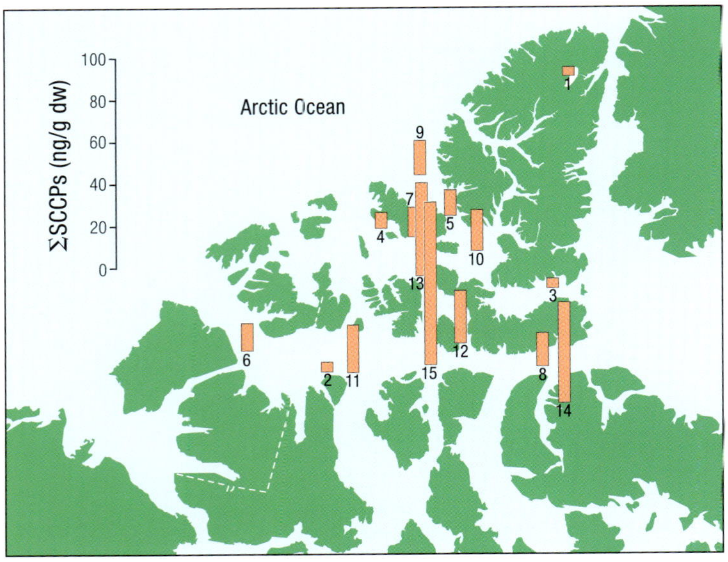

	ΣSCCPs	Site
1	**4.50** ng/g dw	Lake Hazen
2	**4.79** ng/g dw	Melville Sd.
3	**5.14** ng/g dw	Jones Sd.
4	**8.19** ng/g dw	Maclean St.
5	**12.11** ng/g dw	Massey Sd.
6	**12.95** ng/g dw	M'Clure St.
7	**14.15** ng/g dw	Hassel Sd.
8	**16.15** ng/g dw	Lancaster S.
9	**16.64** ng/g dw	Peary Ch.
10	**19.65** ng/g dw	Norwegian Bay
11	**23.51** ng/g dw	Melville Sd.
12	**24.91** ng/g dw	Wellington Ch.
13	**44.07** ng/g dw	Penny St.
14	**47.27** ng/g dw	Nanasivik
15	**77.41** ng/g dw	Barrow St.

FIGURE B.3.3
ΣSCCP concentrations (ng/g) in surface sediment from the Canadian Arctic.

TABLE B.3.1 Organochlorine contaminant concentrations in marine surface sediment samples collected in the Canadian Arctic archipelago and eastern Hudson Bay; all concentrations are in ng/g dw

Site	Date	Depth (m)	Latitude (N)	Longitude (W)	ΣPCB	ΣDDT	ΣCHL	ΣHCH	ΣSCCP	Reference
Beaufort	14-Sept-98	370	74° 17.551	128° 43.175	0.78	0.08	0.13	0.04	n.a.	1
M'Clure St.	19-Sept-98	499	74° 58.646	120° 41.685	0.29	0.05	0.04	0.01	n.a.	1
M'Clure St.	17-Sept-98	500	73° 56.980	115° 4.770	n.a.	n.a.	n.a.	n.a.	12.95	1
V. Melville Sd.	16-Sept-98	550	74° 8.130	108° 2.000	n.a.	n.a.	n.a.	n.a.	4.79	1
V. Melville Sd.	05-Sept-97	203	74° 40.180	104° 4.470	0.28	0.04	0.08	0.02	23.51	1
Maclean St.	11-Sept-98	666	77° 57.994	103° 43.518	n.a.	n.a.	n.a.	n.a.	8.19	1
Barrow St.	14-Sept-98	221	74° 32.930	99° 28.300	1.42	0.28	0.11	0.07	n.a.	1
Peary Channel	23-Aug-98	635	79° 24.670	98° 39.481	0.68	0.08	0.06	0.04	16.64	1
Hassel Sound	10-Sept-98	576	78° 3.008	98° 29.809	n.a.	n.a.	n.a.	n.a.	14.15	1
Penny Strait	21-Aug-98	307	76° 59.820	98° 0.421	0.09	0.01	0.01	0.01	44.07	1
Barrow Strait	05-Sept-97	264	74° 48.490	97° 10.560	2.70	0.23	0.19	0.35	77.41	1
Massey Sd.	21-Aug-98	292	78° 45.381	94° 56.987	0.45	0.13	0.12	0.06	12.11	1
Wellington Chan.	26-Aug-98	252	75° 44.577	93° 30.547	0.36	0.07	0.08	0.15	24.91	1
Norwegian Bay	24-Aug-98	492	79° 44.914	91° 59.028	2.76	0.12	0.20	0.06	19.65	1
Lancaster Sd.	29-Aug-93	550	74° 15.068	85° 59.173	0.75	0.09	0.06	0.01	16.15	1
Strathcona Sd.	01-Sept-98	229	73° 5.905	84° 31.764	2.57	0.09	0.12	0.04	47.27	1
Jones Sd.	08-Sept-98	844	76° 8.104	82° 24.854	2.87	0.44	0.20	0.18	5.14	1
E. Hudson Bay	08-Sept-93	133	58° 0.700	81° 16.680	8.24	0.57	0.35	0.12	n.a.	1
N. Baffin Bay[(1)]	July-98	265–560	75° 15.236–78° 21.107	70° 13.393–77° 22.619	1.80	0.27	0.17	0.16	n.a.	2

[(1)]Mean of 14 samples.
1. Stern (unpublished results).
2. Fisk *et al.* (200[illegible]b).
n.a. = not analyzed.

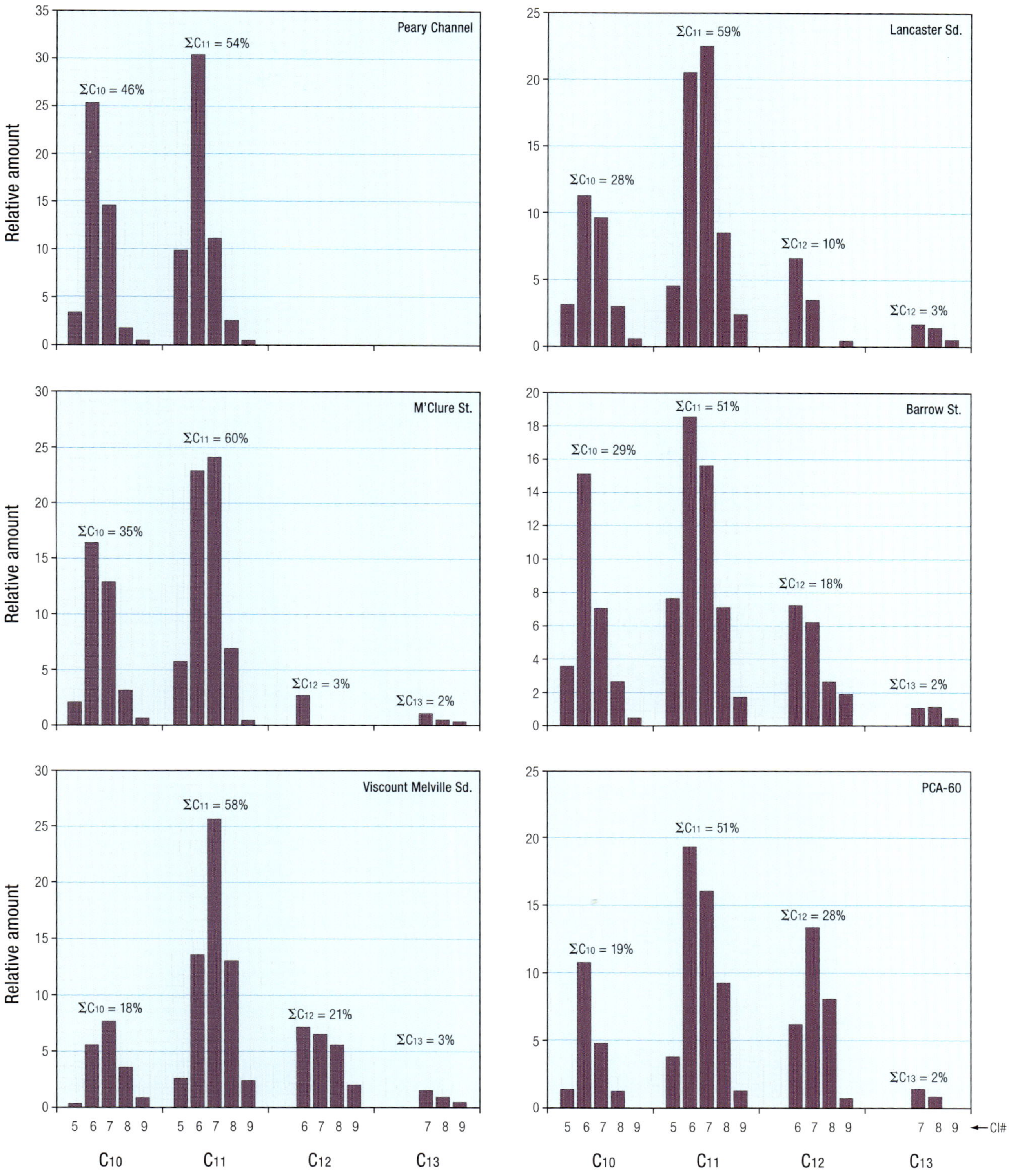

FIGURE B.3.4

ΣSCCP formula group profiles for surface sediment from the Canadian Archipelago and the PCA 60 commercial formulation (C_{10} – C_{13}, 60% chlorine).

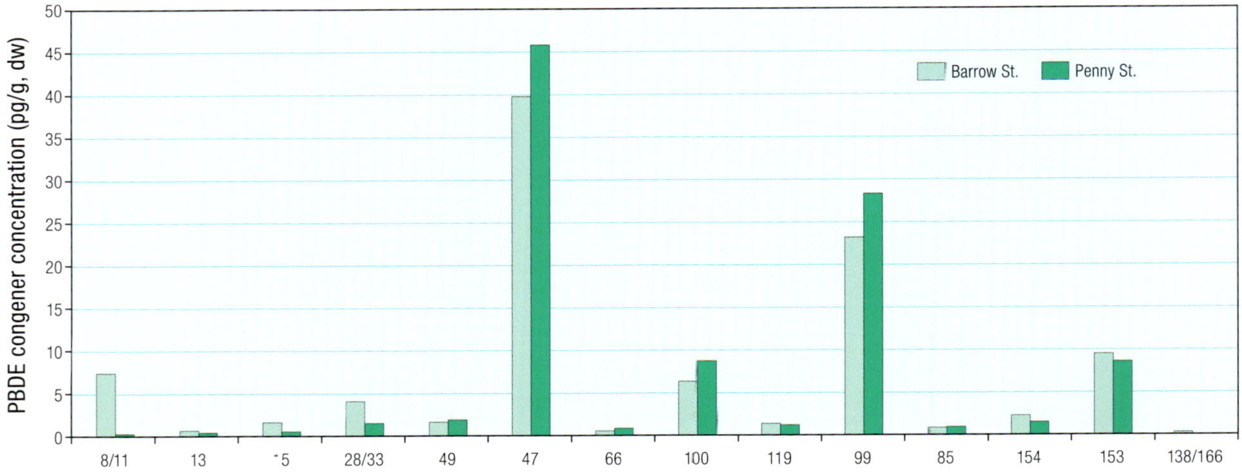

FIGURE B.3.5

PBDE congener concentrations in marine sediment from selected locations in the Canadian Arctic archipelago.

Tomy *et al.* (1999) reported a ΣSCCP concentration of 4.5 ng/g in the surface slice of a sediment core collected from Lake Hazen on Ellesmere Island. Relative to the commercial PCA 60 formulation (C_{10}-C_{13}, 60% Cl) the shorter carbon chain length and lower chlorinated C10 and C11 formula groups are clearly predominant (Figure B.3.4). This result is consistent with the trend of increasing vapour pressures with decreasing carbon chain length and degree of chlorination, reported by Drouillard *et al.* (1998a), and their corresponding susceptibility to long-range atmospheric transport.

Total PBDE concentrations in three of the JOIS sediment samples (Barrow Strait, Penny Strait and Nanisivik) were 106.7, 122.5 and 297.3 pg/g (dw), respectively. PBDE 47, the most abundant PBDE congener present in the marine sediments (Figure B.3.5), had measured concentrations of 39.8, 45.8, and 106.8 pg/g (dw), respectively.

B.3.3 Lake sediments

B.3.3.1 Yukon

Prior to its use as an insecticide in the early 1970s toxaphene was used in both Canada and the USA in fish eradication programs (Nash *et al.*, 1973; Walker *et al.*, 1973). This practice, however, was discontinued in the mid 1960s when it was found that toxaphene was extremely persistent and that lakes could not be successfully restocked for years after treatment (Nash *et al.*, 1973). With the exception of Hanson lake, which was treated with toxaphene as a fish toxin in the early 1960s, no documentation of any past release of toxaphene into Yukon lakes or into the upper Yukon River system exists. Elevated toxaphene (ΣCHB) levels found in fish tissues from several of the large lakes in the system have thus been attributed to an atmospheric source, rather than a local one, and in the case of Lake Laberge a longer than normal food chain (Kidd *et al.*, 1993; 1995b).

The electron capture negative ion selective chromatograms of ΣCHBs in representative slices of sediment cores from five different Yukon lakes are shown in Figure B.3.6. The chromatographic pattern of ΣCHBs in the Hanson Lake sediment is dominated by 2-exo, 3-endo,6-exo,8,9,10-hexachlorobornane (Hx-Sed or B6–923) and 2-endo, 3-exo,5-endo,6-exo,8,9,10-heptachlorobornane (Hp-Sed or B7–1001) (Stern *et al.*, 1995; Andrews *et al.*, 1995). These same two congeners were also predominant in the sediment of two toxaphene-treated lakes located in Alberta (Miskimmin *et al.*, 1995; Stern *et al.*, 1995) but not in remote lakes exposed only to atmospheric sources. B6–923 and B7–1001 are formed through anaerobic reductive dechlorination of other, less stable, CHBs present in the technical mixture but not generally in air (Vetter *et al.*, 1999; Braekevelt *et al.*, 2001).

In the Watson and Fox Lake sediments, Hx- and Hp-Sed are the predominant CHB congeners in both cases suggesting contamination by a non-atmospheric source. In Lake Laberge, Hx- and Hp-Sed are much more prevalent in the core slices representing the years 1972–1976 compared to the period of 1980 to 1991 but, do not predominate nearly to the same extent as observed in the sediments from Hanson, Fox, or Watson Lakes. This result, along with the relatively low ΣCHB levels (Table B.3.2), suggests that atmospheric deposition is the primary source of toxaphene to Lake Laberge but that a small direct contribution, possibly via the Yukon River, may also have occurred before 1964. Hx-Sed is not present in sediment from Kusawa lake and Hp-Sed is only a very minor component of ΣCHB. This result is consistent with inputs associated with atmospheric deposition only.

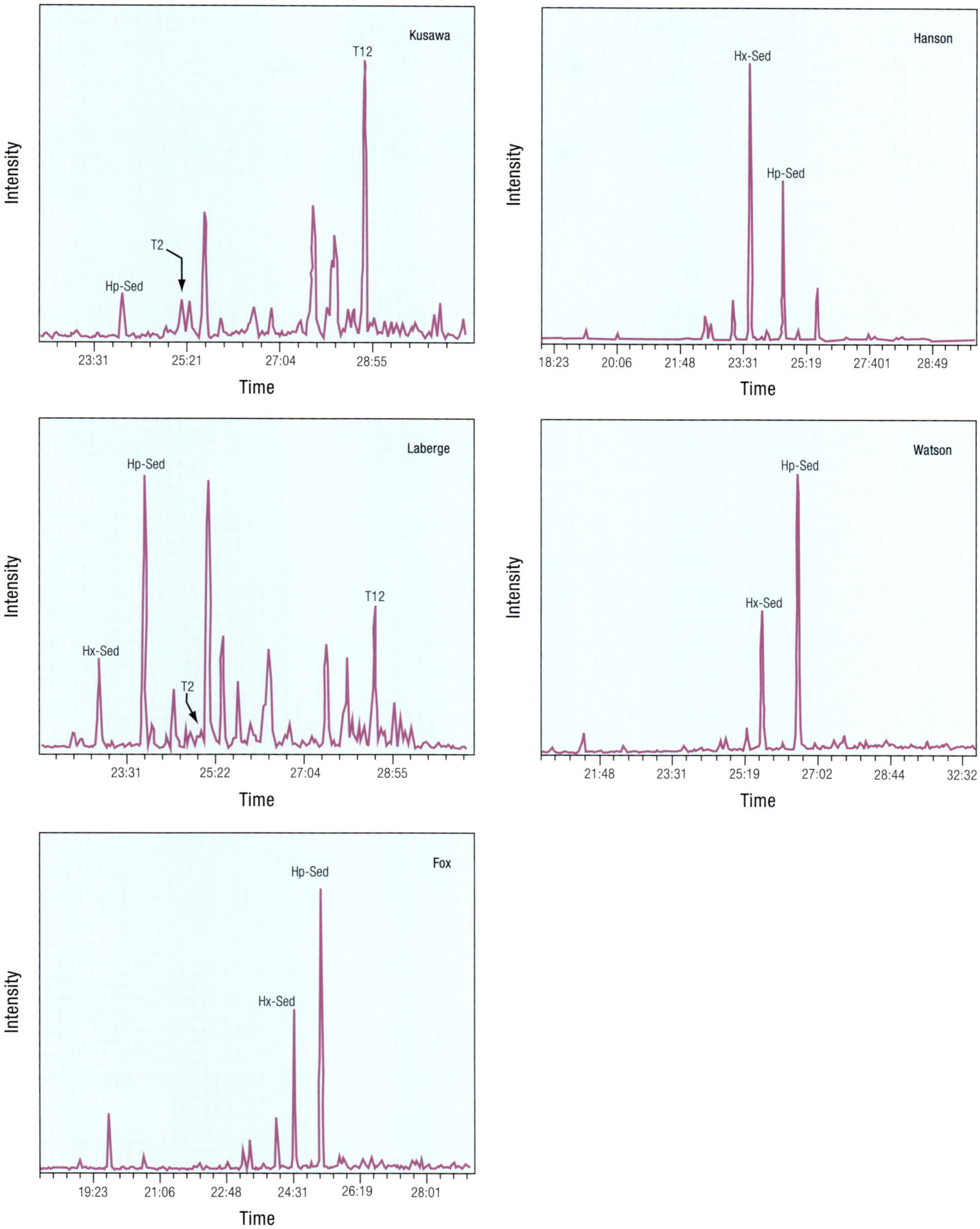

FIGURE B.3.6

ECNI selected ion chromatograms of ΣCHBs in sediment core slices from five Yukon lakes. Hanson Lake was treated with toxaphene as a piscicide in the early 1960s.

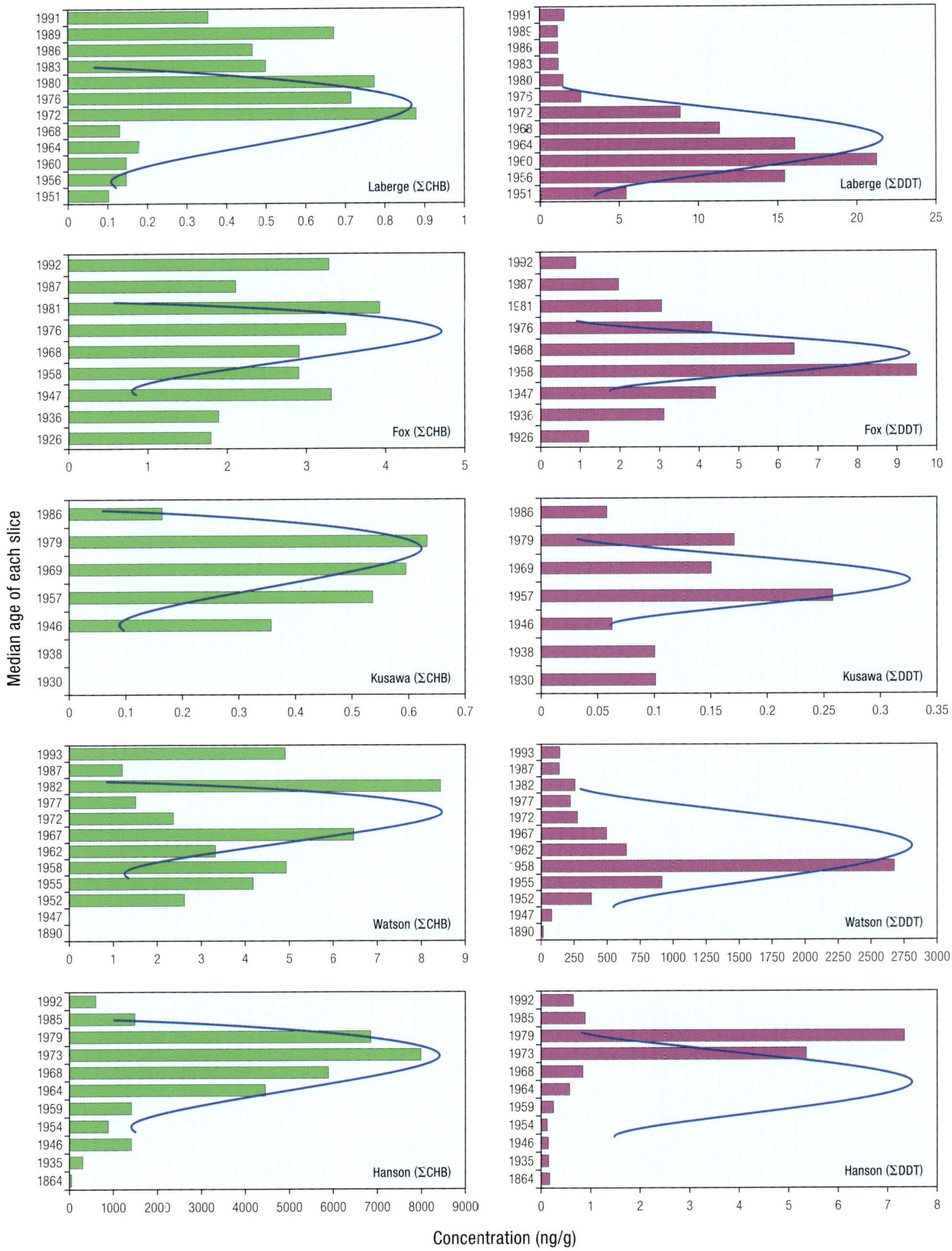

FIGURE B.3.7

Concentration profiles (ng/g, dw) of ΣCHBs and ΣDDTs in sediment cores from four Yukon lakes. The curves correspond to atmospheric input functions of ΣCHBs and ΣDDTs, derived by Rapaport *et al.* (1988).

TABLE B.3.2 Concentrations (ng/g dw) of organochlorine contaminants in lake sediments from Yukon, Nunavut and the Northwest Territories — all toxaphene, Coal Lake and Lake DV09 data from Stern *et al.* (unpublished data) — Great Slave Lake OC data from Evans *et al.* (1996) — all remaining results from Rawn *et al.* (2001)

	ΣCHB		ΣDDT		ΣPCB		ΣCHL		ΣHCH		ΣPeCA	
Lake	**Surface**	**Max**	**Surface**	**Max**	**Surface**	**Max**	**Surface**	**Max**	**Surface**	**Max**	**Surface**	**Max**
Watson	4.92	8.47	114	2680	<MDL	93.6	<MDL	4.35	<MDL	3.18	<MDL	1.72
Hanson	592	8001	1.58	7.20	25.9	37.8	0.56	1.73	1.37	1.57	<MDL	4.52
Fox	3.30	3.94	0.67	8.90	33.5	33.5	0.04	1.30	0.71	0.91	0.51	0.83
Laberge	0.36	0.89	1.60	21.4	8.07	16.1	0.67	0.67	0.54	0.95	0.36	0.60
Little Atlin	–	–	3.47	3.47	28.3	28.3	3.87	3.87	0.33	0.79	0.40	0.62
Kusawa	0.17	0.64	0.06	0.26	2.19	12.7	0.11	0.48	0.13	0.48	0.04	0.33
Lindeman	–	–	0.86	0.86	8.27	11.1	0.26	0.42	0.46	0.46	<MDL	0.36
Coal	–	–	–	–	–	–	–	–	–	–	0.03	0.03
DV09	0.72	1.47	0.07	0.20	2.97	3.00	0.18	0.25	0.23	0.34	0.16	0.44
Great Slave	1.30	3.10	0.41	1.55	8.75	9.95	0.26	0.49	0.31	0.47	0.16	0.38

MDL = Method detection limit.

Collection year, sedimentation rates and focusing factors for a number of Yukon lake sediments cores are shown in Table B.3.3. Total ΣDDT and ΣCHB sediment concentration profiles for a subset of these lakes are shown in Figure B.3.7. Total ΣDDT and ΣCHB levels in the Lake Laberge, Kusawa, and Fox Lake cores peak in the early to mid-1960s and early to mid-1970s, respectively, and coincide with historical usage patterns in North America (Rapaport *et al.*, 1988).

Lindeman Lake, like Kusawa, is a remote glacial-fed lake with a very low maximum ΣDDT sediment concentration (0.86 ng/g, dw); however, no clear temporal trend was obvious from the sediment profile. The small size of this lake and the high sedimentation rate (2050 $g/m^2 \cdot y^{-1}$) point to glacial runoff as being the most significant source of DDT (and other OCs) to the lake (Rawn *et al.*, 2001). The excess ^{210}Pb profile shows an irregular exponential decay down the core which may indicate changes in the sediment characteristics and loadings. This could result from dilution of the bottom sediments with terrestrial materials as a result of weathering, avalanches, or changes in runoff patterns. The ^{137}Cs profile showed a sharp sub-surface maximum that is ^{210}Pb dated to the mid-1960s, the time of maximum bomb fallout, thus reinforcing the ^{210}Pb dating process for this lake.

TABLE B.3.3 Collection year, location, sedimentation rate and focussing factors of Yukon and Nunavut sediment cores

Characteristic	Collection year	Latitude (N)	Longitude (W)	Sedimentation rate ($g/m^2 \cdot y^{-1}$)	Focusing factor	Reference
Watson	1995	60° 06	128° 46	239	3.0	Rawn *et al.* (2001)
Hanson	1995	64° 00	135° 21	76	1.4	Rawn *et al.* (2001)
Fox	1993	61° 15	135° 29	119	1.0	Rawn *et al.* (2001)
Laberge	1992	61° 12	135° 11	999	1.4	Rawn *et al.* (2001)
Little Atlin	1993	60° 16	135° 59	813	1.4	Rawn *et al.* (2001)
Kusawa	1992	60° 25	136° 10	227	0.8	Rawn *et al.* (2001)
Lindeman	1994	59° 47	135° 03	2050	4.9	Rawn *et al.* (2001)
Coal	1998	60° 31	135° 10	102	0.52	Stern *et al.* (2002) unpublished data
DV09	1999	75° 34	89° 19	300	2.4	Stern *et al.* (2002) unpublished data
Great Slave	1994	61° 24	114° 24	692	1.6	Evans (1996)

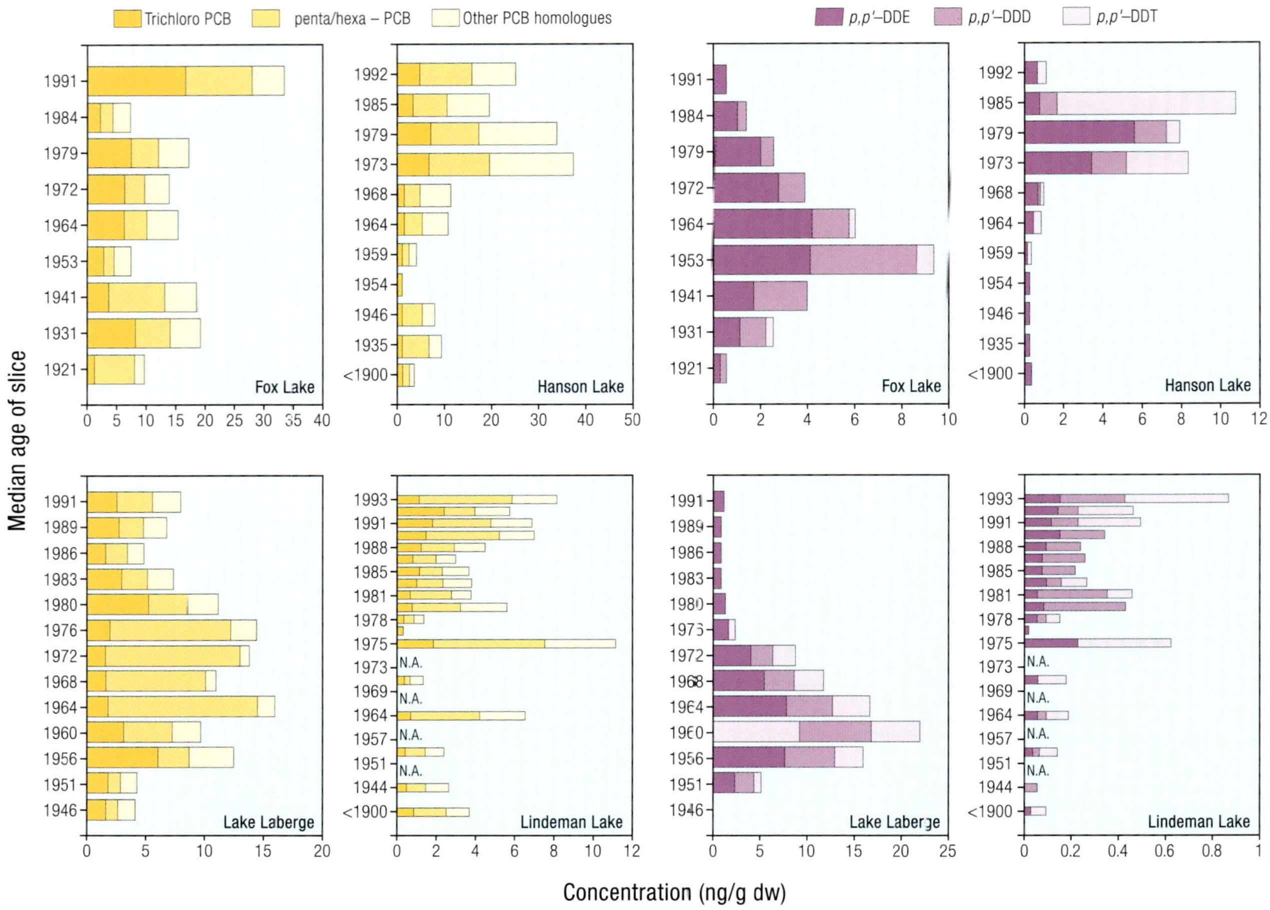

FIGURE B.3.8

Historic profiles of PCBs and DDT-related compounds in Fox Lake, Hanson Lake, Lake Laberge and Lindeman Lake within the Yukon River watershed (Rawn *et al.*, 2001). N.A. – sediment slice not analyzed

In Watson Lake, maximum ΣDDT concentration coincides with the known usage of DDT in the Watson Lake area for mosquito and blackfly control between 1949 and 1963 (Bissett *et al.*, 1995). Levels were approximately two orders of magnitude greater than those measured in other Yukon lakes also known to have been exposed to direct sources of DDT (Hanson, Laberge and Fox). Based on an application rate of 0.20 kg/ha, Rawn *et al.* (2001) estimated that almost 100 kg of DDT may have been applied to the lake on a single occasion. *p,p'*-DDD was the major DDT component throughout the core. Rawn *et al.* (2001) estimated the half-life of the first-order conversion rate for DDT to DDE to be 21 ± 5 y (r^2 = 0.68) for Watson Lake and 32 ± 2 y (r^2 = 0.88) for Lake Laberge sediments. A significant correlation representative of the conversion of DDT to DDD with time, was not observed in any of the other lake sediments studied. The ΣCHB profile in Watson Lake is not consistent with what would be expected based solely on atmospheric deposition (Rapaport *et al.*, 1988) (i.e., peak levels occurring in the early to mid-1970s).

The higher levels of ΣCHBs (Table B.3.2), the predominance of the CHB congeners Hx- and Hp-Sed (Figure B.3.6), and the overlap of the ΣCHB and ΣDDT historical concentration profiles, suggest that toxaphene usage in the Watson Lake area occurred prior to the banning of DDT, possibly for use as a fish toxin or as an additional active ingredient in the insecticidal DDT mixtures. Brief mention of the latter practice is made in the literature (Nordin *et al.*, 1993) but no records documenting the quantities of toxaphene added to the DDT mixtures, or the dates and sites where these mixtures were applied, could be found. The maximum ΣCHB concentration was measured in the combined core slices 8 and 9 with a mean date of 1982.

Concentrations of ΣCHB and ΣDDT in the Hanson Lake sediments are at a maximum in the early and late 1970s, respectively. These results, however, are not consistent with the known application date of toxaphene to the lake (1963) and with the first reported DDT application in the area. DDT applied for insect control around the Hanson Lake area occurred in Keno City and Elsa in 1965 (Nordin *et al.*, 1993). The ^{137}Cs profile showed a sharp sub-surface maximum that is ^{210}Pb-dated to 1974 (linear model) and 1980 (CRS model) suggesting that the sediment in this core is mixed.

In the Yukon, PCB use was very limited before the arrival of the military in 1942. PCBs were used in electrical transformers and radio transmitters, as hydraulic oil and heat exchange fluid, as well as for dust control on Yukon roads. The highest concentration of PCB transformers was thought to be in Whitehorse in connection with the large power demands of the Canol refinery and possibly at Teslin and Watson Lake (Bisset *et al.*, 1995). Total ΣPCB sediment concentration profiles for selected lakes are shown in Figure B.3.8. In the Lake Laberge sediment core, ΣPCB levels above background levels first occurred in the core slice dated to 1956. Levels peaked in the mid-1960s and then start to decline after the early 1980s. Between 1964 and 1976, the penta- and hexa-chlorinated congeners were the largest contributors to ΣPCB (~80%). These results are consistent with the known usage of PCBs in the Yukon and with the banned manufacture and use of PCBs in the US and Canada in the late 1970s.

Of all the Yukon lakes studied, the highest ΣPCB levels were measured in sediment from Watson Lake. Consistent with the arrival of the military, concentrations first started to increase in the mid 1940s. Maximum concentrations were observed in a recent slice, dated at 1984, followed by a decrease in ΣPCB concentrations in the more recent slices. The historical profile, low concentrations, and an even distribution of the homologue contributions to ΣPCB in the Kusawa sediment core are consistent with atmospheric deposition only. As was observed for ΣDDT, ΣPCB concentrations in Lindeman Lake sediments were very low and an irregular historical profile was observed. In Hanson Lake, ΣPCB concentrations peaked from the mid- to late 1970s, much later than one would have expected based historical North American usage patterns and on the known activity during and after the treatment of the lake with toxaphene.

Maximum ΣHCH concentrations in the seven Yukon lakes studied by Rawn *et al.*(2001) ranged from 0.46 ng/g in the slice dated to 1993 in Lindeman Lake to 3.18 ng/g in the slice dated to 1982 in Watson Lake (Table B.3.2). Subsurface maximum concentrations were observed in all but Lindeman Lake. Alpha- and γ-HCH were the dominant isomers contributing, on average, 39 and 52%, respectively, to ΣHCH. Elevated levels of ΣCHL were observed in slices dated to the 1970s and early 1980s in Fox, Kusawa, Hanson, and Lindeman Lakes. Peak concentrations were measured in the surface sediments from Atlin (3.87 ng/g) and Lake Laberge (0.67 ng/g) (Table B.3.2).

As part of a study to assess contamination in the bottom sediments of lakes adjacent to a now dismantled pentachlorophenol (PCP) railway tie treatment plant in Carcross, Yukon (operated from the mid 1930s until its closure in 1975), a sediment core from Coal Lake, a small sub-alpine headwater lake which feeds into the Wolf Creek drainage basin, was collected and used as a control (Stern *et al.* 2002; unpublished data). There is no record of this lake ever being directly exposed to chlorophenols and so the only source of these compounds and their chloroanisole biomethylation degradation products to the lake would be from long-range atmospheric transport.

Coal Lake core data are presented in Table B.3.2 and Figure B.3.9. Total chlorophenol concentrations (ΣCP) started to increase in the 1940s coinciding with the period (1936) when pentachlorophenol (PeCP) was first introduced as a timber preservative (Crosby *et al.*, 1981). A peak concentration of 2.45 ng/g, dw was measured in the core slice dated to 1952 which gradually declined to 1.63 ng/g dw in the surface sediment. While PeCP was

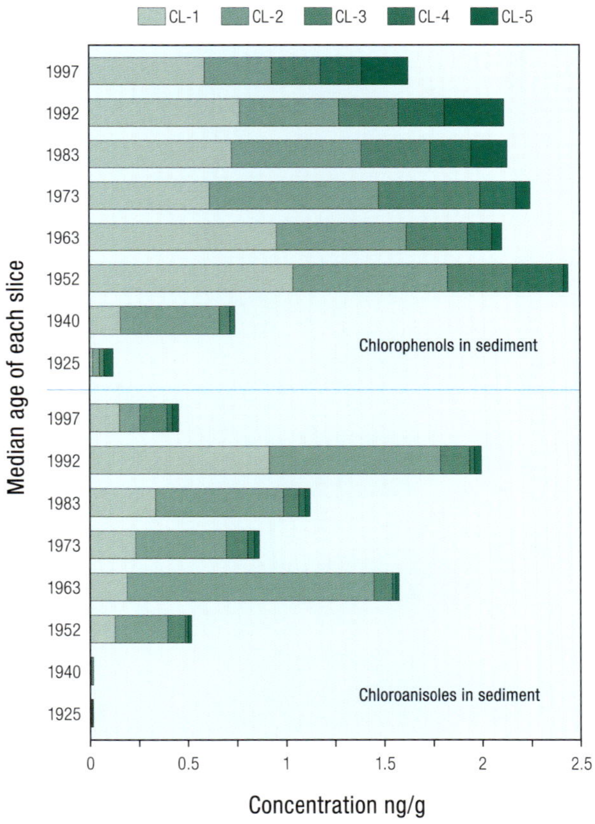

FIGURE B.3.9

Concentration profiles (ng/g dry wt.) of chlorophenols (Cl_{1-5}) and chloroanisoles (Cl_{1-5}) in a sediment core collected from Coal Lake, Yukon Territories, in March 1998.

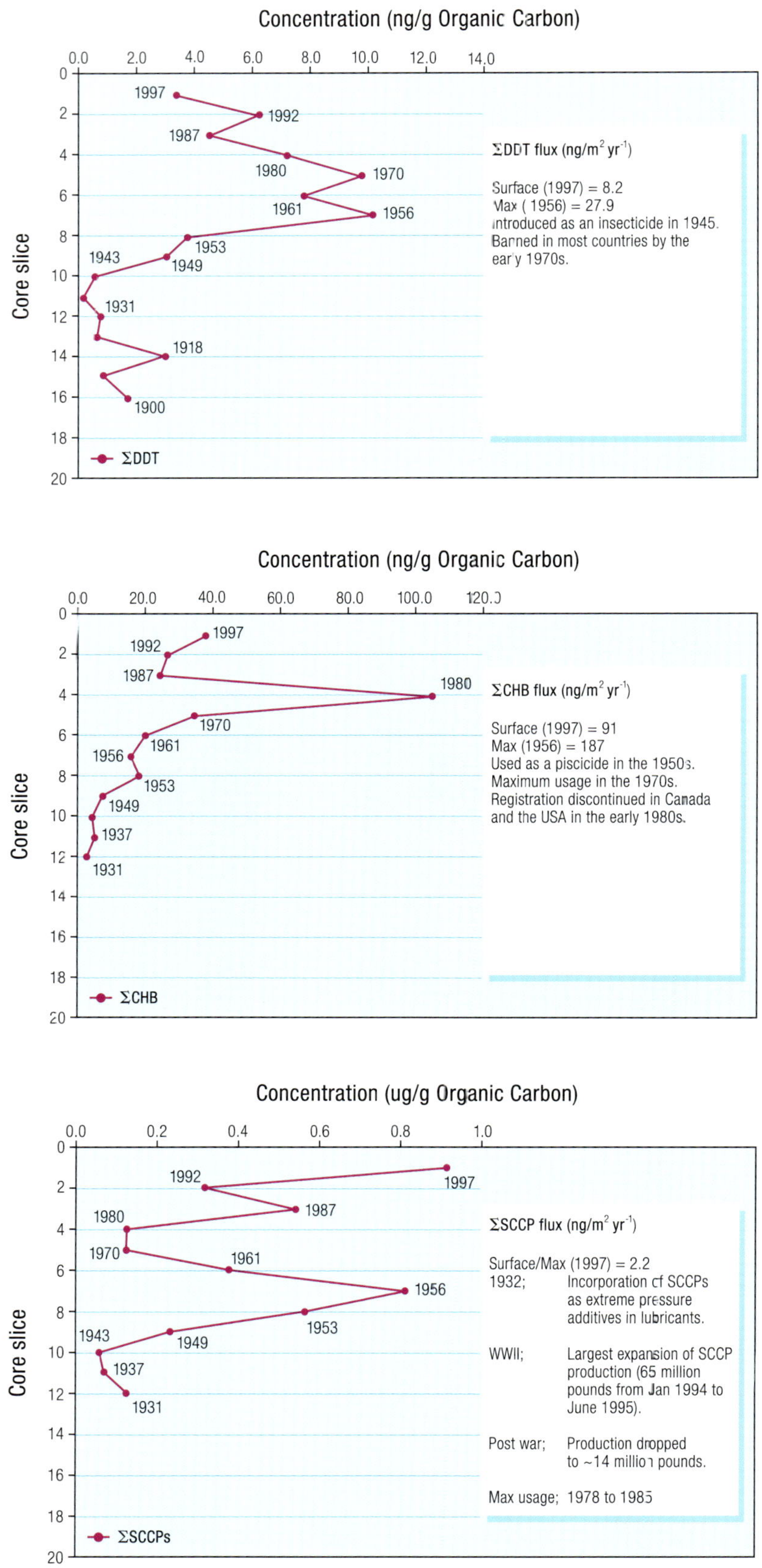

FIGURE B.3.10

Concentration profiles of ΣDDTs (top), ΣCHBs (middle) and ΣSCCPs (bottom) in a dated core from Lake DV09 on Devon Island, Nunavut, collected May, 1999.

designated as a "Restricted Use Pesticide" by the USEPA (USEPA 1991) in the late 1980s, as of yet, it has not been banned outright. In all slices, the mono- and dichlorophenols contribute the greatest percentage to the total, ranging from 89 % in 1940 and gradually declining to 57% in 1997. Lower chlorophenols, in particular the tri- and tetrachlorophenols are present as impurities in the technical formulations but can also be formed by the photochemical reduction of PCP in water (Crosby *et al.*, 1981). The onset of the chloroanisoles occurred in 1992 about 12 years later than the chlorophenols and a peak concentration of 2.0 ng/g dw was measured in the slice dated 1992. As was observed with the chlorophenols, the mono- and dichloroanisoles were predominant in all slices. A maximum pentachloroanisole (PeCA) concentration of 0.03 ng/g was measure in both the Coal Lake surface sediment and in the slice dated to 1983. Rawn *et al.* (2001) reported surface and maximum PeCA concentrations ranging from < MDL to 0.51 ng/g and 0.36–4.52 ng/g, respectively.

B.3.3.2 Nunavut

Sediment cores are subject to several kinds of artifacts that can make interpretation of historical deposition difficult. A rare and valuable type of core is one which has annual layers of sediment with the colour or texture of layers differing to form detectable laminations that can be counted. Such a core has not experienced physical mixing (the laminations would have been obscured) and with such cores, the historical record is more intact. Gajewski *et al.* (1997) reported a lake with laminated sediments on the coastal region of northern Devon Island (Lake DV09, 75°34' N; 89°19' W). Core material from the 1994 samples was not of a sufficient size to allow for contaminant analyses, but the dating with ^{210}Pb and counts of laminae indicated that the laminae were annual.

In May 1999, four sediment cores (approximately 4 m apart) were collected from Lake DV09 at 75°34' N latitude, 89°19' W longitude. The core tops were all undisturbed and distinct laminations were observed down the lengths of each (Lockhart *et al.*, 2000). The exponential decline in the ^{210}Pb profile and the timing of the ^{137}Cs inputs indicate core 1 had a credible historical record. The sedimentation rate calculated from the ^{210}Pb data was 304 $g/m^2{\cdot}y^{-1}$. There was some focusing of sediment to the site because the flux of ^{210}Pb was 65 $Bq/m^2{\cdot}y^{-1}$, a value higher than expected for 75°N latitude. A recently obtained flux of 27 $Bq/m^2{\cdot}y^{-1}$ for a soil profile taken from the north shore of Banks Island in 1998 (74°N) was used to correct for focusing. The flux at Lake DV09 should be similar to that for the Banks Island site since they were at essentially the same latitude. Assuming an expected flux of ^{210}Pb of 27 $Bq/m^2{\cdot}y^{-1}$, the actual flux of 65 $Bq/m^2{\cdot}y^{-1}$ indicates that the measured flux is too high by a factor of 2.4.

Core 1 was analyzed for a wide array of organochlorine compounds, a representative selection of which is shown in Figure B.3.10. These profiles show definitively that materials originating exclusively from human activities have reached the site. Furthermore, in view of the location of the site remote from any local human activity, the source of these synthetic materials must have been from atmospheric deposition. Total ΣDDT concentrations peaked between the mid-1950s and early 1970s. An increase in ΣCHB levels was first observed in the early 1970s but levels dropped down to background levels by the late 1980s. A consistent increase in chlordane concentrations is observed between the early 1950s until the early 1990s and a small decline was observed in the surface sediments. These results are consistent with the known historical usage of DDT, toxaphene, and chlordane in North America (Rapaport *et al.*, 1988; EHP, 1997). The historical resolution of the DV09 sediment core was considerably greater than that reported for other Arctic lake sediments (Muir *et al.*, 1995; 1996a). In addition, unlike the sediment cores collected from Sophia (75°07' N) and Amituk (75°03' N) lakes (Muir *et al.*, 1995; 1996a), concentrations of OCs such as toxaphene and ΣDDT in the core from DV09 were not highest in the most recent slices and were much more representative of known historical usage patterns. Reported sedimentation rates for Sophia and Amituk lakes were 177 and 389 $g/m^2{\cdot}y^{-1}$, respectively, compared to 304 $g/m^2{\cdot}y^{-1}$ for DV09. Thus, the increased resolution and the more historically representative OC profiles in this very unique DV09 core can almost certainly be attributed to the lack of physical mixing in the laminated sediments, and not a difference in sedimentation rate.

The first large volume use of SCCPs, as an extreme pressure additive in lubricants, occurred in 1932. The largest expansion in production, however, occurred during World War II. The output of SCCPs between 1944 and 1945 was estimated to be 29 million kilograms. After the war, production fell drastically reaching about 6.4 million kilograms by 1946. In subsequent years, production slowly increased beyond that of wartime production (Howard *et al.*, 1975). Maximum usage of SCCPs occurred between 1978 and 1985 (Muir *et al.*, 2000a). A pattern consistent with historical usage is observed for the SCCP concentrations in the Devon Island sediment core (Figure B.3.10). Maximum concentrations occur in the surface sediment (17.6 ng/g, dw) and in the core slice dated to 1956 (17.7 ng/g, dw). The latter result occurred approximately 10 years later than expected and can almost certainly be attributed to the complexity of the SCCP industrial mixtures (Tomy *et al*, 1997) and a range in the atmospheric transport and the deposition times to the high Arctic of the individual congeners.

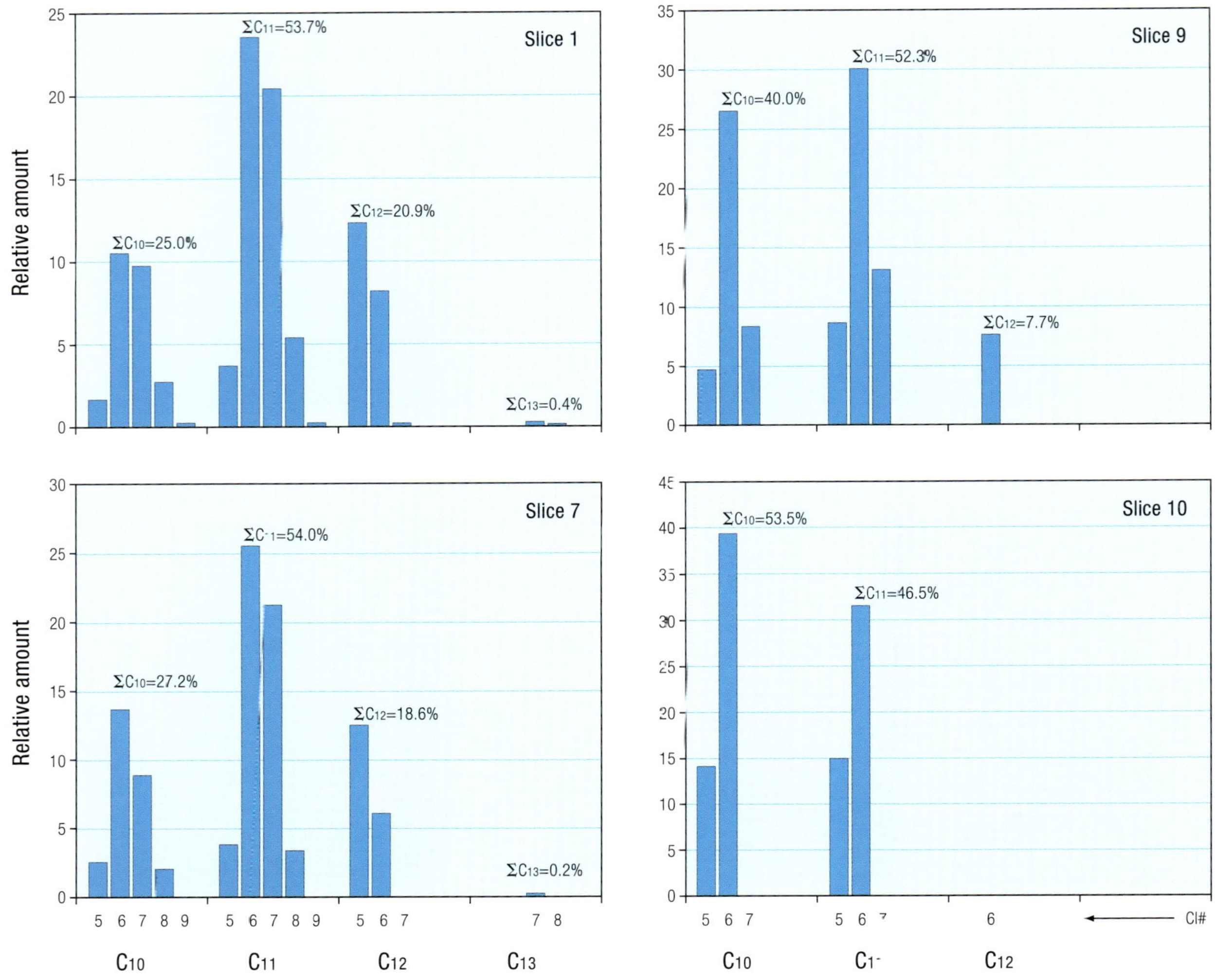

FIGURE B.3.11

SCCP formula group profiles for selected sediment slices from a laminated core collected from Lake DV09 on Devon Island, Nunavut, May, 1999.

Shorter carbon chain length and lower chlorinated C10 and C11 formula groups become more predominant moving down the core suggesting microbial degradation of the longer chain, more highly chlorinated compounds, over time and/or, that in the earlier sediments, the observed profile is representative of the first, more volatile, SCCP congeners to be transported and deposited in the Arctic (Figure B.3.11).

Like core 1, the exponential decline in the ^{210}Pb profile and the timing of the ^{137}Cs inputs indicate core 2 had a credible historical record. The sedimentation rate calculated from the ^{210}Pb data was 290 $g/m^2 \cdot y^{-1}$ with a focusing factor of 2.1. Concentration profiles for total dioxins and furans, normalized to organic carbon, are shown in Figure B.3.12. PCDFs levels start to increase in the early 1950s, peak at 3.02 pg/g (dw) in 1978 and are dominated by the TCDF homologue group, in particular, the 1,2,4,8-TCDF congener. PCDDs levels start to increase about 10 years early but also peak in 1978 (1.88 pg/g, dw). Between 1938 and 1963, sediment core slices were dominated by OCDD (64.32 ± 8.73%) (Figure B.3.13). This homologue profile is consistent with a signature resulting from use of pentachlorophenol as a wood preservative. OCDD is the major impurity in the PeCP technical mixtures and can also be formed by photolytic degradation of PeCP (Crosby *et al.*, 1981). From 1970 to 1992, the PCDD/F profiles are indicative of signatures arising from the combustion of coal and wood (Kjeller *et al.*, 1996).

Using the HYSPLIT air transport model, Commoner *et al.* (2000) estimated the deposition flux of airborne dioxin to Arctic Bay, Nunavut, to be 0.84 pg $TEQ/m^2 \cdot yr^{-1}$. Based on a corrected sedimentation rate of 138 $g/m^2 \cdot y^{-1}$, the surface and maximum deposition flux values for PCDDs to Lake DV09 on Devon Island were calculated to be 0.01 and 0.36 pg $TEQ/m^2 \cdot yr^{-1}$, respectively.

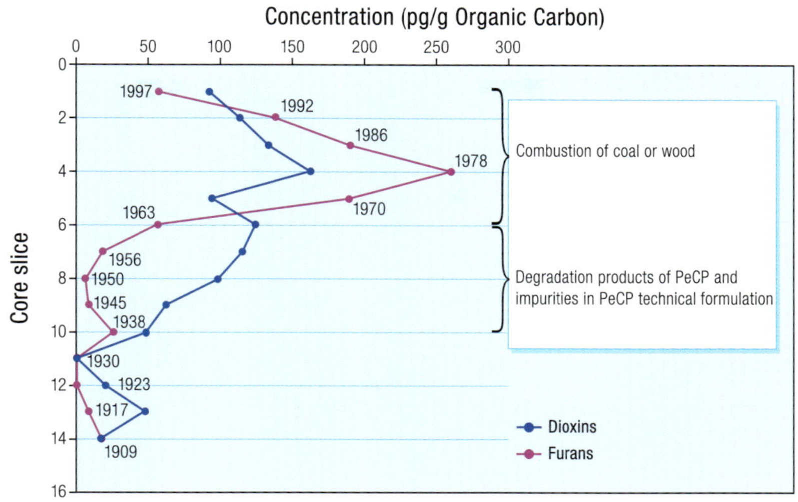

FIGURE B.3.12

Concentration profiles of ΣPCDDs and ΣPCDFs in a dated core from Lake DV09 on Devon Island, Nunavut, May, 1999.

B.3.3.3 Northwest Territories (Great Slave Lake)

Great Slave Lake is one of the most important aquatic ecosystems in northern Canada. In both surface area (27 000 km^2) and volume (2088 km^3), Great Slave Lake is the fourth largest lake in Canada. The Slave River, which receives most of its water from the Peace and Athabasca Rivers, is the major river entering Great Slave Lake and contributes some 87% of the total annual inflow of 135 km^3 (Evans *et al.*, 1996). More than 50% of the population in the Northwest Territories reside along the Great Slave Lake shoreline. Yellowknife, the territorial capital, has a population of ca. 14 000, while Hay River, an important shipping and commercial fishing centre, has a population of ca. 3,000. A commercial fishery, whitefish being the major annual harvest, has been operating on Great Slave Lake since 1945.

Great Slave Lake, because of its remote location, is a relatively pristine ecosystem with low concentrations of organic contaminants in all but the most industrialized areas (Evans *et al.*, 1996). However, given the concern regarding increased development of the forest and oil industries in the Peace and Athabasca drainage basins, and the subsequent possibility of transport of contaminants such as dioxins/furans and PAHs downstream via the Peace and Athabasca Rivers to Great Slave Lake, a study was conducted to try and determine the influence of the Slave River on contaminant loading to Great Slave Lake and to obtain baseline organic contaminant concentration data in the lake sediment.

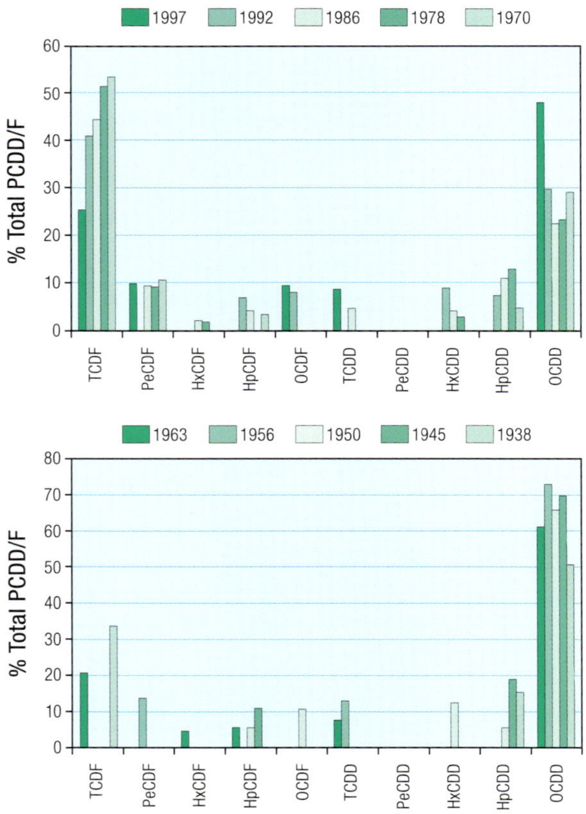

FIGURE B.3.13

PCDD/F mixtures in Lake DV09, Devon Island, sediment core slices.

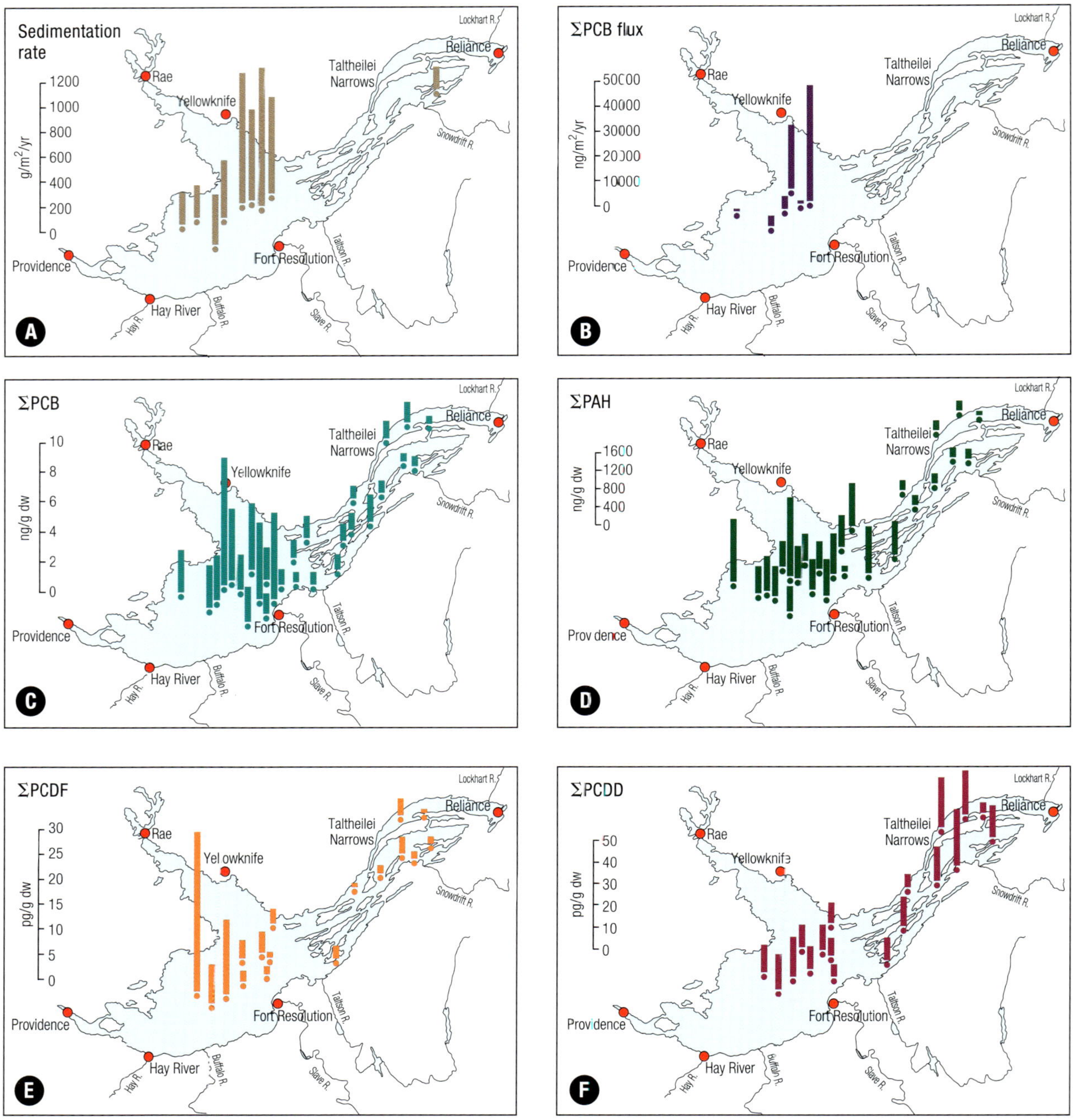

FIGURE B.3.14

Great Slave Lake, 1995; a) sedimentation rate, $g/m^2 \cdot y^{-1}$, b) total PCB flux rate ($ng/m^2 \cdot y^{-1}$) uncorrected for sediment movement, c) total PCB concentration (ng/g, dw), d) total PAH concentration (ng/g, dw), e) total PCDF concentration (pg/g, dw), f) total PCDD concentration (pg/g, dw)

A series of sediment studies were conducted on Great Slave Lake in the mid-1990s, investigating various aspects of these issues. Sediment coring studies, conducted mainly in the West Basin, determined that sedimentation rates were highest nearest the Slave River mouth and decreased with increasing distance from the mouth (Figure B.3.14a). Rates, adjusted for focusing, revealed even more clearly the influence of the Slave River on sediment (and contaminant) loading to the lake. Sedimentation rates were very low at the western most sampled station and roughly comparable to that in the East Arm. Most sediments (and associated contaminants), therefore, appear to be deposited in the deeper regions immediately offshore of the river mouth.

Contaminant concentrations were determined in surface sediments collected in various regions of the lake (Figure B.3.14b-e). Concentrations of PCBs, PAHs, and PCDFs were highest in sediments offshore of the river mouth and decreased with increasing distance. This, combined with the higher sedimentation rates in this region, indicate that contaminant fluxes of these compounds to the sediments were substantially greater offshore of the river mouth than with increasing distance away (Figure B.3.14f). The converse pattern was observed for PCDDs, which tended to occur in high concentrations in the East Arm.

A series of sediment cores was analyzed for contaminant temporal trends. The core analyzed closest to the Slave River mouth and in a region of moderate sediment mobility (Evans *et al.*, 1996) showed increased trends in PCB concentrations with time, with no evidence of a decline even in the surface slice (Figure B.3.15). PAHs also showed a somewhat similar pattern with most of the increase associated with lightweight PAHs of probable petrogenic origin. Concentrations of PCDDs were higher in the late 1950s with a trend of decreasing concentration thereafter. These trends were associated primarily with higher molecular weight PCDDs, possibly reflecting the usage patterns of PeCP. No trend was evident for PCDFs, which occurred in substantially lower concentrations than PCDDs. Overall, these data show the West Basin of Great Slave Lake is a depositional sink for contaminants entering the lake with Slave River inflow (and from the atmosphere) and that relatively small amounts of contaminants associated with sediments are transported out of the lake into the Mackenzie River.

Pat Roach

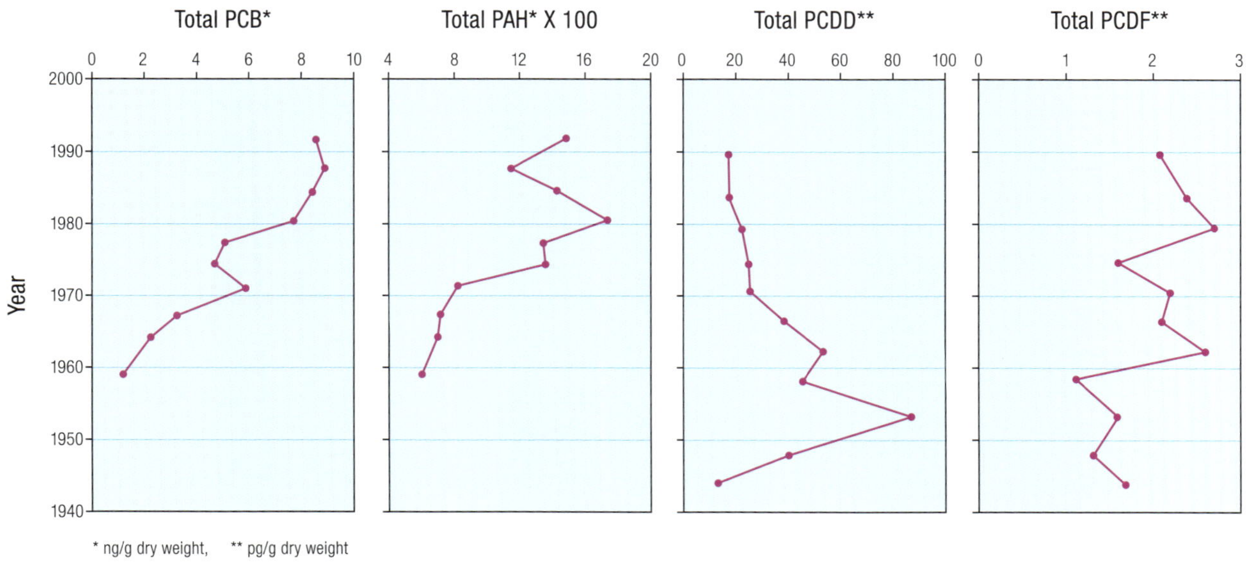

FIGURE B.3.15

Time trends for selected organic contaminants, Great Slave Lake. From Evans *et al.* (1996).

B.4 New persistent chemicals in Arctic air and water

Authors:

M. Alaee (1); D. Muir (1); C. Cannon (1); P. Helm (2); T. Harner (3); and T. Bidleman (4).

Affiliations:

1. Aquatic Ecosystem Protection Research Branch, National Water Research Institute, Canada Centre for Inland Waters, 867 Lakeshore Road, Burlington, ON, L7R 4A6, Canada;
2. Freshwater Institute, Fisheries and Oceans Canada, 501 University Crescent, Winnipeg, MB, R3T 2N6, Canada;
3. Meteorological Service of Canada, 4905 Dufferin Street, Downsview, ON, M3H 5T4, Canada;
4. Centre for Atmospheric Research Experiments, 6248 Eighth Line, R.R. #1, Essa Township, Egbert, ON, L0L 1N0, Canada.

B.4.1 Introduction

This section focuses on "new" chemicals in the Arctic abiotic environment — new because they were not being monitored in the NCP during Phase I, and/or new in the sense of being currently used in Canada, the United States and Europe (e.g., pesticides, BFRs, etc.). From a national and international regulatory perspective, finding new chemicals in the Arctic implies that these compounds are sufficiently persistent in the atmosphere and seawater to be transported long distances from their sources. Both the Canadian Toxic Substances Management Policy and the United Nations Economic Commission for Europe (UN-ECE) draft protocols on POPs include the presence of chemicals in remote environments as evidence of persistence. If these compounds are also found in Arctic food webs, there is the added threat that they are bioaccumulative and are contaminants of human food.

Prior to current studies, reviews of data on contaminants in the Arctic environment found that information was limited or non-existent on levels of certain industrial compounds: polychlorinated naphthalenes (PCNs), coplanar PCBs, polybrominated diphenyl ethers (PBDEs) and chloroparaffins (short-chain chlorinated paraffins, SCCPs) (DeMarch *et al.*, 1998; Muir *et al.*, 1999a). These contaminants have been detected in biota in Svalbard and northern and southern Sweden (Falandysz and Rappe, 1996; Jansson *et al.*, 1993; Järnberg *et al.*, 1993; 1997; Sellström *et al.*, 1993). PCNs have been reported in freshwater biota in the Great Lakes region of North America (Kannan *et al.*, 2000; 2001), and in beluga and ringed seal from the eastern Canadian Arctic (Helm *et al.*, 2002a). They have also been found in human adipose tissue from Ontario municipalities (Williams *et al.*, 1993), in children from Germany and Russia (Witt and Niessen, 2000), and in the breast milk of Swedish women (Lundén and Norén, 1998), demonstrating their ability to be transferred to people. Of the chloroparaffins, the short-chain C_{10}-C_{30} compounds (SCCPs) have the greatest potential for environmental release (EC, 1993) and exhibit the highest toxicity (Willis *et al.*, 1994). Researchers in Sweden have found that the C_{10}-C_{13} SCCPs inhibit intercellular communication in rat liver epithelial cells, a phenomenon which suggests that these chemicals may be acting as tumour promoters (Kato and Kenne, 1996). In Canada, SCCPs are classified as priority toxic substances.

In addition, several currently used pesticides have the potential to be transported to the Arctic, where their persistence is expected to be longer than in temperate regions (Barrie *et al.*, 1997). Evidence of this is provided by identification of several modern agrochemicals in the air, fogwater, and seawater of the Bering and Chukchi seas (Chernyak *et al.*, 1996).

This section presents new information on the occurrence of some industrial chemicals (PCNs, SCCPs, coplanar PCBs, and PBDEs) and modern agrochemicals in Arctic air and water. Results for SCCPs in sediments are presented in Section B.3.

B.4.2 Polychlorinated naphthalenes and coplanar PCBs

B.4.2.1 Usage and sources

Polychlorinated naphthalenes (PCNs) are a group of 75 compounds which have been manufactured as complex technical mixtures. They were used in similar applications to polychlorinated biphenyls (PCBs) and include dielectrics for flame resistance and insulation in capacitors, transformers, cables, and wires. Polychlorinated naphthalenes were also use as lubricants, fungicides and insecticides, and as binding agents and plasticizers (Falandysz, 1998; Jakobsson and Asplund, 1999). Total global production has been estimated to be 150 000 t (Falandysz, 1998). Polychlorinated naphthalenes are also found in PCB formulations (Yamashita *et al.*, 2000) and are produced during combustion processes such as municipal solid waste incineration (Oehme *et al.*, 1987; Abad *et al.*, 1999).

B.4.2.2 Physicochemical and toxicological properties

Polychlorinated naphthalenes have vapour pressures ranging from 6.83×10^{-5} to 5.6 Pa (Lei *et al.*, 1999), and octanol-air partition coefficients (K_{OA}), which range from 1.43×10^{7} to 4.2×10^{10} (Harner and Bidleman, 1998a), that are similar to PCBs. This suggests that PCNs are subject to similar atmospheric transport processes as PCBs and are likely to be found in remote locations. Log K_{OW} values for PCNs range from 4.0–8.5 for mono- to octachloronaphthalenes (monoCNs — octaCNs; Lei *et al.*, 2000), similar to PCBs. The pentaCN and hexaCNs (log K_{OW} = 6.8–7.7) are the most bioaccumulative with bioaccumulation factors up to 23 in herring and > 50 in white-tailed sea eagles were found for some congeners in the Baltic Sea region (Falandysz, 1998).

Polychlorinated naphthalenes exhibit toxic activity similar to 2,3,7,8-tetrachlorodibenzo-*p*-dioxin (TCDD) and planar (mono- and non-*ortho*-substituted) PCBs through an Ah-receptor mediated mechanism. Toxic equivalency factors (TEFs) or relative potencies (REPs) have been determined for several PCN congeners using H4IIE-ethoxyresorufin-*O*-deethylase (EROD) and H4IIE-luciferase induction assays (Hanberg *et al.*, 1990; 1991; Blankenship *et al.*, 2000; Villeneuve *et al.*, 2000). A number of the PCN congeners, particularly hexaCNs and heptaCNs have TEFs similar in magnitude or greater than non- and mono-*ortho* PCBs (Giesy *et al.*, 1997).

B.4.2.3 Occurrence in the environment

Polychlorinated naphthalenes are widespread in the physical environment. In addition to the occurrences in biota and humans previously mentioned, they have been found in sediments and source-related samples in Sweden (Jansson *et al.*, 1993; Järnberg *et al.*, 1997), the Baltic Sea food web (Falandysz and Rappe, 1996; Falandysz *et al.*, 1997; Falandysz, 1998; Ishaq *et al.*, 2000) and in sediments in a UK lake (Gevao *et al.*, 2000). Measurements of PCNs in ambient air have been reported from Germany (Dörr *et al.*, 1996), the UK (Harner *et al.*, 2000b; Lee *et al*, 2000), Great Lakes municipalities and over the Great Lakes (Harner and Bidleman, 1997; Helm *et al.*, 2000a) and in southern and northern Sweden (Wideqvist, 1999).

B.4.2.4 PCNs and coplanar PCBs in Arctic air

Polychlorinated naphthalenes and coplanar PCBs were first measured in Arctic air by Harner *et al.* (1998) in 1996 samples collected from on board ship in the Barents Sea and eastern Arctic Ocean, and in a few archived 1993–94 air sample extracts from Alert and Dunai. Total PCNs were highest in the Barents Sea, followed by the eastern Arctic Ocean, Alert, and Dunai (Table B.4.1). The congener profile followed the order: triCNs > tetraCNs > pentaCNs > hexaCNs. Lower congeners were not determined. Coplanar PCBs ΣCoPCBs, sum of mono-ortho congeners 105, 114, 118 and 156, and non-ortho congeners 77, 81, 126 and 169) had a similar spatial distribution, except that higher values were found at Dunai than Alert (Table B.4.1). The especially high concentrations of ΣPCNs and ΣCoPCBs over the Barents Sea were linked to parcels of air arriving from the UK — central Europe (Harner et al., 1998). Subsequent measurements showed elevated PCNs in UK ambient air (Harner *et al.,* 2000b).

Further investigations were carried out using archived 1994–95 air sample extracts from Alert, Dunai and Tagish (Helm and Bidleman, 2002b). Annual mean ΣPCN concentrations were highest at Dunai, followed by Alert and Tagish (Table B.4.1). Mean ΣCoPCBs at Dunai were more than four times those at Alert and Tagish, and the differences were significant at $p < 0.002$ (Table B.4.1). A seasonal trend for ΣPCNs was observed with higher concentrations occurring during the winter/spring months (Figure B.4.1). This is similar to the seasonality observed for polycyclic aromatic hydrocarbons (PAHs) (Halsall *et al.*, 1997), but quite different from the ΣCoPCB trend (see following).

TABLE B.4.1 Mean concentrations of ΣPCNs and ΣCoPCBs in Arctic air, pg/m³

Location	Dates	ΣPCN	ΣCoPCB[1]	Reference
Alert	1993–94 1994–95	2.5[2] 0.50	0.46 0.084	Harner *et al.*, 1998 Helm *et al.*, 2002 submitted
Dunai	1993 1994–95	0.84 0.65	1.9 0.44	Harner *et al.*, 1998 Helm *et al.*, 2002 submitted
Tagish	1994–95	0.33	0.04	Helm *et al.*, 2002 submitted
Barents Sea	1996	40.4	6.2	Harner *et al.*, 1998
Eastern Arctic Ocean	1996	11.6	0.98	Harner *et al.*, 1998
Norwegian Sea	1996	7.1	not determined	Harner *et al.*, 1998

(1) Sum of mono-ortho congeners 105, 114, 118, and 156, and non-ortho congeners 77, 81, 126, and 169, 81 not included in the Harner *et al.* (1998) samples.
(2) Mean after omitting one high sample is 1.4 pg/m³ ΣPCNs.

Total CoPCBs were dominated by pentaCBs 118 and 105, and contained lower proportions of the other mono- and non-*ortho* congeners (Table B.4.1). A seasonal variation was seen at Dunai with higher concentrations during the spring-summer months (Figure B.4.1); no trend was evident at Alert or Tagish. This agrees with the findings of Stern *et al.*, (1997) that the contribution of pentaCB congeners to ΣPCBs increased in summer, particularly at Dunai. The authors suggested that emissions of Russian PCB mixtures with higher percentages of pentaCBs could account for this observation.

Sources of PCNs in Arctic air include evaporation of PCN commercial mixtures, concomitant release of PCNs with evaporation of PCB fluids, and combustion. Evaporation of Russian PCBs might explain the higher ΣPCNs (as well as ΣCoPCBs) in air at Dunai. A comparative analysis of PCNs in commercial PCB mixtures showed the highest levels to be in Russian Sovols (Yamashita *et al.*, 2000). The higher PCN concentrations in the cold months, similar to PAHs, however, suggests a relationship to Arctic haze and combustion. Evidence of combustion sources is indicated by the presence of several PCN congeners known to be associated with combustion: CNs 13, 26, 44, 29, and 54 (Helm and Bidleman, 2002b). It is unknown if PCNs are lost from the atmosphere during summer through photochemical reaction. Certain congeners have been shown to be removed from methanol-water solutions under solar irradiation (Järnberg *et al.*, 2001).

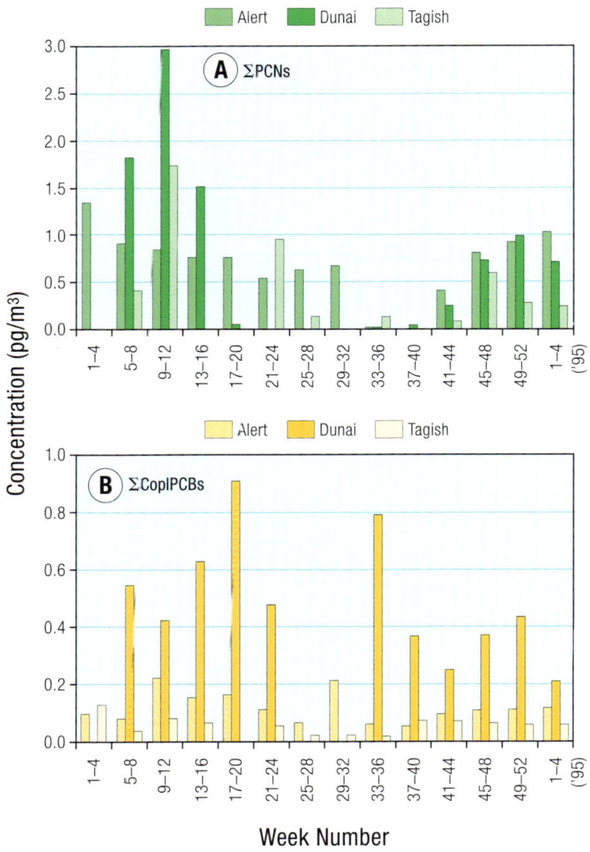

FIGURE B.4.1

Monthly variations in ΣPCNs (top) and ΣCoPCBs (bottom) in air (PUFs + filters) at Alert, Dunai and Tagish from January, 1994–January, 1995 (pg/m^3). Archived extracts of weekly samples were combined into 28-day composites.

Toxic equivalents (TEQs) for PCNs and CoPCBs at the three Arctic stations were estimated (Helm and Bidleman., 2002b) using the TEF values derived from H4IIE induction assays. Average TEQs due to PCNs + CoPCBs ranged from 0.003 fg/m^3 in the warm season to 0.061 fg/m^3 in the cold season, with the highest values observed at Dunai during both periods. In winter, 71–75% of the PCN + CoPCB TEQ was due to PCNs at Alert and Dunai, while only 30% was due to PCNs at Tagish. In summer, CoPCBs accounted for 66–98% of PCN + CoPCB TEQs at all sites, with CB 126 being the major contributor.

A recent report of polychlorinated dibenzodioxins and dibenzofurans (PCDD/Fs) in Arctic air for the 2000-01 winter (Hung *et al.*, 2002b) allowed these compounds to be included in a total TEQ assessment (Helm and Bidleman., 2002b). Using H4IIE TEFs for PCDD/Fs (Giesy *et al.*, 1997), the average concentration on a TEQ basis was 0.8 fg/m^3. Although not directly comparable since six years separate the measurements, the PCN + CoPCB TEQ estimated for 1994–95 was only 2.5–5% of the PCDD/F TEQ. Thus, PCDD/Fs appear to dominate TEQ contributions in air.

B.4.3 Polybrominated diphenyl ethers (PBDEs) and other brominated flame retardants

B.4.3.1 Usage and sources

Fire has been a major cause of property damage and death throughout recorded history and to the present day. In fact, between 1986 and 1995 in Canada, 67 000 fires were reported each year, resulting in a total of 3700 injuries, 465 deaths, and direct damage to properties over $1.125 billion (Council of Fire Marshals and Fire Commissioners, 1997). Similar trends have been observed in other industrialized countries. During the past several decades, modern technology has responded to this challenge by introducing heat resistant chemicals to reduce the chances of ignition and burning of a wide range of textiles, plastics, building materials, and electronic equipment used in commerce and in residential homes. By some estimates, the use of flame-retardant chemicals has saved many lives annually and hundreds of millions of dollars in property damage (Spiegelstein, 2000). By 2000, flame-retardant chemicals had become the second largest additive used by the plastics industry, resulting, in part, in a market value for flame-retardant chemicals estimated at nearly US $2.2 billion (Tullo, 2000).

Modern flame retardants comprise 175 different chemicals, divided into four major groups: inorganic, halogenated organic, organophosphorus, and nitrogen-based compounds and mixtures (EHC, 1997). Halogenated organic flame retardants are generally classified as either chlorinated or brominated flame retardants (BFRs). Brominated flame retardants are further classified as either reactive or additive materials. The reactive BFRs, which include compounds such as the tetrabromobisphenol A (TBBPA) and derivatives, are chemically bonded into plastics (EHC, 1995). The additive BFRs, which include the polybrominated diphenyl ethers (PBDEs) and hexabromocyclododecane (HBCDD), are used as additives in a wide variety of polymers and resins (EHC, 1997). The additive BFRs generally are believed to be more easily released to the environment than the reactive BFRs (Hutzinger *et al.*, 1976; Hutzinger and Thoma, 1987). The annual global production of PBDE in 1999 was 67,000 t, which was the second largest amount of BFRs (after TBBPA) currently in use (BSEF, 2000).

B.4.3.2 Physicochemical and toxicological properties

Polybrominated diphenyl ethers (PBDEs), which are similar to PCBs, are made up of 209 congeners, with the same substitution pattern; therefore, the same IUPAC numbering convention is used for PBDEs. Vapour pressures of PBDEs range between 0.1 Pa for mono-BDE to 10^{-6} Pa for hepta-BDE (Wong *et al.*, 2001); log K_{OW} range between 5.9 for tetra-BDEs and 10 for deca-BDE (Watanabe and Tatsukawa, 1990); and log K_{OA} ranges between 9 for tri-BDE and 12 for hepta-BDE (Harner, 2001). Similar to coplanar PCBs, the PBDEs which lack ortho-bromines have lower vapour pressures than others in the same homologue group (Wong *et al.*, 2001).

Polybrominated diphenyl ethers have a weaker affinity for the Ah receptor compared to tetrachlorodibenzo-*p*-dioxin (TCDD), which indicates low dioxin-like activity, and only limited numbers of BDE congeners induce EROD activity (Chen *et al.*, 2001a). Using ER-CALUX, Meerts *et al.* (2001) showed that selected hydroxylated PBDEs exhibit estrogenic activities and a number of PBDE metabolites were potent competitors for *trans*-thyretin. Ericksson *et al.*, (1998) showed neurotoxicological effects in mice over several months after the administration of a single dose of BDE-47 or BDE-99.

B.4.3.3 Occurrence in the environment

The extensive use of products containing PBDEs has resulted in the release of these compounds into the environment. Polybrominated diphenyl ethers are ubiquitous environmental contaminants and have been detected in all environmental compartments including human blood (Sjödin, 2000), and breast milk (Ryan and Patry, 2000). These are lipophilic compounds and are shown to bioaccumulate through the food web (Sellström *et al.*, 1993). In 1979, the presence of BDE-209 (deca-BDE) was measured in soil and sludge samples collected from areas surrounding PBDE manufacturing facilities in the US (DeCarlo, 1979). Two years later, Anderson and Blomkvist (1981) reported the presence of PBDEs in samples collected along the Visken River in Sweden. Jansson *et al.* (1987) first suggested that PBDEs were global contaminants by demonstrating their presence in tissue samples of fish-eating birds and marine mammals collected from the Baltic Sea, North Sea and Arctic Ocean. Similar reports confirmed the widespread distribution of PBDE congeners in marine fish, shellfish, and sediments collected in the Pacific region and elsewhere (Watanabe *et al.*, 1987). Polybrominated diphenyl ethers were also reported in cod liver and herring from the North Sea (de Boer, 1989), and in fresh water eels in the Netherlands (de Boer, 1990). Stafford (1983) confirmed the presence of PBDEs in North America, reporting elevated concentrations of several congeners in the eggs and tissues of fish-eating birds from six US states and Ontario, Canada. Watanabe *et al.* (1992) were among the first to suggest a global long-range transport process for PBDEs based, in part, on studies of air particulate samples collected from Japan and Taiwan.

B.4.3.4 Polybrominated diphenyl ethers in Arctic air

Archived air samples (116 PUFs, filters, and blanks) from the Arctic air monitoring stations Tagish, Alert, and Dunai were analysed for PBDEs. These samples represented composites of four weekly samples (i.e. weeks 1 to 4, 4 to 8 etc.) from January to December 1994 and January 1995. Polybrominated diphenyl ethers were determined in air samples with high-resolution electron ionization mass spectrometry using the method described by Alaee *et al.* (2001), and Luross *et al.* (2002).

TABLE B.4.2 Concentrations of selected PBDE congeners in monthly composite atmospheric samples from the Arctic (all values are in pg/m³)

	Week #	BDE-47	BDE-99	BDE-100	BDE-153	BDE-154
Tagish	W 1–4	5.5	13	3.4	2.6	2.0
	W 5–8	8.9	58	17	16	12
	W 9–12	36	51	11	8.9	6.0
	W 13–16	65	73	15	7.7	4.8
	W 17–20	40	33	6.8	2.5	2.1
	W 21–24	81	78	15	5.5	4.7
	W 25–28	440	780	120	60	45
	W 29–32	540	1200	170	140	77
	W 33–36	160	330	47	49	22
	W 37–40	79	210	29	36	17
	W 41–44	57	170	20	29	12
	W 45–48	21	87	9.8	14	6.7
	W 49–52	18	79	9.0	19	7.5
	W 1–4	21	87	10	18	7.9
Alert	W 1–4	5.2	21	4.7	3.6	2.5
	W 5–8	23	89	18	14	11
	W 9–12	15	81	18	19	12
	W 13–16	20	99	12	6.4	5.1
	W 17–20	12	110	10	34	16
	W 21–24	160	420	54	59	33
	W 25–28	65	150	25	31	12
	W 29–32	250	270	63	15	14
	W 33–36	52	120	24	9.9	8.4
	W 37–40	6.8	20	3.7	3.1	1.9
	W 41–44	5.5	11	2.0	1.5	1.0
	W 45–48	180	260	38	28	17
	W 49–52	97	140	25	12	7.3
	W 1–4	3.4	5.3	0.85	0.53	0.31
Dunai	W 1–4					
	W 5–8	2.6	7.1	1.7	1.9	
	W 9–12	1.7	3.2	0.72	0.25	
	W 13–16	3.4	4.6	1.0	0.33	
	W 17–20	2.8	2.9	0.94	0.27	
	W 21–24	3.8	3.1	1.1	0.26	
	W 25–28					
	W 29–32					
	W 33–36	9.2	34	11	7.4	
	W 37–40	5.2	12	3.4	2.5	
	W 41–44	5.2	19	5.8	4.4	
	W 45–48	2.2	7.3	2.4	1.7	
	W 49–52	2.3	4.7	1.3	0.66	
	W 1–4	2.2	4.5	0.96	0.57	

Concentrations of ΣPBDEs in air samples from Alert, Dunai and Tagish are presented in Figure B.4.2, and levels for selected BDE congeners are reported in Table B.4.2. Concentrations of PBDEs in air samples from Dunai were the lowest among the three sites, and those from Alert were lower than concentrations in air samples from Tagish. Relatively high levels of mono- to hepta-BDEs were detected at Tagish with maximum concentrations occurring during the summer months. The much higher air temperatures at Tagish compared to Alert and Dunai combined with closer proximity to populated areas in southern Yukon may be the reason for the high levels although local sources cannot be ruled out. Field blanks showed only very low levels of the PBDEs.

The main BDE congeners observed in these samples were BDE 47, 99, 100, 153, and 154. As shown in Table B.4.2, BDE 99 has the highest concentration followed by BDE 47, 100, 153, and 154. The pattern in these samples resembles the commercial products, rather than the pattern reported by Bergander *et al.* (1995) from two sampling sites in Sweden and by Strandberg *et al.* (2001) in samples from the Great Lakes area. An interesting feature in the samples collected at Alert is the presence of mono-, di- and triBDE congeners, which is an indication of photodegradation of PBDEs during long-range atmospheric transport. It is important to note that field blanks showed varying levels of PBDEs. BDE congeners 47 (2,4,2',4'-tetra) and 99 (2,4,2',4',5'-penta) were the major congeners in both PUFs and filters. To resolve a number of issues additional samples will be analyzed in future. In future, additional samples, including blanks, will be analyzed to provide more accurate measurements of PBDEs and to gain a better understanding of their sources in samples collected at Tagish.

The distributions of four PBDEs between the particle and gas phases are presented in Figure B.4.3. The environmental fate of an organic chemical is strongly influenced by its partitioning characteristics. Atmospheric behaviour in particular is determined by the distribution between gas and aerosol phase (Bidleman, 1988). Vapour-particle partitioning for selected PBDE congeners in the environmental temperature range can be estimated using Junge-Pankow relationship. As shown in Figure B.4.3, BDE 47 is predominately in the vapour phase during summer and partitioning to the aerosol increases with drops in the temperature. Significantly higher proportions of BDE 99 and 100 are bound to the particulate phase than BDE 47. BDE 153 is predominately bound to the particulate phase during winter.

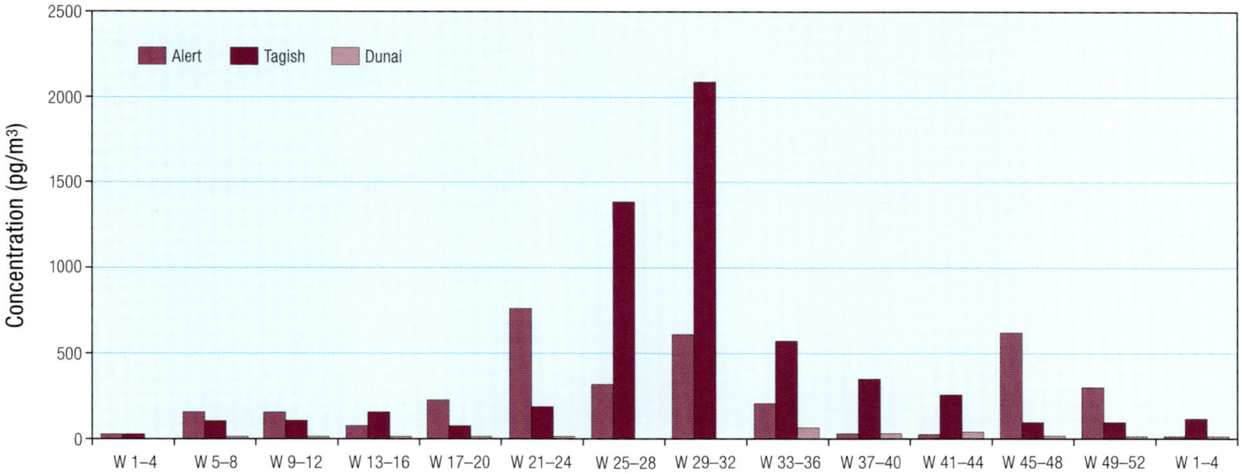

FIGURE B.4.2

Total mono- to heptabromodiphenyl ethers in air (PUFs + filters) in 28-day composites of archived air sample extracts from Tagish, Alert and Dunai 1994/95 (pg/m^3).

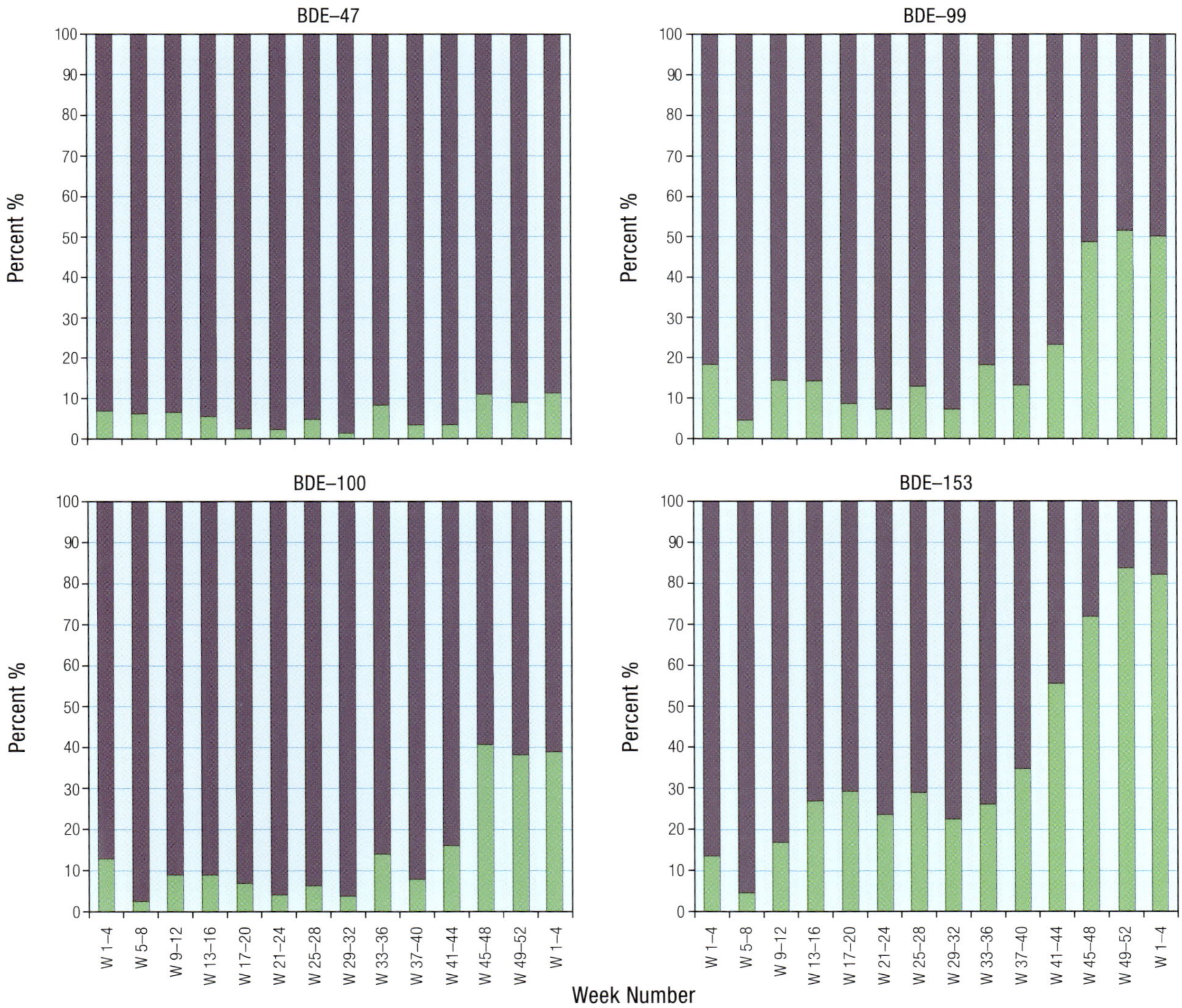

FIGURE B.4.3

Distribution of selected PBDE congeners between the vapour (purple) and particle phases (green) for 28-day composite air samples collected at Tagish.

B.4.3.5 Other brominated flame retardants (BFRs) in Arctic air

In order to investigate occurrence of other BFRs in archived air samples, four samples each from Alert, Tagish and Dunai were analyzed for hexabromocyclododecane (HBCDD), tetrabromo-bisphenol A (TBBPA), and polybrominated biphenyls (PBBs) (Bidleman *et al.*, 2000). Preliminary results are listed in Table B.4.3. One filter sample from Alert (summer) and one from Tagish (winter) appeared to contain some PBB congeners, but levels were very low and further concentration of the subsamples will be required for further analysis. No HBCDD was detected in any of the samples analyzed. TBBPA was detected on the filter of one spring sample from Dunai at a concentration of 70 pg/m^3. Decabromodiphenyl ether was not present on filters at detection limits of about 1 pg/m^3.

B.4.4 Short-chain chlorinated paraffins (SCCPs)

The use, properties, and environmental distribution of SCCPs are discussed in Section B.3, along with measurements of these chemicals in Arctic lake sediments. Short-chain chlorinated paraffins were also determined in 1994–95 archived air samples from Alert, Tagish and Dunai (the same samples used for the analysis of PCNs and BFRs, see previous sections); results are presented in Table B.4.4.

TABLE B.4.3 Preliminary results of analysis of Arctic air sample filters[1] for several brominated flame retardants including decabromodiphenyl ether

Compound	Alert pg/m^3	Tagish pg/m^3	Dunai pg/m^3
Hexabromocyclododecane (HBCDD)	< 1.8	< 1.8	< 1.8
Tetrabromo-bisphenol A (TBBPA)	< 0.07	< 0.07	< 0.07 – 70
Polybrominated biphenyls (PBBs)	< 0.03 – ~0.1	< 0.03 – ~0.1	< 0.03
Decabromodiphenyl ether	< 0.1	< 0.1	< 0.1

[1]Four extracts representing four week composites of filters were analyzed from each location.

B.4.5 Modern pesticides, other biocides, and by-products

B.4.5.1 Occurrence in Arctic air

Halsall *et al.* (1998) reported the presence of the current-use pesticides endosulfan, methoxychlor, and trifluralin, as well as the pentachlorophenol metabolite, pentachloroanisole (PeCA) in air at Alert, Tagish and Dunai during 1993–94 (Table B.4.4), and Hung *et al.* (2002a) reported these chemicals for Alert in 1995–97. Pentachloroanisoles and endosulfan were among the top five biocide-related compounds at all three sites, exceeded only by hexachlorobenzene (HCB) and α-HCH and γ-HCH (itself a current-use pesticide; see Sections A.2.2.3 to A.2.2.7).

TABLE B.4.4 Novel pesticides and other chlorinated semivolatile chemicals in Arctic air (gas phase) during 1993–94,[1] pg/m^3 (Halsall *et al.*, 1998; Bidleman *et al.*, 1999; 2000; 2001)

	Alert			Tagish			Dunai		
Compound	**mean**	**min**	**max**	**mean**	**min**	**max**	**mean**	**min**	**max**
Current use pesticides									
Endosulfan	4.22	0.02	16.2	7.05	0.08	88 6	2.99	0.05	7.18
Methoxychlor	0.27	0.07	1.43	0.36	0.09	6.53	0.41	0.22	0.73
Trifluralin	0.12	0.03	0.64	0.16	0.04	2.92	0.18	0.09	0.13
Byproducts and industrial chemicals									
Pentachloroanisole	3.03	0.1	20.5	3.28	0.04	73.4	2.92	0.95	6.92
Trichloroveratrole	0.93	0.05	10.1	1.85	0.07	20.6	1.49	0.15	3.65
Tetrachloroveratrole	0.19	0.05	0.99	0.25	0.07	4 52	0.44	0.15	2.19
Octachlorostyrene	0.79	0.04	2.96	0.67	0.07	11.9	0.60	0.11	1.74
Chlorinated paraffins (C_{10}-C_{13})	2.0	< 0.4	7.3	–	–	–	–	–	–

[1]Based on retrospective analysis of 28-day composite samples from 1994 and January 1995 only.

Hung *et al.* (2002a) identified an increasing trend in α-endosulfan at Alert from 1993–97, opposite of most OC pesticides which exhibited stable or declining concentrations.

More than 5 x 10^6 kg of trifluralin were applied annually in western Canada and the United States in the mid-1990s for weed control in soybean, cotton and cereal crops. It has been well documented that trifluralin volatilizes from agricultural soils, but it has a very short atmospheric half-life, estimated at about 21–74 minutes (Atkinson *et al.*, 1994) making its appearance in Arctic air at all three monitoring stations surprising. Concentrations at Alert in 1994 (Halsall *et al.*, 1998) were more than 1000 times lower than observed over the Mississippi River by Majewski *et al.* (1998) (http://ca.water.usgs.gov/pnsp/use92/triflrln.html). The high volume of trifluralin use, results in detectable levels of this pesticide, and possibility others such as methoxychlor, even though removal processes in the atmosphere are quite rapid. The presence of similar levels of trifluralin at Dunai Island is more difficult to explain. Welch *et al.* (1991) also detected trifluralin in a "brown snow" event that occurred in the Chesterfield Inlet area in the late 1980s. Trajectories and minerology indicated an Asian source for the particles in the snow, although the trifluralin may have been transported from the central prairie region of North America and deposited in a closely following rain and sleet event. Calculations based solely on photochemical degradation indicate that trifluralin should not reach the Arctic at measurable quantities. Current results indicate that other pathways such as transport on dry particulate or aerosol might be the primary mode of transport for trifluralin.

Calculations based in purely gas-phase reactions with OH radicals, present at average tropospheric concentrations, conclude that several conventional OC pesticides also have short atmospheric lifetimes: chlordane = 8 days, DDTs = 2–4 days, and dieldrin = 1 day (Atkinson, 1990). Nevertheless, there is no doubt that these chemicals are present in Arctic air. Such discrepancies need to be resolved, through continued modelling and research.

Methoxychlor had limited agricultural use on fruit crops (http://ca.water.usgs.gov/pnsp/ use92/mthoxyclr.html) as well as in home gardening products. Like other methoxylated aromatic compounds methoxychlor may have a relatively long atmospheric half-life.

Two other methylated chlorophenolics, trichloroveratrole and tetrachloroveratrole, were reported by Halsall *et al.* (1998) in air from all three sampling locations (Table B.4.4) and by Hung *et al.* (2002a) at Alert in 1995–97 at sub-pg/m^3 concentrations. These compounds may originate from the bleaching of wood pulp and chlorination of wastewaters; however, their widespread detection at all locations suggests other unknown sources. Führer *et al.* (1996) have reported a series of chlorinated anisoles and tetrachlorodimethoxybenzene in air from waste treatment plants and in marine air over the tropical Atlantic Ocean. Schreitmüller and Ballschmiter (1995) measured tetrachlorodimethoxybenzene and PeCA over the north and south Atlantic Ocean, suggesting that these chlorinated methoxy-aromatics are widespread in the troposphere.

Octachlorostyrene (OCS), a by-product of magnesium and chlorine manufacturing, was detected at the three Arctic stations studied by Halsall *et al.* (1998) and by Hung *et al.* (2002a) at Alert during 1995–97. While mean concentrations of OCS were low compared to many OC pesticides, maximum concentrations were up to 15 times higher, especially at Tagish (Table B.4.4), illustrating that long-range transport events from source regions are contributing to the presence of OCS. Actual source regions have not been identified.

B.4.5.2 Occurrence in Arctic water

Chernyak *et al.* (1996) reported modern agrochemicals (endosulfan, chlorpyrifos and atrazine) in surface water of the Bering and Chukchi seas in 1993. In NCP studies, a search was made for several currently used pesticides in the Arctic Ocean, including lindane (γ-HCH), endosulfan, and others (Bidleman *et al.*, 2001). Endosulfan and lindane were the only modern pesticides found, others were below detection limits.

Although lindane is still used in Canada and was deregistered recently in other circumpolar countries (see Section A.2.2.6 and Table A.2.2), it is likely that historical atmospheric and oceanic transport of technical HCH, as well as lindane, from source regions in Eurasia are largely responsible for present-day concentrations in Arctic seawater (Section D.1). Levels of lindane are generally highest in the Canadian archipelago, and southern Beaufort Sea, intermediate in the Bering and Chukchi seas and the northern Canada Basin, and lowest in the eastern Arctic Ocean (Table C.3.2), over the Chukchi Plateau (Macdonald *et al.*, 1999b) and in the White Sea (Muir *et al.*, 2002). Lakaschus *et al.* (2000) found lindane in the North Sea (0.33 ng/L) within the range of concentrations observed in the Greenland Sea (0.072–0.50 ng/L). White Sea waters were characterized by higher γ-HCH levels than in the Barents Sea reflecting use in urban

areas within the watershed. In the White Sea, higher lindane concentrations were found in Kandalaksha Bay (0.19–0.41 ng/L) compared with the central basin (0.16–0.37 ng/L) (Muir *et al.*, 2002). However, in general, γ-HCH levels in the White Sea were lower than those in the southern Beaufort Sea. Alpha/Gamma HCH ratios averaged 1.8 ± 0.05 ng/L in the White Sea, compared with 2.2 ± 0.52 ng/L in the Greenland Sea, 3.4 ± 0.77 ng/L in the northern Barents Sea/central Arctic Ocean, and 2.3 ± 0.77 ng/L in the Beaufort Sea (Hoekstra *et al.*, 2002a; Macdonald *et al.*, unpublished).

Endosulfan is a widespread contaminant in Arctic seawater although present at much lower levels than lindane. Endosulfan (sum of α- and γ- isomers) concentrations in Arctic seawater ranged from < 0.0005 to 0.003 ng/L with no clear spatial trends. The average concentrations of endosulfan along the Tundra Northwest-99 cruise track in the eastern Canadian archipelago ranged from 0.004–0.01 ng/L, with an average of 0.0063 ng/L (Jantunen, unpublished) similar to concentrations found in the northern Canada Basin in 1994 (Jantunen and Bidleman, 1998). Hargrave *et al.* (1997) found endosulfan concentrations were higher during periods of open water in the central Canadian Archipelago, reflecting inputs from gas exchange and freshwater. During their 12-month study in 1993, Hargrave *et al.* (1997) found that endosulfan concentrations increased from 0.002–0.006 ng/L during the late summer and fall months paralleling increases in air concentrations. This was possibly related to summer use of endosulfan in mid-latitude North America.

B.5 Mercury in the Arctic atmosphere

Authors:

A. Steffen (1); W. Schroeder (1); L. Poissant (2); and R. Macdonald (3).

Contributors:

G. Hönninger (4); G. Lawson (5); and C. Banic (1).

Affiliations:

1. Meteorological Service of Canada, Environment Canada, 4905 Dufferin Street, Toronto, Ontario, M3H 5T4, Canada;
2. Meteorological Service of Canada, Environment Canada — Quebec region, 105 rue McGill, 7e étage (Youville) Montréal, Québec, H2Y 2E7, Canada;
3. Institute of Ocean Science, Department of Fisheries and Oceans, 9860 West Saanich Road, Sidney, BC, V8L 4B2, Canada;
4. Gerd Hönninger, Institut fuer Umweltphysik, Universitaet Heidelberg, Im Neuenheimer Feld 229 69120 Heidelberg, Germany;
5. Environment Canada, National Water Research Institute, 867 Lakeshore Rd., Burlington ON, L7R 4A6, Canada.

B.5.1 Introduction

Mercury (Hg) is a toxic metal that has been found at elevated concentrations in the tissues of aquatic animals from the Canadian Arctic (Wagemann *et al.*, 1995, 1997; Muir *et al.*, 1999a; Lockhart and Evans, 2000). Although humans and apex predators are exposed to Hg almost exclusively through the consumption of traditional/country food (Van Oostdam *et al.*, 1999), it is the physical and chemical processes in air and water that ultimately support that exposure. Atmospheric mercury, which is predominantly gaseous elemental mercury [Hg (0)], can undergo long-range transport because it has a global atmospheric residence time of about 6–24 months (Schroeder and Munthe, 1998; Lamborg *et al.*, 2002) allowing it to be carried by air currents far from emission sources to remote regions like the Arctic.

In aquatic environments, inorganic forms of mercury may be transformed (by biotic or abiotic pathways) to the more toxic methylated form (Mason *et al.*, 1994) and biomagnify (Atwell *et al.*, 1998). The major anthropogenic emissions of mercury to the atmosphere include fossil fuel consumption, waste incineration, chlor-alkali plants and metal smelting and processing (Pacyna and Keeler, 1995; Nriagu and Pacyna, 1988). These anthropogenic loadings interfere considerably with the natural biogeochemical Hg cycle. For example, Hg concentrations in global surface waters and the atmosphere are presently thought to be two to three times the natural background (Mason *et al.*, 1994; Fitzgerald *et al.*, 1998). Because Hg also occurs naturally in the environment, and because both natural and anthropogenic Hg are susceptible to atmospheric scavenging, deposition, and re-emission, it is difficult to ascertain the relative contributions of Hg from human activities and natural sources.

Several lines of evidence, including estimates of anthropogenic and natural emission strengths, proxy records from sediments, ice cores, and biological tissue, strongly imply post-industrial enhancement of the Hg cycle, and that Arctic aquatic environments may be particularly vulnerable to these inputs (Macdonald *et al.*, 2000). Anthropogenic emission estimates from Europe and North America have decreased from 1990 to 1995 but emissions in Asia have increased significantly (Pacyna and Pacyna, 2002) during the same time period. These sources are primarily from the combustion of fossil fuels and it is anticipated that the demands of this consumption will only increase (primarily in Asia) in the coming years. Although there is evidence that atmospheric mercury concentrations in Eastern European countries may have begun to decline during the 1990s (Slemr *et al.*, 1995), sediment cores from lakes in the Canadian Arctic do not as yet provide unequivocal evidence of such declines (Cheam *et al.*, 2001; Lockhart *et al.*, 1998) and marine sediments have not yet proven to be reliable recorders of Hg trends in the Arctic (Gobeil *et al.*, 1999).

Partly because the role of the atmosphere in transferring Hg to Arctic ecosystems was not understood and partly because there were no data from which to evaluate Arctic atmospheric trends for Hg, the Northern Contaminants Program (NCP) initiated studies to investigate how atmospheric chemical processes in the Arctic might act to enhance the capture of Hg from the global atmosphere.

B.5.2 NCP-Phase I and II programs

For the synthesis of the NCP-Phase I results [The Canadian Arctic Contaminants Assessment Report (CACAR); Jensen *et al.*, 1997], atmospheric mercury data in the Arctic were limited to a one-year time series of total gaseous mercury (TGM) measurements that were based on a weekly integrated sampling between August 1992 and 1993 at Alert, Nunavut (Figure B.5.1). These data provided the important baseline concentration against which future measurements could be compared to evaluate whether or not global values were increasing (Jensen *et al.*, 1997). The first assessment recommended that more research be undertaken to reduce the uncertainties in Hg budgets in the Arctic and to determine temporal trends of this pollutant in polar regions. At that time, lake sediment core data strongly implied increasing Hg fluxes in the Arctic but there were no atmospheric data to corroborate those findings. Further gaps in our knowledge of Hg were identified including: a lack of data on its spatial distribution in air and water; a poor understanding of its phase-partitioning between environmental media; almost no knowledge of exchanges between the atmosphere and snowpack; and no information on wet deposition.

FIGURE B.5.1
Locations of atmospheric mercury sampling – Alert, Nunavut, and Kuujjuarapik, Quebec.

Significant advances in understanding the behaviour of atmospheric Hg in the Arctic have been made during the past five years, partly due to a revolution in measurement technology (the development of the Canadian-made automated Hg Analyzer, Tekran Inc.), and partly due to the surprising discovery of atmospheric mercury depletion events (MDEs) following polar sunrise at Alert (Schroeder *et al.*, 1998; 1999a). This has sparked interest in the international community to conduct similar research in the United States, Norway, Greenland/Denmark, Russia and the Antarctic (Berg *et al.*, 2001; Lindberg *et al.*, 2001; Schroeder *et al.*, 2001b; Skov *et al.*, 2001; Ebinghaus *et al.*, 2002; Nielsdottir *et al.*, 2002).

As part of Phase II of the Northern Contaminants Program, atmospheric mercury vapour concentration measurements have now been made almost continuously on a half hourly averaged basis at Alert, NT since 1995 and on a 15-minute averaged basis at Kuujjuarapik, Quebec since 1999 (Figure B.5.2a). The main goals of these two measurement programs were to establish atmospheric trends for Hg in Canada's Arctic and sub-arctic, and to reveal the mechanisms unique to cold climates that can transfer Hg from the atmosphere to aquatic and terrestrial surfaces where it then has the potential to enter the food web. It is likely that other valuable insights will emerge from these high temporal-resolution data sets, including the identification of source regions from North America, Europe, and Russia (Poissant, 1998; Cheng and Schroeder, 2000; Lin *et al.*, 2001).

The discovery of MDEs has engendered research at Alert and Kuujjuarapik on the partitioning and exchange of Hg in gas and particle phases as well as the exchange of Hg between snow and air phases (Poissant, 2000, 2001; Lu *et al.*, 2001; Schroeder *et al.*, 2001a; Steffen *et al.*, 2002; in press). The addition of the second monitoring site at Kuujjuarapik has expanded our knowledge of the spatial extent of MDEs in the Canadian North, and is providing temporal trends of Hg in the eastern Canadian sub-arctic, as well as generating an improved understanding of the cycling of Hg between air and snow.

The observations made at Alert have required the development of new terminology to describe atmospheric mercury species. Historically, the term "total gaseous mercury" (TGM), which is predominantly Hg (0), was applied in manual atmospheric mercury measurements (because the sampling procedure was somewhat different). The contemporary Alert data show that gaseous species of Hg other than Hg (0) are present in ambient Arctic air after polar sunrise. Accordingly, in the Arctic atmosphere TGM is now recognized to comprise several gaseous Hg species including: 1) gaseous elemental mercury [GEM or Hg (0)] which is known to be the primary component of atmospheric Hg; and 2) reactive gaseous (oxidized) mercury species [RGM, probably inorganic Hg (II)] currently thought to be mainly, but perhaps not exclusively, $HgCl_2$. Particulate-phase mercury (PM) is also present in the air and is considered to be associated with airborne aerosols.

B.5.3 Continuous gaseous mercury concentration measurements

Continuous hourly averaged gaseous elemental mercury (GEM) concentrations from January, 1995, at Alert (82.5°N; 62.3°W) and August, 1999, at Kuujjuarapik (55.5°N; 77.7°W) are shown in Figure B.5.2a. All times, unless otherwise specified, are in Greenwich Mean Time (GMT). Alert, a high Arctic station, experiences total darkness from October 15 to February 25 with polar sunrise occurring between February 26 and April 6 when 24-hour daylight begins (Figure B.5.2b, top panel). In contrast, the more southerly Kuujjuarapik site experiences no days of complete darkness, exhibiting instead a daylight cycle that drops to less than eight hours between November 20 and January 20, and a maximum length of daylight of greater than 17 hours in June (Figure B.5.2b, lower panel).

Considering the mean global residence time of Hg (0) in the atmosphere, the GEM record from Alert is astonishing, but its annually reoccurring pattern unequivocally signals that something unique happens in the Arctic at polar sunrise. Beginning in early September, GEM concentrations exhibit a small variance about a repeatable background value of approximately 1.5 to 1.6 ng/m^3. The GEM concentration decreases slightly until about the end of September then remains relatively constant until January. As winter progresses, the variance increases until in early March the GEM record becomes dominated by MDEs where, at times, concentrations of GEM drop precipitously to values below the

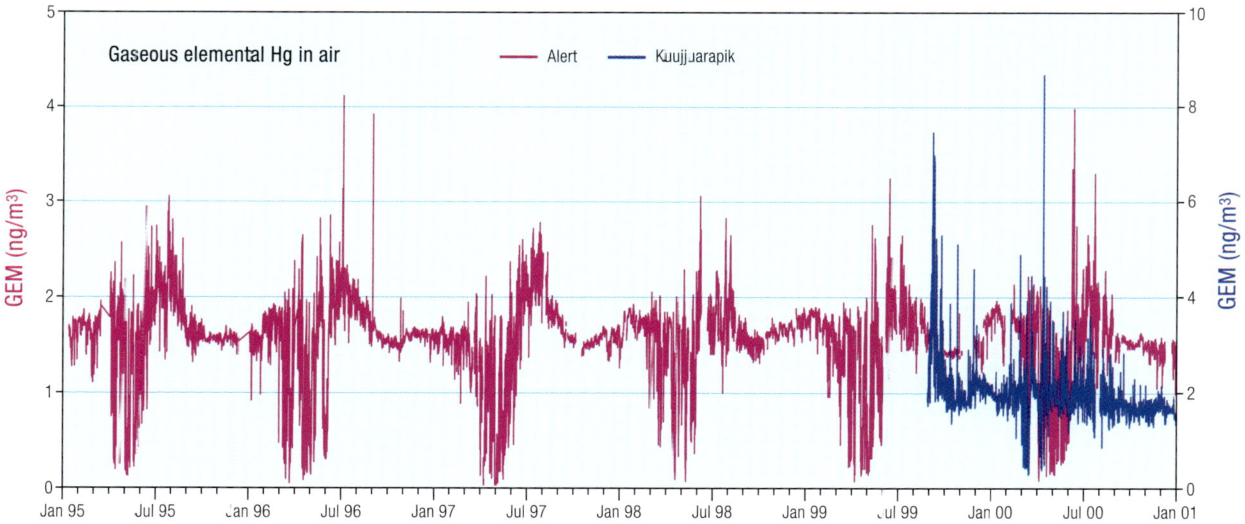

FIGURE B.5.2a

Gaseous Elemental Mercury (GEM) measurements (hourly averages) at Alert, Nunavut and Kuujjuarapik, Quebec.

detection limit of the instrument (0.1 ng/m^3). Despite considerable noise in the signal during summer, an increase in average GEM concentration (1.7 to 1.9 ng/m^3) is evident in the shape of the envelope during that time period. Potential source-contribution-function (PSCF) modelling results of the 1995 Alert data (Lin *et al.*, 2001) indicate that, in the fall and winter, anthropogenic Hg emission source regions include the populated areas in Europe, Eurasia, the United States, and Canada. Spikes of increased GEM concentration during summer may relate to emissions from such sources but the dominant increase in concentration during summer appears to be due to air-surface exchange (re-emission as well as primary emission) of volatile mercury forced by the annual temperature cycle (Figure B.5.2b and see, for example, Xiao *et al.*, 1991). Between 1995 and 2000, annual averages of mercury concentrations at Alert ranged between 1.47 and 1.62 ng/m^3 (Table B.5.1).

The short record of GEM at Kuujjuarapik reveals a seasonal pattern distinct from that observed at Alert (Figure B.5.2a, b). In general, Kuujjuarapik data exhibit higher variability, higher annual average concentration (1.78 ng/m^3 ranging between 0.29 and 3.17 ng/m^3, Table B.5.1) and a less intense seasonal modulation. Since Kuujjuarapik is in the sub-arctic and experiences different climatology, it is not surprising that the behaviour of Hg

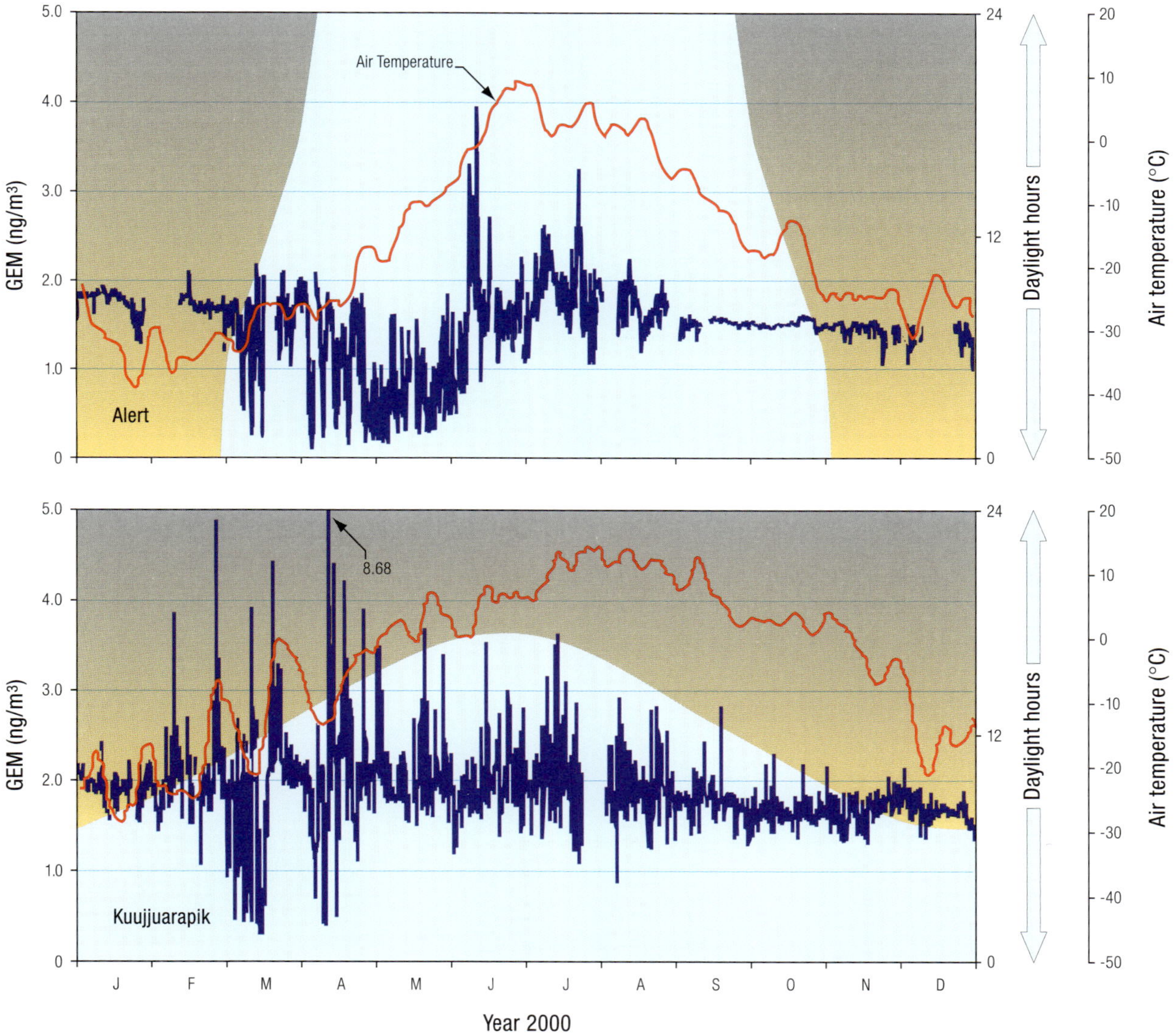

FIGURE B.5.2b

The GEM record at Alert (top panel) and Kuujjuarapik (bottom panel) for the year 2000. Also shown are the number of daylight hours and the temperature record (moving 7-day average) at both locations.

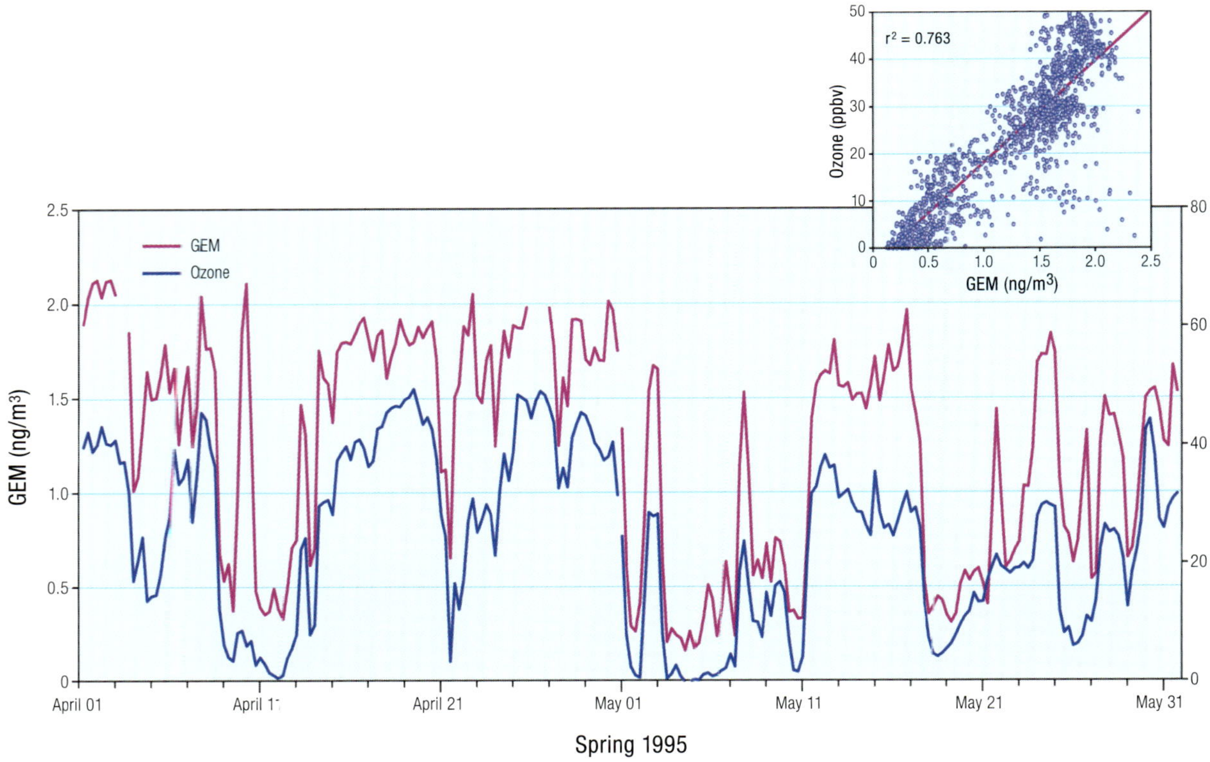

FIGURE B.5.3a

GEM and ozone hourly averaged concentration measurements between April 1 and May 31, 1995 at Alert. Insert: correlation between GEM and ozone.

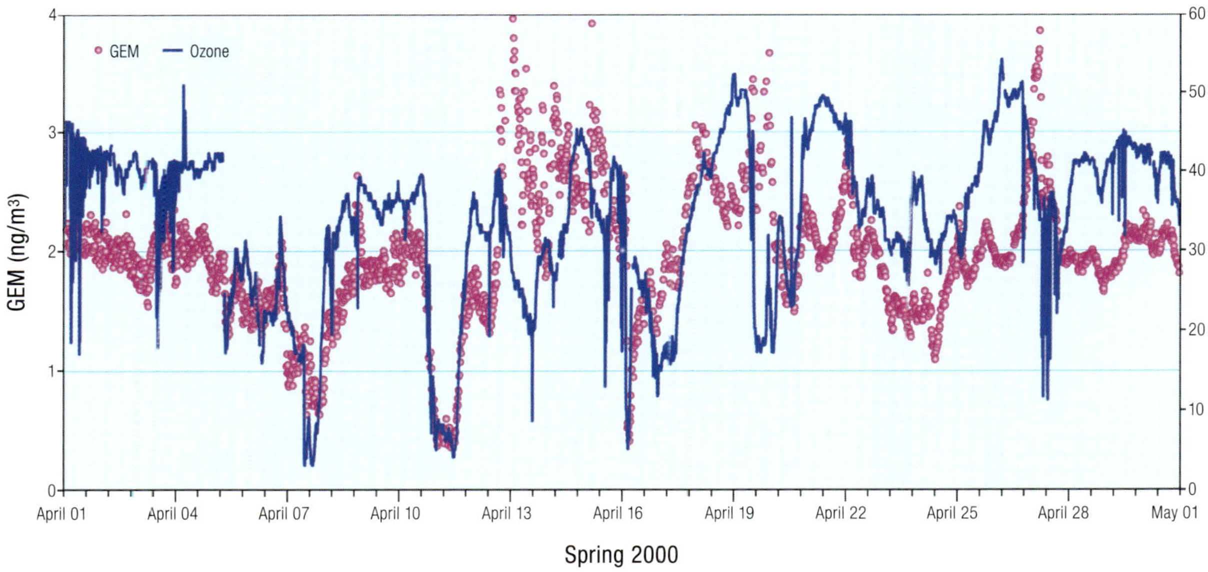

FIGURE B.5.3b

GEM and ozone hourly averaged concentration measurements between April 1 and May 1, 2000 at Kuujjuarapik (15 minute averaged data for GEM, 5-minute averaged data for ozone).

TABLE B.5.1 Annual averages and seasonal ranges of GEM (TGM) ng/m³ concentrations at Alert and Kuujjuarapik

Site	1995	1996	1997	1998	1999	2000	2001
Alert Annual Average (s.d.)	1.62 (0.41)	1.56 (0.42)	1.52 (0.44)	1.58 (0.31)	1.50 (0.43)	1.47 (0.44)	n/a
Alert Seasonal Average Range	1.41–1.88	1.40–1.79	1.21–1.88	1.38–1.67	1.26–1.70	1.13–1.47	n/a
Kuujjuarapik Annual Average (s.d.)	n/a	n/a	n/a	n/a	2.24* (0.72)	1.84 (0.38)	1.62 (0.38)
Kuujjuarapik Seasonal Average Range	n/a	n/a	n/a	n/a	2.22–2.26	1.83–1.84	1.62–1.63

n/a – data not available

* August–December

(as seen in the annual pattern) differs from that found in the high Arctic. However, these data clearly demonstrate that MDEs are not restricted to the high Arctic but may also occur in sub-arctic regions. At Alert, the onset of MDEs occurs sharply with polar sunrise and these then continue to be observed well into June (Figure B.5.2b, top panel). At Kuujjuarapik, there is evidence of weaker MDEs earlier in winter (January–February), but in the year 2000, intense MDEs were initiated at about the same time at both Kuujjuarapik and Alert (comparing the record in top and bottom panels in Figure B.5.2b). Intense MDEs were not observed later than April at Kuujjuarapik. Although there is a hint of increased GEM concentration during the warmest period at Kuujjuarapik (June–August) the evidence of this is not as clear as it is at Alert.

B.5.4 Discovery of mercury depletion events (MDEs) in the Arctic.

B.5.4.1 Gaseous elemental mercury and ozone during mercury depletion events

With the first annual time series of high-resolution atmospheric mercury vapour data collected in the Arctic in 1995, the GEM concentrations in the fall and early winter period exhibited, as expected, a fairly constant background value of ~1.5 ng/m^3. This concentration is consistent with measurements at other background monitoring sites for GEM, such as Macehead, Ireland and northern Wisconsin, USA (Ebinghaus *et al.*, 1999; Schroeder *et al.*, 1995). In regions that experience polar sunrise, however, GEM concentrations underwent extraordinary fluctuations, decreasing at times to values < 0.1 ng/m^3 within periods of 24 hours or less. This behaviour runs counter to what is expected for an air pollutant characterized by a long atmospheric residence time (Schroeder and Munthe, 1998a).

The unique environmental condition at Alert that appeared to initiate this unusual behaviour was the sudden exposure to solar radiation in early March after approximately five months of total darkness. The measurements at Alert in 1996 corroborated the distinctive behaviour of GEM after polar sunrise, as seen in the previous year, and revealed a strong correlation between GEM and ground level ozone concentrations (Schroeder *et al.*, 1998). Springtime tropospheric ozone depletion events were observed for the first time at Alert in 1985 (Bottenheim *et al.*, 1986; Barrie *et al.*, 1988). During and after polar sunrise, GEM and ozone concentrations were found to deplete at the same time (Figure B.5.3a) with excellent correlations during the period between late March and mid-June in the high Arctic [correlation coefficient (r^2) between GEM and O_3 is ~0.8]. The relationship between ozone and GEM appears endemic to northern regions as shown by the results from Kuujjuarapik in April 2000 [a correlation coefficient (r^2) between GEM and O_3 was determined to be ~0.6 for the data displayed in Figure B.5.3b; Poissant, 2001]. At this lower latitude, MDEs are less frequent and less intense than those in the high Arctic and are observed between February and April (Figure B.5.2b). It is interesting that MDEs occur at the lower latitude of Kuujjuarapik even though the climate differs significantly from Alert. Furthermore, MDEs are found to occur in Kuujjuarapik during both the day and night time, whereas at Alert, the events are primarily observed during 24-hour daylight periods. Yet, during the spring of 2000, in mid-March, a large MDE occurred at both Alert and Kuujjuarapik (Figure B.5.2b), suggesting a large spatial extent for these events.

B.5.4.2 Particulate-phase mercury during mercury depletion events

Particulate-phase mercury (PM) measurements were first made at Alert intermittently between May, 1995, and December, 1996. These weekly-integrated samples showed the spring PM concentrations (~5 to 30 pg/m^3) to exceed those measured in the fall (< 0.5 to 3.5 pg/m^3) by almost an order of magnitude (Schroeder *et al.*, 1999b). The increase in PM during spring was the first indication that chemical processes may be converting GEM to inorganic Hg (II) species which are much less volatile and thus more likely to associate with airborne particulate matter.

A newly developed technique for determining PM (described in Lu *et al.*, 1998) showed that the decrease in GEM concentration during MDEs at Alert in April and May, 1997, was accompanied by an increase in PM concentration [Figure B.5.4 (inset)]. From these data, dry depositional fluxes to the Canadian Arctic were estimated to be 3.10 and 3.24 ng/m^2·h^{-1} for the springs of 1995 and 1996, respectively. These fluxes imply that MDEs have the potential to deliver significant amounts of Hg to Arctic ecosystems, estimated by Schroeder *et al.* (1999b) at ~50 to ~60 tonnes deposited in spring on the Arctic Ocean and an additional 10 to 15 tonnes deposited over the eastern Canadian Arctic.

Particulate-phase mercury samples collected at higher temporal resolution in the spring of 1998 at Alert confirmed the strong anti-correlation between GEM and PM concentrations during MDEs (Figure B.5.4) and showed the conversion to occur rapidly. From sampling at Alert between March, 1998 and February, 1999, PM was found to average 83 pg/m^3 during the polar sunrise MDE period in contrast to 7.4 pg/m^3 during the rest of the year.

B.5.4.3 Conversion of mercury during mercury depletion events

To confirm that GEM was being converted to a more reactive Hg species during MDEs, the total atmospheric mercury (TAM) concentration in the ambient air at Alert during a MDE was measured using a Cold Regions Pyrolysis Unit (CRPU). This unit, placed at the front end of the measurement instrument, heats incoming air to 900°C (pyrolysing the air) thereby converting any inorganic and organo-metallic mercury forms to GEM [Hg (0)] which can then be detected by the mercury measurement instrument (Schroeder and Jackson, 1984). Parallel instruments, one sampling ambient (unheated) air, the other sampling the ambient air after pyrolysis (heated) were used to determine changes in mercury speciation during depletion events (Figure B.5.5).

From April 1 to April 29, 1998 at Alert, unfiltered ambient air entered the CRPU to allow all airborne Hg species into the system. During the times when MDEs were not occurring (e.g., April 6 to April 11, Figure B.5.5), both the heated and unheated ambient air streams measured identical concentrations indicating that most (if not all) of the background mercury (~1.5 ng/m^3) in the air was Hg (0). With the initiation of a depletion event (e.g., April 11 in Figure B.5.5) the Hg concentrations in unheated ambient air exhibited a typical rapid decrease (blue circles, Figure B.5.5). The magnitude of the decrease of mercury in the ambient air stream was only partially matched in the heated (pyrolysed) air stream (red circles, Figure B.5.5) signaling the presence of mercury species (~0.5 to 1 ng/m^3) other than Hg (0) which are converted to Hg (0) by pyrolysis. These results confirm the presence of oxidized mercury species [previously converted from Hg (0)] during depletion events (Schroeder *et al.*, 1999c; Banic *et al.*, 2001).

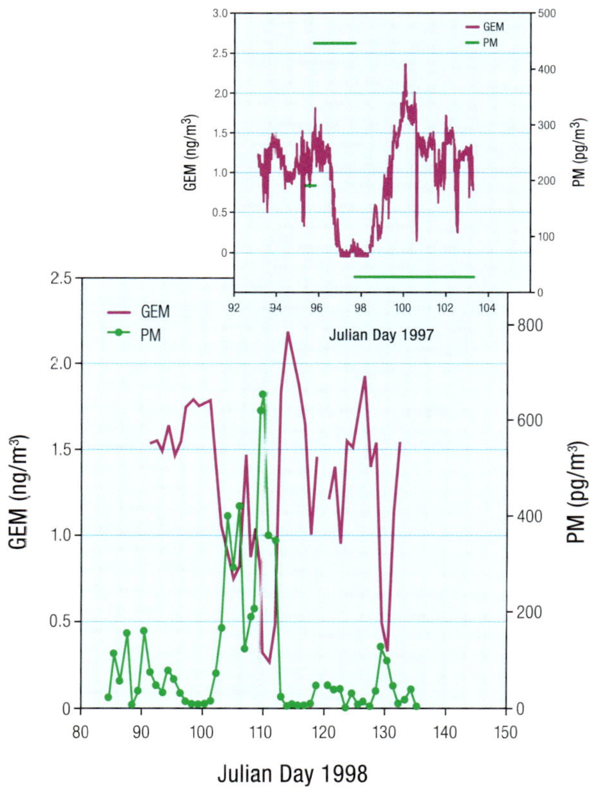

FIGURE B.5.4

Daily GEM and particulate mercury (PM) concentration measurements in the springtime at Alert, 1998. Insert: GEM and PM concentration at Alert, 1997.

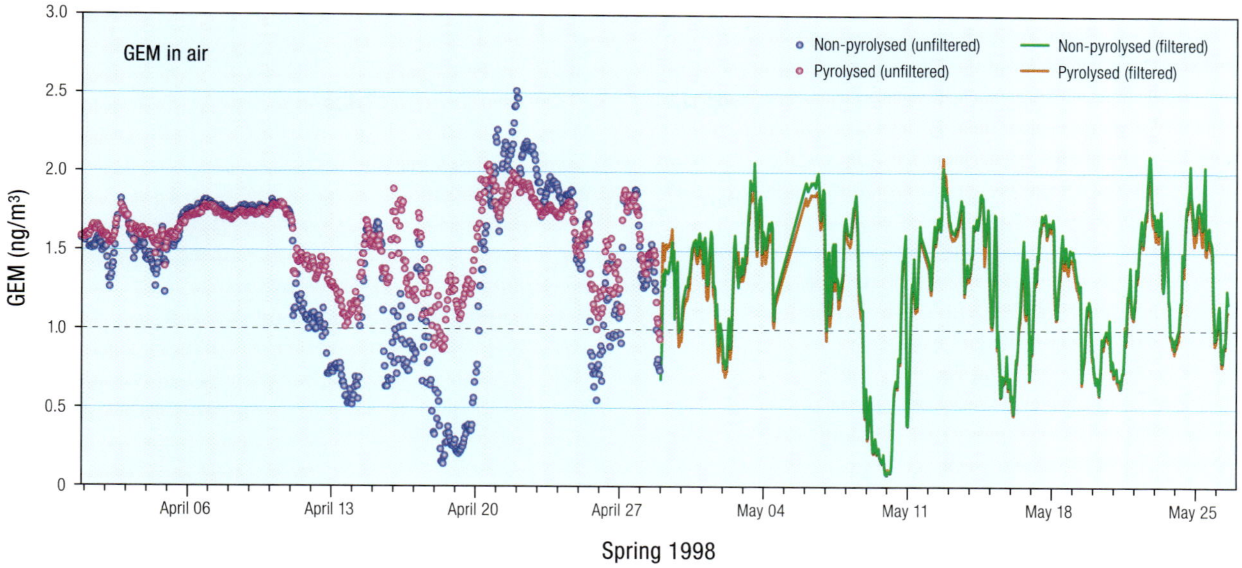

FIGURE B.5.5

Cold Regions Pyrolysis Unit GEM hourly measurements between April 4 to May 25, 1998 at Alert, confirming atmospheric mercury speciation during mercury depletion events.

After April 29, air entering the two parallel sampling streams was filtered, removing all oxidized (PM and RGM) Hg species. Immediately, the two filtered air streams exhibited identical mercury concentrations [Figure B.5.5, right hand side (pink and light blue lines)] whether or not a MDE was occurring. This result clearly shows that the Hg (II) species present in the air during MDEs (difference between blue and red circles on Figure B.5.5, left hand side) are captured by filtration. It is important to note that pyrolysis does not recover all of the converted Hg (0) vapour as shown by the incomplete recovery of Hg in pyrolysed, unfiltered air (red circles; Figure B.5.5). This suggests that some of the oxidized mercury has been deposited from the atmosphere on snow and ice surface en-route to the monitoring site. Further work at Alert in 2000 confirmed the data in Figure B.5.5, finding that, on average, 48% (ranging from 11 to 87%) of the converted GEM was recovered through pyrolysis with the remaining component removed from air, presumably having been deposited to the frozen surface (to be mobilized at a later time).

B.5.4.4 The effect of mercury depletion events on mercury in snow

The important presumption that some of the reactive mercury [Hg (II)] produced during MDEs deposits on snow and ice was investigated directly by collecting snow, up to, during and after depletion events, from a wide range of locations throughout the Arctic over several years. In general, snow samples taken during and after depletion events exhibit elevated Hg concentration. At Alert in 1998, freshly fallen snow was collected on a Teflon-coated “snow table”. From the few snowfall events that occurred, the deposition flux of mercury was found to increase about 3.5-fold after a MDE (Schroeder *et al.*, 2000), primarily due to an increase in Hg concentration in the falling snow. Similar results have been obtained in subsequent years with, in one case during 2000, an observed 20-fold increase in Hg concentration in the snow from before a MDE to after a MDE (Steffen *et al.*, 2002).

The spatial distribution of Hg in snow over sea-ice in the Arctic was determined from samples collected by students from northern communities between April and May 1997 (Figure B.5.6; Lu *et al.*, 2001). Mercury concentrations in surface snow were found to be generally higher toward the east in the Canadian Island Archipelago and at coastal sites on Hudson Bay (25–160 ng/L). Snow samples from Tuktoyaktuk, the eastern shore of Greenland and the south shore of Lake Winnipeg were found to contain much lower concentrations (~2 ng/L). Sites were revisited in 1998, yielding similar results and confirming the spatial distribution of Hg in snow implied in Figure B.5.6.

The drift of the CCG icebreaker Des Groselliers across the Beaufort Sea between October 1997 and October 1998, as part of the SHEBA program (Surface Heat Budget of the Arctic Ocean), provided an unique opportunity to evaluate the seasonal cycle of mercury in snow at a remote, high Arctic site (Lu *et al.*, 2001). As predicted from studies at Alert, the transition from winter darkness to sunlight in the spring months was accompanied by a four-fold increase in Hg concentration in surface snow. From an average of 7.8 ng/L between November 1997 and January 1998, Hg concentration in snow increased to 34 ng/L between February and May 1998. However, much of the enhanced concentration of

Site	Number
North of Alert	1
Alert	2
Arctic Bay	3
Arviat	4
Cambridge Bay	5
Cape Dorset	6
Chesterfield Inlet	7
Churchill	8
Clyde River	9
Eureka	10
Gjoa Haven	11
Greenland (East)	12
Hall Beach	13
Igloolik	14
Iqaluit	15
Kimmirut	16
Kugluktuk	17
Lake Winnipeg	18
Mould Bay-buoy	19
Mould Bay	20
Pangnirtung	21
Pelly Bay	22
Pond Inlet	23
Quaqtaq	24
Repulse Bay	25
Resolute	26
Sachs Harbour	27
Sanikiluaq	28
Taloyoak	29
Tuktoyaktuk	30
Whale Cove	31

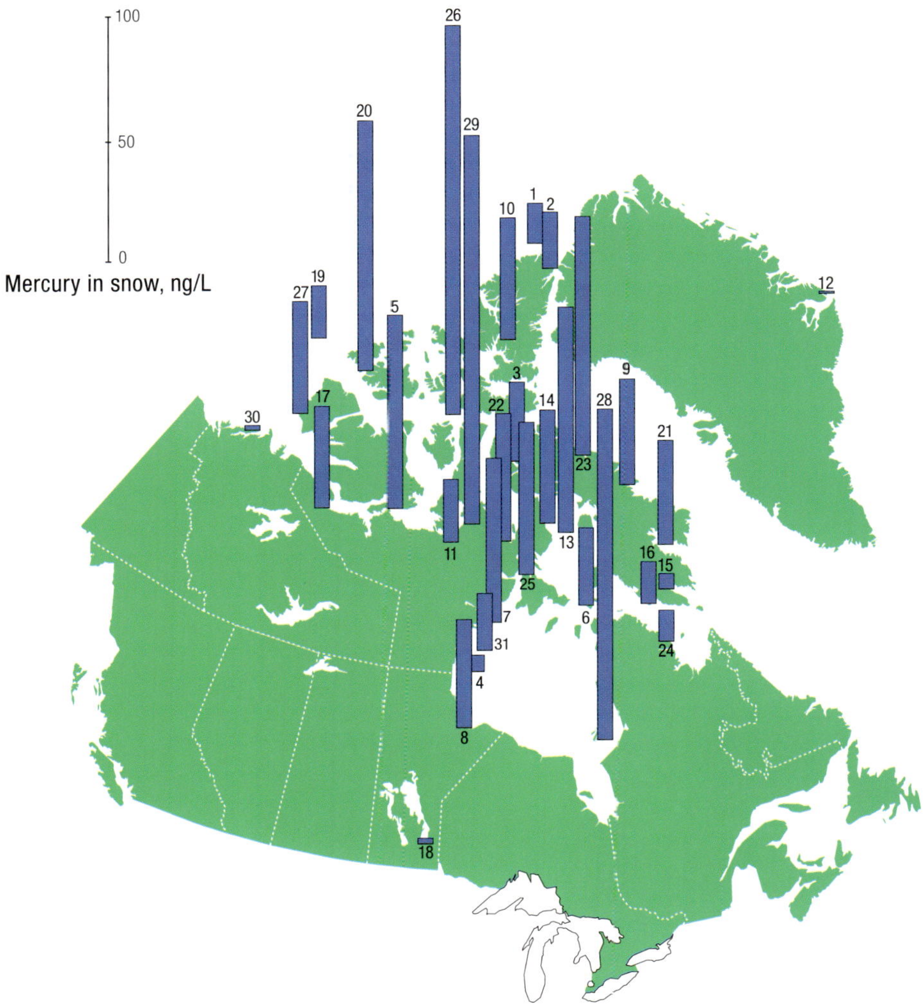

FIGURE B.5.6

Mercury snow survey across Canada, 1997.

Hg observed in snow during late winter to early spring appears to be transient. A dramatic plunge in the surface snow concentrations was recorded in May 1998 perhaps due perhaps to: (i) melting of the surface snow and dissolution of the highly soluble Hg (II) compounds (air temperatures at this time of the year hover around 0°C); (ii) volatilization of Hg (0) produced from reduction of Hg (II) by photolysis in the snowpack (Poissant *et al.*, 2001; Lalonde *et al.*, 2002); or, (iii) leaching of easily soluble Hg (II) species by meltwater percolating through the snowpack at the onset of summer. Furthermore, if the dramatic decrease in the incidence of MDEs that occurs at Alert during June (as seen Figure B.5.2b) also occurred along the SHEBA drift track, it would imply a dramatic decrease in the supply of reactive Hg at the snow surface at this time of the year.

The process of MDEs leading to enhanced surface snow concentrations in spring appears to be endemic to large regions of the Arctic, being witnessed in the Beaufort Sea and at Alert (Lu *et al.*, 2001) and in the sub-arctic at Kuujjuarapik (Poissant *et al.*, 2001). At this latter location a survey in the spring of 2000 showed a 16-fold increase in Hg concentration (from ~5 ng/L to 82 ng/L) in snow within 24 hours after a MDE. Similar results were obtained in 2001 in Kuujjuarapik, as well. However, strong evidence was found that much of the deposited (oxidized) mercury was transient, escaping back to the atmosphere within a few hours following the MDE. Approximately 50% of the deposited mercury appears to have been lost from the snowpack within 12 hours (see Figure B.5.7). The process by which mercury is lost from the snow is inferred to depend on environmental conditions such as air temperature and solar radiation intensity. If the air temperature is > 0°C, then total Hg may be removed into surface snow meltwater which would then leach through the snow to become available to aquatic or soil ecosystems. A fraction of the total Hg is removed from the snowpack [as Hg (0)] via solar radiation following photo-reduction from Hg (II) and this process will become relatively more important if air temperature is < 0°C. Such a process would explain the findings in the high Arctic where, at the end of the MDE "season", a dramatic plunge in Hg concentrations in snow is observed. Kuujjuarapik, however, exhibits loss of Hg from snow in the middle of the MDE "season" suggesting that coupling between MDEs and the cycle of Hg in snow differs between locations due to the difference in climate (temperature, radiation).

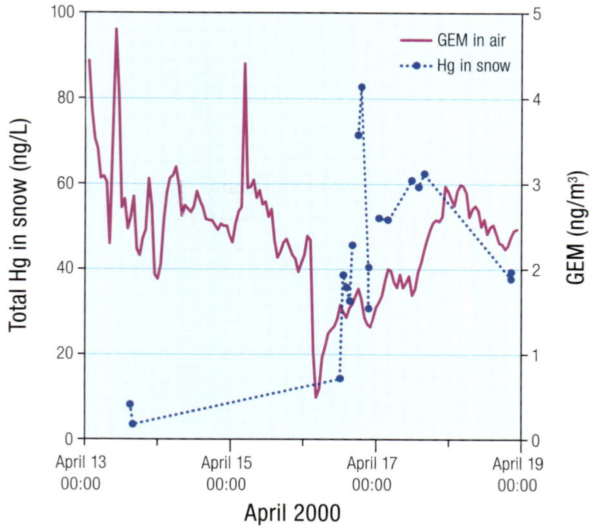

FIGURE B.5.7

Spike in mercury concentration in snow at Kuujjuarapik following mercury depletion event in 2000.

B.5.4.5 The chemistry of mercury depletion events — halogens and mercury

Although the oxidation of elemental gaseous mercury during MDEs is strongly correlated with ozone depletion in the Arctic marine boundary layer, the two processes are not necessarily linked directly in a cause-effect relationship. Rather, it is thought that Hg (0) is oxidized to Hg (II) by reactive halogen species (e.g., Br, BrO, BrCl) produced through photochemically initiated autocatalytic reactions (Macdonald *et al.*, 2000; Lu *et al.*, 2001; Lindberg *et al.*, 2001; Ebinghaus *et al.*, 2002). The high Arctic is an ideal location for such environmental chemistry to occur due to long dark winters (Figure B.5.2b) followed by rapid emergence of sunlight in spring. In essence, it is like turning on the UV radiation after a long dark incubation period. Unlike the high Arctic, the sub-arctic around Kuujjuarapik, does not experience true polar night nor a rapid emergence of the sunlight. Nevertheless, MDEs have been observed at this location where important reactive halogen species derive from sea-salt aerosols previously deposited on frozen ocean, aquatic and terrestrial surfaces (Barrie and Platt, 1997). It is now generally accepted that reactive halogen species cause polar tropospheric ozone depletions (Bottenheim *et al.*, 1986; Barrie *et al.*, 1988) but they can also be effective oxidants for mercury (Schroeder

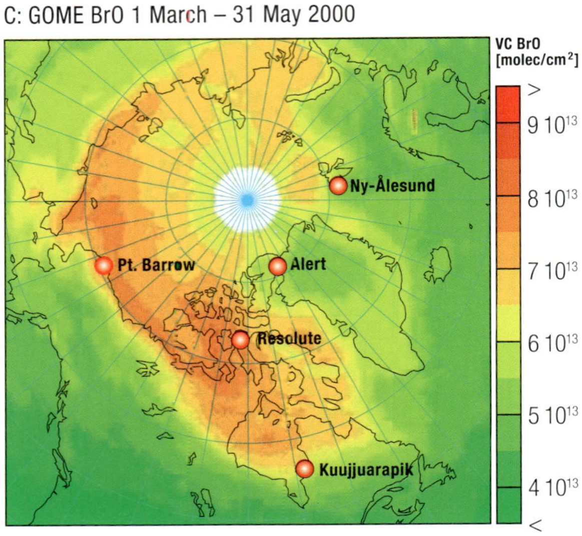

FIGURE B.5.8

GOME satellite measurements of BrO over northern Canada in spring 2000. (*Source*: A. Richter, University of Bremen)

et al., 1991; Lu *et al.*, 2001). Elevated concentrations of BrO have been observed in the air column (by satellite) and in the boundary layer (*in situ* measurements) over wide areas of the Arctic, including Alert and Ny-Ålesund and further south over Hudson Bay (Figure B.5.8; Richter *et al.*, 1998; Lu *et al.*, 2001). Elevated BrO levels have been found more recently in the free troposphere as well (McElroy *et al.*, 1999). Because the inventory of BrO (molecules/cm^2) in the air column can now be measured from satellite we have a clear idea of its spatial distribution throughout the Arctic in spring — as exemplified for the year 2000 in Figure B.5.8. The pattern varies from year to year, but elevated inventories of BrO in the air column during spring are seen in the marginal seas — including the SHEBA drift site, in the Canadian Arctic Archipelago and, especially, in the Hudson Bay area. Although the bromine in BrO is derived predominantly from sea salt, it is clear that air column concentrations of BrO can be elevated over land as well.

A detailed examination of the BrO chemistry accompanying a depletion event was conducted from an ice camp near Alert during the spring of 2000 (Hönninger and Platt, 2002a,b) and at Kuujjuarapik in 2001. Near Alert, observations of mixing ratios of BrO (using Multi-Axis DOAS), GEM, ozone and meteorological parameters made at ground level during the onset of an MDE (Figure B.5.9) show that decreases in GEM [measured at 1 cm from the ice surface] (blue dots) and ozone (light blue line) are matched by an increase in BrO (red line). From vertical profiles, it was ascertained that most of the BrO was present in a layer near the earth's surface (1 ± 0.5 km), reaching concentrations of 20 to 30 ppt in the boundary layer (Hönninger and Platt, 2002a). These data imply that much of the BrO inventory, mapped through satellite measurements, probably resides close to the earth's surface and within the boundary layer. The same technique was later used to measure BrO *in-situ* at Kuujjuarapik in the spring of 2001 and results showed BrO levels up to 30 ppt (Hönninger, 2002b). *in-situ* measurements of GEM, BrO and ozone revealed that during depletion events, GEM and ozone concentrations were inversely correlated with BrO. Interestingly, these events were recorded from air masses containing high amounts of BrO that had originated from the north and had been in contact with the Hudsons Bay sea ice for several days prior to arrival at the measurement site.

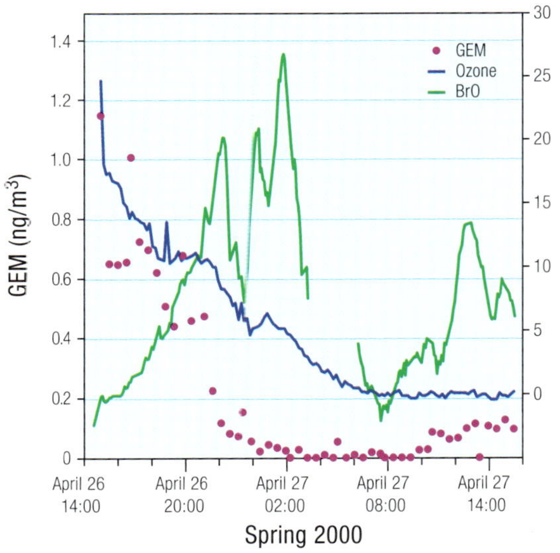

FIGURE B.5.9

GEM, BrO and ozone measurements at an ice camp near Alert, April, 2000 (GEM, 15–20-minute averaged data; BrO, 5-minute averaged data; ozone, 10-minute averaged data)

B.5.4.6 The role of the ocean in mercury depletion event chemistry

An important question yet to be resolved is the geographical location of GEM depletion chemistry. Spatially, MDEs have been observed across the Arctic and sub-arctic, including two areas of Canada (Lu *et al.*, 2001; Poissant *et al.*, 2000), Point Barrow, Alaska (Lindberg *et al.*, 2001), Ny-Ålesund, Spitsbergen (Berg *et al.*, 2001), Faroe Islands (Nielsdottir *et al.*, 2002) and, recently, in the Antarctic (Ebinghaus *et al.*, 2002). Clearly, air masses exhibiting mercury depletion are widely distributed, but where, actually, does the depletion chemistry occur? This point is crucial to understanding the likely impact of MDEs on Arctic and sub-arctic ecosystems and their vulnerability to climate change, because the locus of the depletion is likely to be the locus of highest air to ground fluxes of mercury.

Back trajectories of air masses from Alert in 1998 show that, during February and May, 46–78% of the air masses arrive from sectors over the Arctic ocean. Furthermore, MDEs usually occur at Alert when the wind direction is from the north (between 300° and 90°). These observations could indicate either that air masses containing depleted mercury concentrations are arriving from over the Arctic Ocean where the conversion chemistry occurred, or that the air masses from the northern sector supply the halogenated species to allow Hg (0) oxidation conversion to occur at or near Alert. Cross correlation function analysis of mercury and ozone data from the Antarctic (Ebinghaus *et al.*, 2002) suggests two possibilities for the reaction between ozone and Br atoms and the resulting BrO–Hg reaction. Either Br atoms react with ozone and subsequently or concurrently with GEM at such a rapid rate that the reaction is complete on site (or complete within 15 minutes of the sample site) or the reaction has already occurred before the depleted air parcels are advected to the sample site. Similarly, at Kuujjuarapik MDEs were most predominant when winds had passed over Hudson Bay.

During April 1998, Canadian aircraft flights at altitudes from 100 m above sea level up to 7 km were made over the Beaufort Sea as well as thousands of kilometres from land over the Arctic Ocean to measure the vertical distribution of mercury (Banic *et al.*, 2002). The vertical distribution of Hg was a uniform 1.7 ng/(standard)m^3 on days when mercury was not observed to be depleted from the atmosphere at the lowest altitude flown (Figure B.5.10, red circles). During days when mercury was observed to be depleted at the lowest altitude flown, the mercury concentrations remain at the 1.7 ng/sm^3 at altitudes above about 1 km (Figure B.5.10, blue circles). Below about 1 km in altitude, however, lower mercury concentrations were observed, with values < 1 ng/sm^3 seen below 500 m. The frequency distribution of the spot measurements of Hg made with the aircraft at altitudes up to 1 km showed the same range of values and remarkably similar shape to the Hg concentration data collected at ground level at Alert for the month of April 1998. These results demonstrate that the depletion events observed at Alert are part of a regional pattern that includes large areas of the Arctic Ocean and the Beaufort Sea. Furthermore, the depletion mechanism affects the concentration of mercury up to an altitude of 1 km. These studies did not, however, make observations closer than 100 m from the snow and ice surfaces as discussed in the next section.

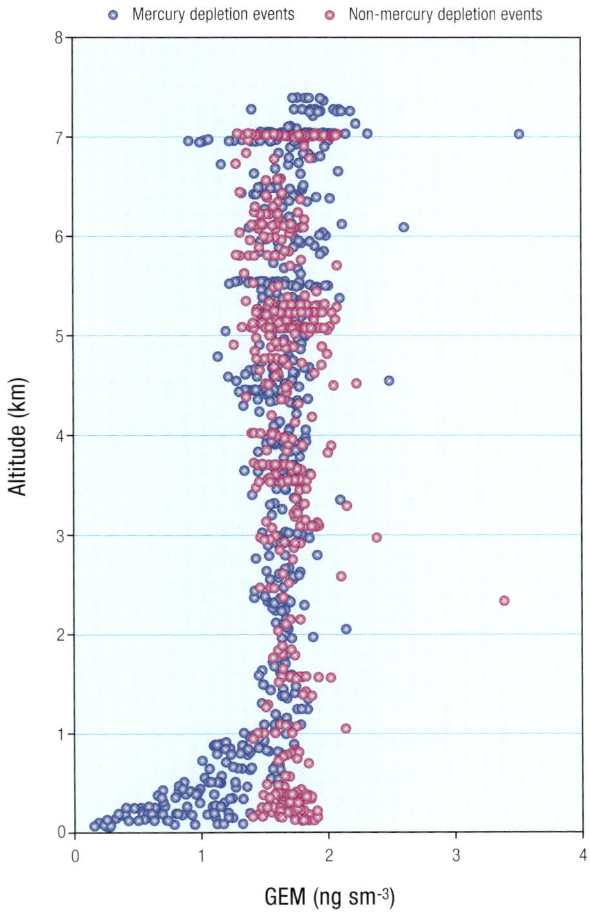

FIGURE B.5.10

Aircraft GEM profile measurements made over 15 days from over the Beaufort Sea and Arctic Ocean, April 1998, during non-mercury depletion events (red circles) and during mercury depletion events (blue circles) (GEM concentrations are normalized at all altitudes to 0°C and 1 atm for reference to the measurements made with the instrument at sea level).

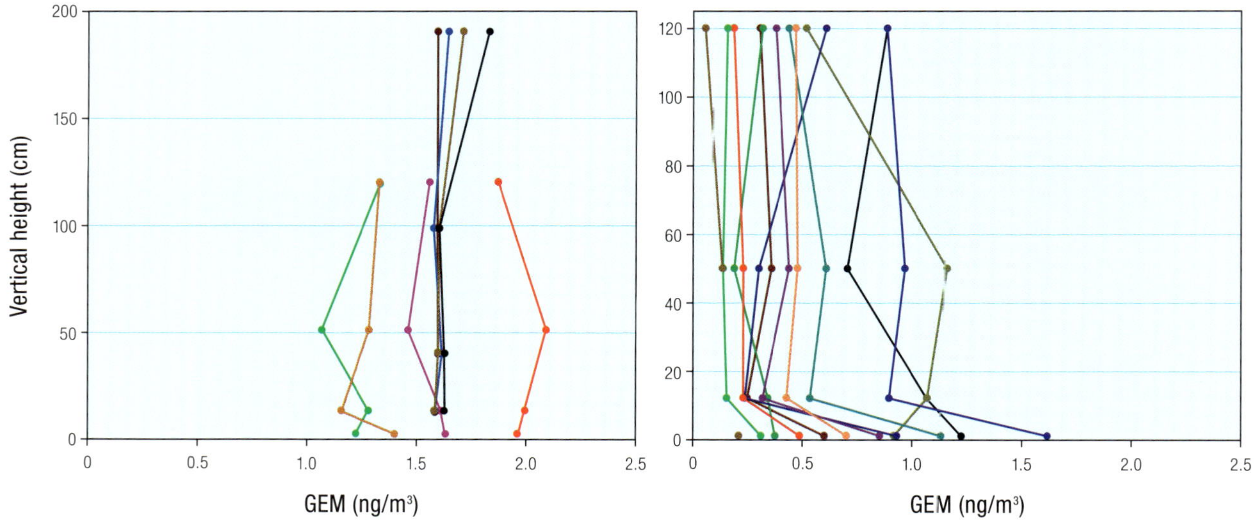

FIGURE B.5.11a

Four level GEM concentration profiles at Alert in spring 2000 during non-mercury depletion events (left panel) and during mercury depletion events (right panel).

B.5.4.7 Profiles and fluxes of mercury in air near the snow surface during mercury depletion events

Considerable research has been undertaken on the behaviour of mercury during depletion events near ground level and aloft, but very little has been undertaken directly at the snow surface where key deposition and re-emission processes occur. Vertical concentration gradients of mercury in interstitial air within the snowpack and in air above the snow surface, therefore, were measured at Alert in the spring of 2000. Ambient air profiling measurements were conducted up to two metres above the snow surface, during both light and dark periods, and in the presence and absence of MDEs (Figure B.5.11a). When no MDE was occurring, GEM concentrations were essentially invariant in the air column between the snow surface and 2 m above the snow (Figure B.5.11a; left hand

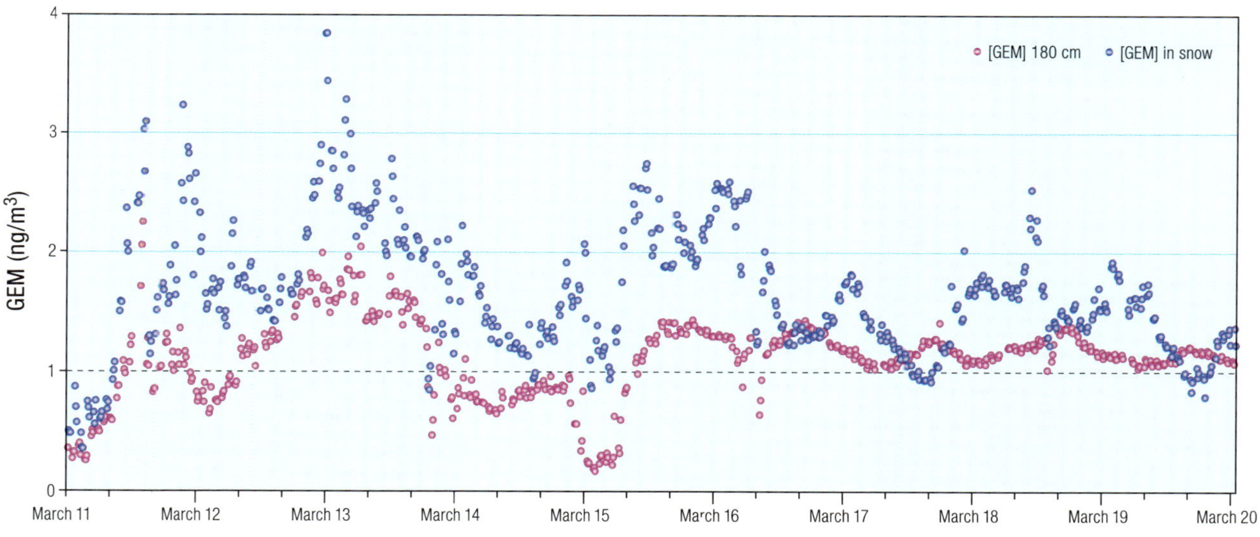

FIGURE B.5.11b

GEM concentrations from within the interstitial snowpack and at 180 cm above the snowpack surface measured in March 2000 at Alert.

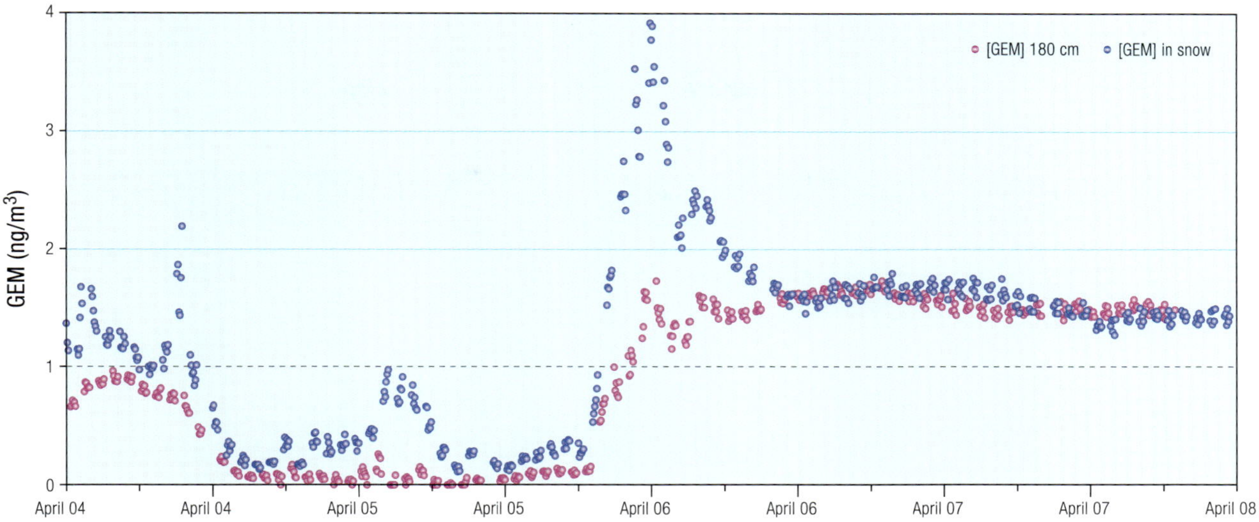

FIGURE B.5.11c

Two level GEM concentration profile measurements showing spikes in interstitial snowpack air during and after a mercury depletion event at Alert, April, 2000.

panel). During MDEs, GEM concentrations were also found to be invariant with height above the snow except right at the snow surface where they increased noticeably (Figure B.5.11a; right hand panel). The negative gradient between the snow surface and 10 cm above the snow, shown in Figure B.5.11a (right hand panel), implies a net flux of GEM from the snow surface to the atmosphere.

Measurements conducted by burying one sample inlet in the snow (red dots on Figure B.5.11b) and maintaining a second inlet at 180 cm above the snow surface (blue dots on Figure B.5.11b) reveal a gradient that would require GEM to diffuse out of the snow almost continuously through the period sampled (March 11–20, 2000). A detailed examination of the GEM concentration of interstitial air in the snow during a MDE in early April, 2000 (Figure B.5.11c) shows that with the onset of the MDE (April 4) both the air within the snow and ambient air above the snow exhibit a GEM concentration decrease. At the end of the MDE, however, the GEM in the snowpack air overshoots the ambient air concentration implying that a pulse of GEM must have been emitted to the atmosphere over a period of about 12–15 hours.

Flux measurements were made at Kuujjuarapik also in 2000 (Poissant *et al.*, 2000; 2001) showing GEM fluxes out of the snow to be four times larger after a MDE than before. When no depletion event was occurring, diel cycles were observed for GEM surface evasion with a maximum flux out of the snow estimated to be ~0.7 $ng/m^2 \cdot h^{-1}$, observed during periods of increasing solar radiation (Figure B.5.12). Following a MDE the flux of GEM out of the snow was greater, increasing to almost 2.5 $ng/m^2 \cdot h^{-1}$.

The data collected at Alert and Kuujjuarapik unequivocally show that Hg concentration in snowfall increases in response to MDEs. Most of the Hg in the snowfall is presumed to be Hg (II), either as RGM or PM, both of which are susceptible to photochemical reduction back to Hg (0) (Lalonde *et al.*, 2002). The observations at Alert and Kuujjuarapik can be understood as an enhanced flux of mercury (RGM or PM) from the atmosphere to the snow surface, possibly mediated by snowfall, followed by reduction at or near the snow surface mediated by photolysis, and then emission of GEM back to the atmosphere. However, as suggested in B.5.4.4, if the air temperature is > 0°C, the loss mechanism of GEM from the snow surface may not be entirely due to photoreduction but may involve meltwater leaching of Hg (II) from surface snow.

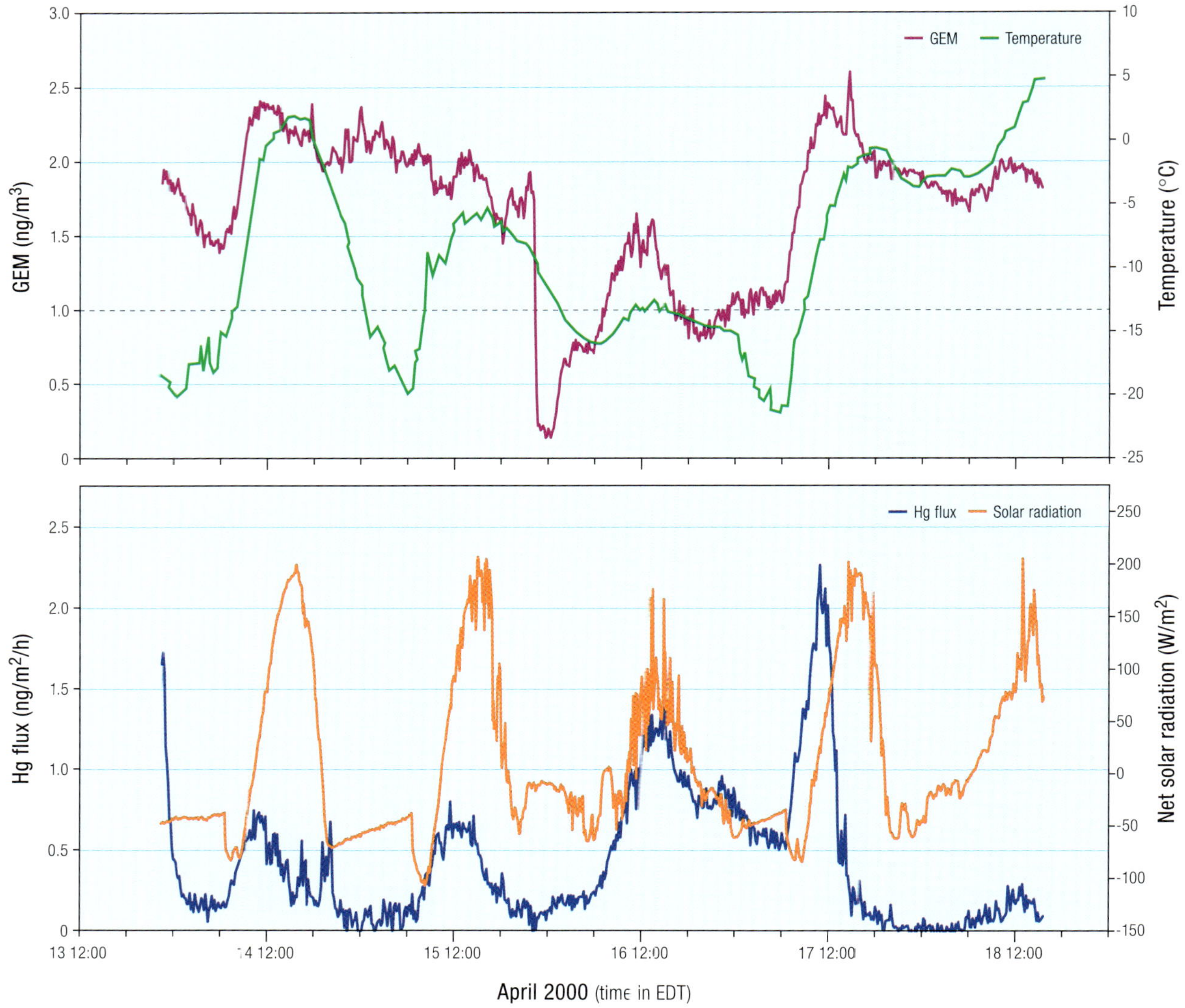

FIGURE B.5.12

Measurements of mercury flux at Kuujjuarapik, April, 2000 shown with solar radiation, ambient GEM concentrations and air temperature.

At Alert, spikes of GEM emitted from the snow surface were observed not only following MDEs when GEM concentrations were returning back to normal, but also during an extended MDE in May 2000 when GEM concentrations remained below 1 ng/m^3 for over eight days. At 1 cm above the snow surface, GEM was observed to spike to concentrations over twice those measured at ~2 metres above the snow surface, as shown in Figure B.5.13. In this latter example from Alert the solar radiation pattern matched well with the GEM spikes, corroborating the finding at Kuujjuarapik that photochemical reduction may at times play an important role in recycling previously deposited reactive mercury from the snow back to the atmosphere. Direct correlation between radiation and GEM spikes was not always observed at Alert; therefore, the following mechanism for GEM spikes is proposed: Hg (II) is deposited on the snow surface during a depletion event to be subsequently photochemically reduced back to Hg (0); as snowpack temperature increases during the day, exchange of GEM out of the snowpack is stimulated, producing spikes. The relationship between the GEM flux and solar radiation appears more direct at Kuujjuarapik, which could be the result of higher snow and air temperatures at this location. These findings show that there is a redox-reaction cycle of mercury at the snow surface and in the interstitial snowpack air during and after MDEs.

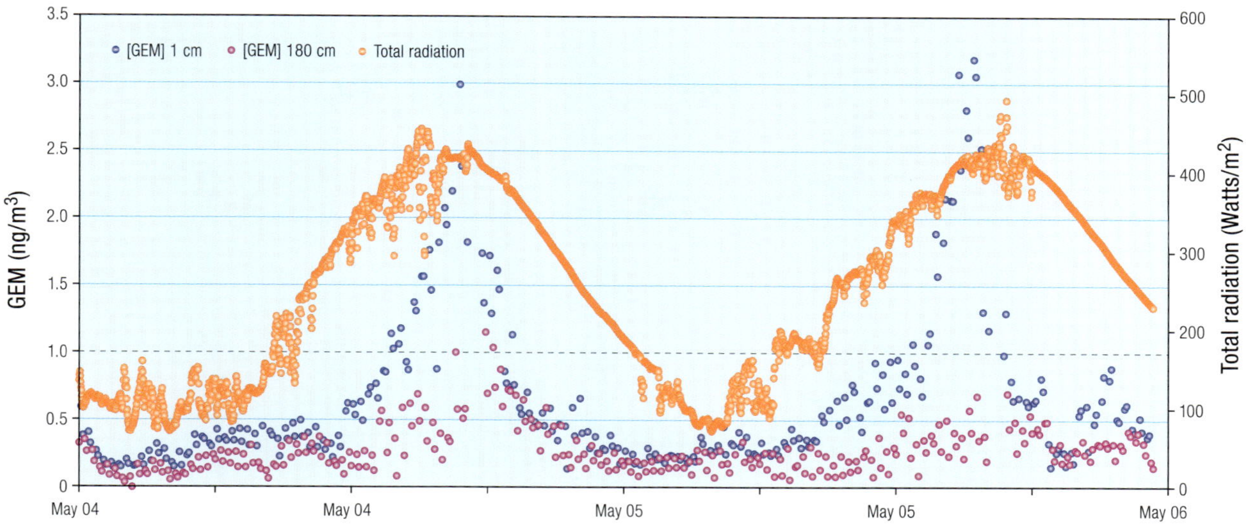

FIGURE B.5.13

GEM concentration measurements from the snowpack surface to 180 cm above the snow pack surface at Alert, May, 2000 during an extended mercury depletion event, showing back reduction of mercury correlated with solar radiation.

B.5.4.8 *Reactive gaseous mercury during mercury depletion events*

Studies at Point Barrow, Alaska (Lindberg *et al.*, 2001, 2002) demonstrated that, during MDEs, surprisingly large amounts of RGM (postulated to be mainly $HgCl_2$) were present in the ambient air (as high as 900 pg/m^3 compared to < 2 pg/m^3 non-MDE concentrations) providing the first direct measurement of this gaseous oxidized Hg species in the Arctic air during a MDE. Sharp increases in total Hg concentration in the surface snowpack were also found, coincident with peaks in RGM concentrations, similar to findings at Alert and Kuujjuarapik.

Continuous measurements of GEM, RGM, and PM, initiated at Alert and Kuujjuarapik in the spring of 2001, reveal a complex interaction between these forms of mercury (Figures B.5.14a and 14b, respectively). During MDEs, decreases in GEM concentration are accompanied by an increase in RGM and PM. The inverse correlation between GEM and its reactive forms is expected, but an important observation is that PM concentrations are significantly higher than RGM (i.e., PM:RGM ratio is > 1) at Alert contrary to those from Alaska (Pt. Barrow) where the PM:RGM ratio was found to be < 1 (Lindberg *et al.*, 2002). At Kuujjuarapik in 2001 the PM:RGM ratio was initially < 1 but as further MDEs occurred, the PM:RGM ratio increased to > 1 (an overall average PM:RGM ratio was observed to be ~1).

At present, evidence suggests that most of the springtime oxidative chemistry for mercury occurs over the ocean. RGM (i.e., $HgCl_2$) is a very reactive ("sticky") species and will readily associate to particles. It is hypothesized, therefore, that the RGM concentrations exceed those of PM at Pt. Barrow because the monitoring site is close to the Arctic Ocean (2 km) where the conversion of GEM occurs and thus more RGM remains in the air sampled at Pt. Barrow. The differing PM:RGM ratios found at Kuujjuarapik are not well understood but could relate to the chemistry, meteorology, or climatology at this location which is sub-arctic and not adjacent to the Arctic Ocean (but is along Hudson Bay which is a large reservoir of halogens). Alert is in the high Arctic but farther inland from the ocean (6 km) than the other two sites giving more time for RGM to associate with airborne particles and thus be removed (scavenged or deposited) during the transport of the air mass to the monitoring site. If true, this hypothesis suggests that the observed PM:RGM ratio may provide an estimate of the age and origin of a particular air mass exhibiting depleted GEM concentrations.

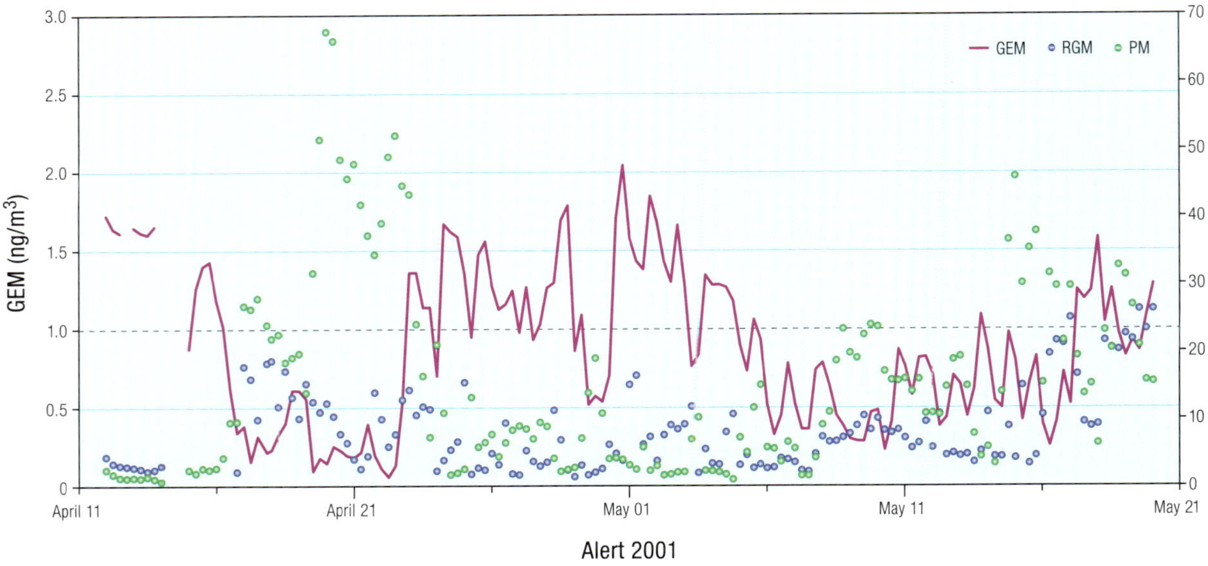

FIGURE B.5.14a

First simultaneous measurements of 6-hour averaged GEM, RGM, and PM concentrations at Alert, spring 2001.

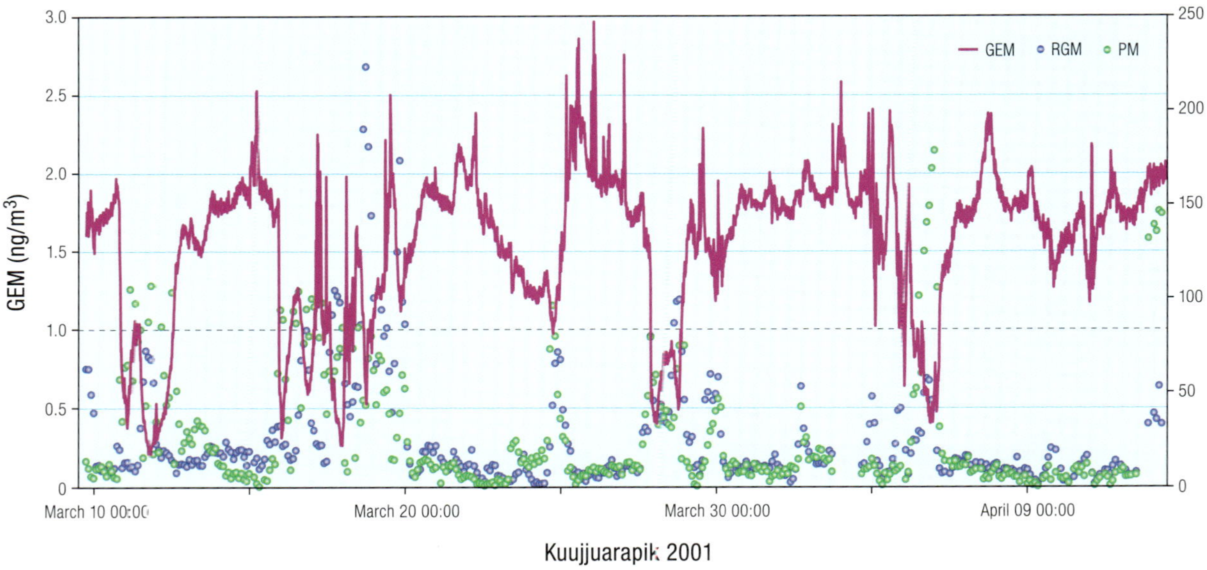

FIGURE B.5.14b

First simultaneous measurements of 6-hour averaged GEM, RGM and PM concentrations at Kuujjuarapik, spring 2001 (GEM is 15-minute average; RGM and PM are 2-hour averages).

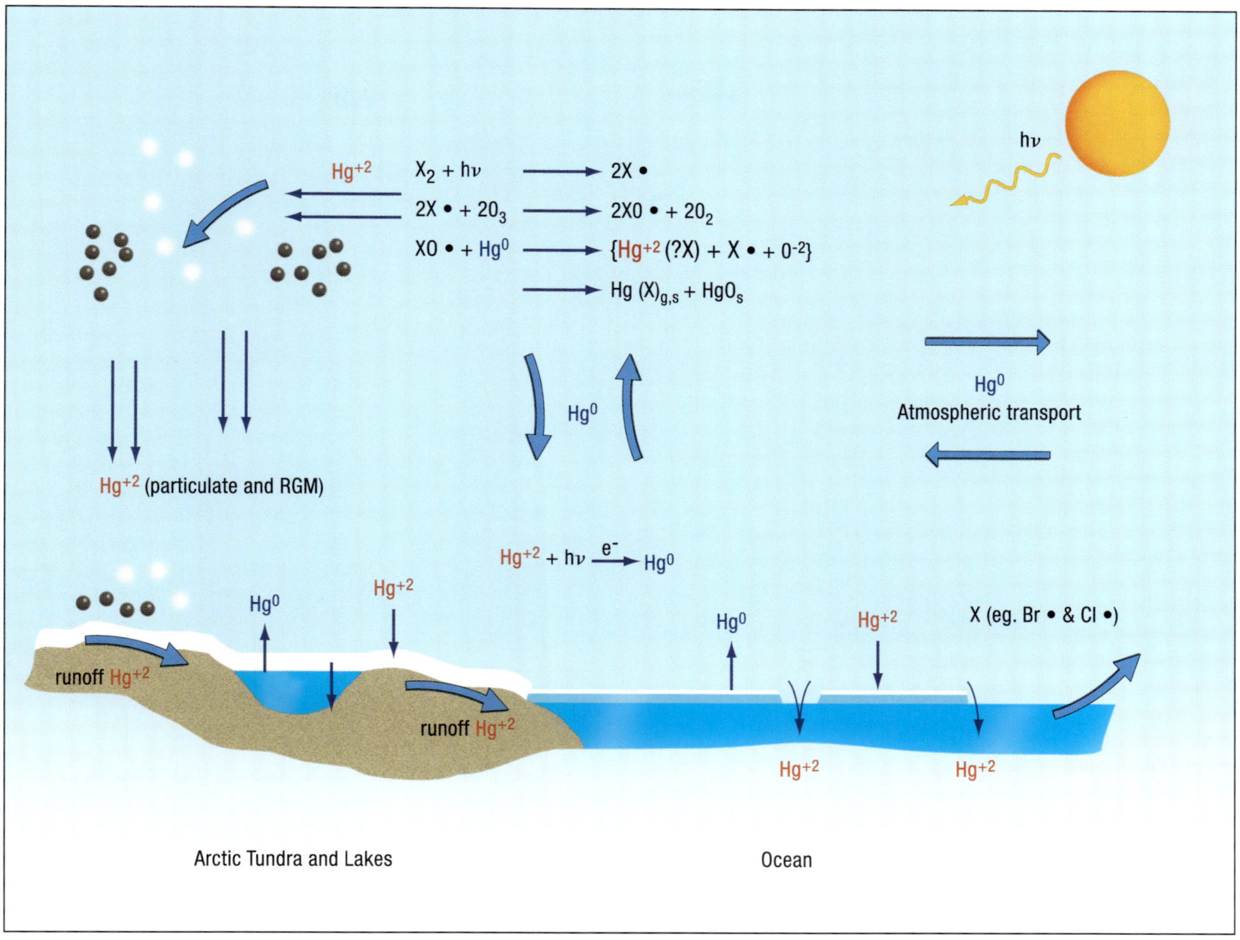

FIGURE B.5.15

Simplified schematic diagram of mercury cycling in the Arctic environment.

B.5.5 The cycle of atmospheric mercury in the Arctic

Despite a long history of study, only recently has the scientific understanding of the global mercury cycle undergone a revolution, partly due to the development of new instrumentation by a Canadian company (Tekran Inc.), and partly due to a focus on contaminant pathways in the Arctic under the NCP. We may expect that understanding to continue to evolve over the next few years from both field observations and laboratory experiments. New insights are likely to emerge from current and prospective studies in polar regions.

The current understanding of atmospheric environmental mercury cycling in the Arctic is shown schematically in Figure B.5.15. Before the measurements were taken at Alert, it was thought that mercury was distributed ubiquitously in the atmosphere of the Northern Hemisphere at a concentration of ~1.5 ng/m^3 and that Arctic air would consistently exhibit this background value. Concentrations of elemental mercury vapour above this background would then constitute contamination delivered by long-range atmospheric transport. These views were not incorrect but were, however, incomplete. During the past seven years, depletion events, wherein almost all of the Hg (0) in Arctic air near the surface is converted to a less volatile form, have occurred consistently at Alert and also the Kuujjuarapik site. Approximately 50% of the converted Hg at Alert remains in the air during these depletion events, with the remainder thought to be deposited on the snow surfaces. Direct measurements of total mercury content in snow have confirmed this process. The reactive Hg (II) deposited on the snow has been found to be reduced back to elemental mercury and then partly diffuses back into the air. Important components of this system remain to be understood. For example, how much of the mercury deposited during MDEs follows pathways that allow it to enter the food web? What are the factors that control those pathways? How might they change with time?

B.5.6 Summary

Significant advances in understanding the cycling and behaviour of mercury in the Arctic environment have been made since 1995. Field measurements at both Alert and Kuujjuarapik have been undertaken to understand the chemical processes leading to MDEs. It was discovered at Alert in 1995 that, during and after polar sunrise, atmospheric mercury episodically depletes to concentrations well below annual averages (at times to lower than the detection limit of the instrumentation). Ensuing research has found that these mercury depletion events occur throughout the high and sub-arctic in Canada and other polar regions (both in the Arctic and Antarctic). In the form of GEM, atmospheric mercury is expected to have a long atmospheric residence time, and for that reason, these significant decreases in the atmospheric concentrations were surprising. During MDEs mercury is depleting at the same time as tropospheric ozone, a phenomenon that is by now well known to occur annually in the Arctic. It has been concluded that gaseous elemental mercury is converted to more reactive species of mercury either in reactive gaseous form or associated with aerosols. Studies at Alert have shown that approximately half of these gaseous species can be recovered in the air during and after MDEs. The remainder are assumed to be deposited on the snow and ice surfaces during transit to the monitoring site. Studies have revealed increases in mercury concentration in snow during and after MDEs demonstrating that these converted mercury species are deposited onto the snow surface. In Kuujjuarapik, it was found that only about half the initial post-MDE mercury pulse remained in the snow within 12 hours of the depletion event, suggesting a reduction and volatilization loss of mercury at the surface after a MDE.

The location where MDEs occur is still to be determined; however, various studies have been undertaken to understand this question. Aircraft studies showed depleted levels of mercury in the first 1 km from the surface over wide areas of the Arctic ocean and the Beaufort Sea. Air concentration profiles and flux measurements conducted at the snow surface and within the snowpack show that depletion does occur at these locations. It was also found in these measurements that there is another cycle of mercury occurring at the snow surface (and within the snowpack) involving reduction of oxidized mercury. It is thought that a fraction of the oxidized mercury that is deposited during MDEs is photoreduced back to elemental mercury vapour immediately at the snow surface.

The chemistry causing these MDEs is yet to be explained, but there have been some significant advances in the understanding of mercury oxidation in the Arctic. In the presence of sunlight the combination of halogen species derived from the Arctic ocean are thought to contribute to the depletion of ozone at this time of the year. It is thought that Hg undergoes oxidation with similar halogens (oxidants) derived from sea salt. BrO has been measured both by satellite and *in-situ* by DOAS at both Alert and Kuujjuarapik and exhibits elevated concentrations during MDEs. This suggests that halogens play an integral role in the oxidation of mercury in the Arctic. Research into the oxidation kinetics and reaction products following MDEs is on-going.

Over the past several years the understanding of the behaviour of mercury in the Arctic atmosphere has significantly changed and evolved. What was once thought to be a pollutant known for its long residence time in the atmosphere is now known to be capable of being removed from the air within a period of hours during and after polar sunrise. Additionally, it is postulated that once the mercury is deposited to the snow and ice surfaces it can move further into the environment. Although researchers have made considerable progress in the study of atmospheric mercury behaviour in the Arctic and Antarctic, a complete understanding of the cause, effects, and implications of this discovery is still pending.

B.6 Mercury in sediments

Author: W. Lockhart

Affiliation: National Water Research Institute, Canada Centre for Inland Waters, 867 Lakeshore Road, Burlington, ON, L7R 4A6, Canada.

B.6.1 Introduction

Studies of mercury and in sediments have been included in the Northern Contaminants Program because of their ability to contribute insights to three issues:

1. the geographic variability of mercury and other elements in aquatic settings throughout northern Canada;
2. changes in inputs of metals to aquatic systems over time;
3. the relative contributions of natural and anthropogenic sources to amounts measured in present-day samples.

B.6.2 Geographic distribution of mercury in sediments from northern Canada

Several investigators have reported concentrations of mercury and/or other elements in samples of sediments from aquatic settings in northern Canada. The earliest survey was that of Atkins Baker (1979) who reported mercury in sediments from lakes and streams in Yukon.

The beginnings of a national survey of lake and stream sediments by the Geological Survey of Canada has been described by Painter *et al.* (1994). This study included sections of southern Yukon, central Quebec, and Labrador and several smaller areas. Painter *et al.* (1994) published colour-coded maps describing ranges of mercury found in sediments. The purpose of that study was not environmental but rather geochemical prospecting for potentially economic mineral resources. The samples were collected in a way to exclude the surface material in order to get a better geological perspective in the absence of interference from recent human activities.

A more recent survey of mercury and other elements in sediments from the Slave Structural Province has been completed by the Department of Indian Affairs and Northern Development, Yellowknife, (Puznicki, 1997) and raw data from that survey has been provided for this report. Other sources of data included individual research projects usually with a small number of samples (D.C.G. Muir, unpublished data on cores from several lakes; G.A. Stern, core data from lakes in the Yukon; E.G. Pannatier, (1997), thesis on sedimentation in the Mackenzie Delta; Gubala *et al.* (1995), cores from two lakes in central and northern Alaska).

Most samples have been grabs collected using a dredge with upper layers consisting of a few centimetres of sediment nearest the sediment-water interface. Other samples have been cores sliced into layers that were analyzed separately to construct depth profiles. For the description of geographic variation, cores have been treated as grab samples by taking the average concentration of the top few slices, usually to a depth of 3–5 cm.

In some instances, concentrations have been obtained directly from raw data but in others the concentrations have been estimated from graphical presentations in the original sources. Similarly, the geographic positions were usually obtained as numeric data but sometimes the positions of the sampling points were estimated from maps.

The values obtained from these sources have been tabulated and presented as maps with different ranges in concentrations shown as points of different colours (Figure B.6.1). The colours were selected based on Canadian sediment quality guidelines (Table B.6.1) for mercury: green = Hg < 170 ng/g; yellow 170–486 ng/g; red > 486 ng/g. The map depicts 368 points from Alaska to the Quebec/Labrador border, with the most detailed coverage in the area north of Great Slave Lake based on the data of Puznicki (1997). It shows 302 sites in green, 43 in yellow and 23 in red, although overlaps frequently obscure points when plotted at a scale required to cover so large an area. There are several red points in the area of Strathcona Sound (from Fallis, 1982), and in the region north of Great Slave Lake. In addition to the points shown on the map (Figure B.6.1), Geological Survey data on lake and stream sediments were illustrated by Painter *et al.* (1994), who provided excellent coverage for a few areas but contained no data for many large areas, especially in the North. The report did not give the number of samples in each range for each province or territory but it did give a cumulative distribution figure for all 161 228 analyses nationally. The 95th percentile figure was 190 ng/g suggesting that only a small percentage of sediments exceeded the Canadian sediment quality guideline of 170 ng/g for mercury. An area of the south-eastern Yukon with high mercury levels in sediments is presented at a finer scale by Painter *et al.* (1994) showing points near the Yukon/NWT border with results mostly above 189 ng/g.

TABLE B.6.1 Canadian sediment quality guidelines for mercury

Element	Sediment quality guideline (ng/g)	Probable effect level (ng/g)
Mercury	170	486

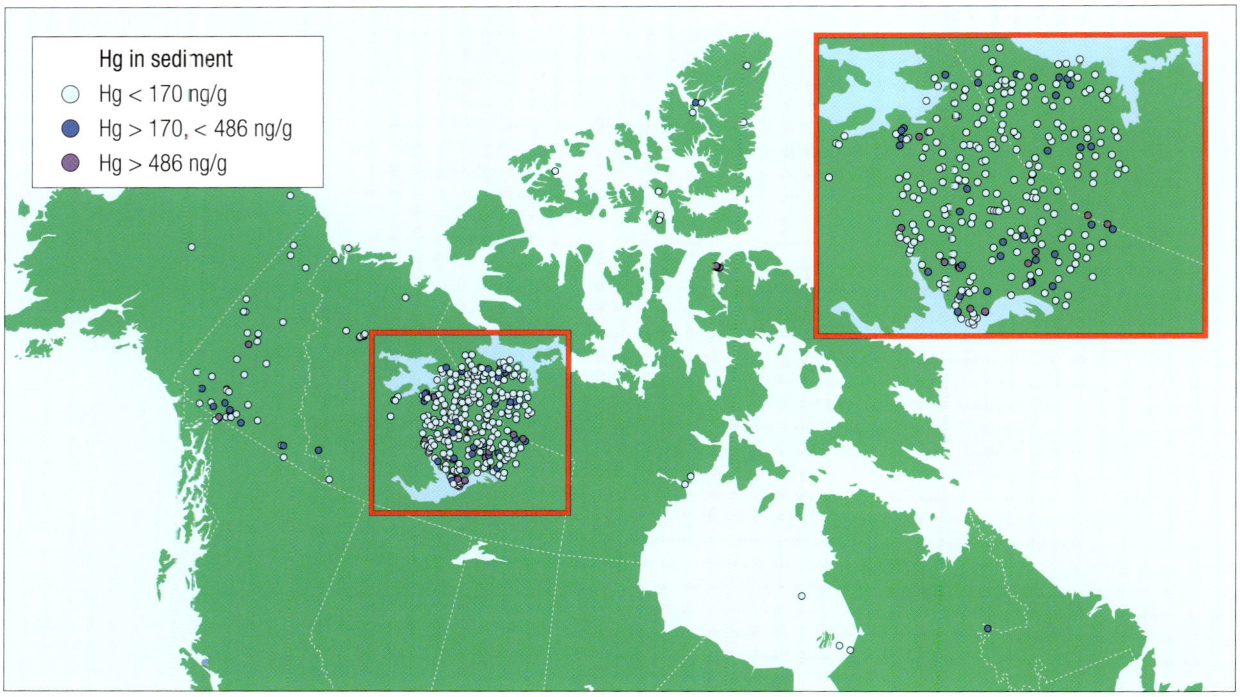

FIGURE B.6.1

Levels of mercury in surface sediment collected from lakes and rivers in northern Canada.

In some instances, high levels of mercury have been explained by local, industrial sources, usually metal mines. For example, Moore and Sutherland (1981) reported mercury at high levels in sediments from the east end of Great Bear lake near an operating silver mine (Echo Bay Mines) and an abandoned uranium mine (Eldorado mine). Concentrations in sediment from different sampling stations reached as high as 3 mg/kg (3000 ng/g) and decreased with increasing distance from the mining site. Moore and Sutherland also noted a negative correlation between the mercury content and the organic content of sediments.

Considering the map in Figure B.6.1, and the similar presentation from Painter *et al.* (1994), mercury levels exceed the guideline level of 170 ng/g at only a small proportion of sites — approximately 10%. Levels exceed the probable effect level of 486 ng/g at even fewer sites. In spite of this, fish frequently exceed human consumption guidelines for mercury in human food and high levels in the fish appear to bear little relationship to levels in the sediments (see fish section of biology chapter). Perhaps conditions within each lake regulate the flow of mercury to the fish as much or more than basal geological levels and mineralogy.

B.6.3 Changes in inputs of mercury over time

Perhaps the most helpful insight derived from sediments are down-core profiles when they provide historical records. The use of ^{210}Pb to date layers of sediment has become commonplace and has been used to date a number of Arctic cores. With the detection limits for unsupported ^{210}Pb, deposition over a period of about 100–150 years can be measured frequently. ^{210}Pb by itself, however, is inadequate and an independent test of the dates is required. ^{137}Cs from the atmospheric testing of nuclear bombs is probably the most frequently used independent test of ^{210}Pb dates but there are others such as synthetic organic compounds and biological remains. The profile of ^{137}Cs should show peak inputs in the early 1960s and ^{210}Pb layer dates are more credible if they place the ^{137}Cs peak at the correct time. Mercury is often well preserved in lake cores over times datable by ^{210}Pb, notably when organic carbon levels are high. There remains scientific debate over the interpretation of some core profiles, however, and ambiguities are commonly observed.

Essentially cores either reflect the history of the site, or they are artifacts of vertical redistribution of mercury, or they are some combination of these. The two artifacts most likely to cause vertical redistribution of mercury are particle mixing by physical disturbance in the sediments and exchanges between sediment particles and pore water with vertical diffusion during the periods spent in solution.

One northern study, bearing on the fidelity of core records of inputs of mercury, was that of Giauque Lake (63°10' N 113°50' W), possibly the most highly contaminated site in northern Canada (Lockhart *et al.*, 2000). Giauque Lake was the site of the Discovery gold mine which operated from the mid-1940s until 1968. Mercury amalgamation and cyanidation were both used to recover the gold (Moore *et al.*, 1978). Two cores were taken in March, 1995 at different distances from the mine site. The records of inputs of mercury to the core sites provided accurate records of the history of mine operations. In this case, and in other temperate sites described in the same publication, core records of inputs of mercury were faithful to the known histories of the sites.

Most studies of mercury in sediment cores were completed during NCP-Phase I and were reviewed in the previous report; however, a few additional core studies have been reported during Phase 2.

TABLE B.6.2 Summary of recent Hg in sediment studies

Location	Reference	Observations
Mackenzie Delta lakes	Pannatier, 1997	Very shallow lakes; one lake with no physical mixing indicated by the presence of varves; sedimentation rates high; dating based on ^{137}Cs; no change in Hg over depth of cores; no indication of recent contamination with Hg
Arctic Ocean basins	Gobeil *et al.*, 1999	Deep, marine sites; sedimentation rates extremely low with several thousand years required to deposit 1 cm of sediment; sedimentation too slow to use ^{210}Pb dating; ^{210}Pb profiles have excess in top 3 cm only indicating a mixed layer to that depth; Hg-enriched in surface sediments extending back several thousand years; positive correlations of Hg with 'reactive' Mn and Fe; vertical redistribution of Hg was indicated; probably enrichment of Hg was generated by chemical gradients in the sediment, not by contamination
Lake DV09, Devon Island	Lockhart *et al.*, 2000	Laminated core, not mixed physically, dating by ^{210}Pb with ^{137}Cs confirmation and also with synthetic organic compounds; Hg-enriched in surface layers, negative correlations between Hg and both total Mn and Fe; positive correlation with C; probably surface enrichment of Hg is at least partially the result of contamination with increased inputs of Hg; given the remote setting, the atmosphere is the only credible source
Several lakes and one marine core from Baird Inlet	Cheam *et al.*, 2001	Mostly ^{210}Pb dating; ^{137}Cs available for three cores, moderate sedimentation rates in lake cores; some discrepancies between ^{210}Pb dates and Cs profiles; no relationships between Hg and 'reactive' Fe or Mn; probably surface enrichment of Hg is at least partially the result of contamination with Hg
Lake Laberge	Stern *et al.*, 2001	^{210}Pb dating with ^{137}Cs confirmation; high sedimentation rates; cores appear unmixed because they all indicate declines in inputs of mercury at about the same time, possibly due to Yukon River dredging; Hg-enriched in surface layers; other metals not reported; probably enrichment of Hg is at least partially the result of contamination with Hg

An elegant series of studies on sedimentation in the Mackenzie Delta region has been reported by Pannatier (1997). In this study, cores from lakes in the area showed no enrichment of mercury in upper layers and hence no indication of increased inputs of mercury from any source. This observation is consistent with cores from Yaya Lake in the Mackenzie Delta reported by Lockhart *et al.* (1998) and also with results from Schrader Lake in Alaska north of the Brooks range of mountains (Gubala *et al.*, 1995). The sediments from lakes in the Mackenzie Delta and the North Slope of Alaska indicated little or no increase in inputs of mercury. Rummy Lake (69°05' N, 123°38' W) east of the Mackenzie Delta, however, did show an increase in mercury in surface sediment layers (Cheam *et al.*, 2001).

Gobeil *et al.* (1999) reported a series of ocean cores taken on a cruise from Bering Strait to the North Pole. The down-core profiles of mercury showed increased concentrations in the uppermost layers but the sedimentation rates were so low at these deep ocean sites that the mercury inputs would have to have increased several thousand years ago and to have been preserved in order to interpret the profiles as histories of pollution. Instead, the correlations of mercury with 'reactive' iron and manganese suggested that the mercury profiles were generated by vertical redistribution within the sediment over very long periods of time in response to oxidation-reduction gradients. That is, the mercury profiles in these cores cannot represent histories of pollution. With the scope of hemispheric or planetary circulation of mercury it is difficult to imagine that even remote sites like the deep Arctic Ocean basins have escaped contamination completely. Most core studies interpreted as histories of contamination have suggested that the contamination has increased during the last century. If that applied to the deep Arctic basins, and if the mercury were perfectly stable in the core, it would be confined to a layer only a fraction of 1 mm in thickness. Other studies have suggested that mercury in sediments has doubled or tripled over the last century and so a very thin layer at double or triple the basal levels would not likely be detectable when mixed over 1–3 cm, even if it were present. It remains possible that some component of contaminant mercury may be confined to a very thin layer and be obscured by natural amounts; however, the most northerly freshwater core reported to date, that from Lake Hazen (81°45' N, 71°30' W) on Ellesmere Island, showed little or no indication of recent contamination (Lockhart *et al.*, 1998).

A laminated core from northern Devon Island was described by Lockhart *et al.* (2000). Excess ^{210}Pb measurements were used to date the slices, and these dates agreed well with inputs of ^{137}Cs. Mercury was enriched in the upper slices spanning about 100 years and was positively correlated with both (total) manganese and iron suggesting the possibility that some or all of the enrichment of mercury in upper slices could be an artifact. The rate of movement of mercury, however, would have to be very high in order to generate the observed profile in less than a century. It seems likely that at least some of the enrichment in mercury at this site reflects increased inputs during the past century.

Cheam *et al.* (2001) studied lake cores from diverse sites across northern Canada. The ^{210}Pb dates for three of these sites were compared with ^{137}Cs profiles and the results indicated relatively poor agreement in two cores (Rummy and Oksana), and extensive surface mixing in the third (TK-54). Two of these lakes are in the high Arctic, Baird Inlet Lake (79°20' N, 74°00' W) and Lake AXAJ (80°00' N, 87°00' W). Baird Inlet Lake had a clear increase in mercury in slices with ^{210}Pb dates after about 1900 with calculated fluxes about three times those of the deepest layer. The fluxes of iron indicated by the slices showed an apparent negative correlation between these two elements. Lake AXAJ had a variable profile for mercury with a surface and a sub-surface peak and an apparent positive correlation between mercury and iron. Considering these two lakes and the two other extreme northern lakes (Hazen and Buchanan) reported by Lockhart *et al.* (2000), there are inconsistencies that will require further study. Mercury concentrations were elevated in slices at or near the surface relative to deep slices in Lake Oksana where total and reactive iron were measured but no relationship was evident. It seems that the value of reactive iron to predict the profile of mercury may vary from site to site and that additional efforts to understand these relationships will be required.

Stern *et al.* (2001) obtained three cores from Lake Laberge, Yukon, in March, 2000. The ^{210}Pb profiles showed near-ideal exponential decline with depth and provided quite good dates for the ^{137}Cs peaks. The mercury profiles in these cores, however, were quite variable showing large variations among slices. All three of them showed a large drop in mercury in the 1890s and again in the 1930s, possibly associated with historical events in the watershed, namely dredging the Yukon River starting in the 1890s and construction of the Marsh Lake dam in 1924. The cores all suggested some increase in loading with mercury relative to pre-1850 levels and that peak inputs occurred in the 1970s or 1980s. Present-day inputs have fallen since then, but are still above historical inputs.

On balance, the cores suggest that different regions of the Canadian Arctic may be receiving different inputs of mercury. The cores suggest that inputs to the eastern Arctic south of about 80°N have increased over about the past century by a factor of about two, but that the area north of there has been affected little. Progressing southward, the signal from recent inputs becomes clearer. In the western Arctic, a similar gradient seems to exist with cores from the Mackenzie Delta and the north slope of Alaska indicating little recent enrichment of mercury, but cores further south indicating it consistently.

Measurements of mercury in air and snow are reported separately (See B.5 and C.2). The consistent detection of mercury in these samples indicates the import of mercury to the Arctic and its deposition to the surface. Some, hopefully most, of this mercury may revolatilize during spring thaw, but some is probably retained in Arctic ecosystems. It has been assumed that the import of atmospheric mercury is a source that contributes to the food chain and ultimately to fish and the people who consume them. This chain of events remains somewhat tentative until future work can test it more rigorously.

B.6.4 Natural and anthropogenic sources

This is the most speculative of the three sections in this report. While mercury has a number of stable isotopes like lead, mercury isotopes are not radiogenic in origin and so different deposits of mercury do not differ as strikingly as they often do with lead. Consequently, mercury isotopes have not been used to discriminate environmental sources until very recent improvements in analytical methods. It is unknown if these emerging technologies have been applied to Arctic samples. Lacking isotopic evidence, arguments about natural or anthropogenic sources are based on sediment core profiles and a few cases in which biological samples have been preserved. The lake sediment cores generally suggest that inputs of mercury have increased, at least in some parts of the Canadian Arctic, over the past century. The only obvious mechanism to explain this is the known emission of mercury to the atmosphere as a result of industrial processes at lower latitudes, followed by transport with moving air masses and deposition in the Arctic. If this mechanism applies, then mercury should be found in air and snow, which is the case. The amount of increase is generally consistent with up to approximately twice the mercury reaching newly deposited sediments as that which reached sediments prior to 1900. Several authors have inferred that this increase is the result of anthropogenic activities (e.g., Hermanson, 1993), notably burning fuels or garbage, or smelting certain ores containing trace amounts of mercury. In addition, several cases of biological samples have suggested increased levels in recent time (e.g., Hansen *et al.* 1991).

Perhaps it is even more speculative, but it seems appropriate to suggest a possible link between the dynamics of mercury and climate warming. That is, warming may produce something like a reservoir effect but at a regional scale. Destabilization of permafrost may be expected to produce increased erosion of mercury-containing soils and sediments, while at the same time providing better conditions (warmer, more organic matter) for the methylation of mercury. Increased sediment transport might be anticipated in a large, uncontrolled river like the Mackenzie and in some Russian rivers, and with increased particle transport, increased mercury of particle-bound mercury. Whether this is 'natural' or 'anthropogenic' mercury does not matter much; the result is the same if it becomes methylated and moves up food chains into animals and people.

B.7 Other heavy metals in the Arctic atmosphere

Authors:

S. Gong (1); L. Barrie (2)

Affiliations:

1. Meteorological Service of Canada, 4905 Dufferin Street, Downsview, ON, M3H 5T4, Canada;
2. Atmospheric Research and Environment Programme, World Meteorological Organization, 7 bis avenue de la Paix, CH-1211 Geneva 2, Switzerland

B.7.1 Introduction

Heavy metals are natural constituents of the Earth's crust and are present at varying concentrations in all ecosystems. Human activity has drastically changed the biogeochemical cycles and balance of some heavy metals. Between 1850 and 1990, production of copper, lead and zinc increased nearly 10-fold, with emissions rising in tandem (Nriagu, 1996). Excessive levels of metals in the marine environment can affect marine biota and pose risk to human consumers of seafood. Today, much more is known about the health effects of heavy metals. Environmental exposure to high levels of heavy metals has been linked with developmental effects, various cancers, and kidney damage.

The main anthropogenic sources of heavy metals are various industrial sources, including present and former mining activities, foundries and smelters, and diffuse sources such as piping, constituents of products, combustion by-products, traffic, etc. Relatively volatile heavy metals and those that become attached to airborne particles can be widely dispersed on very large scales.

Measurements of metal concentrations in the Arctic aerosols have been carried out since 1980 at Alert, Canada and seasonal variations and long-term trends were reported by Sirois and Barrie (1999). Selected metal air concentrations and trends at Alert up to 1995 have been summarized in the previous CACAR report (Macdonald *et al.*, 2000). This section presents the most up-to-date data set for more metals in air at Alert with the analysis of seasonal variations and trends.

B.7.2 Experimental methods

The detailed descriptions of sample collection and chemical analyses have been given by Sirois and Barrie (1999). Seven-day aerosol samples were collected on 20 x 25 cm Whatman 41 filters using a high-volume sampler. Once samples were back in the laboratory, they were cut into eight aliquots to be used for chemical analysis. Three analytical methods were used, liquid ion chromatography (IC), instrumental neutron activation analysis (INAA), and inductively coupled plasma emission or mass spectroscopy (ICP/MS). Most metal analysis was done with

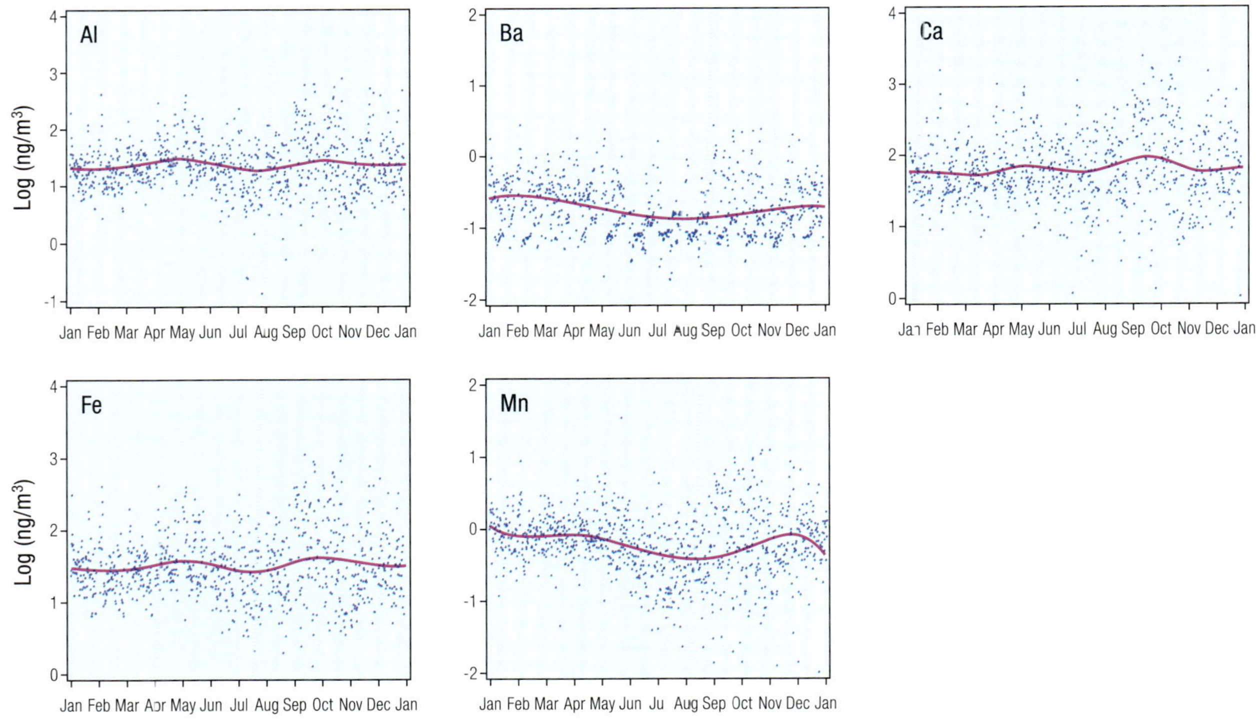

FIGURE B.7.1
Metals without significant seasonal variations, concentration scale is log (ng/m^3).

INAA or ICP/MS. There are some overlaps of metal analysis between INAA and ICP/MS; however, INAA data were preferentially used for trend and seasonal variation analysis as it yields the total metal contents in a sample.

B.7.3 Seasonal variations

Trends and seasonal variations in the weekly mean concentrations of 15 metal elements in aerosols collected on filters at Alert, Nunavut, Canada over a 20-year period (1980–1999) are summarized in this section. Part of the first 15-year data has been published by Sirois and Barrie (1999).

Seasonal variations of metal components in Alert aerosols are well established from the 20-year data set. No significant seasonal variations were found for soil related metals such as Al, Ba, Ca, Fe and Mn (Figure B.7.1). There are two weak maxima during the year; however, one in spring (April–May) from long-range transport and one in fall (September–October) from local soil emissions.

The anthropogenic-derived metals such as Pb, Zn, and Cu and sea-salt components Na and Mg exhibit a maximum in winter and minimum in summer (Figure B.7.2). Similar variations were observed for non-soil fractions of vanadium and manganese (Sirois and Barrie, 1999). The strong seasonality of aerosols for the aerosol components is associated with the lifetime of aerosols in the Arctic air mass. Because of stable thermal stratification in the atmospheric surface boundary layer and less than 10-mm H_2O/month of precipitation in the cold half of the year from October to May, aerosol residence times are much longer in winter (~3 to 7 weeks) than in

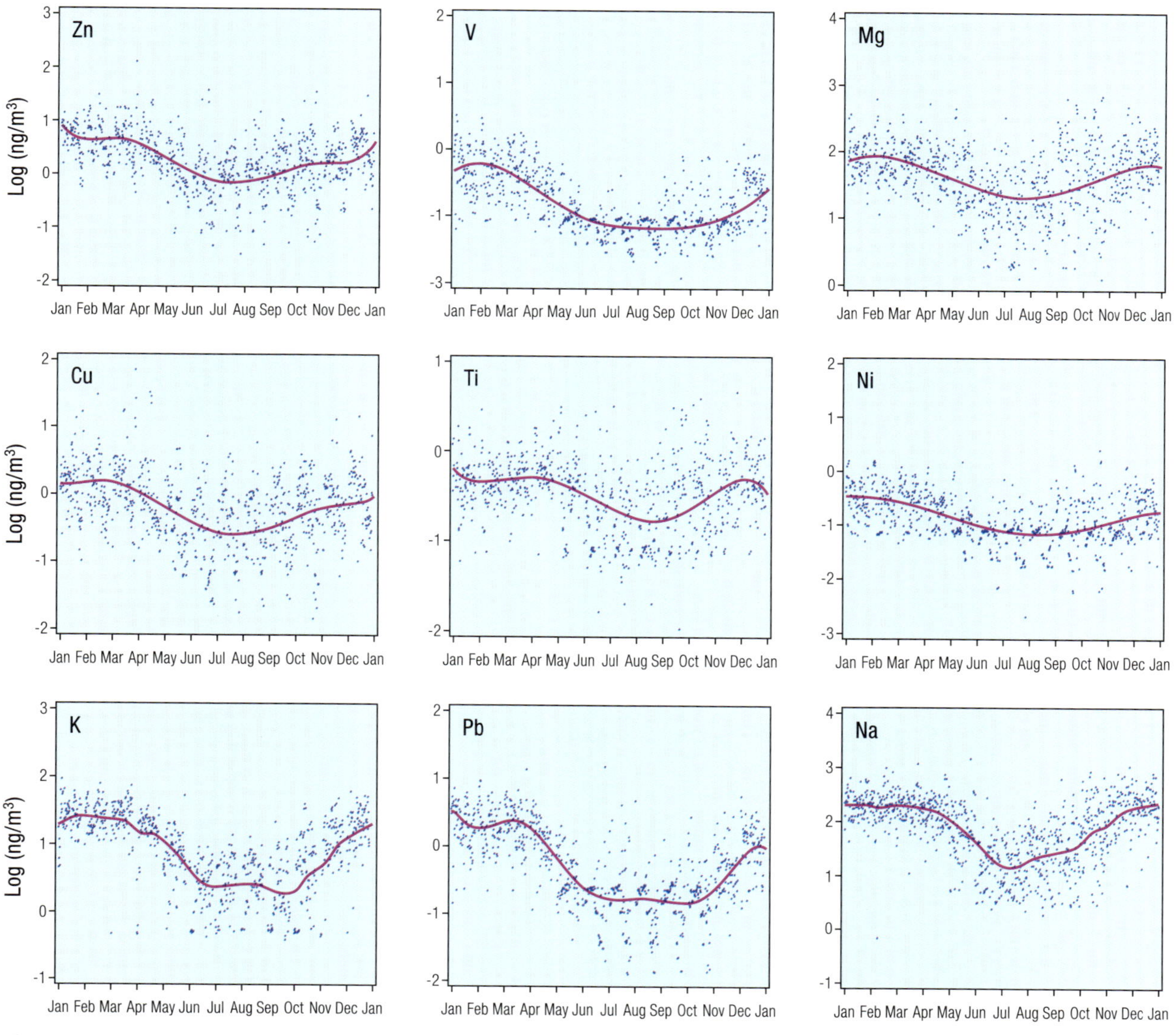

FIGURE B.7.2

Metals with significant seasonal variations, concentration scale is log (ng/m³).

summer (~3 to 7 days). The seasonality of any species at Alert, therefore, is a result of factors such as meteorology, transport, and removal processes specific to the Arctic environment as no such variations were observed in the source regions of these metals.

B.7.4 Long-term trends

The long-term trends were analyzed from all data for 12 metal components and from winter data only for selected metals. For all data, no long-term trends were detected for sodium and calcium. Zinc and copper showed a slight increase from 1980 to 1985 and then thereafter decreased. For most other metals, a slight decreasing trend was observed (Figure B.7.3). Lead showed a decreasing trend from 1985 to 1993 and then increased. Since there are too many BDLs (below detection limits) for all the data of Pb, it may not be statistically significant to resolve a long-term trend from them. Trends, however, were obtained from winter data only as shown in Figure B.7.4. Results from trend analyses on the more polluted winter period alone confirmed earlier observations and also showed long-term decreases in lead and zinc as well as reductions in the concentrations of vanadium and manganese not seen in the larger data set (Sirois and Barrie, 1999). The analyses indicate that there are no obvious increasing trends for anthropogenic components, but rather, a slight decrease.

The long-term trends observed at Alert for the anthropogenic components such as Cu, Pb, and Zn are governed by trends in the emissions of the anthropogenic components surrounding the Arctic basin as well as the transport/circulation patterns that influence the air masses that reach Alert. Emissions data indicated that changes have been taking place from 1983 (Nriagu and Pacyna, 1988) to the mid-1990s (Pacyna and Pacyna, 2002) on a hemispheric scale. For trace metals emitted mostly during coal combustion, namely Hg, As, Cd, Cr, Cu, Mn, Mo, Pb, Sb, Sn, and Zn, emissions in 1995 are either similar or lower than in they were 1983. Some of the changes were consistent with the trends observed while others did not correlate at all, reflecting the influences of circulation pattern changes to the long-term trends.

Kahl *et al.* (1999) studied the decadal-scale variability in long-range atmospheric transport to the summit of the Greenland ice sheet from the past 44 years of modeled 10-day air trajectories arriving daily at the summit. Trajectory residence times over specific upwind

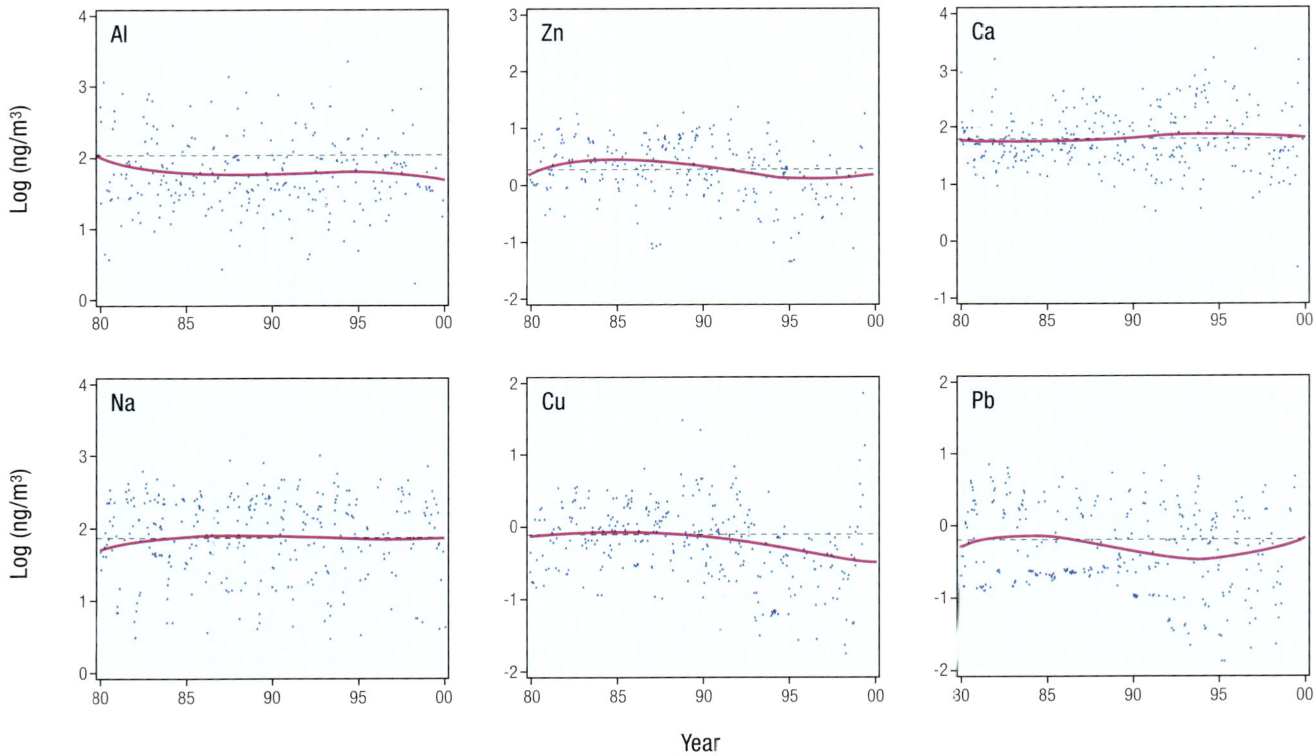

FIGURE B.7.3

Estimated long-term trends of weekly concentrations of 12 metals at Alert. Long-term trend curves are fits to the observations displayed here after taking into account the seasonal variation. A redhorizontal grid line is included for reference. Concentration scale is log (ng/m³).

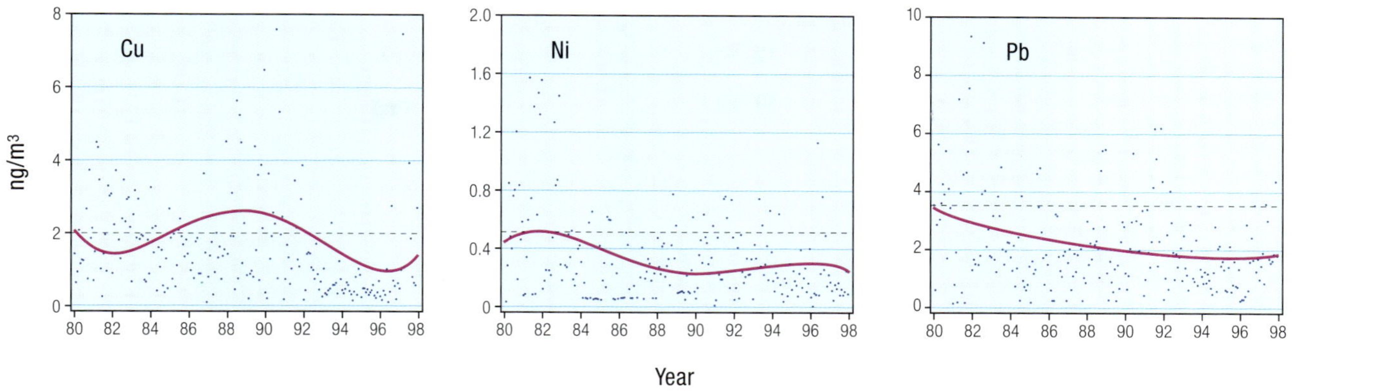

FIGURE B.7.4
Estimated long-term and seasonal trends of Pb, Cu, and Ni (ng/m^3) in air at Alert (weekly concentrations from winter only).

regions reveal inter-annual oscillations with periods of 4–14 years. This implies that even a region with a constant emission could result in an oscillating contribution to the Arctic in a 4–14 year cycle, which may either weaken or amplify contributions of the changes in the anthropogenic emissions to the Alert observations. To realistically interpret the long-term trends of observed species at Alert, changes in the emission inventories, circulation patterns, and removal processes need to be considered simultaneously in a coherent way. A 3-D chemical transport model would be an ideal tool to link them together.

B.7.5 Principal component analysis

The lead isotope ratios of $^{206}Pb/^{207}Pb$ are analyzed from the long-term measurements, which are indicative of origins of atmospheric pollutants to the Arctic (Sturges and Barrie, 1989). The stable isotopic ratio of $^{206}Pb/^{207}Pb$ at Alert was found to be 1.16 ± 0.01 and is characteristic of air originating over northern parts of the former USSR. Recent analysis of 1994–1995 lead isotope data $^{206}Pb/^{204}Pb$, $^{208}Pb/^{204}Pb$, and $^{206}Pb/^{207}Pb$ (Mercier *et al.*, 1999) suggests three major sources of atmospheric aerosols reaching the Canadian High Arctic: i) a radiogenic "natural" end-member from Canadian Arctic islands and the coastal areas of west Greenland in September; ii) an unradiogenic end-member originated from west and/or northwest Europe in later fall and winter and iii) an anthropogenic end-member with $^{206}Pb/^{207}Pb$ = 1.16, originating from Asia and Russia during late spring and early summer.

Consideration of associations between the various metals yielded further information on the trends and sources of the metals. Application of a Principal Component Analysis (PCA) approach for winter/spring data showed that ten factors (Table B.7.1) explained 93% of the variance. Metals such as Pb, V, Mn, and Zn, for the most part, were linked with a factor (ANTHRO) which peaked January to March and included non-acidic salts of sulphate with ammonia as the dominant cation. This factor, which reflects anthropogenic sources in general, has shown a gradual decline over the 20-year period (Table B.7.1). Another factor dominated by zinc and copper (SMELTER), is associated with smelter emissions and the long-term trend includes a slight increase in the 1980s and then beginning in 1991 has decreased to the end of the current measurement period. A marked seasonal variation is also shown with a peak in November to March and lower concentrations in April and May. This factor, however, does not include a significant contribution from sulphate (Table B.7.1) as most sulphate is formed in the Arctic atmosphere and the sulphur associated with this factor reflects primary emissions from smelters in the Arctic air shed. Metals associated mainly with the SOIL factor (Mg, Al, Ca, Ba, Ti, and Fe) show little long-term trends but weak seasonal variations. The weaker seasonality than the anthropogenic components are associated with the fact that there are two contributions to the soil components: long-range transport in spring and local contributions in fall. Similar seasonality to the SEASALT components (Na, K, and Mg) and ANTHRO components would occur if there were no local soil sources.

The analyses also confirm that the major source regions influencing the concentrations seen at Alert are primarily from Eurasia, although other regions adjacent to northern Canada but at higher elevations record a different aerosol mix and may be more significantly influenced by emissions from North America (Sirios and Barrie, 1999). The average concentration of a series of chemical elements in the surface air of the eastern part of the Russian Arctic was measured in five consecutive springs (1985–1989) (Vinogradova, 1996). The concentrations of Ca, Al, Si,

Sc, Fe, Sm, Th, Ni, and Cd detected in the eastern part of the Russian Arctic are the highest among all published data on the Arctic atmosphere. For comparison purposes, atmospheric concentrations of the selected metals in the Arctic regions are listed in Table B.7.2 which includes Barrow and Alert. Averaged concentrations at Alert are comparable to those at Barrow but are about 3–4 times lower than those at the eastern Russian Arctic. This indirectly confirms our analysis that the major pollution sources of metals to the Canadian Arctic are from Eurasia.

TABLE B.7.1 PCA rotated factor patterns from winter/spring data

Component	Soil	Sea-salt	Photo-S	Anthro	Smelter	Iodine	MSA	Broine	Manganese	Nitrate
INAA_Cl	-0.02	**0.92**	-0.17	-0.01	-0.05	-0.02	-0.09	-0.01	-0.02	-0.02
IC_NO3_	0.04	**0.66**	-0.05	0.19	0.04	0.05	0.08	-0.14	0.04	**0.57**
IC_SO4_	0.05	0.07	**0.91**	**0.23**	0.15	0.01	0.01	0.18	-0.02	-0.01
IC_H_	-0.26	-0.29	**0.82**	0.05	-0.02	0.05	-0.03	0.14	-0.04	-0.08
INAA_Na	0.02	**0.90**	-0.04	0.07	0.03	0.21	0.08	0.06	0.01	-0.01
IC_K_	-0.09	**0.85**	**0.27**	0.18	0.09	0.00	-0.14	0.03	-0.02	0.03
ICP_Pb	-0.06	0.15	**0.45**	**0.70**	**0.27**	-0.07	-0.18	-0.01	0.00	0.20
ICP_Mg	**0.52**	**0.74**	-0.04	0.17	0.06	-0.15	-0.15	0.14	0.00	0.01
IC_NH4_	-0.06	0.09	**0.84**	**0.33**	0.19	0.05	0.00	-0.06	0.00	0.10
INAA_V	0.19	0.16	0.25	**0.82**	0.05	0.17	0.12	-0.16	0.09	-0.05
INAA_Mn	0.24	-0.01	-0.04	0.08	0.00	0.04	0.02	0.03	**0.96**	0.04
ICP_Cu	0.02	0.02	0.07	0.01	**0.96**	0.06	-0.03	0.07	-0.01	0.00
INAA_Al	**0.90**	0.00	-0.14	0.06	0.01	0.19	0.14	-0.08	0.11	-0.05
ICP_Ca	**0.91**	0.10	-0.10	0.11	0.02	-0.11	-0.07	0.01	0.06	0.02
INAA_Br	-0.08	0.08	0.20	-0.07	0.04	0.21	0.14	**0.90**	0.03	-0.03
INAA_I	0.08	0.11	0.07	0.04	0.04	**0.92**	0.17	**0.20**	0.04	0.06
IC_MSA	0.07	-0.13	-0.02	-0.04	-0.11	0.17	**0.95**	0.13	0.02	0.03
ICP_Zn	0.09	0.07	0.23	**0.34**	**0.84**	-0.03	-0.10	-0.05	0.01	0.03
ICP_Ba	**0.57**	0.00	0.03	**0.53**	0.01	0.04	-0.02	0.13	0.03	**0.48**
ICP_Ni	0.27	0.16	0.18	**0.86**	0.13	-0.05	-0.06	0.03	0.03	0.06
ICP_Ti	**0.70**	0.01	0.12	0.08	0.02	**0.26**	0.04	-0.07	0.12	**0.48**
ICP_Fe	**0.96**	-0.02	-0.06	0.14	0.06	-0.04	0.04	-0.03	0.08	-0.02

IC – Ion Chromatography, CP – Inductively Coupled Plasma – MS, INAA – Instrumental Neutron Activation Analysis

TABLE B.7.2 Comparisons of average concentrations (standard deviation) of selected metals in spring Arctic air (ng/m^3)

Place	Wrangel Island		SP-28	Eastern Arctic	Central Arctic	Alaska		Alert
Season	**1986**	**1987**	**1987**	**1985–1989**	**1985, 1986, 1988**	**Winter–Spring**	**1986–1987**	**1986 Winter**
Ca		1130 (720)	250 (110)	600	180	53	95	88
Al		690 (270)	540 (180)	360	78	30–114	25	17
Mn		4.5 (3.0)	5.0 (2.8)	3.8	1.4	0.5–1.1	0.9	0.8
Cu	3.3 (2.0)	8.3 (2.1)	7.0 (3.8)	6.4	3.8	4	0.2–1.0	1.4
Zn		28 (20)	14 (9)	16	5.6	4.8–11	1.8–3.0	4.2
Pb		5.6 (8.8)	13 (14)	10	3.5	0.1	1.8	1.9

Except for Alert, data source: Vinogradova, 1996.

B.8 Radionuclides: local and long-range sources

Authors:

J. N. Smith

Affiliation:

Bedford Institute of Oceanography, Fisheries and Oceans Canada, PO Box 1006, Dartmouth, NS, B2Y 4A2, Canada.

B.8.1 Introduction

Transport of radioactivity into the Arctic Ocean occurs by long-range atmospheric and ocean-current pathways and by direct injection into shelf seas from local sources. Although artificial radioactivity, mainly ^{137}Cs, had been used for some years as a circulation tracer in the Arctic Ocean (eg., Smith *et al.*, 1990; Smith and Ellis, 1995), interest in actual radioactivity pollution of the Arctic Ocean was heightened by the release in 1993 of the Yablokov Report (Yablokov *et al.*, 1993) which outlined the large quantities of radioactive materials that had been discharged into the Barents and Kara Seas since the early 1950s by the former Soviet Union. The radiological effects associated with sources in the Arctic Ocean have been recently summarized in Strand *et al.* (2002), and some of the more recent work on sources and pathways are reviewed in the following sections.

B.8.2 Long-range transport

Recent investigations have revealed that the Arctic Ocean is undergoing major changes in contaminant concentrations and rates of water circulation on time scales of the order of years (Carmack *et al.*, 1995; 1997; Morison *et al.*, 1998; Steele and Boyd, 1998). Results from icebreaker cruises in 1993–1995 in the Western Arctic Ocean indicated that a major thermohaline disturbance was underway, characterized by an increase in the transport and/or temperature of Atlantic water flowing into the Eurasian Basin and a corresponding increase in flow past the Lomonosov Ridge into the Canada Basin (McLaughlin *et al.*, 1996; Carmack *et al.*, 1997; Swift *et al.*, 1997). This direct intrusion of water from the North Atlantic along the continental margin of the Eurasian Basin into the Canada Basin was also carrying elevated inventories of the radioactive tracers, ^{129}I and ^{137}Cs, which during the past 30 years have been released in large quantities from the nuclear fuel reprocessing plants at Sellafield, UK and La Hague, France, which discharge into the Irish Sea and English Channel, respectively. Changes in the reprocessing plant release rates of ^{129}I and ^{137}Cs are propagated through the Norwegian and Atlantic Coastal Currents into the marginal seas and interior of the Arctic Ocean on time scales of 4–5 years (Livingston, 1988; Smith *et al.*, 1990; Kershaw and Baxter, 1995) and measurements of these signals can be used to constrain water circulation time-scales throughout the Arctic Ocean (Smith *et al.*, 1998).

There are two basic tracer applications of these radionuclides in the Arctic Ocean: (1) measurements of ^{129}I and ^{137}Cs, separately provide evidence for Atlantic-origin water labelled by discharges from European reprocessing plants; and (2) measurements of ^{129}I and ^{137}Cs, together can be used to identify a given year of transport through the Norwegian Coastal Current (NCC) thereby permitting the determination of a transit time from the NCC to the sampling location (Smith *et al.*, 1998). In contrast to ventilation tracers such as chlorofluorocarbon compounds (CFCs), which are introduced from the atmosphere, ^{129}I and ^{137}Cs are laterally injected into the water column, mainly from two point sources (Sellafield and La Hague) and, as a result, their input functions to the Arctic Ocean can be constrained in time and space through knowledge of the historical record of reprocessing plant discharges. In addition, both the release rate of ^{129}I from La Hague and the dynamic range of the tracer pair for transit time measurements is rapidly increasing while the input functions for many ventilation tracers are levelling off or decreasing owing to global constraints placed on their production and usage.

The distribution of ^{129}I and ^{137}Cs throughout the Arctic Ocean during the 1990s was characterized by samples collected during the five-year Science Ice Experiment (SCICEX) program to support unclassified oceanographic research in the Arctic Ocean. These expeditions employing US Navy nuclear powered submarines were used to determine the synoptic, large-scale distribution of these tracers in the Arctic Ocean. Stations occupied during the first two expeditions, the USS Cavalla in 1995 and the USS Pogy in 1996 are illustrated in Figure B.8.1. The advantage of nuclear submarines as oceanographic platforms is illustrated by: (1) the Cavalla set of stations showing the wide geographical coverage available on a single expedition owing to the speed of the vessel (25 knots) under the ice; and (2) the Pogy section across the centre of the Canada Basin through previously unsampled regions of hard, multi-year ice not easily penetrated by surface ships.

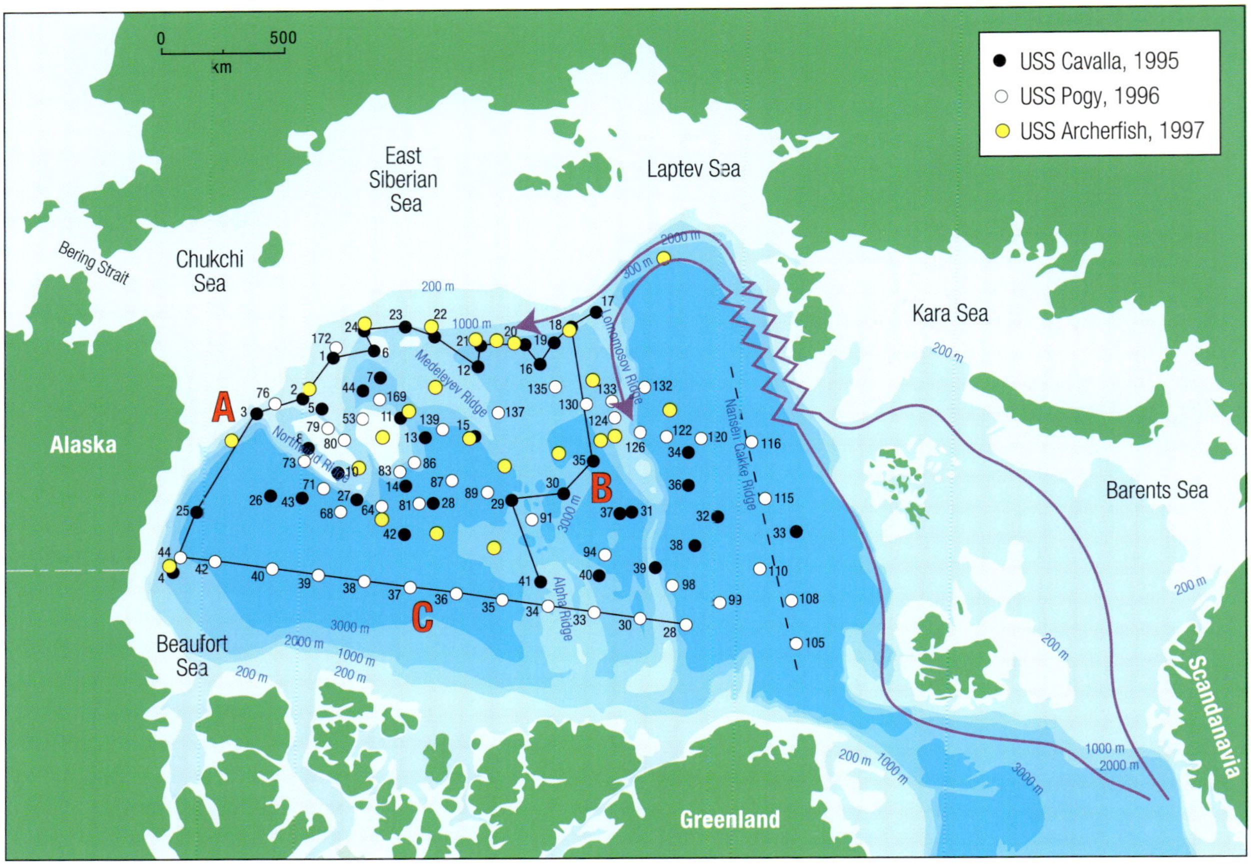

FIGURE B.8.1

^{129}I and ^{137}Cs tracer plumes from Sellafield and La Hague enter the Arctic Ocean with Atlantic Water via Fram Strait and the Barents Sea following the circulation of Atlantic Water as indicated by arrows in inset (Rudels *et al.*, 1994; McLaughlin *et al.*, 1996). Closed and open circles indicate stations occupied in the Arctic Ocean by the USS Cavalla in 1995 and USS Pogy in 1996, respectively. Solid lines identify sections on the USS Cavalla (A and B) and USS Pogy (C) cruises.

The spatial distribution of ^{129}I in halocline and upper Atlantic water is delineated by contours at the three standard submarine sampling depths of 59 m, 134 m, and 240 m in Figure B.8.2a. The highest levels of ^{129}I (> 120 x 10^7 at/l) were measured at 59 m at Stations 17 and 18 proximal to the Lomonosov Ridge. These results reflect the pathway for the entry of ^{129}I labelled, Atlantic halocline water into the Makarov Basin along the continental margin of the Laptev Sea (Rudels *et al.*, 1994; McLaughlin *et al.*, 1996; Frank *et al.*, 1998). The most prominent feature in Figure B.8.2a is a steep ^{129}I concentration gradient aligned with the Mendeleyev Ridge that delineates a "front" between Pacific-origin and Atlantic-origin halocline water that has been observed in the same general location since 1993 (Carmack *et al.*, 1995; 1997; Morison *et al.*, 1998; Smith *et al.*, 1998). Atlantic-origin water on the western side of the front is labelled by elevated levels of ^{129}I from releases from Sellafield and La Hague, while Pacific-origin water in the Canada Basin is distinguished by much lower ^{129}I concentrations (< 5 x 10^7 at/l) associated mainly with nuclear weapons fallout (Smith *et al.*, 1998). There is also a broad decrease in ^{129}I concentrations which occurs northward across the Makarov Basin and along the Mendeleyev and Lomonosov Ridges. This gradient in ^{129}I concentrations is consistent with the "downstream", return flow of Atlantic-origin, halocline water along the Mendeleyev and Lomonosov Ridges, similar to the flow predicted by Rudels *et al.*, (1994) for Atlantic intermediate water.

At a water depth of 134 m (Figure B.8.2b), the ^{129}I gradient has shifted from the Mendeleyev Ridge to a position over the Chukchi Plateau and Mendeleyev Abyssal Plain. These results illustrate the eastward displacement of the Pacific-Atlantic front towards the Canada Basin with increasing water depth (McLaughlin *et al.*, 1996). The alignment of this front between Stations 14 and 28 suggests that halocline water may be topographically steered along the northern slope of the Chukchi Plateau. There is also a northward decrease in ^{129}I concentrations

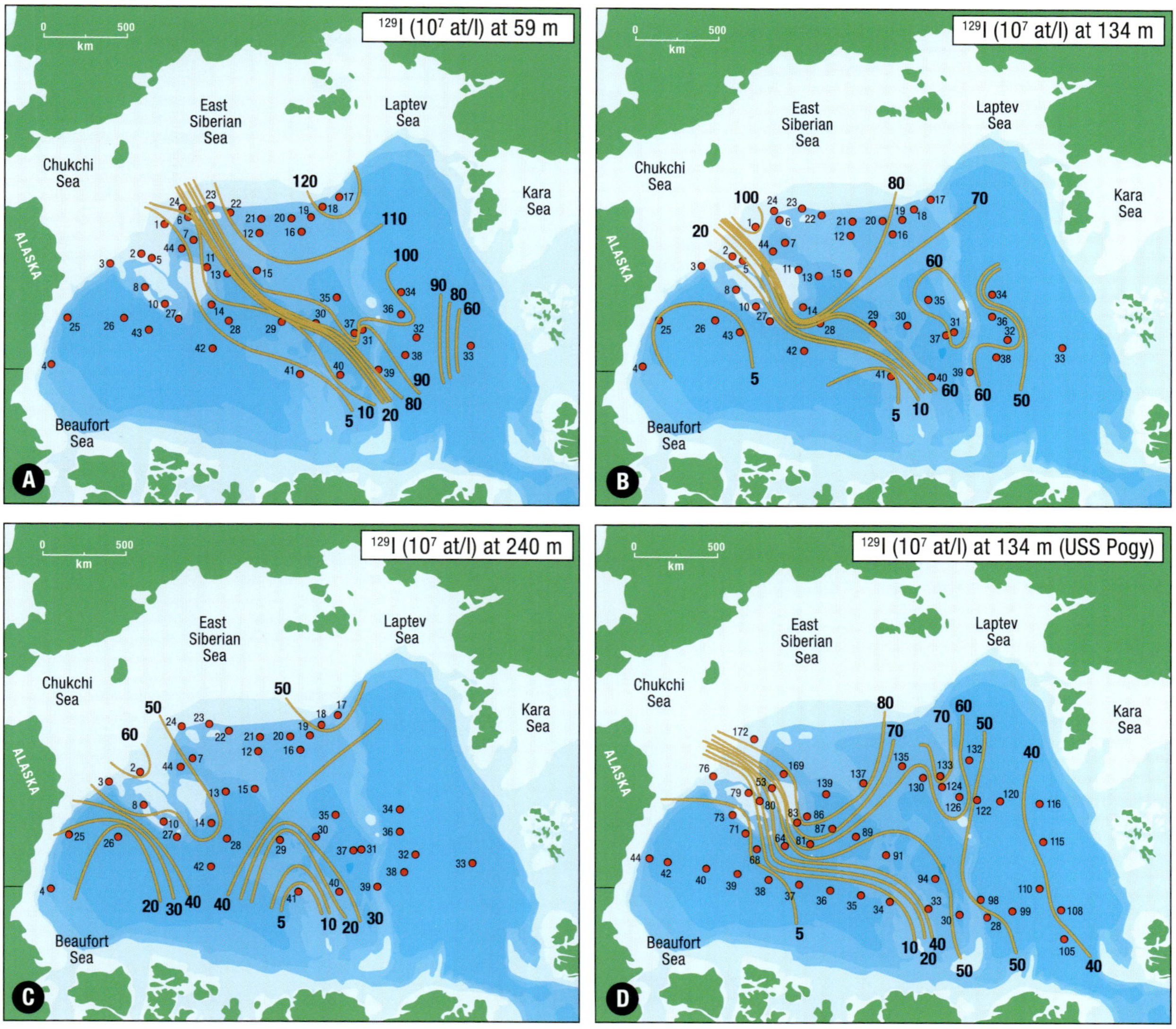

Aug./10/98/02c

FIGURE B.8.2

(a) ^{129}I contours at the 59 m depth level from the USS Cavalla cruise delineate the front between Atlantic-origin, halocline water having elevated ^{129}I levels and Pacific-origin, upper halocline water labelled solely by fallout levels of ^{129}I that is oriented along the Alpha-Mendeleyev Ridge. (b) The ^{129}I gradient marking the Atlantic-Pacific halocline front is shifted towards the Canada Basin at a water depth of 134 m and, north of the Chukchi Plateau, shows some evidence for topographically steering of halocline water into the interior of the Canada Basin. (c) ^{129}I contours at the 240 m depth level show uniform levels in the interiors of the Makarov and Canada Basins, but extremely low levels over the Alpha Ridge. (d) ^{129}I contours at the 134 m depth level from the 1996 USS Pogy cruise are similar to those from the 1995 USS Cavalla cruise.

across the Makarov Basin and along the Mendeleyev and Lomonosov Ridges, similar to that observed at the 59-m water depth. The possible steering of high ^{129}I water north of the Chukchi Plateau is even more evident at a water depth of 240 m (Figure B.8.2c), where elevated ^{129}I levels (> 40 x 10^7 at/l) extend along the continental margin of the Makarov Basin and across the Mendeleyev Ridge to stations 28 and 42 in the interior of the Canada Basin. Much lower ^{129}I levels were measured both at stations 4, 25 and 26 in the southern part of the Canada Basin and at stations 29 and 41 over the northern Mendeleyev and Alpha Ridges. The elevated ^{129}I levels at a water depth of 240 m at station 42 indicate that lower halocline/upper Atlantic layer water in the interior of the Canada Basin is relatively well ventilated. In contrast, the extremely low ^{129}I levels (< 5 x 10^7 at/l) measured at the 240 m water depth at Station 41 give little indication that Atlantic origin water contaminated by ^{129}I from

European reprocessing inputs had reached the Alpha Ridge by 1995 and suggest that ventilation rates in this region are extremely low.

The ^{129}I distribution at a water depth of 134 m (Figure B.8.2d), measured during the USS Pogy cruise is similar to that measured during the USS Cavalla cruise indicating that only small changes occurred in the ^{129}I distribution between 1995 and 1996. This will certainly not be the case in future years as the leading edge of a large, post-1991, ^{129}I input signal enters the Arctic Ocean, probably resulting in annual increases in ^{129}I concentrations of the order of 10–20% at locations over the continental margins of the Nansen and Makarov Basins.

In summary, ^{129}I measurements on water samples collected from the 1995 USS Cavalla, and 1996 USS Pogy nuclear submarine (SCICEX) cruises in the Arctic Ocean clearly delineate the front between Atlantic-origin and Pacific-origin halocline water that is presently aligned over the Mendeleyev Ridge. The Atlantic-Pacific halocline front is displaced eastward towards the Canada Basin with increasing water depth and has exhibited little variance in its spatial configuration since the 1995 and 1996 SCICEX cruises. Extremely low levels ($< 2 \times 10^7$ at/l) of ^{129}I measured in Atlantic intermediate water over the Northern Canada basin and Alpha Ridge reflect minimal contamination from European reprocessing plants and are consistent with ventilation ages in excess of 25 years for this region of the northern Canada Basin. In contrast, higher ^{129}I levels ($> 20 \times 10^7$ at/l) were measured in the interior of the Canada Basin indicating that ventilation of intermediate layer water in this region is significantly more rapid, probably by lateral mixing from topographically-steered boundary currents flowing over the continental margin north of the Chukchi Plateau.

B.8.3 Local sources

Local contamination of Arctic seas, such as the highly productive Barents Sea which supports a large fishery, is an issue of considerable concern. Investigations addressing local inputs of radioactivity include studies of radioactive waste dumpsites in the Kara Sea (Salbu *et al.*, 1997), downstream transport of radioactivity from nuclear facilities on the Ob and Yenesey Rivers (Sayles *et al.*, 1997; Kenna and Sayles, 2002), leakage of radioactivity from underground nuclear test sites at Amchitka, Alaska (Dasher *et al.*, 2002) and offshore transport of radioactivity into the Barents Sea from nuclear test sites on the southern coastline of the island of Novaya Zemlya (Smith *et al.*, 1995; 2000). These latter investigations revealed that some of the highest levels of ^{239,240}Pu ever measured in the marine environment can be found in the sediments of Chernaya Bay, a small (1–6 km-wide, 15 km-long) fjord on the south-western coast of Novaya Zemlya. Chernaya Bay is a former Soviet Union nuclear weapons test site where at least two underwater nuclear tests were conducted in 1955 and 1957. The high level of ^{239,240}Pu (> 8000 Bq/kg) measured on a single surface sediment sample collected from Chernaya Bay in 1992 indicated that a large inventory of ^{239,240}Pu from the nuclear tests may have been sequestered in bottom sediments (Smith *et al.*, 1995). Chernaya Bay plutonium was also shown to be distinguished by a low ^{240}Pu/^{239}Pu isotope atom ratio of 0.03 (compared to typical fallout ratios of 0.18) indicative of low-yield nuclear tests. In addition, elevated levels of ^{239,240}Pu measured in box cores collected within 50 km of the approaches to Chernaya Bay indicated that ^{239,240}Pu transport had occurred from this embayment into the Eastern Barents Sea (also known as the Pechora Sea) and could represent a potential radiological threat to it's important commercial fishery.

More extensive oceanographic investigations were undertaken aboard the R/V Geolog Fersman in 1993 and 1996 to collect seawater, sediment, and biota samples in Chernaya Bay and its surrounding approaches in the eastern Barents Sea (Figure B.8.3). These studies revealed elevated levels of artificial radionuclides, as high as 15000 Bq/kg for ^{239,240}Pu, 250 Bq/kg for ^{137}Cs, and 100 Bq/kg for ^{60}Co in sediment cores. Significant uptake of ^{239,240}Pu has occurred in Chernaya Bay from contaminated sediments into benthic invertebrates (1292 Bq/kg and 104 Bq/kg in polychaete and Macoma, respectively) and from seawater into brown algae (5.1 Bq/kg and 14.7 Bq/kg in Laminaria and Fucus, respectively). The total inventories of ^{137}Cs and ^{239,240}Pu in Eastern Barents Sea sediments are contoured in Figures B.8.3a and 3b. The fractional contribution from nuclear tests in Chernaya Bay compared to fallout in Barents Sea sediments was determined from the ^{240}Pu/^{239}Pu results. The inventory of ^{239,240}Pu derived from Chernaya Bay is contoured in Figure B.8.3c while the percent of the total ^{239,240}Pu inventory derived from Chernaya Bay is contoured in Figure B.8.3d. The influence of Chernaya Bay discharges decreases rapidly with distance seaward into the eastern Barents Sea, but extends for much greater distances (> 100 km) northwestward along the Novaya Zemlya coastline. This is consistent with circulation patterns which indicate that longshore sediment transport occurs northward along the southern Novaya Zemlya coastline, mainly as a result of the Litke Current flowing westward through the Strait of Karskiye Vorota or Kara Gate (Pfirman *et al.*, 1997). Although the actual magnitude of plutonium transport southeastward into the shallower, coarser-grained sediments of the Barents Sea is relatively small, Chernaya Bay $^{239\ 240}$Pu still represents more than

40% of the total ^{239,240}Pu inventory (Figure B.8.3d) in much of the region off the southern coast of Novaya Zemlya. The inventory of Chernaya Bay plutonium in the eastern Barents Sea, estimated by linear interpolation between adjacent stations from the contour plots in Figure B.8.3c, is 2.0 TBq. This can be compared to the 11.2 TBq of ^{239,240}Pu still retained in the sediments of Chernaya Bay. The decreasing ^{239,240}Pu inventory gradient, northwestward along the Novaya Zemlya coastline, is relatively small, however, suggesting that a significant additional quantity of ^{239,240}Pu could have been transported farther along the Novaya Zemlya coastline by the Litke Current and into the Central Arctic Ocean. It would only require an additional 0.2–0.3 TBq of Chernaya Bay plutonium, or approximately 10–15% of the inventory already measured in the eastern Barents Sea, to account for the low ^{240}Pu/^{239}Pu ratio, plutonium inventories recently measured in the Eurasian and Makarov Basins (Huh *et al.*, 1997) .

In addition to contamination from local sources, the Russian Arctic shelves also receive contaminants by long-range transport from European nuclear fuel reprocessing plants (Kershaw and Baxter, 1995; Smith *et al.*, 1998). This was made abundantly clear in recent studies by Matishov *et al.*, (2002) of possible contamination from the sunken Russian submarine, "Kursk". Following the vessel's sinking in September, 2000, measurements of radioactivity were conducted on seawater, sediment and biota samples collected near the "Kursk" to determine whether leakage had occurred from the submarines two nuclear reactors (Figure B.8.4) and to assess the impact on one of the most productive fishing areas in the world. Contamination of the marine environment could have occurred during the accident or during the recovery of the submarine and subsequent attempts to defuel the vessel at naval facilities in Kola Bay, close to the city of Murmansk. Levels of radioactivity in surface sediments and biota were found to be within the range of values

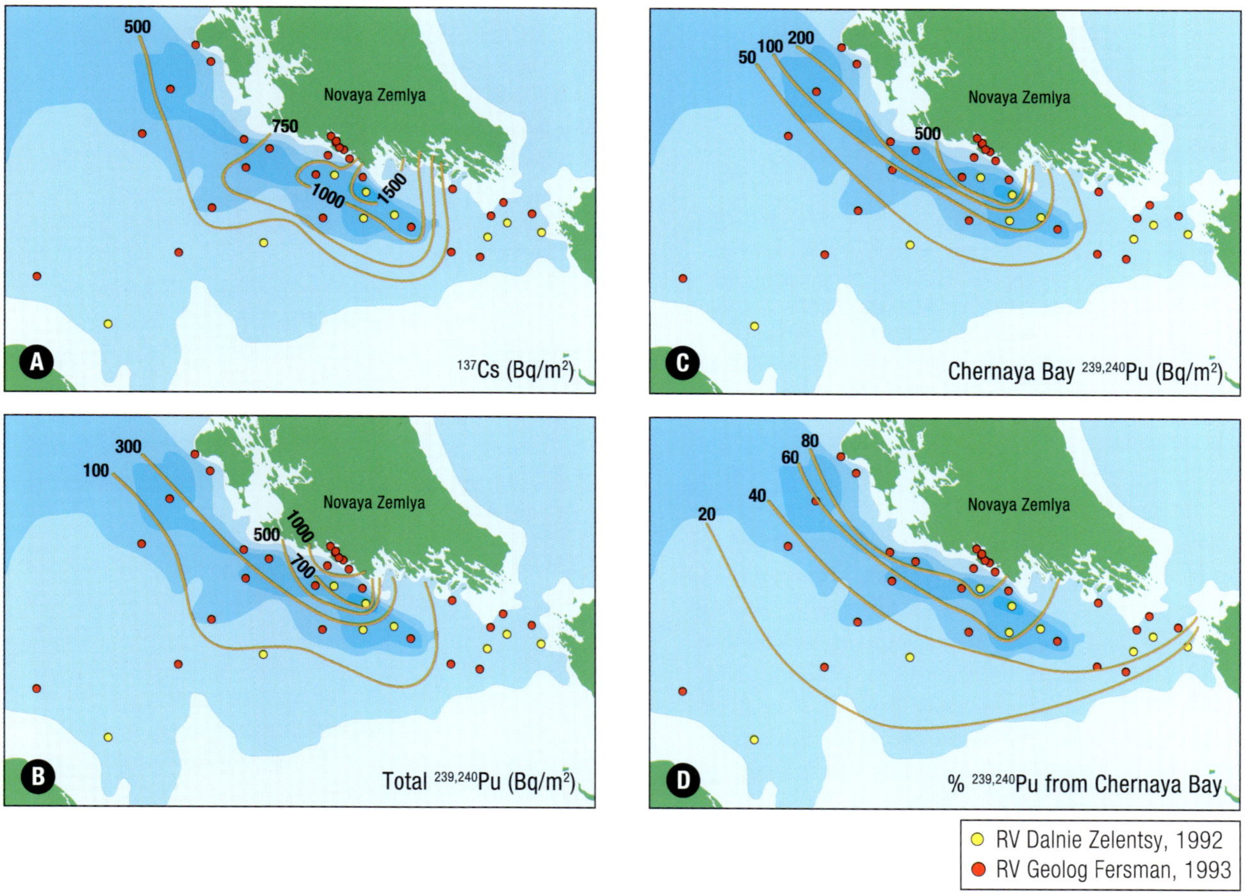

FIGURE B.8.3

(a) Contours of ^{137}Cs sediment inventories (Bq/m^2) conform to the bathymetry of the Eastern Barents Sea with highest values associated with finer-grained, basin sediments. (b) ^{239,240}Pu inventory (Bq/m^2) contours tend to be more centred around the entrance to Chernaya Bay. (c) Contours of the ^{239,240}Pu inventory derived exclusively from transport from Chernaya Bay illustrate the northwestward transport of ^{239,240}Pu carried by the Litke Current. (d) Contours illustrating the percentage of the total ^{239,240}Pu inventory derived from Chernaya Bay have been calculated from the ^{240}Pu/^{239}Pu atom ratio data for each core.

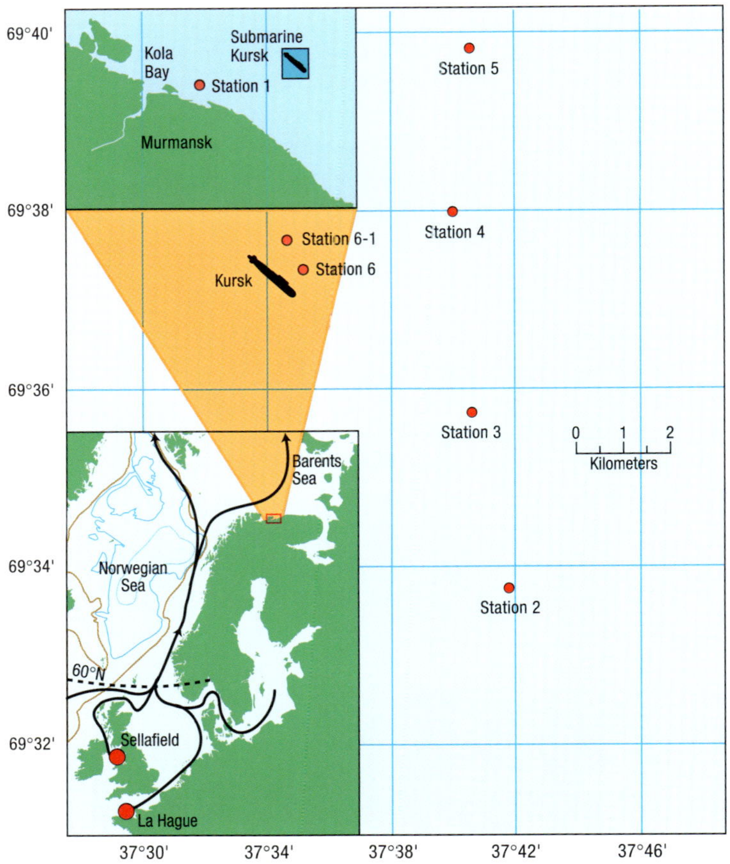

FIGURE B.8.4

Sediment, seawater, and biota samples were collected at station locations, proximal to the site of the sunken submarine, Kursk, located northeast of Murmansk and Kola Bay. The principal source of ^{129}I contamination at the site is radioactivity discharged from European nuclear fuel reprocessing plants at Sellafield (UK) and La Hague (France) followed by transport through the Norwegian Sea into the Barents Sea.

previously measured in the Barents Sea and can be ascribed to inputs from global fallout, European nuclear fuel reprocessing facilities and the Chernobyl accident. Levels of ^{129}I in seawater in the Southern Barents Sea, however, increased by 500% between 1992 and 2000 and the $^{129}I/^{137}Cs$ ratio increased by more than an order of magnitude during this time, owing to long-range transport of releases from reprocessing facilities at Sellafield (UK) and La Hague (France). Although these results indicate that at the time of sampling, leakage from the "Kursk" had a negligible impact on the environment, they also show that regional background levels of artificial radioactivity are varying rapidly on annual time scales and that Europe's nuclear reprocessing facilities are a leading contributor of anthropogenic radioactivity to the region.

B.9 Local sources of contaminants in the Canadian Arctic

Authors:

D. Gregor (1); J. Stow (2); D. Kennedy (2); K. Reimer (3); and C. Ollson (3).

Contributors:

Z. Kuzyk (3); I. Koch (3); C. Hough (4); N. Burgess (5); S. Solomon (6). P. Roach (7); and G. Stern (8)

Affiliations:

1. MDA Consulting, Campbellville, ON, L0P 1B0, Canada;
2. Indian and Northern Affairs Canada, Ottawa, ON, K1A 0H4, Canada;
3. Environmental Sciences Group, Royal Military College, Kingston, ON, K7K 7B4, Canada;
4. Department of Nation Defence, National Defence Headquarters, Ottawa ON, K1A 0H4, Canada;
5. Canadian Wildelife Service, Environment Canada, Mount Pearl Newfoundland and Labrador, A1N 4T3, Canada;
6. Geological Survey of Canada (Atlantic), Natural Resources Canada, Dartmouth, NS, B2Y 4A2, Canada;
7. Department of Indian Affairs and Northern Development, 345–300 Main Street, Whitehorse, Yukon Territory, Canada;
8. Freshwater Institute, Department of Fisheries and Oceans, 501 University Crescent, Winnipeg, Manitoba, Canada, R3T 2N6.

B.9.1 Introduction

B.9.1.1 Local sources of contaminants in CACAR-I

In order to address local sources of contaminants to the Arctic, CACAR-I provided a brief comparison of PCBs and radionuclides originating from local and long-range sources. Abandoned Distant Early Warning Line (DEW-line) military radar stations were assessed as sources of PCBs and various sites, including: uranium processing plants; waste disposal sites; and, nuclear weapons test sites, were assessed as local sources of radionuclides.

Studies conducted by the Environmental Sciences Group, Royal Military College, concluded that PCB contamination from DEW-line stations generally contributed to a 'halo' of elevated PCB concentrations in soil and plants within at least a 10 km radius of each site. While the DEW-line stations were considered the most important source of PCB contamination (accounting for 85%) at a local level, long-range atmospheric transportation was considered to be the most important source on an Arctic-wide basis (accounting for 99%) (Macdonald *et al.*, 2000)

Local sources of radionuclide contamination in Russia, including waste disposal sites and nuclear weapons test sites, were identified as the major sources of contamination to the Arctic Ocean. By comparison, uranium processing plants and long-range atmospheric sources of anthropogenic radionuclides were considered to have a negligible contribution to Arctic contamination. Radionuclides from the effluent of European nuclear fuel reprocessing plants have provided valuable tracers for the study of ocean-currents.

B.9.1.2 Local source of contaminants in CACAR-II

Local sources of contamination in Canada are being assessed by other oganizations within Indian and Northern Affairs Canada (INAC) as well as other federal and territorial departments and are not within the mandate of the NCP; despite this, there are several reasons to address local sources in this report. For example, the NCP program is interested in any information concerning contaminant levels in the North, especially as it relates to traditional/country foods and spatial trends. Local sources of contamination, in particular contaminated sites, raise serious concerns at a local level. These concerns are often heightened because many of these sites, including DEW-line stations, mines, and other industrial sites, are located close to communities. When exchanging contaminants-related information with Northerners the subject of local contaminants concerns is consistently raised and the NCP must be prepared to address these issues.

It is also important to account for local sources of contamination from the perspective of their contribution to global sources. As the direct emissions of many "past use" POPs declines, or are terminated completely, the remaining global sources of these pollutants are becoming more dominated by soil residues and contaminated sites (Li *et al.*, Section A.3, Blanchard *et al.*, Section B.1, Muir *et al.*, Section B.2). While the contribution of local sources within the Canadian Arctic may be relatively minor to overall Arctic-wide contamination, their contribution to this remaining global source cannot be ignored. A need to assess these sites with a view to changes that may arise as a result of climate change has also been identified. These changes relate primarily to the destabilization of contaminated materials currently sequestered in permafrost (Macdonald *et al*, Section D.1).

Under NCP-Phase II, funding was provided to Territorial Contaminants Committees for local contaminants concerns (LCC) projects. These relatively small projects were designed and executed by local people and were intended to address concerns that arise at a community level. The LCC funding envelope was developed to meet a need that was not otherwise being dealt with by other federal or territorial agencies. While these projects will not be discussed in any detail in this section, where appropriate, data collected from LCC projects has been incorporated into technical sections of CACAR-II. These projects are discussed in more detail in Chapter 3 of the Knowledge in Action Report.

The Canadian government has undertaken several programs to assess and if necessary remediate contaminated sites in the Canadian Arctic, most notably the DEW-line cleanup project, through the Department of National Defence (DND), and the Arctic Environmental Strategy — Action on Waste program, through INAC. These programs have led to a comprehensive inventory of contaminated sites in the Canadian North that can be broken down into three main categories: abandoned military sites (eg., DEW-line stations); abandoned mines; and, other industrial sites. This section will briefly describe this inventory of contaminated sites and the state of assessment and will then discuss several case studies in more detail. Case studies have been chosen from each of the main categories to provide examples of sites that contain relatively large amounts of contamination and that have seen substantial investigation and, in some cases, remediation. The extent of contamination at these sites is not representative of all contaminated sites in the North — indeed, in several cases they represent extreme examples. Sites chosen for case studies include: a INAC DEW-line site in Nunavut (Resolution Island); a DND abandoned Polevault radar station in Newfoundland and Labrador (Saglek); two INAC mine sites in the Northwest Territories (Giant gold mine and Port Radium uranium mine); and, a former industrial site in the Yukon (Carcross railway tie treatment plant). These sites, in particular Saglek and Giant mine, have provided excellent opportunities to conduct research that integrates both abiotic and biotic processes over a wide range of contaminant concentrations.

B.9.2 Contaminated sites overview

For the purposes of this discussion, a site is considered to be contaminated if a contaminant has been identified at concentrations in excess of an appropriate guideline or criteria [eg., Canadian Council of Ministers of the Environment (CCME) Guidelines and Canadian Environmental Protection Act (CEPA) regulations]. The contaminants under consideration include PCBs and other organochlorines, heavy metals, arsenic, and radionuclides. Sites that have contaminant issues relating to petroleum hydrocarbons are not considered herein.

Over two thousand waste sites in the Canadian North have been catalogued in a database prepared by INAC (INAC, 2002). Many of these sites contain refuse and

TABLE B.9.1 Status of cleanup by site for DEW-line sites administered by DND

Site	Designator	Main contaminant	Status
BAR-1	Komokuk Beach	PCB, Cu, Pb, Zn	Complete
BAR-2	Shingle Point	PCB, Cu, Pb, Zn	Complete
BAR-4	Nicholson Peninsula	PCB, Cu, Pb, Zn	Complete
CAM-M	Cambridge Bay	PCB, Cu, Pb, Zn	Complete
FOX-4	Cape Hooper	PCB, Cu, Pb, Zn	Complete
PIN-M	Cape Parry	PCB, Cu, Pb, Zn	Complete
BAR-3	Tuktoyaktuk	PCB, Cu, Pb, Zn	Complete
CAM-4	Pelly Bay	PCB, Cu, Pb, Zn	In progress
PIN-1	Clinton Point	PCB, Cu, Pb, Zn	In progress
CAM-1	Jenny Lind Island	PCB, Cu, Pb, Zn	Not yet started*
CAM-2	Gladman Point	PCB, Cu, Pb, Zn	Not yet started*
CAM-3	Shepherd Bay	PCB, Cu, Pb, Zn	Not yet started*
CAM-5	Mackar Inlet	PCB, Cu, Pb, Zn	Not yet started*
DYE-M	Cape Dyer	PCB, Cu, Pb, Zn	Not yet started*
FOX-2	Longstaff Bluff	PCB, Cu, Pb, Zn	Not yet started
FOX-3	Dewar Lake	PCB, Cu, Pb, Zn	Not yet started
FOX-5	Broughton Island	PCB, Cu, Pb, Zn	Not yet started*
FOX-M	Hall Beach	PCB, Cu, Pb, Zn	In progress
PIN-2	Cape Young	PCB, Cu, Pb, Zn	Not yet started
PIN-3	Lady Franklin Point	PCB, Cu, Pb, Zn	In progress
PIN-4	Byron Bay	PCB, Cu, Pb, Zn	Not yet started

*Detailed site investigation work has been completed and design/specifications for cleanup are being developed. Actual site remediation work has not begun (as of December 2002).

debris but have not necessarily been identified as being chemically contaminated. Contaminated sites included in this database as well as other that are administered by DND generally break down into the following categories: DEW-line sites that are the responsibility of INAC; DEW-line and other radar sites that are the responsibility of DND; and, abandoned mines and industrial sites that are the responsibility of INAC.

In the 1950s, the Canadian and US governments built a large number of military radar bases at remote locations across the North American Arctic. These sites spanned from Greenland to Alaska in an arc just north of the Arctic Circle. These radar stations, known collectively as the Distant Early Warning (DEW) Line, were part of the North American air defence system designed to detect attack from the former Soviet Union. The total network of DEW-line stations in Canada consisted of 42 sites of three types: main, auxiliary, and intermediate (Holtz *et al.*, 1986). The DEW-line was complemented by systems further south referred to as: the Mid-Canada Line at approximately 55°N; the Pinetree Line located at about 50°N; and, the Pole Vault Line which ran north-south along the coast of Newfoundland and Labrador.

In the 1980s, 21 DEW-line sites were upgraded as part of the North American Air Defence Modernization project and were incorporated into the North Warning System (NWS) which is jointly operated by the Canadian Department of National Defence (DND) and the United States Air Force. The Department of National Defence retains responsibility for the former DEW-line facilities and associated contamination that remain at these sites. In an effort to decommission those parts of the DEW-line that are not required for the operation of the North Warning System, DND has undertaken a massive cleanup effort known as the DEW-line Cleanup Project (DLCU). Through a series of site assessments that were carried out between 1989 and 1993, the DLCU identified contaminants of concern and developed a set of risk-based criteria for cleanup. The DEW-line Cleanup criteria (DCC) for PCBs in soil were developed to correspond to different standards of disposal: DCC-I = 1–5 µg/g; and, DCC-II = 5–50 µg/g. Above 50 µg/g, PCB contaminated soil is regulated under the Canadian Environmental Protection Act (CEPA). DEW-line cleanup criteria have also been developed for metals (Cu, Pb, Zn, Hg, Ni, Cr, Cd, Co, and As). The general state of these

DND administered sites is summarized in Table B.9.1 and their locations are illustrated in Figure B.9.1.

In the 1970s, INAC took charge of 21 "I" or intermediate sites that had been abandoned in the 1960s. Following investigations by Environment Canada and INAC in the mid-1980s (Holtz *et al.*, 1986) a partial cleanup of these sites was undertaken during which site contents were identified and many sources of PCBs were removed. Additional work in the 1990s undertaken on behalf of INAC further characterized the sites and documented additional PCB contamination. Sites were also assessed for for a full range of potential contaminants including metals, hydrocarbons, asbestos and in some cases chlorinated solvents.

Table B.9.2 provides a brief summary of the former DEW-line sites that are administered by INAC and have documented PCB contamination. The PCBs generally

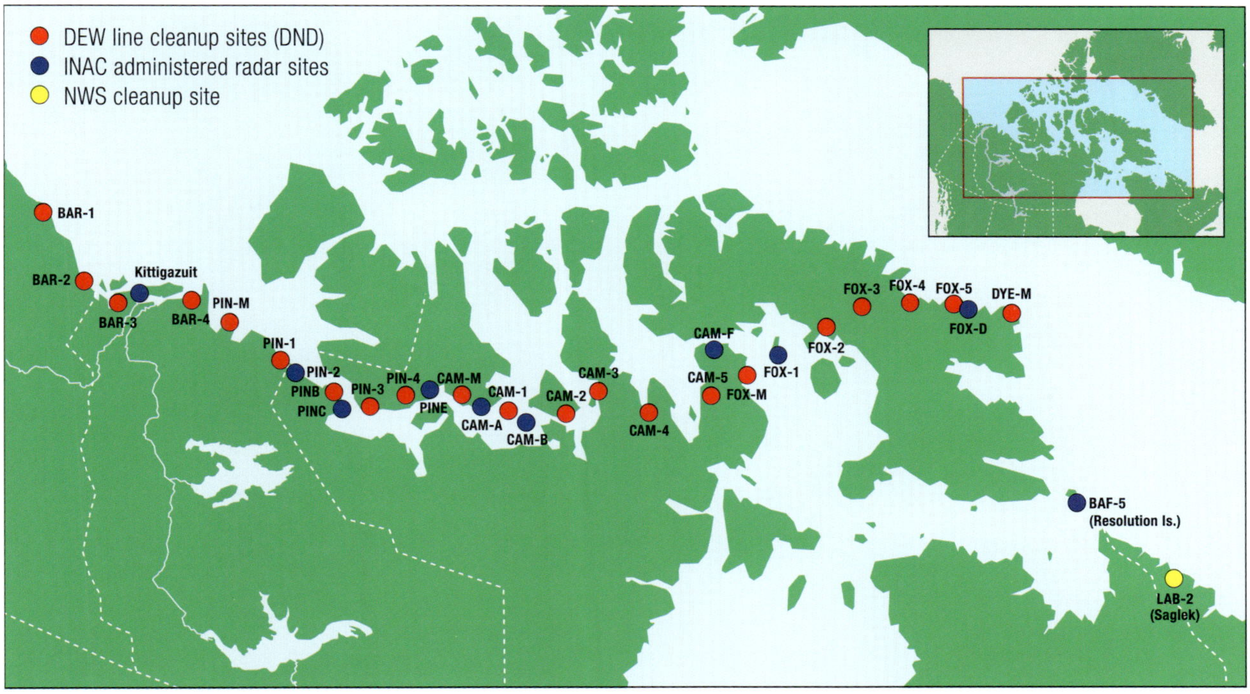

FIGURE B.9.1

Map of abandoned military sites in the Canadian Arctic that are managed by INAC and DND.

TABLE B.9.2 INAC administered DEW-line sites with residual PCB contamination (DIAND, 2002)

Site #	Site name	Main contaminants	General description of conditions
NM080	Kittigazuit Bay	DDT, PCBs	320 m^3 of DDT-contaminated soil, small quantity of PCB-contaminated soil
EK005	CAM F – Sarcpa Lake	PCBs	76 m^3 of soil with PCBs > 50 ppm excavated and stored in drums, 301 m^3 of soil with PCBs > 5–50 ppm (DCC-II) excavated and removed in 1995/96
SB030	FOX 1 – Rowley Island	PCBs	20 m^3 of soils with PCBs > 50 ppm and 70 m^3 of DCC-II excavated and removed in 1995
SB033	FOX D – Kivitoo	PCBs, As	"As" higher than background in marine sediments, PCB and metal concentrations pose risk in terrestrial ecosystem, partially cleaned between 1973 and 1985, abandoned buildings contaminated with PCBs
WK141	CAM A – Stuart Point	PCBs	Contaminated soils with PCBs between 5 and 50 ppm (~8 m^3)
WK142	CAM B – Hat Island	PCBs	Estimated 16 and 6 m^3 of DCC-I and DCC-II soils respectively
WK150	PIN C – Bernard Harbour	PCBs	Estimated 23 and < 1 m^3 of DCC-I and DCC-II soils respectively, buildings contaminated with PCBs
WK152	PIN E – Cape Peel	PCBs	Estimated 135 and 19 m^3 of DCC-I and DCC-II soils respectively
WK163	PIN B – Clifton Point	PCBs	72 m^3 of DCC-I soil for PCBs and lead, DCC-II soils for PCBs stored here from another site since 1995, buildings contaminated with PCBs

remain as low levels of contamination in soils [DCC-I (1–5 μg/g) and DCC-II (5–50 μg/g)]. Soil contamination that exceeds CEPA criteria (> 50 μg/g) has been or is in the process of being removed from these sites. Table B.9.2 does not include all 21 "I" sites as not all of these sites have PCB contamination. The location of these sites is illustrated in Figure B.9.1.

Table B.9.2 includes site number NM080, Kittigazuit Bay, which although it is not a DEW-line site, is a former military site on the Mackenzie Delta. This site is unique in that it is contaminated with DDT, which was used to control flies and mosquitoes during the operation of the site. An ecological risk assessment concluded that the DDT-contaminated soil posed minimal risks

TABLE B.9.3 Summary information for key abandoned mine sites in the Canadian North (DIAND, 2002)

Site #	Site name	Main contaminant	General description of conditions
NM182	Terra #2 (Northrim Mine)	As	45 000 tonnes of waste rock releasing As into soils and Camsell River
NM183	Terra #3 (Norex Mine)	As	100 000 tonnes of waste rock with elevated concentrations of As and moderate potential for acid mine drainage
SM221	Giant Mine – Yellowknife	As	Mine is still in operation but DIAND is responsible for pre-1999 conditions
NM181	Terra #1 (Northrim Mine)	As, Hg	Over 500 000 tonnes of waste rock could be releasing As and other metals into Camsel River during high runoff
SM214	Discovery Mine	Hg	Over 1 Mt of tailings, high levels of Hg in runoff into Giaugue Lake, acid drainage contributes to high metals concentration in runoff
SM122	Hay River Dene Reserve	Radionuclides	9 dump sites around the reserve, radiation levels 10 times background, other contaminants
SM275	Rayrock Mine	Radionuclides	2 tailings locations with high radioactivity, local hunters avoid area because of radiation
NM180	Sawmill Bay/Great Bear Lake	Radionuclides	22 m^3 of uranium contaminated soil (> 500 ppm), larger quantity of less contaminated soil
NM203	Contact Lake Mine	Radionuclides, As	61 000 tonnes of waste rock with low level radioactivity and elevated As in tailings pond

TABLE B.9.4 Selected INAC administered sites that have been remediated (DIAND, 2002)

Site #	Site name	Main contaminant(s)	General description of conditions
KW019	Rankin Inlet	As	Abandoned site contained 297 000 tonnes of tailings which were being eroded by wind, waves and runoff; site was cleaned up in 1995, environmental monitoring continues
TA006	Venus Tailings/Mine Site	As	Abandoned site required containment of leachate entering lake and stopped erosion by means of sealed sheet pile wall and capping of tailings, completed in 1995
BC002	Snag Airstrip	PCBs	Military site constructed in 1942 and abandoned in 1964, high PCBs in powerhouse and pump station, PCB-contaminated material removed or buried as per criteria
HJ003	Aishihik Airstrip	PCBs, Hg, Pb	Military airstrip from 1942–1966, high Hg concentration in soils, very high PCBs in old floor of power house and paint, material with PCBs > 50 ppm removed, Pb contaminated soils removed
HJ028	Border Pump Station/ Rainy Hollow	PCBs, Hg, DDT	Military/construction camp from 1950s–1979, groundwater contaminated with DDT and Hg, DDT contaminated soil and buried canisters of DDT from Rainy Hollow contained in 1994, continued ground and surface water monitoring
TA033	Granger	PCBs	PCB-contaminated soil removed
SB003	Iqaluit 6 North 40 Dump Site	PCBs	Military dump from 1950 to 1963, site cleaned up in 1996
SB004	Iqaluit 7 – Upper Base	PCBs	Military dump from 1951 to 1974, 80 m^3 of contaminated soil removed in 1995, 89 m^3 of soil > 50 ppm and 257 m^3 of DCC-II soils remain
SB028	Iqaluit West 40 Dump Site #1	PCBs	Active military dump to 1973, partial cleanup between 1986–89 removed some PCBs, PCB-contaminated soils removed in 1995
NM041	BAR E – Horton River	PCBs	Contaminated soils, site remediated in 1995, monitoring to be maintained to ensure capped landfill does not fail

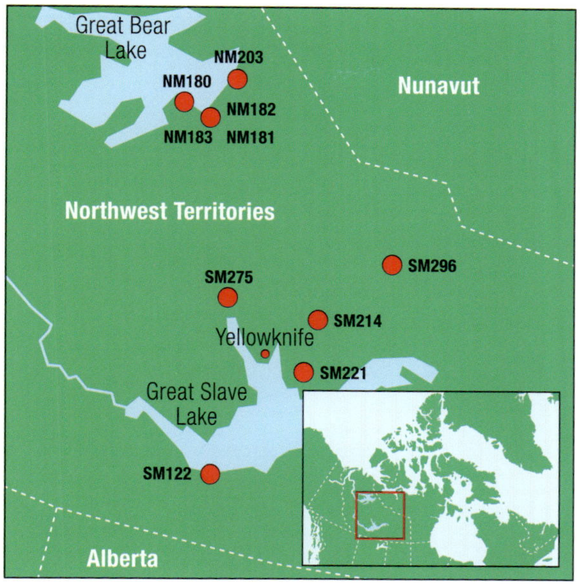

FIGURE B.9.2

Map of abandoned mine and industrial sites in the Canadian Arctic that are managed by INAC.

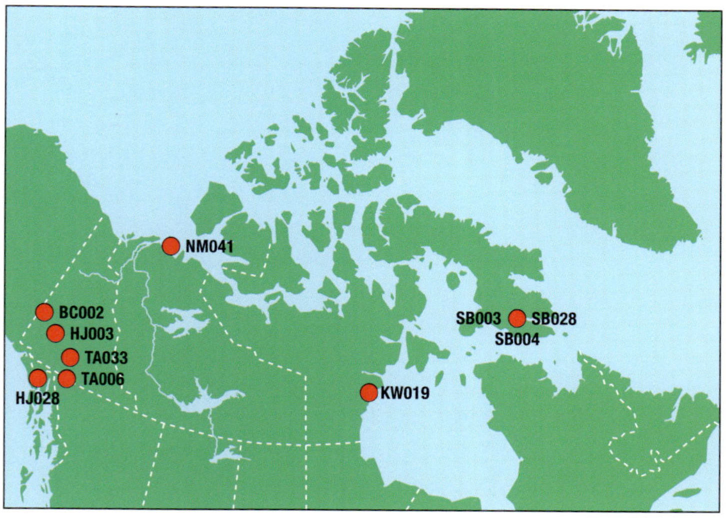

FIGURE B.9.3

Map of INAC managed sites that have been remediated.

and it was recommended that it be left in place in order to avoid disturbance of the vegetation that has re-colonized the site (ESG 2000a, Nirwal *et al.*, 2000).

The INAC waste sites database also contains a large number of sites that are abandoned mines or exploration sites. Many of these are small, and past assessments do not indicate the presence of contamination. Table B.9.3 summarizes information related to a number of the largest abandoned mines where local contamination is known to be extensive. The locations of these sites are indicated in Figure B.9.2.

Several contaminated military, mine, and industrial sites that are included in the INAC waste sites database have been remediated over the past decade. Some of these sites along with basic information about the remediation are listed in Table B.9.4 and located in Figure B.9.3.

B.9.3 Resolution Island, Nunavut

B.9.3.1 History and introduction

Resolution Island is located offshore of the southeast tip of Baffin Island at 61°35' N and 65°0' W (see Figure B.9.1), approximately 310 km from Iqaluit. The surface of the island has been scoured by glacial ice and generally there is only a thin veneer of soil over the bedrock if the rock is not exposed at the surface. With little soil development the plant community is limited to lichens, mosses, grasses, sedges, rushes, willows, saxifrage, crowberries, heather, and Labrador tea. Polar bears are frequently observed traversing the island, however, no other terrestrial mammals have been observed. Seals (bearded and ring) walrus and a healthy nearshore fishery have been documented as well as killer whales, narwhal, beluga, and bowhead whales migrating through the adjacent waters. The closest documented traditional fisheries occurs approximately 100 km to the northeast. During the summer there is an active commercial shrimp fishery approximately 20 km off the coast (INAC, 2000).

The radar base at Resolutions Island, designated as BAF-5, was opened in 1953, two years after construction began. At its peak, the base accommodated up to 200 military personnel and included 40 permanent buildings. It is accessible by air and sea when the weather permits, which, due to ice in the winter time and poor weather and fog throughout the year, is at best limited and hazardous.

The US Air Force (USAF) abandoned the base in 1972 and it was eventually placed under the administrative control of INAC. In 1985, the DND negotiated the use of a small part of the original base for the construction of an unmanned Short Range Radar Station as part of the new North Warning System. Prior to construction of the new site in the late 1980s and early 1990s DND conducted a cleanup of the small area that was required for the new station.

B.9.3.2 Site investigations

The remaining area of the site, not included in the footprint of the North Warning System facility, was the subject of several site investigations in the early 1990s. Investigations of the site, which included most of the abandoned infrastructure as well and at least eight dump sites, focused on the assessment and delineation of soil contamination for the purposes of cleanup. These investigations identified approximately 5000 m^3 of soil contaminated with over 50 µg/g PCBs. Contamination originating from site equipment was found adjacent to buildings and in and around dumps. Polychlorinated biphenyls were also found to be migrating from source areas along drainage channels to the ocean, conceivably in violation of the Fisheries Act (INAC, 2000).

Polychlorinated biphenyl contamination at Resolution Island was the subject of an ecological risk assessment that was undertaken by Golder and Associates in 1997. The objective of this work was to assess potential risks to humans and/or ecological receptors using standard models and surrogates for Arctic species. The Canada goose, a migratory bird known to stop on the island during migration periods, was selected as the primary ecological receptor for the assessment. Marine receptors were not evaluated on the assumption that migration of PCBs to the ocean had been mitigated by the installation of adsorbent barriers in 1994. Ecological risk was found to be minimal with respect to the Canada goose and since the area is virtually uninhabited by humans and seldom used for hunting, human exposure was also found to be negligible. The report concluded that the site does not present unacceptable risks to humans or terrestrial and marine wildlife.

B.9.3.3 Site remediation

In 1997 INAC developed a plan for site remediation that included the excavation and containerization of soil contaminated with over 50 µg/g PCB as required under CEPA. Long-term storage of the containerized soil was considered but eventually dismissed because year-round site conditions do not allow for frequent site inspections as required under PCB storage regulations. Several disposal options including on-site remedial technologies as well as off-site shipment and disposal were considered. After careful consideration it was determined that shipping the containerized soil to an established PCB destruction facility in southern Canada was the best disposal option. While off-site shipping and destruction was relatively expensive compared to other options, project related risks were minimized by avoiding the transportation to and operation of sophisticated equipment at Resolution Island. This option was also considered acceptable to the local population in Nunavut (INAC, 2000).

The basic elements of the overall site remediation plan include:

- excavation, containerization and temporary storage of soil contaminated with PCBs at concentrations above 50 µg/g (ongoing);
- construction and maintenance of adsorption barriers to prevent PCB migration into the ocean;
- demolition of buildings deemed structurally unsafe or contaminated with PCBs (ongoing);
- collection and removal of PCB contaminated liquids from the site (completed);
- removal of other materials contaminated with PCBs (e.g., floor tiles, capacitors — completed);
- excavation and containerization of soil contaminated with cobalt, lead and mercury (ongoing);
- disposal of various solvents, additives, and fuels;
- removal of asbestos from buildings and burial in a landfill; and,
- collection and disposal of debris and barrels.

Since most of the PCBs at Resolution Island are concentrated in a relatively small volume of highly contaminated soil, the current remedial plan is expected to remove 97% of the PCBs from contact with the surrounding environment. This constitutes approximately 3.9 tonnes of PCBs (ASU, 1999).

B.9.4 Saglek, Newfoundland and Labrador

B.9.4.1 History and introduction

Saglek is located on the northeast coast of Labrador (58°23' N, 62°35' W) at the Southern extent of the Torngat Mountains (Figure B.9.1). This mountainous and coastal region of northern Labrador is ecologically characterized as alpine tundra. The region supports a rich community of terrestrial and marine wildlife and is an important traditional harvesting area for Labrador Inuit.

The radar station at Saglek, designated as LAB-2, was operated as a Polevault Line radar station by the US Air Force between 1953 and 1971. At the height of operation Saglek was the largest site on the Polevault line and supported in excess of 200 people. In the late 1980s the abandoned facilities were demolished to make way for a new long-range radar (LRR) station that was constructed as part of the current NWS.

B.9.4.2 Site investigations

In 1996 the North Warning System Office (NWSO) commissioned an assessment of their current operations. This investigation revealed extensive PCB contamination in areas associated with the abandoned radar station, including soil with PCB concentrations greater than 50 µg/g, the criteria established under CEPA (ESG, 1997). Elevated concentrations of PCB in sculpin samples collected directly off shore of a contaminated beach area also revealed that contamination was migrating into the marine environment, potentially in violation of the Fisheries Act. In response to these findings, DND funded further study into: the redistribution of PCBs to surrounding soil, plants, and freshwater sediments; uptake and accumulation of PCBs by terrestrial and marine biota; and, an ecological risk assessment of contaminated marine sediments.

B.9.4.2.1 Redistribution of PCBs to area soils, plants, and freshwater sediments

Paired soil and plant samples were collected at increasing distances along transects that radiated from the three main areas of contamination. Samples were collected at 100–200 m intervals and up to 1600 m from the contaminated areas. Large diameter sediment cores were also collected from 16 lakes that radiated away from the site to a distance of 27 km. Paired soil and plant samples were also collected from beside each of these lakes.

Not surprisingly, ΣPCB concentrations in soil, plant, and sediment samples collected within a 3-km radius of contaminated areas at Saglek are significantly more contaminated than those samples collected from beyond 3 km (Table B.9.5). Soils from within the 3 km zone ranged from 0.69–7842 ng/g ΣPCB while those beyond this zone ranged from 0.146–79 ng/g. Plants from within the 3-km zone had 242 times the average concentration of plants beyond the direct influence of the site (> 3 km). A willow, growing within the PCB-contaminated area of the beach, had the highest concentrations at 12 380 ng/g. Concentrations of ΣPCB in the surface sediments from lakes close to the site were 48 times the mean concentrations of sediments from lakes beyond the 3-km zone (Table B.9.5) (Pier *et al.*, 2002a). The PCB-congener profiles for the samples collected closest to the site tended to resemble Aroclor 1260 the most, while increasing distance from the site led to decreases in the higher chlorinated congeners. PCB-congener signatures in plants tended to illustrate this pattern most clearly.

Although the ΣPCB concentrations beyond the 3-km halo of Saglek are much lower than observed close to the contaminated sites, the congener pattern typical of the Aroclor 1260 commercial mixture used at Saglek can be observed in soils, plants, and lake sediments up to 27 km from the site (Pier *et al.*, 2002a). It was concluded that the halo of PCB contamination around Saglek could extend to approximately 30 km, significantly larger than the area of impact assumed in the previous CACAR report.

B.9.4.2.2 Bioaccumulation of PCBs in herbivores

Uptake and accumulation of PCBs in terrestrial herbivores was assessed in deer mice and caribou from Saglek. Deer mice were chosen as a monitoring species because of their abundance throughout the site and their year-round residence in a relatively small area. Caribou, which have also been frequently observed at the site, are not year-round residents, but were also collected to address local concern over the safety of this very important traditional food species. Deer mice were analyzed as whole-body samples (less livers), while muscle, liver and fat were analyzed for caribou.

Deer mice collected within one of the most contaminated areas of the site had ΣPCB concentrations ranging from 25 000 ng/g to 64 000 ng/g, ww. Deer mice collected adjacent to two other contaminated areas had mean ΣPCB concentrations of 11 000 ng/g and 5000 ng/g, ww. At a site approximately 6 km from any contaminated area ΣPCB concentrations in deer mice had a mean of 65 ng/g, ww (ESG, 1999).

TABLE B.9.5 Summary of ΣPCB (69 congeners) concentrations in lake sediments, soil and plants within 3 km of Saglek, and from locations > 3 to 27 km from the contaminated sites at Saglek (Pier *et al.*, 2002a) [geometric mean (95% confidence interval)]

Distance	Soil (n)	Soil ΣPCB (ng/g dw)	Plants (n)	Plants ΣPCB (ng/g lipid)	Sediment (n)	Sediment ΣPCB (ng/g dw)
≤ 3 km from Saglek	46	100 (59–185)	52	63 (33–120)	8	82 (40–170)
> 3 km from Saglek	8	0.44 (0.11–1.7)	10	0.26 (0.06–1.14)	8	1.7 (0.8–3.5)

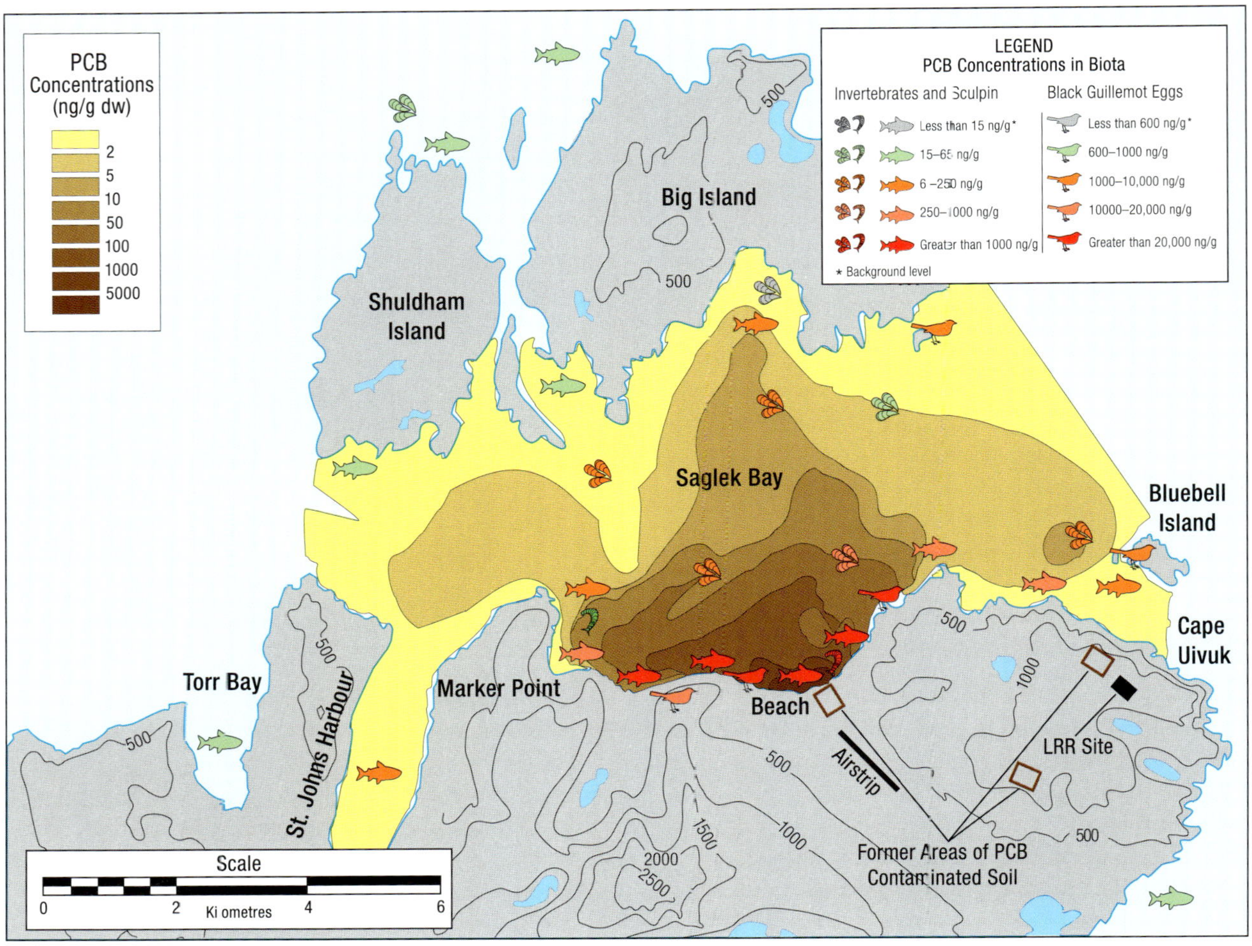

FIGURE B.9.4

Map of PCB contamination in marine sediments and the benthic based food chain of Saglek Bay (ESG 2001).

Eight caribou were collected during the assessment, four from the site itself and two each from Big Island and St. John's Harbour, both approximately 6 km away. Both of these sites are separated from the radar station by substantial physical barriers, a large stretch of open water and a small mountain range. The ΣPCB concentrations from caribou muscle taken at the site were consistently higher than the levels from both Big Island and St. John's Harbour (2.9–17 ng/g, ww versus 0.14–1.9 ng/g ww, respectively). The contrast was even greater for fat (approximately 71% lipid) with 68–365 ng/g (ΣPCB ww) at the site versus 7.8–13 ng/g (ΣPCB ww) for St. John's Harbour and Big Island. Caribou were also analyzed for a suit of organochlorine pesticides for which Saglek does not represent a local source. There was no significant difference in pesticide concentrations between caribou, from the three locations. These results indicate that caribou, which reside at Saglek during the summer are accumulating PCBs as a result of this local source.

B.9.4.2.3 Redistribution of PCBs to marine sediments

One of the three contaminated areas at Saglek was situated on the shores of Saglek Bay and was subject to annual erosion events as a result of spring runoff. As part of the 1996 investigation, sculpin, a non-migratory bottom-feeding fish, were collected from the near shore areas of the contaminated beach. This investigation revealed highly elevated concentrations of PCB in sculpin and indicated that PCB-contaminated materials had migrated into the marine environment, potentially in violation of the Fisheries Act. In response to these findings DND set out to remediate the site making excavation of the contaminated beach and removal of the source to the ocean the highest priority. Coincident with the excavation, DND also commissioned a detailed investigation of PCB contamination in sediment and biota of Saglek Bay.

The delineation of PCBs in marine sediments revealed that PCB contamination greater than 1 µg/g extends up to 1 km from the contaminated beach and covers an area of 1 km^2 (Figure B.9.4). Contaminant levels decrease with increasing distance from the contaminated beach and gradually approach background concentrations at the limits of a natural basin that is formed between the south shore of Saglek Bay and Big Island to the north. The western extremity of site influence can be detected up to 10 km from the source. Approximately 10 km^2 of sediment exceed the interim sediment quality guideline of 21.5 ng/g (Solomon, 2001).

B.9.4.2.4 Ecological risk assessment of PCB-contaminated marine sediments

Uptake and accumulation of PCBs in the marine food web was examined in benthic invertebrates, bottom-feeding fish (shorthorn sculpin), pelagic fish (Arctic char), seabirds (great black-backed gulls and black guillemots), and ringed seals. The results of this study indicated a high degree of PCB bioaccumulation and biomagnification in the local benthic-based food chain (Figure B.9.4). More pelagic and wider ranging species such as Arctic char, ringed seal and great black-backed gulls did not display the same site-influenced effect on tissue PCB concentrations (Table B.9.6). The findings of this study resulted in the establishment of a conceptual model for PCB uptake in the marine food web (Figure B.9.5).

An ecological risk assessment was designed to evaluate PCB-associated risks to benthic invertebrates, shorthorn sculpin, and black guillemots. Risks to benthic invertebrates were evaluated using sediment toxicity tests (10-day amphipod survival and Microctox™ solid-phase test) and benthic community surveys that were accompanied by chemical analysis of paired sediment samples. This assessment did not measure any adverse effects on benthic invertebrates at maximal sediment ΣPCB concentrations as high as 7 µg/g (ESG, 2001a). These results were surprising given the interim sediment quality guideline for PCBs of 21.5 ng/g.

TABLE B.9.6 PCB concentrations in marine biota sampled in the vicinity of Saglek compared to distant locations (ESG, 1999)

Species/Location (distance)	N	Tissue	ΣPCB (ng/g ww) range or mean (standard deviation)
Snails and scuds/intertidal area (0 km)	–	Whole	8000–49,000
Sea urchins/intertidal area (0 km)	–	Whole	5200
Sea urchins/Branagin I. and east of beach (20 km)	–	Whole	3.3–79
Mussels/Big Island and White Point (6 km)	–	Whole muscle	7.3–11.2
Clams/Saglek Bay (1–5 km)	–	Whole muscle	46–285
Shorthorn sculpin/near contaminated beach (< 1 km)	–	Whole body minus liver	1000–10,000
Shorthorn sculpin (up to 6 km away)	–	Whole body minus liver	65–1000
Shorthorn sculpin (6 to 10 km away)	–	Whole body minus liver	15–65
Arctic char/Saglek Bay near Big Island (5 km)	10 4	Whole Liver	37.6 (10.1) 90.2 (42.4)
Arctic char/Saglek Fjord (50 km away)	11 5	Whole Liver	28.2 (12.3) 26.1 (10.8)
Ringed seal/Saglek Bay (1–5 km)	5	Blubber	2480 (3397)
Ringed seal/Nain, northern Labrador (200 km)	4	Blubber	480 (116)
Great black-backed gull/Saglek Bay (1–5 km)	8 5	Muscle Eggs	10,090 (18,285) 1180 (513)
Great black-backed gull/Nain (200 km)	10 6	Muscle Eggs	3430 (5060) 844 (456)
Guillemot/Saglek Bay (1–5 km)	10	Muscle	2570 (2230)
Guillemot/Nain (200 km)	8	Muscle	120 (35.2)
Guillemot (1–2 km from Saglek)	8	Eggs	18,300 (17,773)
Guillemot (16–18 km W. of Saglek)	6	Eggs	435 (176)

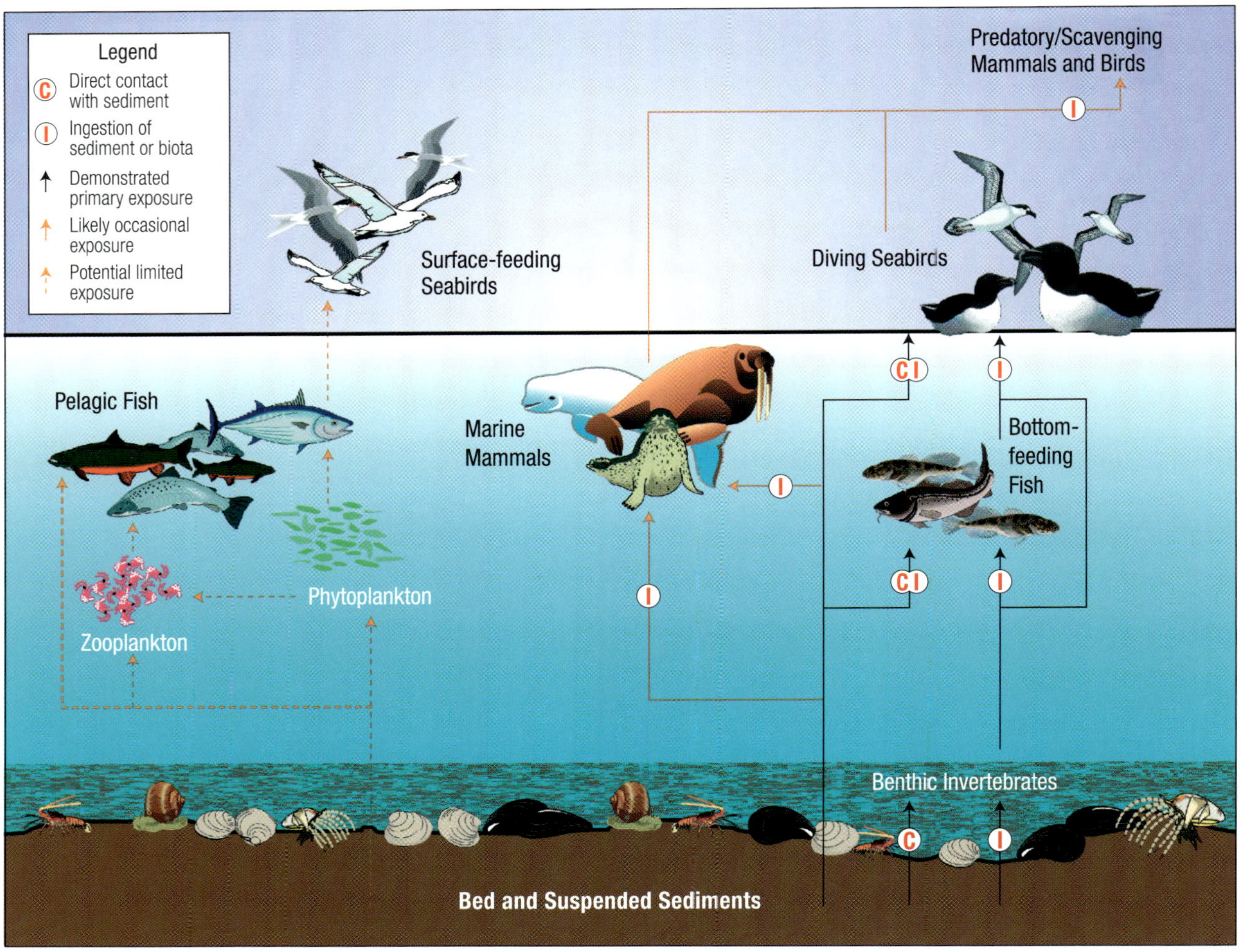

FIGURE B.9.5

Conceptual model of PCB uptake in the benthic based food chain of Saglek Bay (ESG 2001).

Risks to shorthorn sculpin were assessed as a function of coplanar PCB and PCDD/F toxic equivalence quotients (TEQs). Sculpin TEQs in the most contaminated areas are as high as 2.2 pg TEQ/g. Dioxin-like TEQs for shorthorn sculpin were compared to other reported studies that observed survival and reproductive effects associated with dioxin like toxicity. This assessment concluded that shorthorn sculpin in the most highly contaminated sections of the bay were at risk of decreased survival and reproduction as a result of the sediment PCB contamination (ESG, 2001a).

The next step in evaluating effects on the benthic-based food chain was the black guillemot. Black guillemots are diving seabirds that feed on fish and invertebrates at or near the bottom. During the ice-free summer months black guillemots are known to occupy relatively small foraging ranges and therefore make excellent subjects for the study of localized contaminant related effects. At Saglek, the foraging range for the nesting guillemots was estimated at 500 m (ESG, 2001a). Risks to black guillemot survival and reproduction were assessed through a series of 13 toxicological endpoints that ranged from observations of hatching success to measures of biochemical indicators. The endpoints that were evaluated fell into four general categories based on biological systems: reproductive, immune, metabolic, and endocrine. The study was carried out on 31 guillemot nestlings that were evenly distributed among three exposure groups: beach (most exposed, n = 11); Islands (moderately exposed, n = 10); and, reference (background exposure, n = 10). Mean liver ΣPCB concentrations differed significantly between each group and were 830, 73, and 25 ng/g, ww, respectively. Results of the toxicological study indicated significant PCB-associated effects on endpoints associated with reproductive, immune and endocrine systems (ESG 2001a, Kuzyk *et al.*, 2003). These effects, were considered to represent a significant risk to guillemot survival and reproduction, were associated with liver PCB concentrations as low as 73 ng/g, ww. The Saglek guillemot study is unique in that PCB-associated effects were measured on wild seabirds in the virtual absence of other contaminants (Kuzyk *et al.*, 2003).

B.9.4.3 Site Remediation

In 1997 DND undertook a cleanup of PCB-contaminated soil at the site in response to the discovery of soil PCB concentrations greater than 50 µg/g and to initial evidence of PCB migration into the marine environment. The priorities for cleanup were: the excavation and containment of PCB contaminated soil in excess of 5 µg/g PCB from areas that were subject to erosion and migration to the marine environment; excavation and containment of soil with PCBs in excess of 50 µg/g; and, the *in-situ* containment of remaining soil in excess of 5 µg/g PCBs. A site specific cleanup criteria of 5 µg/g was arrived at by consensus among a diverse group of stakeholders that included representatives of DND, Environment Canada, the Department of Fisheries and Oceans, the Newfoundland Department of Environment and Labour, and the Labrador Inuit Association. One of the main considerations in this decision was the optimal reduction in PCB mass. Cleaning up the site up to a 5 µg/g criteria effectively removed 99% of the PCBs at Saglek from contact with the environment (ESG, 2000b).

The excavation and *in-situ* stabilization of contaminated soil was completed in 1999. Approximately 20 000 m^3 of soil with over 50 µg/g of PCBs was stockpiled in a secure staging area and covered with impermeable geomembranes. The amount of PCBs contained in these stockpiles is estimated at 4.2 tonnes (ESG, 2000b). The removal and/or destruction of the PCBs contained in these stockpiles is scheduled to begin in 2003 and to be completed by the end of 2004 (Kuzyk Z.A., pers. comm.).

The fate of contaminated marine sediments is currently being considered by the Saglek stakeholder group. Aside from the ecological risk assessment, the stakeholders must also consider: the findings of a sediment transport study that provides evidence as to how ecological risk might change in the future under a variety of remedial scenarios; a technical and financial analysis of remedial options; and, an assessment of risks to Inuit under several harvesting scenarios. DND will make a decision on how to manage the contaminated marine sediments based on recommendations from the stakeholder group in 2003 (Kuzyk Z.A., pers. comm.).

B.9.5 Giant Gold Mine, Northwest Territories

B.9.5.1 Introduction

Yellowknife, NWT has an extensive terrestrial and aquatic arsenic contaminant problem as a result of 60 years of gold mining activity. The Miramar Giant Mine is located five kilometres north of the city and the Miramar Con Mine borders the southern city limit (see Figure B.9.6). The Giant Mine mineral rights were recently purchased by Miramar Mining Corporation, however, as part of the agreement of purchase, INAC retained the environmental legacy issues of the site. Between 1997 and 2002, a committee comprised of local residents, First Nations, and government regulators was formed to assess the potential risk posed to human and ecological health from arsenic contamination. It was called the Yellowknife Arsenic Soil Remediation Committee (YASRC).

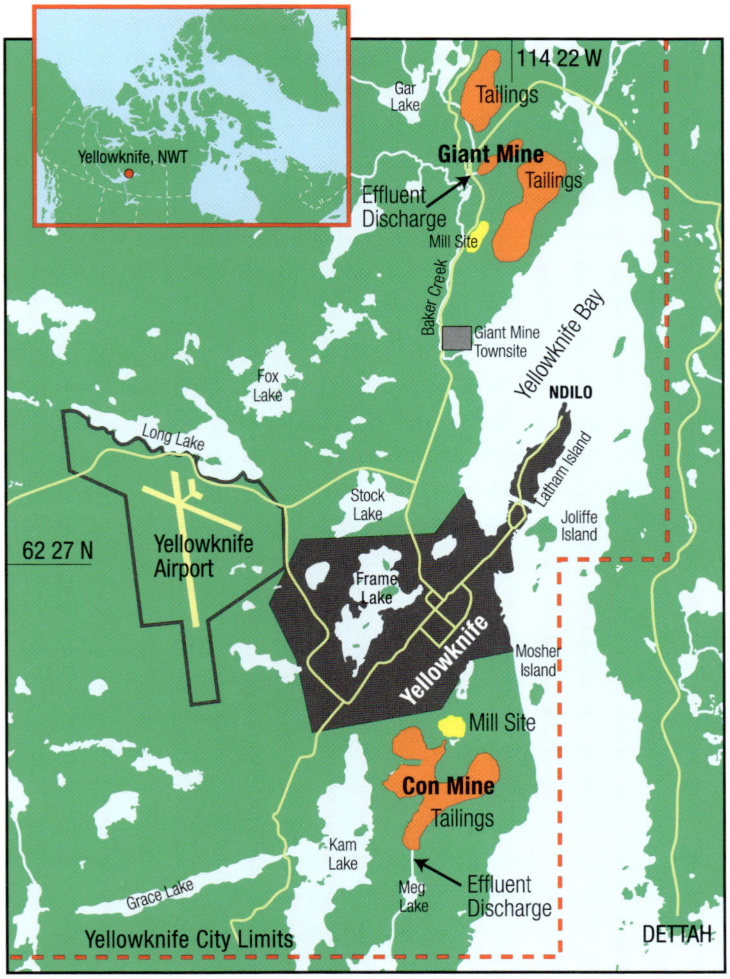

FIGURE B.9.6

Map of gold mining activity in the Yellowknife area.

For several years it has been known that gold mining has increased the levels of arsenic in several environmental media in the Yellowknife area. The gold ore mined in Yellowknife is associated with the arsenic-bearing mineral sulphide, arsenopyrite (FeAsS). Over the years, there have been several modifications and updates to the milling and waste disposal practices at both mine sites. Currently, ore is only extracted from the Giant Mine site and is trucked to the Con Mine, where it is milled, with Con Mine ore in an autoclave process. Historically, the gold separation process at both mine sites consisted of cyanidation and roasting. The resulting waste streams included: arsenic trioxide dust (As_2O_3), which at the Giant Mine was collected and is currently stored in various underground vaults and mined-out stopes; fugitive emissions of arsenic trioxide from the roaster stacks; solid and liquid mill tailings that have been deposited in tailings ponds; and, liquid effluent from the tailings ponds. Until the mid-1980s untreated liquid effluent from the tailings ponds was discharged into the Baker Creek watershed (Giant) and the Meg-Keg-Peg Lake system (Con) (Ollson, 1999). Consequently, there are two arsenic issues in Yellowknife. The management of the stored arsenic trioxide dust at Giant Mine and the environmental contamination of the land and water system in the general vicinity.

Arsenic is a natural and ubiquitous element, which ranks twentieth in abundance in the earth's crust. Traditional risk assessment methodology incorporates only measurements of total arsenic in exposure calculations; however, arsenic's toxicity depends on both its chemical form and potential bioavailability. In fact, many forms of arsenic are not biologically active and safely pass through the body (e.g., arsenobetaine; an arsenic compound commonly found in seafoods), but inorganic arsenic compounds, a group that includes arsenic trioxide, are toxic. The bioavailability of arsenic differs depending on the matrix ingested (Ruby *et al.*, 1996, Rodriguez *et al.*, 1999). For example, arsenic in water is believed to be 100% bioavailable, while it has been shown that only a fraction (mean 30%) may be potentially bioavailable from soils. For this reason an assessment of environmental risks associated with arsenic must take the chemical form and bioavailability into consideration and not simply total arsenic concentrations.

B.9.5.2 Environmental assessment of arsenic contamination in the Yellowknife area

The Environmental Sciences Group, Royal Military College (ESG, RMC), has investigated the spatial distribution and environmental impact of arsenic in soil, sediment, and water throughout the Yellowknife area. These studies have considered: total arsenic in different environmental media (soil, water, sediment, plants, biota); the relative contributions of the different forms of arsenic in soils based on sequential selective extractions (SSE); an assessment of bio-availability based on simulated gastric fluid extractions (GFE); ecological risk assessment based on a short food chain (soil-plant-deer mouse); and, a human health risk assessment that incorporates both total and biologically available arsenic concentrations from all exposure pathways.

B.9.5.2.1 Distribution of arsenic in soil and sediment

The typical Canadian background soil concentration of arsenic ranges from 5 to 15 mg/kg but high natural background concentrations can be expected in areas with arsenic-rich rocks or soil, such as Yellowknife (ESG, 2001b). As noted previously, gold mining activities in this area have introduced arsenic to the environment over the last 50 to 60 years. The primary sources of arsenic originating from the mine have historically been condensed arsenic dust from the roaster stack and effluent, liquid, and solid (tailings), from the gold extraction process. Scrubbers were introduced to the roasting process in the early 1950s to remove most of the arsenic trioxide from the stack gases. In the case of Giant Mine, arsenic trioxide was collected as baghouse dust and stored underground. In recent years, the Con Mine has successfully managed to treat its arsenic trioxide contaminated sludge ponds.

One of the challenges associated with assessing surface arsenic contamination in the Yellowknife area was determining whether sources of elevated soil concentrations were natural (geologic in nature) or anthropogenic (having resulted from mining activities). It was very important to establish this difference to set a site-specific guideline for what might be considered "background" concentrations. While the CCME has established a guideline of 12 µg/g for all land-use categories, it recognizes that local geological conditions may result in an elevated background level of arsenic and endorses the establishment of site-specific guidelines in these situations (CCME, 1999).

TABLE B.9.7 Arsenic concentrations in soils in the Yellowknife area[1] (SD = standard deviation)

Location	n	Minimum concentration (μg/g)	Maximum concentration (μg/g)	Median concentration (μg/g)	Mean concentration (μg/g ± SD)
Giant Mine					
Giant Mine Tailings	23	738	4840	2650	2460 ± 1330
Giant Mine Mill/roadways	8	1890	89 000	4350	16 000 ± 29 200
Giant Ore	3	5420	6050	5460	5650 ± 350
Giant Surrounding Property	30	7	3940	52	257 ± 724
Con Mine					
Con Mine Tailings	31	491	25 000	3430	4790 ± 5200
Con Mine Mill/roadways	9	383	3140	1800	1660 ± 995
Con Mine Ore	3	1020	1410	1150	1190 ± 199
Con Mine Surrounding Property	59	10	1450	264	338 ± 323
Residential					
City of Yellowknife	67	3	307	24	36 ± 47
Ndilo	50	4	885	127	254 ± 194
Rat Lake	42	24	2330	290	475 ± 583
Dettah	8	7	144	24	47 ± 53
Background Area	4	54	110	81	81 ± 30

[1]From Ollson, 2002.

To establish background concentrations in the Yellowknife area, samples of soil and sediment from various locations were analyzed for a suite of metals thereby producing a geological fingerprint for the various soil and sediment types in the region. These fingerprints were then compared using various statistical tools including principal components analysis (PCA). This use of PCA allowed for the soil and sediment samples to be classified as either mine waste, anthropogenically impacted, or as naturally elevated (ESG, 2001b; Reimer *et al.*, 2003). Samples that were considered to be representative of background conditions had arsenic concentrations ranging from < 10 to 150 μg/g. Such concentrations are typical of the soils in most of the residential areas of Yellowknife.

Total concentrations of arsenic in Yellowknife soil samples are found in Table B.9.7. Not surprisingly, the highest concentrations were found at the mill sites and tailings ponds on both mine properties. The highest concentration of arsenic measured in a Yellowknife soil sample was 87 000 μg/g, collected at the base of the Giant Mine roaster stack. The levels of arsenic in the Giant ore (5650 ± 350 μg/g) were found to be higher than those from the Con Mine ore (1190 ± 199 μg/g).

Concentration of arsenic in organic soils surrounding the tailings ponds were 257 ± 724 μg/g on the Giant Mine, and 338 ± 323 μg/g on the Con Mine property. Although these concentrations are over 20 times the CCME arsenic in soil guideline, a vibrant forest ecosystem exists. Elevated concentration of arsenic in soil were also found in the First Nations community of Ndilo, located on the northern tip of Latham Island. Rat Lake, an area close to the city of Yellowknife, also contained elevated levels of arsenic in soils as a result of tailings being deposited in the area during the early years of milling at the Con Mine.

In the early years of milling, tailings from the Giant Mine were deposited on the shores of Yellowknife Bay (Beach Tailings). The average concentration of arsenic in these tailings was 909 ± 150 μg/g. Over time, the tailings in this area have eroded into Yellowknife Bay with sediments 300 m offshore having arsenic concentrations as high as 400 μg/g. Recently, the Giant Mine in cooperation with INAC, has addressed the Beach Tailings issue by stabilizing the material in place.

Elevated concentrations of arsenic were also found in surficial sediment samples from lakes throughout the Yellowknife area. Giant Mine tailings pond effluent is discharged to Baker Creek, which has an average arsenic concentration of 2024 ± 1101 µg/g. The effects of the arsenic discharge can be seen as far as 1 km away in Back Bay (1193 µg/g) (ESG, 2001b). The Con Mine discharges its tailings pond effluent into the Meg-Keg-Peg Lake watershed, which drains into Great Slave Lake. Arsenic concentrations ranged from 865 µg/g in Meg Lake, peaked at 5550 µg/g in Peg Lake and declined by several orders of magnitude at the confluence with Great Slave Lake. The average arsenic concentration in Kam Lake was 891 ± 491 µg/g resulting from its location west of the Con Mine, and the historical tailings discharges, which occurred in the early years of mining. With the exception of Rat Lake, which was once used for waste disposal by Con Mine, the lakes in the city of Yellowknife were uninfluenced by mining activities and had sediment arsenic concentrations typically < 100 µg/g (ESG, 2001b).

B.9.5.2.2 Biological availability of soil arsenic

One of the highest risk pathways of exposure in environmental and human health risk assessment is the inadvertent ingestion of soils. An adult human (20 years and older) inadvertently consumes, on average, 0.02 g of soil a day; however, relatively little is known about the bioavailability and/or toxicity of arsenic from this exposure pathway. If after oral ingestion, the arsenic in the soil passes through the gastrointestinal (GI) tract without being solublized and then absorbed into the bloodstream, it does not pose a toxicological risk. *In vitro* testing of Yellowknife soils was conducted to assess the potential bioaccessibility (% of arsenic extracted) of arsenic in human and mammalian gastrointestinal tracts using a simulated gastric fluid extraction (GFE). Sequential selective extractions (SSE) were used to determine the fraction of arsenic that was bound in soils and to assess the potential for arsenic to be environmentally available from soils (Ollson, 2003; Reimer *et al.*, 2003).

Based on the results of SSE analysis, it was determined that a large percentage of arsenic in topsoil (organic soils) was in the environmentally available fractions. Most of the arsenic in tailings and rock, however, was bound as arsenic-sulphides and was deemed to be relatively insoluble (i.e., not environmentally available) under ambient environmental conditions (ESG, 2001c). These results were supported by GFE analysis. Rock and tailings samples had low potential bioaccessibility, while arsenic contaminated mine organic soils and residential organic soils were determined to be 20–30% bioaccessible (Table B.9.8). The arsenic bioaccessibility results were incorporated into human and ecological risk assessments.

TABLE B.9.8 Arsenic concentration in the gastric fluid extraction

Sample category	Arsenic (µg/g)	GFE arsenic (µg/g)	% Bioaccessible[1]
Rock	3170 ± 5220	110 ± 127	5.0 ± 3.6
Tailings	4570 ± 4420	116 ± 112	2.9 ± 1.7
Mine organic	850 ± 1020	164 ± 211	20 ± 11
Residential	142 ± 128	72 ± 112	31 ± 28

[1]The amount of arsenic extracted by GFE is expressed as a percent of total arsenic.

B.9.5.2.3 Yellowknife ecological risk assessment

A field-based ecological risk assessment of a short terrestrial food chain — soil → plant → deer mouse, was conducted in the Yellowknife area (Hough, 2001). Mice were found to be abundant in burrows that they had established in tailings. Such mice were found to have higher concentrations of arsenic in various tissues than mice living in areas well removed from mine operations, but there was evidence of arsenic transformation in various tissue types — suggestive of a biochemical response to this contamination. Hazard quotients calculated using bioaccessibility data (from gastric fluid extraction) approached "1" — indicating that the risk was much less than would have been expected based on the total arsenic concentrations (Ollson, 2003). Deer mice captured from areas outside the tailings ponds (mine organic soil) and from background areas all had hazard quotients below "1", which suggests that there would be little effect on mice living in these areas.

B.9.5.2.4 Human exposure to arsenic in Yellowknife

The YASRC committee commissioned a screening level, deterministic human health-risk assessment in 2001 (Risklogic, 2002). This study used total concentrations of arsenic in environmental exposure pathways. Results of this study indicated that it was "safe" to live in the city of Yellowknife; however, the mine properties were not assessed.

A recent human health-risk assessment conducted by the ESG (Ollson, 2003), included field data on exposure pathways, such as concentrations of arsenic in vegetable gardens, soils, sediments, and water. The domestic water supply for the city of Yellowknife is drawn upstream of potential anthropogenic impacts of arsenic. Regular testing of drinking water consistently indicates arsenic concentrations are below 1 µg/L, which is well below both the Canadian drinking water guideline of 25 µg/L. Therefore, drinking water was excluded as a route of exposure that warranted concern.

To evaluate potential health risks associated with Yellowknife garden produce, ESG collected vegetables from 10 city gardens and three abandoned gardens in the Giant Mine town site. Soil samples from these gardens revealed arsenic concentrations ranging from 11 to 56 µg/g in Yellowknife and 81 to 351 µg/g in the town site. The results of this study indicated that green leafy vegetables and greens such as beet greens and celery leaves had the highest concentrations of arsenic, which were approximately 10-fold higher than results of a survey conducted on supermarket vegetables in 1993. Despite the elevated concentrations of arsenic in certain vegetables, the ESG report concluded that there is likely no potential health risks from consuming these vegetables (ESG, 2001d; Koch *et al.*, 2002).

The ESG human health-risk assessment investigated the human health risks associated with all arsenic exposure pathways. This assessment, which included bioaccessibility results, also determined that it was "safe" to live in the city of Yellowknife. An investigation of different exposure scenarios, and the inclusion of bioaccessibility results, also indicated that the mine properties (except tailings ponds) could potentially be used as limited use recreational areas (i.e., hiking and biking) after mine closure.

B.9.5.3 Subsurface stored arsenic trioxide

Approximately 237 000 tonnes of As_2O_3 dust has been stored in mined-out stopes or purpose-built chambers in the mine (SRK, 2001). The storage of the dust below ground at the mine site was considered to be an acceptable, permanent disposal mechanism for handling this waste. This conclusion was based on a 1977 report by the Canadian Public Health Association on the basis that once the Giant mine was closed, permafrost would re-establish around the storage vaults and seal in the arsenic trioxide. In recent years, the improved understanding of permafrost and groundwater in Arctic regions has forced the realization that permafrost would not naturally re-establish in the Yellowknife area. Consequently, unless the pumps were kept running to de-water the mine in perpetuity, groundwater could result in the migration of the arsenic from the subsurface vaults (SRK, 2001).

Recent inspections of the lower chamber bulkheads have found that water is seeping through the face of some of the bulkheads and the rock surrounding them, indicating that at least some portion of some of the chambers and stopes may currently be saturated with water.

Sampling to characterize mine water quality has continued since 1999 with the focus being to identify and characterize the principal sources of arsenic contamination of the mine water. The main sources of water entering the mine are direct infiltration of snowmelt and rainfall, inflow from Baker Creek, service water, and seepage from tailings ponds. Surface infiltration is predominant (SRK, 2001).

Subsurface water at the extremities of the mine have relatively low arsenic concentrations ranging from 0.018 to 63 µg/L. Interaction with the mine workings increases As concentrations to 500 µg/L while seeps from As_2O_3 dust storage chambers are in the range of 3.5×10^6 µg/L. Calculations indicate that seepage from the dust chambers accounts for approximately half of the arsenic in water pumped out of the mine. The remainder is due to contact with tailings, waste rock backfill and, to a lesser extent, seepage from above-ground tailings ponds. Currently, water from the mine is pumped to the tailings ponds where it is treated using hydrogen peroxide oxidation and ferric iron precipitation and then discharged to Baker Creek. The total As concentration in the water pumped to the surface is of the order of 10 000 to 20 000 µg/L while the regulated criterion for effluent discharged to Baker Creek is 500 µg/L (SRK, 2001).

B.9.5.3.1 Management of stored arsenic trioxide dust

The preparation of a long-term arsenic management plan for the mine is being developed by a team of consultants lead by SRK Consulting. To date, the team has agreed to a minimum level of remediation that is considered to be essential for the safe closure of the site. Beyond these measures the team has also identified several management options which are currently being evaluated.

The minimum level of remediation includes:

- removal of spilled tailings and As_2O_3 dust from unsealed tunnels and access points;
- removal of piping used to deliver As_2O_3 dust to the storage chambers and stopes;
- reinforcement of bulkheads around the stopes and chambers containing the As_2O_3 dust to serve as permanent barriers to water; and,
- providing the Northwest Tailings Pond with an impermeable cover to prevent continued infiltration of water from the pond into the mine.

Under the minimum level of remediation described, the mine would be allowed to flood thereby mobilizing an unknown amount of arsenic trioxide into the groundwater that would eventually make its way into the surrounding aquatic environment. A risk assessment was carried out to evaluate this scenario and found that it could result in annual arsenic discharges as high as 16 000 kg/year. It was concluded that the associated ecological and human health risks were unacceptable (SRK, 2001).

Options for the further remediation of the stored arsenic trioxide dust fall under two basic approaches: *in-situ* stabilization; and removal and further processing of the dust to either remove residual gold and/or arsenic or to produce stabilized waste. Evaluation of numerous alternatives that fall under each of these basic approaches is ongoing. Interactive public consultations are integral to this process.

One of the promising management options being discussed is *in-situ* stabilization with active ground freezing. Since it is already known that the underground storage areas will not freeze naturally a technology known as a thermosyphon is being proposed. A thermosyphon is a proven cold regions technology consisting of a sealed tube filled with pressurized carbon dioxide (CO_2) that penetrates beneath the ground surface and has a radiator extending into the air. The CO_2 becomes liquid when the radiator is sufficiently cool and, as a liquid, drains down the thermosyphon into the warmer ground. The CO_2 takes heat from the ground — thereby freezing it, changes back to a gas, and rises back into the radiator to complete the cycle. The process continues as long as the air temperature is sufficiently cold; but once the air temperature rises, the cycling of the CO_2 stops, precluding the transfer of heat into the ground.

B.9.6 Port Radium Mine, Northwest Territories

B.9.6.1 Radioactivity sources in the Arctic

Following the reduction in atmospheric nuclear testing in the early 1960s, and eventual global ban on atmospheric and underground nuclear testing in 1996, the contribution of anthropogenic fission products in Arctic ecosystem has markedly decreased, most importantly in caribou which are a major food source for northern Aboriginal populations (Jensen, *et al.*, 1997). Research contained in CACAR I showed that long-range atmospheric and oceanic transport are not currently pathways of concern for radionuclides entering the Canadian Arctic at present. However local sources of metals and radionuclides, such as uranium mines, continue to be a concern for northern residents.

The Canadian Shield contains pockets of uranium-rich rock. Two uranium mines, Port Radium and Rayrock, were operational in the Northwest Territories in the mid 1900s and the mine sites continue to be a concern to local communities. Through inclusion in site assessment and management processes, communities are being made aware of the risks associated with closed mines, and about radiation exposure in general. Port Radium was Canada's first uranium mine, and one which is viewed by the Dene community of Déline on Great Bear Lake as a potential local source of radiation exposure to those Dene who worked there, those who continue to use the eastern shore of Great Bear Lake for traditional purposes and the residents of Déline itself. The federal government and the community of Déline, in partnership, are undertaking studies to answer questions about historical and present day exposures, and of the effects of the former mine site on the environment.

B.9.6.2 History of the site

The abandoned Port Radium mine site is located on a peninsula on the eastern shore of Great Bear Lake in the Northwest Territories (66°05' N; 118°02' W). Figure B.9.7 is a picture of the site in 1972 when the mill and town site facilities were still in place.

Mining and milling took place at Port Radium almost continuously from 1930 to 1982. From 1942 to 1960, a uranium mine and mill was operated by a Crown-owned mining company. The Port Radium town site was closed in 1960 and subsequently reopened in 1964 for a silver operation. The mine site was decommissioned in 1982 as a silver mine at which time time the mine surface openings were sealed and most of the buildings demolished. During the various mining operations, tailings were deposited directly in Great Bear Lake as well as in a number of on-land lakes and depressions. It is estimated that approximately 100,000 m^3 of tailings are deposited in depressions around the site, most of which were covered by waste rock. The remaining tailings are deposited in Great Bear Lake and McDonough Lake.

The mining activities and practices have generated concern from Déline residents regarding historical exposure, contamination of their environment and the potential for present day exposure to radiation.

Several environmental studies have been conducted at the site since the 1970s, however it was difficult to compare the data from the different reports and identify any trends or draw meaningful conclusions. In 2001, a human health and environmental risk assessment was commissioned, and the report is currently being finalized (SENES 2002a).

FIGURE B.9.7
Photograph of the Port Radium site. Source: Government of Canada archives.

B.9.6.3 Site assessment

The recent mine site assessments (SENES 2001, SENES 2002b) include air, water, soil, rock, sediment and tailings sampling, as well as limited biotic sampling including vegetation, fish, grouse, squirrel and hare. The quantitative human health and ecological risk assessment predict exposures based on these environmental media. Where predicted levels and measured levels of contaminants are different (e.g., in fish), the higher value is used in the risk assessment. More field work is required to characterize the in-lake transfer of contaminants from tailings into the aquatic food chain, and further sampling work must be done on terrestrial biota to obtain a more complete data set. Analysis of the scientific data will be correlated to information currently being gathered using traditional knowledge approaches.

The site assessment data collected to date are presented here, starting with the metals which may be a concern, followed by the radionuclide data and the gamma radiation data from the site.

It is estimated that 750 000 tonnes of uranium tailings remain in Great Bear Lake. These tailings exceed many guidelines for sediment quality; however measurements of water quality near the site undertaken so far do not indicate an effect on water quality in Great Bear Lake. Table B.9.9 shows levels of selected metals in water around the site (SENES 2002a).

SENES (2002b) reported results of leachate tests for on-land tailings suggesting that the tailings are not generating substantial quantities of acid. Metal leaching, however, is occurring and some tailings have been observed to contain leachable arsenic, molybdenum and zinc. Samples of spilled tailings on the surface were also observed to have high levels of leachable metals. The presence of leachable metals in the tailings are a concern in terms of potential contamination of surface runoff from these areas, however it has been suggested that these impacts could be localized.

TABLE B.9.9 Concentrations of metals in water (mg/L)

Contaminant	Great Bear Lake[1]	McDonough tailings basin	Ponded tailings water	Maximum of seep concentrations
Aluminum	0.06	0.03	0.05	1.96
Antimony	0.0008	0.008	0.283	0.014
Arsenic	0.0032	0.122	3.72	0.574
Cadmium	0.00003	0.00006	0.0011	0.0047
Chromium	0.00022	0.00027	0.0014	0.0069
Cobalt	0.0015	0.0011	0.187	1.24
Copper	0.0012	0.011	0.009	1.11
Lead	0.0012	0.0011	0.0046	0.0628
Molybdenum	0.0004	0.013	0.267	0.0117
Nickel	0.0009	0.008	0.938	0.817
Selenium	0.0006	0.00005	0.0027	0.0058
Silver	0.00005	0.00005	0.0007	0.0051
Uranium	0.0014	0.0998	0.978	0.813
Zinc	0.0083	0.038	0.332	1.69

[1]The average of the Great Bear Lake surface water concentration is also considered as the background water concentration.

TABLE B.9.10 Concentrations of metals in sediments (μg/g)

Contaminant	Average Inner Labine Bay[1]	Average Great Bear Lake[2]	McDonough tailings basin[3]	Background
Antimony	< 20	27	< 20	0.34[5]
Arsenic	95	1370	1100	1.9[4]
Cadmium	< 30	< 30	< 30	0.2[4]
Chromium	261	82	62	65.1[4]
Cobalt	64.5	513	190	10.7[5]
Copper	211.5	3700	1900	20[4]
Lead	84	1187	980	6[4]
Molybdenum	9 5	14	5	9.5[5]
Nickel	53	270	350	15[4]
Selenium	< 50	< 50	< 50	0.069[5]
Silver	6	210	100	0.26[5]
Uranium	< 60	213	87	2.8[4]
Zinc	183	597	570	69[4]

[1]Average measured values from two samples for Inner Labine Bay.
[2]These sediments are obtained from all Great Bear Lake sediment sampling sites with the exception of Inner Labine Bay.
[3]Measured from one sample from McDonough Lake Tailings Basin.
[4]Mean typical background sediment concentration presented in Painter et al. (1994).
[5]From U.S. ACE (2001).

Table B.9.10 shows the concentration of metals in sediments around the site. The background levels are taken from the literature cited. Low sample numbers (n) are also noted.

Concentrations of metals in on-site soil are generally elevated, and reflect the local agregate material, including waste rock, which was used to cover the area during decommissioning. Table B.9.11 compares on-site and background metal concentrations. Soil concentrations are used in the risk assessment to predict uptake in plants of impacted areas.

TABLE B.9.11 Concentrations of metals in soil (μg/g)

Contaminant	On-site(1)	Background(2)
Antimony	10.0	5.3
Arsenic	180.0	10.0
Cadmium	15.0	3.5
Chromium	87.6	102.0
Cobalt	80.4	15.5
Copper	397.2	21.5
Lead	48.4	18.0
Molybdenum	9.0	2.6
Nickel	93.6	19.5
Selenium	25.0	1.1
Silver	3.6	1.0
Uranium	284.0	5.0
Zinc	250.0	225.0

(1)On-site concentrations are reflective of the whole mine site area (weighted mean of impacted and un-impacted areas).
(2)Background concentration were measured in soil from un-impacted areas.

TABLE B.9.12 Concentration of radionuclides in water (Bq/L)

Radionuclide	Maximum of McDonough and Great Bear Lake	Maximum of surface seeps	Ponded tailings water
U-238+	1.9	11.3	12
Th-230	0.02(1)	1.52	7(1)
Ra-226+	0.02	2.26	4
Pb-210+	0.02(2)	0.67(2)	3(1)
Po-210	0.02	0.67	3(3)

(1)Measured as less than the detection limit, therefore the detection limit was used.
(2)Concentration of Pb-210 was assumed to be equal to PB-210; supported by additional alpha spectrometry.
(3)Assumed equal to Pb-210+

Radionuclides of concern from the ^{238}U decay series were measured at the site in water, soil and sediment. Along with air sampling of radon and dust, radionuclide levels in food and water are used to calculate a total potential uptake of radionuclides by human or ecological receptors through breathing and ingestion while at the former mine site. Tables B.9.12 and B.9.13 show the levels of radionuclides of concern in the each medium.

A comprehensive gamma survey of the mine site and surrounding area has been undertaken. Over 20,000 gamma radiation measurements were taken across the site in 2001–2002, and the results are summarized in Table B.9.14. Background gamma measurements in nearby areas uninfluenced by the site are typically 10 to 20 μR/h.

B.9.6.4 Risk assessment

A quantitative human health and ecological risk assessment is nearly complete and although there remain some information gaps, these can be addressed with the inclusion of traditional knowledge from the community of Déline and more sampling and analysis of aquatic and terrestrial biota. Some preliminary findings have been proposed.

TABLE B.9.13 Concentrations of radionuclides in sediment and soil (Bq/g)

Radionuclide	Maximum measured sediment concentration	Measured soil concentration(1)
U-238+	4.2	20.8
Th-230	6	8
Ra-226+	10.8	3.9
Pb-210+	10.9	4.7
Po-210+	10.9(2)	4.7

(1)Measured at one location assumed to have radionuclide levels higher than the site average.
(2)Po-210 is assumed to be equal to Pb-210.

TABLE B.9.14 Gamma radiation at the Port Radium mine site

Classification (micro Roentgen per hour)	Measured area (ha)
Less than 20 μR/h	25.8
Between 20 and 50 μR/h	17.2
Between 50 and 100 μR/h	10.1
Between 100 and 250 μR/h	8.0
Over 250 μR/h	2.2
TOTAL	**63.3**

ITK/Eric Loring

The contaminants of potential concern were identified using measured and modeled values from the site and applying a screening of environmental quality and health guidelines and benchmarks. Since these toxicity benchmarks are not site-specific, further toxicity testing is planned for the in-lake tailings. The radionuclides of concern used in the risk assessment are uranium-238 decay series radionuclides (the thorium-232 decay series was considered however proved negligible in dose calculations). The metals of concern that were identified for the human health risk assessment (HHRA) are antimony, arsenic, cobalt, copper, lead, molybdenum, nickel, and uranium. The metals of concern used in the ecological risk assessment (ERA) are aluminum, arsenic, cadmium, chromium, cobalt, copper, lead, molybdenum, nickel, selenium, silver, uranium, and zinc.

The human exposure assessment is based on five occupancy scenarios outlined in Table B.9.15. The likelihood of permanent residences was considered very low at this remote location therefore this option was not included in the calculations. The radiation doses estimated in Table B.9.15 represent the amount of radiation in excess of that which would be attained from background levels. For all of these occupancy scenarios local sources of food and water were used exclusively for exposure calculations. Whereas several local food choices exist, the modelling used only the most potentially affected animals and fish in order to have a conservative dose estimate.

Based on these occupancy scenarios, the radionuclides and metals levels at the site are considered by the risk assessment authors not to be a cause of concern from a human health perspective. The toxicity benchmarks were not reached and the total incremental radiation doses predicted for these exposure scenarios is under 1 mSv/year, which is the Canadian Nuclear Safety Commission reference dose for the general public.

The ERA shows potential impacts to some aquatic and terrestrial receptors (Tables B.9.16 and B.9.17).

The traditional knowledge component still needs to be incorporated into the risk assessment for it to be considered valid by the community. Traditional knowledge will be used for advising the pathways and receptors models, and to identify any other potential effects on the environment. At present, based on existing information and modelling, no toxicity benchmarks for aquatic life should be exceeded in Great Bear Lake, but impacts on McDonough Lake aquatic organisms is expected.

TABLE B.9.15 Radiation doses received at Port Radium for selected exposure scenarios

Receptor	Total incremental dose in milliSieverts (mSv)
Camper for 3 months/yr adult	0.4
Camper for 3 months/yr child	0.7
Fishing lodge worker 2 month/yr	0.7
Fisher/hunter 1 week/yr, 10 hr/d	0.1
Inspector for 2 days/yr	0.03

TABLE B.9.16 Potential impacts to aquatic biota

	Metals exceeding toxicity benchmarks	
Aquatic biota	McDonough tailings basin	Great Bear Lake
Pond weed	None	None
Primary producers	Arsenic, copper, uranium	None
Benthic invertebrates	None	None
Zooplankton	Uranium	None
Lake trout	Copper	None
Lake whitefish	Copper	None
White sucker	None	None

TABLE B.9.17 Potential impacts to terrestrial biota

Terrestrial biota	Metals exceeding toxicity benchmarks
Bear	Arsenic
Ducks: Mallard Merganser Scaup	Arsenic, selenium, uranium None Arsenic, selenium, uranium
Fox	Arsenic
Hare	Arsenic, uranium
Moose	Arsenic
Ptarmigan	None

The contaminants of potential concern from a toxicity perspective are arsenic, selenium and uranium. The key pathways for the terrestrial receptors which may exceed toxicity benchmarks are uptake from contaminated sediments in McDonough Lake and the transfer of arsenic from the soil to the surface vegetation. The remediation options to be developed for the Port Radium mine site will need to address the identified environmental risks that are posed by current conditions.

This site assessment was designed in collaboration with a scientific advisor employed by the Déline Dene Band Council, and several community members participated in the field work. The inclusion of the local community in this risk assessment process is intended to promote ownership of the results of the studies and a common understanding of risk information leading to better and more supported risk decision-making. The community is insistent that the traditional knowledge component of the environmental risk assessment should be completed before any final conclusions are drawn. The community's questions regarding historical exposures to Dene workers at the mine and along the transportation route are being addressed in parallel with the assessment of present day risks posed by the site.

B.9.7 Carcross railway tie treatment plant, Yukon Territory

B.9.7.1 Background

Carcross, is located on the White Pass and Yukon Route (WP&YR) railroad at the point where the line crosses the Nares River. Nares River is a short, narrow channel linking Lake Bennet to Nares Lake. The WP&YR was constructed during the Yukon gold rush to facilitate the flow of goods to the gold fields by connecting from Skagway, Alaska to the Yukon River ferries at Whitehorse via the White Pass, Fraser, B.C., Bennet and Carcross YT. Although construction of the line commenced in 1898, by the time it opened in 1900 the gold rush had ended. Nevertheless the WP&YR remained an important freight route to the coast until it closed in the earlier 1980s. WP&YR re-opened a portion of the route in 1988 and continues to provide tourists with an opportunity to travel from the ocean port to the summit at White Pass during the summer season.

For the first 40+ years of operation, the WP&YR imported pre-treated railway ties from outside sources but in the late 1940s it was decided to cut and treat the ties locally. The tie-treating plant was constructed on the north side of Nares River just as it entered Nares Lake. Operations ceased in 1975 and the plant was dismantled in 1979 (Roach *et al.*, 1999). During its operation, the plant used a non-pressurized process consisting of a hot and cold bath treatment. Diesel fuel and pentachlorophenol (PCP) were used as the preservatives at the plant.

A preliminary assessment of the tie plant site was undertaken in 1997 with the collection of soil samples and the installation of monitoring wells which confirmed the high levels of PCPs in the soil as well as dioxin and furan contamination.

B.9.7.2 *Contaminants of concern*

Pentachlorophenol is known to be toxic to fish at low concentrations. Pentachloroanisoles (PeCAs) and other anisoles are methylated degradation products of chlorophenols. Anisoles are relatively volatile and therefore their presence would suggest recent inputs of chlorophenols. Chlorophenol residues in fish muscle have a reported half-life of 6.9 hours while anisoles have significantly longer half-lives at 6.3 days (Glickman, 1977 as cited in Roach *et al.*, 1999).

As noted previously, dioxins and furans were a major contaminant of concern at the Carcross tie plant. Although polychlorinated dibenzo-*p*-dioxins and furans (PCDD/Fs) are generated from a wide range of combustion sources where carbon and chlorine are present, they likely existed as contaminants in the chlorophenols from

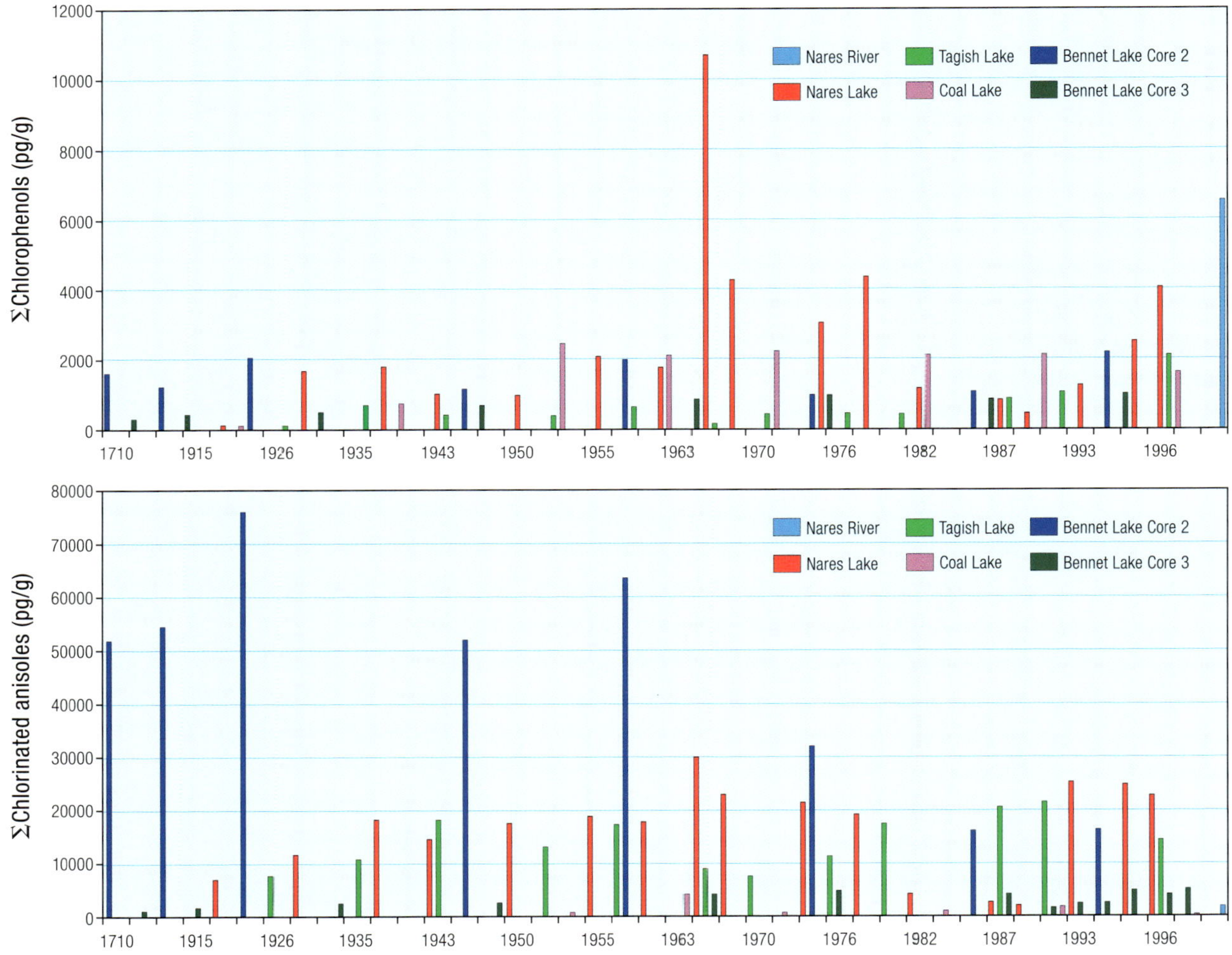

FIGURE B.9.8

Plot of chlorophenols and chlorinated anisoles in sediments from lakes in the vicinity of the Carcross railway tie treatment plant.

the process used in the production of PCP and as degradation products as a result of the chlorophenols being exposed to ultraviolet light (Roach *et al.*, 1999).

B.9.7.3 Investigation of off-site contamination

Sediment cores, sediment grab, and fish samples were collected upstream and downstream and in the immediate vicinity of the site as part of a local contaminants of concern project funded by the NCP. A total of five sediment sites were sampled. Coal Lake was selected as a background site as this location has not been exposed to chlorophenols and anisols except through long-range atmopheric transport. A second background sample was taken from the West Arm of Bennett Lake which is well upstream from the tie treatment site. Bennett Lake-Mid Lake is the closest sample site upstream while the Nares Lake core is the closest downstream from the site. The Tagish Lake core is the farthest downstream from the site. A total of seven grab samples were collected from the sediment of the Nares River immediately adjacent to the tie treatment facility. Grab samples 1, 2, and 3 were taken 47 m offshore and along the shore in a downstream direction, samples 4, 5, and 6 were taken 90 m offshore in alignment with the first three samples, and sample 7 was taken 160 m off the northshore of the Nares River.

Fish (trout and pike) were collected from Nares Lake, downstream from the former tie treatment facility. Comparison or background fish tissue samples were not collected as part of this study, as an extensive survey of fish contaminant burdens taken from across the Yukon provide excellent comparative data. In this case, trout from Lake Laberge and pike from Hanson Lake were used.

Chlorophenols and anisoles in sediments

The results reported by Roach *et al.* (1999) indicated contamination of sediments both upstream and downstream of the former railway tie plant as illustrated for anisoles in Figure B.9.8. All PCP concentrations are below the Yukon Aquatic Life Standard of 20 ppb. The presence of these contaminants in the sediments dated prior to the existence of the tie treatment facility, especially in Bennett Lake core 2, cannot be explained fully, although it may be due to migration in the sediments, or earlier contamination from an unknown source, or smearing in the core. For this reason, Bennett Lake core 2 has been excluded from Figure B.9.8.

The data for Coal Lake are indicative of airborne deposition based on the predominance of the lighter or more volatile chlorophenols and anisoles. The data for Bennett Lake may suggest more local airborne contamination due to the closer proximity of these sites to the tie treatment facility. Tagish sediments, the most downstream site, are higher than those observed for Bennett Lake but are well below those of Nares Lake.

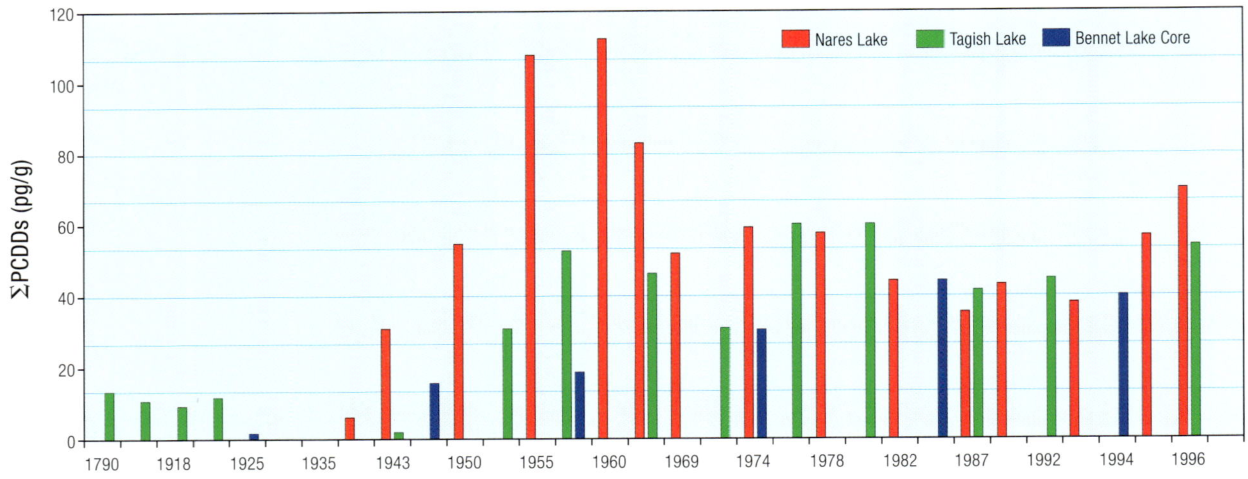

FIGURE B.9.9

Comparison of dioxin concentrations in sediments for lakes in the vicinity of the Carcross railway tie treatment plant.

PCDD/Fs in sediments

There are no Canadian or Yukon standards for PCDD/Fs in sediments, but the Yukon standard for soils is 0.35 ppb. Nares and Bennett Lake cores have the highest concentrations but both are still well below the standard for soils (Figure B.9.9).

Chlorophenols in fish

Interestingly, chlorophenols in the Nares Lake fish were found to be considerably lower than for Hanson and Laberge Lakes. Whether this is due to the short half-lives of chlorophenols in the fish or to some other factor could not be determined (Roach *et al.*, 1999). The data are compared in Figure B.9.10.

Anisoles in fish

Nares Lake pike and trout had chlorinated anisole concentrations an order of magnitude greater than reported for fish from Laberge and Hanson Lakes (Figure B.9.10). There are no known consumption guidelines for anisoles in fish but based on the very low concentrations it is unlikely to be a concern in the short-term (Roach *et al.*, 1999).

As the source of these contaminants have now been isolated from the ecosystem and are being remediated in a treatment facility away from the lake, it is reasonable to expect that these concentrations represent the maximum. Re-sampling at some appropriate time in the future is required to assess whether or not contamination is decreasing in the sediments and fish as would be expected.

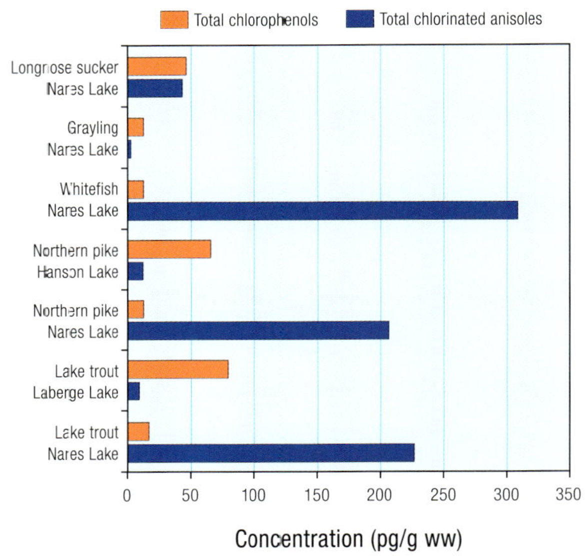

FIGURE B.9.10

Chlorinated phenols and anisoles in fish from Yukon lakes.

B.9.8 Summary

Local sources of contamination in the Canadian Arctic largely consist of abandoned mine, military, and industrial sites. The assessment and remediation of these sites is the responsibility of other groups within INAC and the federal and territorial governments and therefore not covered by the NCP. During the course of investigating these sites, however, the information produced pertaining to contaminant levels in the surrounding environment is of interest to the NCP. There is great concern over local contaminant sources in the Arctic and the NCP must be aware of contaminant levels and related effects associated with these sources and prepared to address local concerns.

The influence of military radar stations on local PCB contamination was examined at a worst-case (exceptionally contaminated) site in Saglek, Labrador. It was concluded that PCB contamination in soil and plants was influenced by site contamination in a halo that extends up to 30 km from the site. Marine sediment PCB contamination displayed a site influence at up to 10 km from the site. Since these studies were initiated, the terrestrial source of PCBs at Saglek has been isolated. The source of PCB contamination at a similar exceptionally contaminated radar site on Resolution Island, Nunavut has also been removed.

GNWT/RWED/Brett Elkin

A detailed quantitative ecological risk assessment conducted on PCB-contaminated marine sediments at Saglek identified significant risks to the survival and reproduction of shorthorn sculpin and black guillemots residing close to site. This study incorporated detailed toxicological effects study on black guillemots and significant PCB-associated effects in the virtual absence of other contaminants were identified. Potential options for managing sediment PCB contamination at Saglek are under review by the Department of National Defence.

Studies into anthropogenic arsenic contamination resulting from mining activities in Yellowknife, NWT, have revealed the importance of identifying site-specific background conditions. The bioavailability of arsenic in tailings and contaminated soils has also been studied and the results used to assess ecological and human health risks. It was concluded that despite high localized arsenic contamination, there was no ecological or human health risks associated with surface soil contamination. Evaluation of the most acceptable way to manage the large quantities of arsenic trioxide dust stored below ground is ongoing.

Radionuclide and heavy metal contamination at Port Radium has been delineated and assessed for ecological and human health risks. It was concluded that based on site occupancy scenarios of less than three months, there was no increased risk to ecological or human health due to the contamination. However, there remain some risks to ecological receptors. The results of these risk assessments will be used to identify priorities for site remediation.

Sediment core samples collected in the vicinity of a railway tie treatment facility in Carcross, Yukon, reflect past use of PCP as a wood preservative at the site. This study provided the opportunity to compare contaminant levels influenced by the site to those strictly resulting from long-range atmospheric transport.

Local sources of contamination have provided opportunities for researchers to study contaminants in both abiotic and biotic media and have allowed intercompartmental processes to be examined. Two such cases are presented, a PCB-contaminated radar station at Saglek, Labrador, and arsenic contamination associated with the Giant mine in Yellowknife, NWT. Both of these sites provided ideal locations for the study of uptake, accumulation, and biological effects over a range of contaminant levels in the virtual absence of other contaminants. These studies have been applied to ecological and human health risk assessments that have produced conclusions at reduced levels of uncertainty compared to traditional probabilistic methods.

Pat Roach

Process Studies and Modelling

C.1 Air/snow exchange of persistent organic pollutants

Authors:

Frank Wania (1) and Crispin Halsall (2).

Affiliations:

1. Division of Physical Sciences, University of Toronto at Scarborough, 1265 Military Trail, Toronto, Ontario M1C 1A4, Canada;
2. Department of Environmental Science, Lancaster University, I.E.N.S., Lancaster, LA1 4YQ, England

C.1.1 Introduction

Snow and ice have a strong impact on how organic pollutants reach Arctic terrestrial and marine ecosystems. Much of the precipitation in the Arctic falls in the form of snow, and the dry forms of atmospheric pollutant deposition occur on snow and ice-covered surfaces for large parts of the year. Contaminant release to soils, vegetation and water bodies takes place during snow and ice melt influencing both location and timing of delivery. A thorough discussion of how snow and ice influence the environmental fate, and specifically air-surface exchange of organic contaminants, can be found in two reviews published within the last five years (Franz *et al.*, 1997; Wania *et al.*, 1998). Here, we focus on the most recent progress as it relates to air-snow exchange of persistent organic pollutants in Arctic regions.

C.1.2 Field measurements of contaminants in snow and ice

Very few measurements of persistent organic contaminant concentrations in Arctic snow fall, snowpack, glacier ice, and sea ice have been reported since CACAR-I (Macdonald *et al.*, 2000). In particular, no recent measurements of organic contaminants in snow or ice from the Canadian Arctic are known. With respect to other polar locations, PAHs were determined in particles filtered from surface snow from the Greenland ice cap (Masclet *et al.*, 2000), revealing the presence of PAHs attributable to fossil fuel combustion, vehicular exhaust, and forest fires. The PAH concentrations in the snow layers showed a distinct seasonal variability and were strongly correlated with black carbon. Masclet *et al.* (2000) thus confirmed major conclusions drawn from a similar study conducted on the Agassiz ice cap on Ellesmere Island several years earlier (Peters *et al.*, 1995).

Halogenated acetic acids are being detected more frequently in the environment, including in the polar region. The presence of trichloroacetate and dibromoacetate in deeper firn layers from Antarctica were interpreted as suggesting significant natural sources of these two compounds (von Sydow *et al.*, 2000). Trifluoroacetate was detected in lake water samples from the Canadian Arctic (D. Muir, EC. 2001), but a significant concentration gradient across Canada with much higher levels in the south suggests predominance of anthropogenic sources for this compound.

A number of studies reported concentrations of organic contaminants in seasonal snow and glacier ice from several sub-arctic Eurasian locations. For example, concentrations of PCBs were reported for snow sampled in northern Norway (Enge *et al.*, 1998), and the presence of numerous organic compounds, including POPs, has been observed in snowpack from northern Russia and Finland (Poliakova *et al.*, 2000). Snowpack samples further served to map the spatial extension of a regional PCB source in the Lake Baikal region of Siberia (Mamontova *et al.*, 1997). A fairly high concentration of tris(2-chloroethyl)phosphate in a sample of glacier ice from northern Sweden suggests that this flame retardant is subject to long-range transport (Laniewski *et al.*, 1998).

Most of the recent field work on POPs in snow has been conducted at high elevations at mid-latitudes. Numerous POPs were detected in snow from the Canadian Rocky Mountains (Blais *et al.*, 1998), from various mountain regions of Europe (Carrera *et al.*, 2001; Gröllert *et al.*, 1997) and in the Californian Sierra Nevadas (Seiber *et al.*, 2001). Relatively high concentrations of selected POPs in snow from high elevations were attributed to vertical temperature differences (Carrera *et al.*, 2001; Seiber *et al.*, 2001), which create the necessary conditions for a "cold condensation" phenomenon on a regional scale. High rates of precipitation further increase the rate of pollutant deposition at higher altitudes (Carrera *et al.*, 2001). This is the case in the Arctic, the presence of significant amounts of POPs in snow is linked with fairly high concentrations of these contaminants in fish from mountain lakes in Canada (Campbell *et al.*, 2000), Europe (Grimalt *et al.*, 2001; Hofer *et al.*, 2001) and the western US (Datta *et al.*, 1998). A depth profile from an ice cap in the Canadian Rocky Mountains was interpreted as a providing an archive of pollutant deposition (Donald *et al.*, 2000), although the validity of this historical record has been questioned (Gregor and Peters, 2000).

C.1.3 Mechanistic understanding of snow-related processes of organic contaminants

One of the key issues of the air/surface exchange of POPs is a quantitative understanding of the atmospheric deposition with and to snow. During a snow event, the falling snowflakes may scavenge chemicals both in the vapour phase and sorbed to particles. The former is believed to occur by sorption to the ice surface, whereas the latter depends on the capture efficiency for the particles that hold the contaminants. After the snow fall, contaminants may further enter the snowpack by dry deposition as gaseous and particle-bound species.

Theoretical approaches regarding the role of snow as an efficient scavenger of airborne POPs have been based on the interpretation, and in some cases re-analysis, of field observations. Wania *et al.* (1999a) demonstrated that falling snow will serve as an extremely efficient scavenger of both organic vapours and atmospheric particulate matter, with vapour scavenging of the lighter PCBs and PAHs being an important, if not the dominating, snow-scavenging process. This finding was based on the re-interpretation of field observations made by Franz and Eisenreich (1998), who examined snowfall events in Minnesota. These authors assumed that the distribution of a chemical between the particle and dissolved phase measured in the snow meltwater, reflected the state of the chemical during the actual scavenging process. Wania *et al.* (1999a) suggested an alternative interpretation, whereby repartitioning was likely to occur during the melting resulting in a positive bias of particle-sorbed contaminants — in effect, overemphasizing the amount of particle-sorbed chemical actually scavenged from the atmosphere. Wania *et al.* (1999a) using their interpretation of the data, found that the gas-scavenging ratio (WG) was strongly negatively correlated to a chemical's vapour pressure, whereby decreasing vapour pressure results in increasing WG. This indicates the importance of air/ice interfacial partitioning for the scavenging of vapour phase semi-volatile organic compounds, and hence the need to understand the physical snow properties (e.g., snow density, snow surface area) to predict contaminant behaviour.

The specific snow surface area (SSSA) is of particular importance, is as it determines the capacity of the snow phase for non-polar organic contaminants. In falling snow, the SSSA controls the extent of vapour scavenging, whereas in the snowpack it controls the diffusive snow-atmosphere exchange, in particular the potential for re-evaporation from the aging snowpack. In both cases, it strongly influences the extent of contaminant delivery to the ecosystem.

In the past five years the work of two groups has considerably increased the quantitative knowledge of the SSSA of freshly fallen snow and aged snowpacks (Cabanes *et al.*, 2002; Chaix *et al.*, 1996; Chaix and Dominé, 1997; Dominé *et al.*, 2000; 2001; 2002; Hanot and Dominé, 1999; Hoff *et al.*, 1998; 1999; Legagneux *et al.*, 2002). Most of the studies derived the SSSA from measured sorption isotherms of methane or nitrogen on the surface of artificial or natural ice; however, one study actually relied on measuring the sorption of a higher molecular weight organic compound onto ice (Hoff *et al.*, 1999). The SSSA derived from sorption measurements were found to be consistent with estimates of surface area based on optical and scanning electron microscopy (Dominé *et al.*, 2001). Hoff *et al.* (1998) obtained SSSA measurements of 0.06 to 0.37 m^2/g for fresh snow sampled in Southern Ontario, and 0.01 m^2/g for aged snow from the Canadian Rocky Mountains (Hoff *et al.*, 1999). Another sample of snow from southern Ontario was aged at -4°C, causing the SSSA to

ITK/Eric Loring

decrease from an initial value of approximately 0.055 to 0.07 m^2/g to 0.03 m^2/g within a few days (Hoff *et al.*, 1999). Hanot and Dominé (1999) observed much higher values of up to 2.25 m^2/g in snow freshly fallen in the French Alps and noted a large decrease of SSSA to levels around 0.25 m^2/g within two weeks. More recent studies by this group gave lower SSSA between 0.01 and 0.15 m^2/g for various types of snow, including graupel, hoar, diamond dust and rimed snow (Dominé *et al.*, 2001). They attributed the higher values obtained in earlier measurements (Hanot and Dominé, 1999; Hoff *et al.*, 1998; 1999) to sampling artifacts and low accuracy of the measurement techniques (Hoff *et al.*, 1995). The first SSSA of Arctic snow, including snow from Ellesmere Island and Svalbard, were reported recently, confirming the range between 0.01 and 0.15 m^2/g (Cabanes *et al.*, 2002; Dominé *et al.*, 2002; Legagneaux *et al.*, 2002). In summary, most of these studies suggest that the SSSA of snow varies between 0.01 to 0.2 m^2/g and whereas freshly fallen snow may on occasion have a SSSA in excess of 0.2 m^2/g, the SSSA typically decreases over a number of days as the snow ages.

For a quantitative treatment of snow-related chemical fate processes, the sorption coefficients of organic contaminants on the snow surface need to be known in addition to the SSSA (Hoff *et al.*, 1995). Little progress has been made within the past five years in quantifying the sorption of organic substances onto the ice surface (Hoff *et al.*, 1999), however, there has been some recent work related to the sorption of organic compounds on the liquid water surface (Donaldson *et al.*, 1999; Mmereki *et al.*, 2000). As the surface of ice close to the melting point is made up of a liquid-like layer, the sorption of organic chemicals onto ice may be estimated by extrapolation of the known sorption to the liquid water surface. Empirical equations that are of particular practical value as they allow the coefficients and heats of sorption on the water and ice surface to be estimated for a large variety of organic substances (Goss, 1997; Goss and Schwarzenbach, 1999).

TABLE C.1.1 Physicochemical properties for estimating sorption of POPs to water, snow and ice

		PCB-52	α-HCH
H	Henry's law constant at 25°C in Pa m³/mol	15.5[(1)]	0.60[(3)]
ΔU_{AW}	Internal energy of air water phase transfer in kJ/mol	51.2[(1)]	65.1[(3)]
P_L	Vapour pressure of the supercooled liquid at 2°C in Pa	0.00776[(2)]	0.224[(4)]
ΔH_{VAP}	Enthalpy of vaporisation of the supercooled liquid in kJ/mol	-82.0[(2)]	-68.44[(4)]

[(1)]Calculated from a regression of log HLC vs 1/T using values from Murphy *et al.* (1987); Brunner *et al.* (1990); ten Hulscher *et al.* (1992); Burkhard *et al.* (1985); and Dunnivant *et al.* (1992).

[(2)]Falconer and Bidleman (1994).

[(3)]Average of value reported by Jantunen and Bidleman (2000) and Sahsuvar *et al.* (submitted).

[(4)]Hinckley *et al.* (1990).

C.1.4 Snow scavenging of persistent organic pollutants

The crucial prerequisites for a quantitative description of the interaction of organic pollutants with snow are now emerging with this significant increase in knowledge of (i) the surface area of snow and ice in the environment, and (ii) the sorption of organic substances onto the ice surface. As an example of the ability to quantitatively describe snow-related fate processes, estimates of rain and snow-scavenging ratios for two persistent organic pollutants (PCB-52 and α-HCH) at different temperatures are presented herin. As the atmospheric temperature decreases, precipitation scavenging is influenced by several interacting processes:

1. The partitioning of semi-volatile chemicals between the gas phase and the aerosol phase shifts towards the aerosol phase favouring particle scavenging over vapour scavenging.
2. As rain becomes snow, the vapours are no longer predominantly scavenged by dissolution in the rain drop, but by adsorption to the ice surface of the snow flake. The water-air partition coefficient K_{WA} is replaced by the ice-air interfacial sorption coefficient K_{IA} as the physicochemical property controlling vapour scavenging.
3. Both K_{WA} and K_{IA} strongly increase at lower temperatures, thus increasing the potential for vapour scavenging.
4. As rain becomes snow, the efficiency of particle scavenging is likely to increase.

The concerted effect of these changes on the efficiency with which organic contaminant chemicals are scavenged under different climatic circumstances was evaluated by estimating the precipitation scavenging ratio for vapours WG, for the total compound in the atmosphere WT, and the percent contribution to WT that stems from gas scavenging within the temperature range of -25 to +25°C. In each case, the influence of variable assumptions concerning the particle scavenging ratio WP and the specific snow surface area A was explored. Table C.1.1 gives the physicochemical properties that were used in the estimation. The interfacial sorption coefficient onto the ice surface K_{IA} in metres was estimated from PL using the relationship developed by Goss (1997), and the internal energy of sorption onto the ice was estimated from K_{IA} using the empirical equation by Goss and Schwarzenbach (1999).

The particle-sorbed fraction Φ in the atmosphere was estimated using a temperature-dependent PL and the Junge-Pankow equation:

$$\Phi = c \bullet S_T / (P_L + c \bullet S_T) \qquad (1)$$

where the constant "c" and the total aerosol surface area "S_T" were assumed to be 0.172 Pa·m and 0.00015 m^2/m^3, respectively. The following equation for the scavenging ratios (Wania *et al.*, 1998):

$$\text{rain } (T > 0°C) \quad W_G = K_{WA} = R \cdot T / H \qquad (2)$$

$$\text{snow } (T < 0°C) \quad W_G = K_{IA} \cdot A \cdot \rho \qquad (3)$$

$$W_T = (1-\Phi) \cdot W_G + \Phi \cdot W_P \qquad (4)$$

where "A" is specific snow surface area in "m^2/g" snow and "ρ" is the density of the snow meltwater in "g/m^3". The results are displayed in Figure C.1.1.

Scavenging ratios for the same compound are highly variable as a result of both the different mechanisms operating at temperatures above and below 0°C, and the temperature influence on the partitioning between gas phase, particle phase, and the (liquid or frozen) aqueous phase. The latter influence always leads to higher scavenging ratios at lower temperatures, typically resulting in an order of magnitude increase of both W_G and W_T over a 25 K° temperature range. The former influence, however, can result in higher and lower scavenging ratios when the hydrometers freeze. The calculations suggest that the relatively water-soluble α-HCH is scavenged more efficiently by rain than by snow, no matter what assumptions were made concerning the snow surface area "A" or the particle scavenging ratio "W_P". These latter assumptions, however, have a tremendous impact on the relative importance of gas and particle scavenging of α-HCH at temperatures below 0°C.

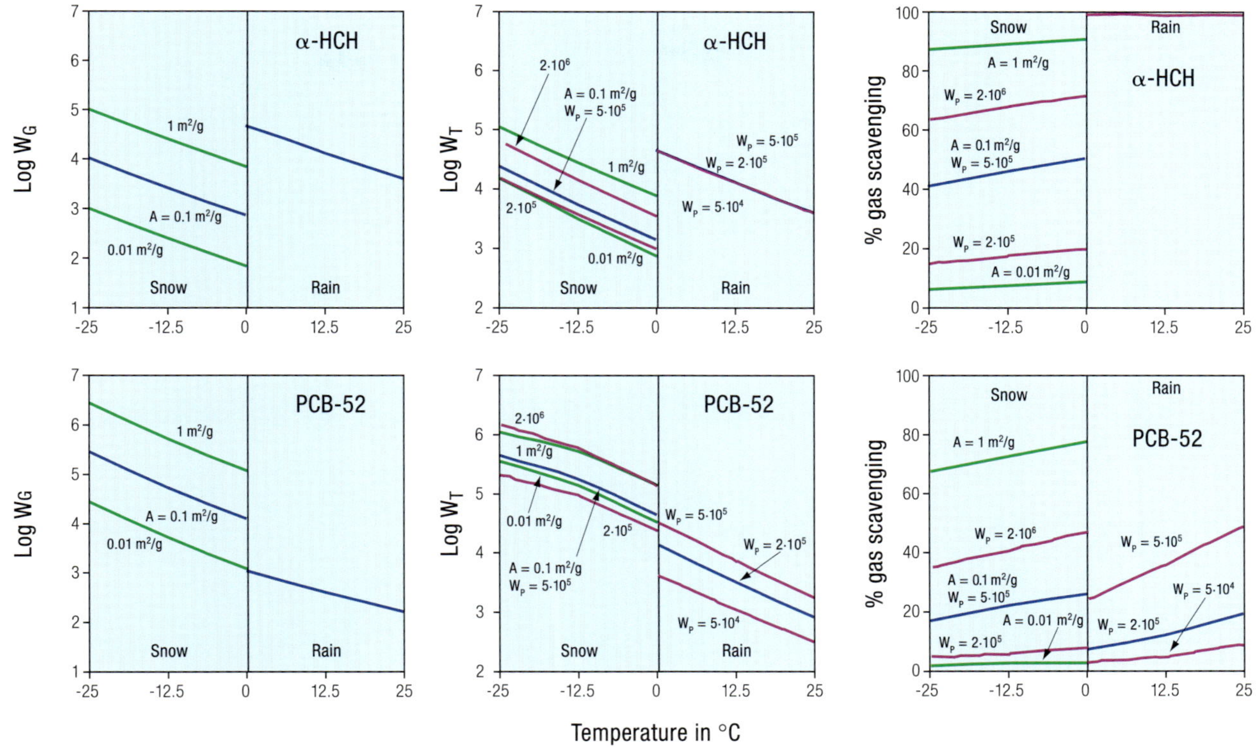

FIGURE C.1.1

Precipitation scavenging ratio for vapours, W_g; the precipitation scavenging ratio for the total compound in the atmosphere, W_T; and the percent contribution to W_T that stems from gas scavenging as estimated for PCB-52 and α-HCH in the temperature range -25°C to +25°C.

In the case of PCB-52, the calculations predict that scavenging by snow is considerably more efficient than by rain. Especially falling snow with a high specific surface area may scavenge PCB-52 more than an order of magnitude more efficiently than rain. As was the case for α-HCH, the relative importance of gas and particle scavenging depends greatly on the assumptions concerning A and W_P.

C.1.5 Fate of persistent organic pollutants in snowpack

The amount of contaminant scavenged from the atmosphere may be very different from the amount delivered to the Arctic ecosystem. In particular, as a snowpack ages and snow metamorphosis leads to a decrease in the surface area of the snow, a considerable fraction of the deposited contaminant may re-volatilize to the atmosphere before and during snowmelt. A chemical that does not re-evaporate will be delivered to the underlying surface during snow melt, but the timing and nature of release from the melting snowpack will depend on the contaminant's distribution between the dissolved and particulate phase. Sufficiently water soluble organic chemicals may be subject to a first flush phenomenon as is commonly observed for inorganic ions. Organic contaminant behaviour in an aging snowpack is too complex to be treated quantitatively in this overview, but relatively simple models to describe these processes have been developed (Wania, 1997). One such model, in particular, was applied to simulate the behaviour of selected POPs observed during snow melt in the Amituk Lake watershed on Cornwallis Island in the Canadian High Arctic (Wania *et al.*, 1999d). Simulated and measured concentrations of four POPs in the water of a creek during snow melt are compared in Figure C.1.2. The model reproduced the observed preferential elution of hexachlorocyclohexanes in the first meltwater fractions, and the relative retention of PCBs within the snowpack (Wania *et al.*, 1999d). More advanced future models of contaminant fate in snowpacks may need to include a dynamic description of the water balance of a snowpack and rely on surface snow routines developed for climate models.

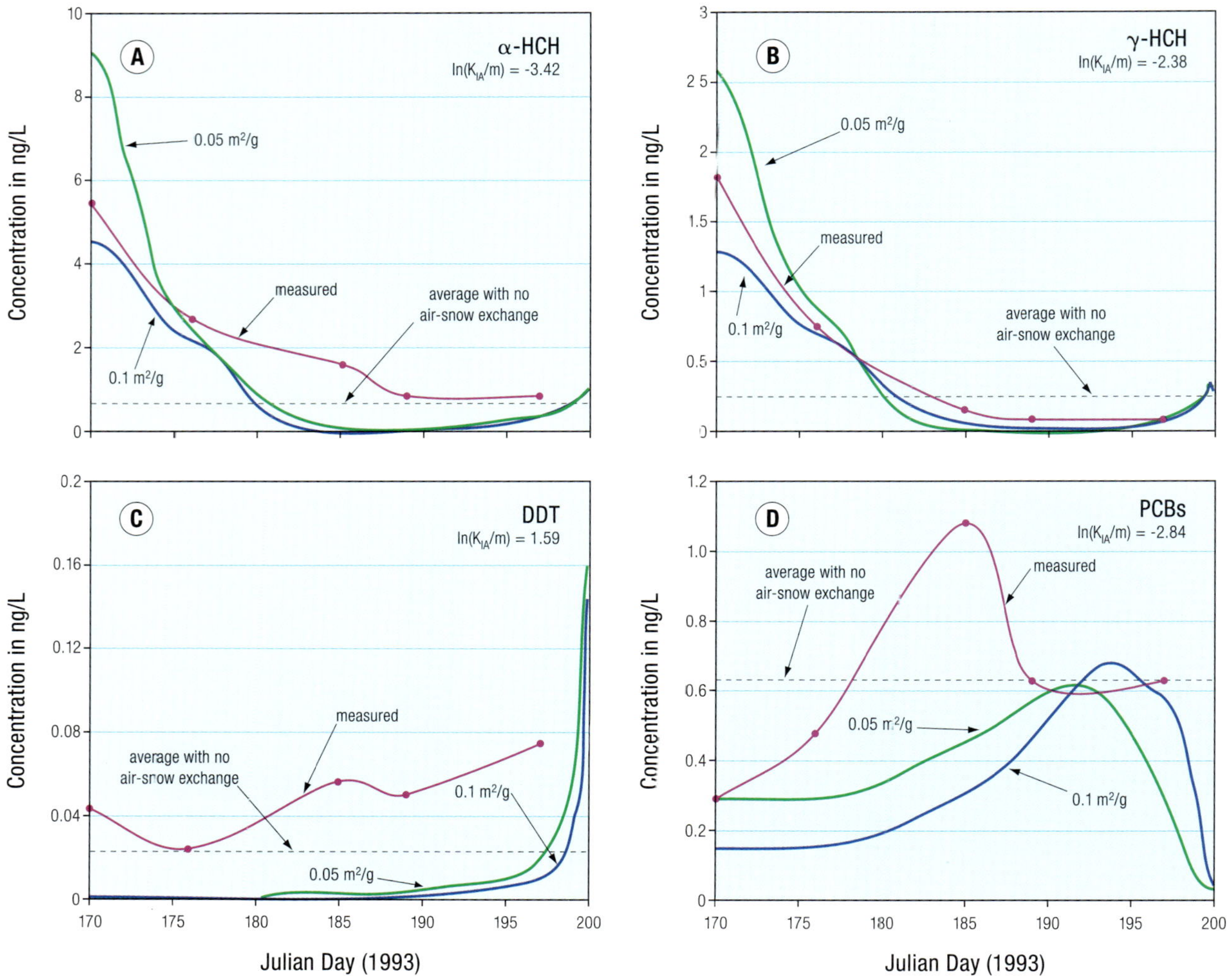

FIGURE C.1.2

Measured and simulated concentrations of four organochlorine chemicals in water from Gorge Creek during the snowmelt period in 1993 on Cornwallis Island, Nunavut. Measured data are represented by pink points. The specific snow surface area was assumed to be either 0.05 or 0.1 m^2/g. The straight line indicates the average snow concentration calculated from the measured initial snowpack burden. Reprinted from Wania *et al.* (1999) with permission by John Wiley & Sons, Ltd.

The emerging quantitative understanding of snow scavenging of organic pollutants, of the diffusive gas exchange between atmosphere and surface snow, and of the fate of organic pollutants during snow melt will eventually contribute to a more complete understanding of organic pollutant delivery to Arctic ecosystems. Growing interest of the atmospheric chemistry community in the exchange of organic substances between the Arctic atmosphere and surface snow may lead to an advancement of this field in the next few years. It was speculated that the observed reduction in specific snow surface area after initial snowfall, would result in a considerable increase in the atmospheric concentration of volatile trace gases initially sorbed to the snow surface (Hanot and Dominé, 1999). Specifically, increased levels of formaldehyde and NO_2 in the air above a snow surface have been observed and may be related to changes in the physical snow structure (Honrath *et al.*, 1999; Hutterli *et al.*, 1999; Sumner *et al.*, 1999; Couch *et al.*, 2000). For semi-volatile pollutants such as OC pesticides and PCBs, it is likely that a proportion may re-enter the atmosphere during a reduction of snow surface area with the aging process, resulting in their availability for further transport in the atmosphere. Unlike trace gases, however, this will largely depend on the volatility of the compound in question as well as other physicochemical parameters, and the state of the snow in which the chemical resides (i.e., particle loading, stratigraphy).

C.2 Mass balance and modelling of contaminants in lakes

Authors:

M. Diamond (1), P. Helm (2), R. Semkin (3), S. Law (4)

Affiliations:

1. Department of Geography, University of Toronto, 100 St. George St., Toronto, Ontario M5S 3G3, Canada;
2. Freshwater Institute, Department of Fisheries and Oceans, 501 University Crescent, Winnipeg, Manitoba R3T 2N6, Canada;
3. National Water Research Institute, Environment Canada, 867 Lakeshore Road, P.O. Box 5050, Burlington, Ontario L7R 4A6, Canada;
4. Exponent Environmental Inc., 15375 SE 30th Place, Suite 250, Bellevue, Washington 98007, USA.

C.2.1 Introduction

Lakes and rivers are dominant features of the Arctic and subarctic landscape. In the first CACAR report results of studies conducted on large rivers indicated that contaminant concentrations were generally low and contribute less than 5% of total loadings of α- and γ-HCH and cadmium to the Arctic Ocean. In the high Arctic, the Amituk Lake study and studies of lakes in the Upper Yukon River Basin, concluded that these systems received chemical loadings from long-range atmospheric transport and that chemical fate was largely controlled by hydrology.

The next phase of studies have further probed the role of Arctic watersheds and lakes as chemical sinks and sources. A major aim of the studies and modelling efforts has been to improve our understanding of contaminant fate processes that result in water and sediment concentrations that, in turn, influence contaminant concentrations in biota. A second aim has been to develop models that allow interpolation of contaminant concentrations in systems that have not been extensively studied and similarly, to extrapolate from past and present conditions to predict future conditions based on estimates of contaminant loadings.

The Amituk study was conducted on the Amituk Lake watershed and lake on Cornwallis Island at latitude 75°02' 57" N and longitude 93°45' 51" W. Field studies spanned the spring and summers from 1992 to 1994 during which the catchment hydrology was estimated and samples taken for chemical analysis (Semkin, 1996; 1997). The study focussed on inorganic compounds and POPs and was designed to allow estimates to be made of contaminant budgets and the development of a mass balance model. CACAR (1997) contains summaries of snowpack, stream and lake chemistry, budgets for inorganic and organic constituents, and results of a mass balance model for organochlorines. Wania *et al.* (1999d) used data on the snowpack, melt regime and contaminant concentrations in snow and stream water to construct a general contaminant runoff model. Freitas *et al.* (1997) developed a general mass balance model of contaminant fate in a high Arctic lake and illustrated its use with data from Amituk Lake. Helm *et al.* (2002b) extended the mass balance model of Freitas *et al.* (1997) to investigate a range of POPs in Amituk Lake.

Studies continue on the Upper Yukon Lake system, notably Lake Laberge. Field studies conducted in 2000 and 2001 are intended to be used to obtain a lake-wide budget and for mass balance modelling. Mass balance modelling is also being concluded on Atlin, Tagish, Bennett and Marsh Lakes (Kawai, 1995). The results of these studies are not reported herein.

As part of the effort to improve our understanding of contaminant dynamics in freshwater systems, several studies have investigated the degradation half-lives and relative age of selected POPs in Arctic and subarctic systems by means of isomer and enantiomer ratios (Helm *et al.*, 2000b; Law *et al.*, 2001). Helm *et al.* (2002) used the mass balance model to estimate degradation half-lives in Amituk Lake, which provides the first estimates for these compounds in Arctic freshwaters.

This section summarizes contaminant budget results compiled for Amituk Lake for Hg and POPs, and estimates of degradation half-lives and the relative age of selected POPs in Amituk and Yukon Lakes.

C.2.2 Amituk Lake

Amituk Lake is located on the east coast of Cornwallis Island in Nunavut. The watershed area is 26 km^2 and drains into the relatively narrow, bow-shaped Amituk Lake. The surface area of Amituk Lake is 37.8 ha and it has a mean and maximum depths of 19.4 and 43 m, respectively. The lake is representative, in terms of volume and morphology, of the many small lakes that dot the Arctic landscape. There are six sub-basins that drain into Amituk Lake, five of which were instrumented. The basin is underlain by limestone, dolomite, and calcareous shale of Ordovician and Silurian age. There is minimal soil development in this area of continuous permafrost and during summer, the active layer is less than 0.5 m deep.

The climate is typical of that in the High Arctic with an annual mean temperature of -16.6°C and an annual precipitation of 131.4 mm. The amount precipitation at Amituk qualifies the area as an Arctic desert. About half the precipitation falls as snow in two peak periods in fall and late spring.

C.2.2.1 Mercury budget for Amituk Lake

From June to August 1994, snow and surface water samples were collected for total mercury analysis. Details of sampling and analytical methods are reported by Semkin (1997) and Semkin and Mierle (in prep). The following discussion summarizes an annual mercury budget developed for Amituk Lake for 1994. The goal of developing the budget was to discriminate the contribution of Hg from geologic sources within the watershed relative to contributions from atmospheric deposition which are derived from natural and anthropogenic sources (Pacyna and Keeler, 1995).

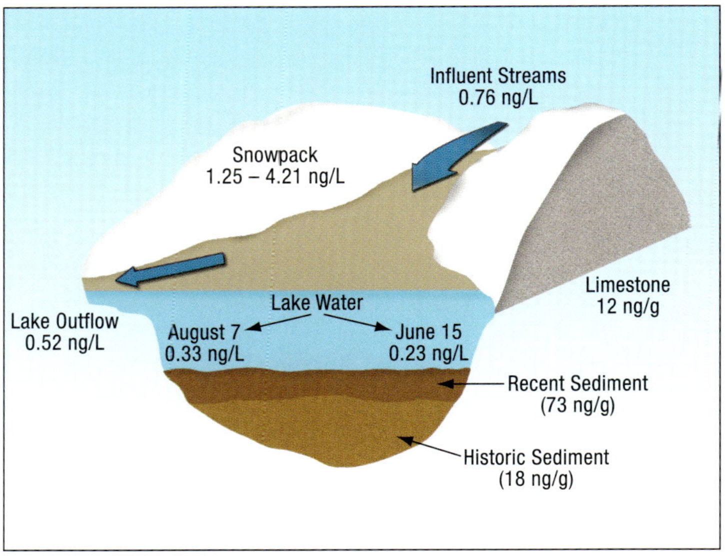

FIGURE C.2.1

Concentrations of mercury in the Amituk Lake basin.

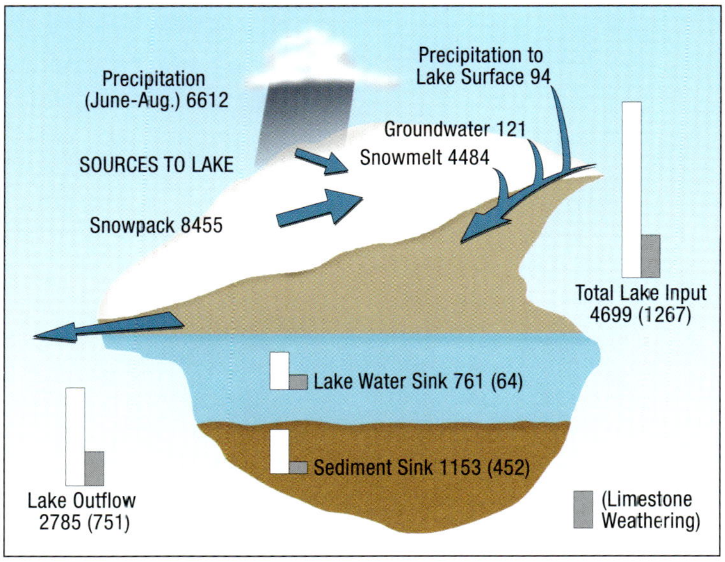

FIGURE C.2.2

Total mercury budget for the Amituk Lake basin (mg).

The snowpack in the Amituk watershed is heterogeneous with respect to water content and chemistry. The shallow snowpack that was found on upland plateau areas and the lake surface, had a mean depth of 0.47 m and a snow-water equivalence (SWE) of 204 mm. It is hypothesized that this snowpack is formed from early fall precipitation. In the late spring, new snowfall is deposited on this slick, frozen snowpack and is readily redistributed by wind scour into topographical depressions and into the heavily incised stream valleys. The deeper snowpack that forms in such areas had an average depth of 2.76 m and SWE of 1803 mm in June. Total Hg was higher in the valley snowpack with a mean concentration of 4.25 ng/L compared to an average value of 1.25 ng/L measured for the shallow snowpack (Figure C.2.1).

Time of sampling must be recorded to compare these concentrations of Hg in snow with those reported in the literature. Atmospheric mercury is depleted for up to three months after polar sunrise in April (Schroeder *et al.*, 1998), with a concomitant increase in snow concentrations by up to 1600% (Lu *et al.*, 2001; Poissant, 2001). For example, snow concentrations measured during SHEBA increased from 7.8 ng/L from Nov. 1997 to Jan. 1998, to 34 ng/L from Feb. to May 1998. Lu *et al.* (2001) reported snow concentrations measured in April and May, during the mercury depletion events, of ~2 ng/L in the western Canadian Arctic and eastern Greenland to 25–160 ng/L at coastal sites on Hudson Bay. Poissant (2000) reported a change in snow concentrations at Kuujjuarapik on Hudson Bay, from 5 ng/L on April 13 to a maximum of 82 ng/L on April 16 of 2000. Poissant *et al.* (2001), however, also reported that ~50% of this burden could be lost within 12h, presumably due to photoreduction followed by volatilization. In this context, the concentrations measured in the Amituk snowpack in late June may reflect the Hg burden after the time of snow enrichment and loss due to on-going photoreduction and volatilization.

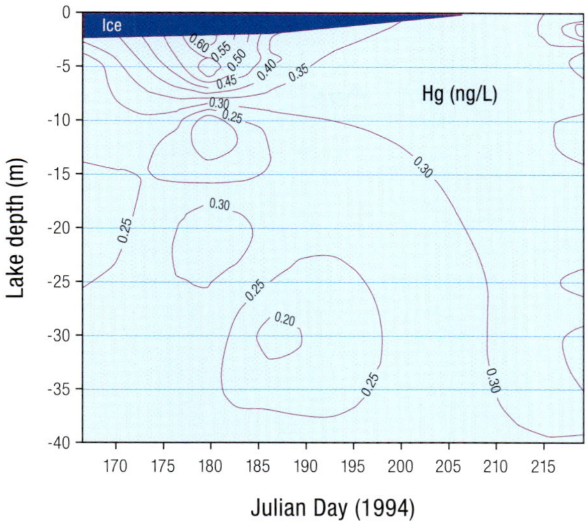

FIGURE C.2.3

Isopleths of mercury in the Amituk Lake basin.

Based on the concentrations measured in late June and estimates of the spatial extent of these snow types, the snowpack contained 8455 mg of total Hg over the 26.5 km^2 watershed, to which direct precipitation to the watershed added 6612 mg (Figure C.2.2). The mass accumulation is equivalent to an atmospheric deposition rate of 0.57 μg/m^2·y^{-1} for both snow and precipitation over the watershed. This atmospheric deposition rate for the Amituk watershed is within the range reported by Schroeder (1996) of 0.1–4.7 μg/m^2·y^{-1} for sites in the Northern Hemisphere, but is considerably less than the range of 2.4–7.64 μg/m^2·y^{-1} reported by Lu *et al.* (2001) across the Arctic. As noted previously, the discrepancy with the values reported by Lu *et al.* (2001) may be attributable to the different sampling times; the measurements of Lu *et al.* (2001) were taken mostly at polar sunrise in comparison to the late June measurements taken by Semkin and Mierle (in preparation). These values are considerably lower than estimates of 10–15 μg/m^2·y^{-1} reported for several remote temperate lakes (e.g., Laurier and Driscoll, 1999; Egstrom *et al.*, 1994).

Snowmelt is the major hydrologic event at Amituk Lake at which time peak stream flow reaches approximately 5 m^3/s. Snowmelt contributes about 95% of stream runoff while groundwater contributes about 5% (Kinney, 1997). In terms of runoff, Semkin and Mierle (in preparation) estimated that snowmelt contributed 4484 mg of total Hg, groundwater 121 mg, and precipitation to the lake surface another 94 mg to give a combined loading of 4699 mg Hg to Amituk Lake in 1994. This is equivalent to 0.18 μg/m^2·y^{-1} over the full watershed. The latter value contrasts with the total atmospheric contribution of Hg to the watershed of 15067 mg or 0.57 μg/m^2·y^{-1} based on snow and precipitation measurements. The difference between loadings to the snowpack and that measured from runoff is 10368 mg or 0.39 μg/m^2·y^{-1}. The discrepancy of almost 70% between loading estimates can be attributed to: (1) loss due to volatilization of Hg. (Poissant, 2000; 2001), (2) retention in the watershed (e.g., Laurier and Driscoll, 1999) since there is much fine-grained rock flour from mechanical weathering that could effectively sorb Hg; and (3) uncertainties and variability in field measurements and calculations.

It is important to note that loadings of 0.18 μg/m^2·y^{-1} over the full watershed represents contributions from atmospheric deposition as well as the weathering of bedrock within the watershed. Thus, the contribution from atmospheric deposition to lake loadings is reduced from 0.57 μg/m^2·y^{-1} (deposition to the snowpack), to 0.18 μg/m^2·y^{-1} (loadings calculated from runoff), to 0.13 μg/m^2·y^{-1}, when bedrock weathering is considered. This implies a loss of 0.05 to 0.52 μg/m^2·y^{-1}, or up to 77%, of Hg deposited to the snowpack, from volatilization and retention in the watershed.

Snowmelt occurs over a three-week period in mid-June at a time when the lake is mostly ice-covered. The ice cover prevents lake mixing and results in cold, dilute meltwater flowing over and just under the ice. Because of the density-induced stratification, the meltwater incursion is stable in the lake to a depth of 4–5 m. This stability results in approximately half of the meltwater and its chemical burden moving through the lake to the outflow without mixing with the water column. The lake mixes after the ice melts in July at which time there is minimal stream flow.

This melt regime has important implications for chemical fate. Freitas *et al.* (1997) and Helm *et al.* (2002b) estimated that about half of the chemical loads in meltwater exit the lake without mixing. This can be seen from isopleths of Hg concentrations (Figure C.2.3). As a result of loadings from runoff, Hg reached a maximum concentration of 0.6 ng/L just below the ice as the meltwater enters the lake. This influx, however, has a limited effect on concentrations at depth and the whole lake after mixing. The depth-averaged concentration measured in the lake while still ice-covered was 0.23 ng/L and increased to 0.33 ng/L after mixing in August (Figure C.2.1).

Of the total runoff loadings of 4699 mg, 761 mg or 16% was estimated to reside in the water column assuming average concentrations of 0.23 and 0.33 ng/L in mid-June (ice-on) and August (ice-off), respectively. Sedimentation or burial accounted for the loss of 1153 mg or 25% of loadings and 2785 mg or 59% left the lake through the single outflow. Retention of 25% of total Hg loadings in

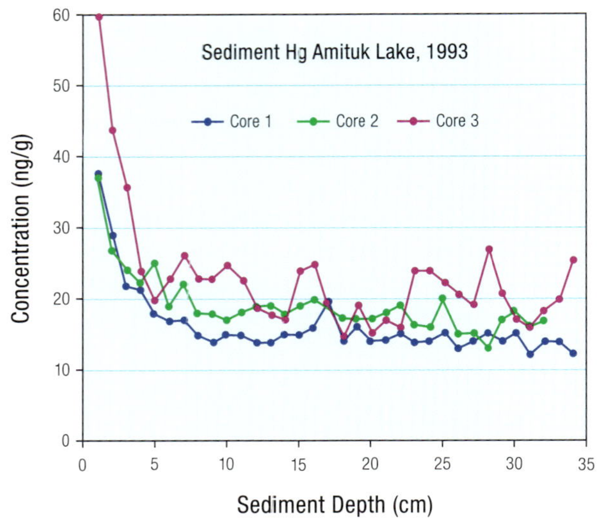

FIGURE C.2.4

Depth profiles of mercury in Amituk Lake sediments.

sediments is much higher than those estimated for organochlorines in Amituk Lake, that ranged from 2–3.5% for the more hydrophobic compounds (DDT and PCBs) to <1% for the less hydrophobic compounds such as HCHs (Helm *et al.*, 2002b). This considerable difference in retention may be due to different origin and aqueous chemistry of Hg versus POPs. As discussed in the following text, 27% of Hg loadings are estimated to originate from bedrock weathering and thus are likely in the form of particulate Hg that would have a high sedimentation rate. In contrast, virtually all POPs are derived from atmospheric deposition to the snowpack and subsequent runoff. Due to the extremely low particulate organic carbon content, more than 90% of POPs loadings are in the dissolved phase that is subject to minimal deposition.

The retention rate of 25% for Amituk is less than that of 65% estimated from a Hg budget for Lakes 373 and 375 in the Experimental Lakes Area (Lockhart *et al.*, 1998). Similarly, Diamond (1999) estimated a retention rate of 60% based on a whole lake mass balance for Hg in a "typical" temperate lake located on the Canadian Shield. While this estimate was based entirely on modelling results, estimated Hg concentrations did correspond well with measured values. Thus, Arctic lakes are much less efficient sinks of Hg than temperate lakes, a conclusion similarly drawn for POPs.

As mentioned previously, the contribution of Hg from bedrock weathering versus atmospheric deposition can be calculated. The Hg content of the limestone and dolomite bedrock in the Amituk watershed averages 12 ng/g. This natural or geologic Hg, along with a host of inorganic elements and minerals, is liberated from the host rocks in the watershed during surface runoff. The contribution of geologic Hg to the total Hg budget can be estimated assuming similar rates of chemical weathering of Hg and other inorganic elements. This calculation was based on a Ca^{2+} budget, assuming that Hg and Ca weather at equal rates. The assumption is supported by Temmam *et al.* (2000) who reported that Hg^{2+} can substitute for Ca^{2+} and Mg^{2+} in the formation of the limestone/dolomite crystal lattice structure and Hg^{2+} can be liberated upon weathering of the host rock. The use of Ca^{2+} is also advantageous since almost all Ca is derived from weathering and a negligible amount comes from atmospheric deposition.

Based on the Ca budget, 1267 mg out of total runoff loadings of 4699 mg, or 27% is derived from bedrock weathering (Figure C.2.2). This leaves about 0.13 $\mu g/m^2 \cdot y$-1 of the estimated 0.18 $\mu g/m^2 \cdot y^{-1}$ calculated from watershed runoff or 0.57 $\mu g/m^2 \cdot y^{-1}$ atmospherically deposited Hg to the snowpack, to actually enter the lake. Conversely, about 77% of the atmospherically deposited Hg in the snowpack did not enter the lake. Thus, measurements of the Hg burden in the snowpack can greatly exceed loadings to aquatic and marine systems during runoff.

As snowmelt and surface runoff channels Hg from the entire watershed to the lake basin, a "watershed funneling effect" enhances Hg deposition to the sediments by five times (3.05 $\mu g/m^2$) compared with atmospheric deposition (0.57 $\mu g/m^2$). This implies that lakes with large watershed-to-lake ratios will have higher chemical concentrations in the water and sediments relative to those with lower ratios, given similar rates of bedrock weathering and retention within the watershed (Laurier and Driscoll, 1999).

The mass of Hg that is deposited annually in the sediments that can be accounted for by bedrock weathering is 452 mg. The ratio of total loadings to that derived from bedrock weathering is a factor of 2.6 (1153/452). This ratio compares closely with that obtained by comparing recent and older sediment concentrations (Figure C.2.4). Below 8 cm sediment depth, which would correspond to before the 1950s, the concentration of Hg in sediment profiles is stable at approximately 18 ng/g (see Figure C.2.4). This value can be interpreted as a background concentration with minimal recent anthropogenic influence. In contrast, the average concentration in the most recent (0–1 cm) sediments is 45 ng/g and reflects geologic as well as anthropogenic influences. The ratio of these concentrations is 2.5, which is similar to a value of 2.1 obtained Lockhart *et al.* (1998) for Amituk Lake and a range of 0.3 to 5.4 for other northern lakes. For temperate lakes, this ratio averages 3.5 (Swain *et al.*, 1992; Laurier and Driscoll, 1999). The data from Amituk Lake support the observation that recent anthropogenic activities have increased Hg inputs to this Arctic system by 2–2.5 times. This conclusion assumes that the rate of sediment accumulation and the preservation of the historical record of atmospheric deposition supercedes changes to the sedimentary record due to post-depositional diagensis.

C.2.2.2 Persistent organic pollutants in Amituk Lake

Chemical budgets, such as that for Hg, are very useful for understanding contributions from various sources and the overall fate of chemicals in lakes. However, budgets can be of limited use when all process rates are known, as occurs with POPs. For POPs, these unknown rates include air-water exchange during the short ice-free period and chemical degradation or transformation. A general mass balance model was developed with this in mind, to track the fate of POPs in high Arctic lakes and to quantify the unknown process rates (Freitas *et al.*, 1997; Helm *et al.*, 2002b). In addition to gaining knowledge of chemical fate processes, the model was to predict lake response given a cessation of loadings.

The model developed by Freitas *et al.* (1997) and Helm *et al.* (2002b) is adapted from the level III fugacity/equivalence model of Mackay (1991) and Mackay and Diamond (1989). Based on a simple mass balance framework, the mechanistic model includes features specific to Arctic lakes such as the through flow of a fraction of loadings, as explained for Hg, and cryoconcentration or the exclusion of chemicals from ice and their consequent concentration in the remaining water column. The models have been developed to describe steady-state conditions where only the ice-free season is considered, and an unsteady-state version that tracks ice development and melting, and hydrologic inputs as a function of time. Freitas *et al.* (1997) and Helm *et al.* (2002b) provide details of the model.

Helm *et al.* (2002b) used the Amituk data set from 1992 to 1994 to test and extend the model of Freitas *et al.* (1997). Tables C.2.1, C.2.2 and C.2.3 summarize parameter values describing Amituk Lake that are used in the model, measured OC concentrations, and physicochemical properties of the chemicals modelled, respectively. Modelled bulk water column and outflow concentrations generally agreed to within one standard deviation of the measured mean for each of the eight OC groups or chemicals considered.

Previous work established, and the present work confirmed, that northern lakes convey most of the POPs loadings received from snowmelt to a downstream sink such as the Arctic Ocean, rather than retaining the loadings in the lake water column or sediments. This feature is largely due to the hydrologic regime discussed previously, whereby most snowmelt and chemical loadings occur while the lakes are ice-covered and thermally stratified, which minimizes the mixing of runoff with the water column (e.g., Bergmann and Welch, 1985). The second reason for the minimal retention is the extremely low suspended particle concentrations (e.g., Schindler, 1994) that scavenge chemicals from the water column.

TABLE C.2.1 Physical and limnological characteristics of Amituk Lake used in the steady- and unsteady-state models

Parameter	Value	Reference
Summer Air Temperature (July) (°C)	4	a
Summer Water Temperature (°C)	2	b
Annual Precipitation (m y^{-1})	0.131	a
Water Surface Area (m^2)	3.78 x 10^5	c
Mean Depth (m)	19.4	c
Active Sediment Depth (m)	0.01	d
Active Sediment Area (m^2)	1.42 x 10^5	e
Sediment Deposition Rate (g m^{-2} y^{-1})	45.4	f, g
Sediment Resuspension Rate (g m^{-2} y^{-1})	0	f
Sediment Burial Rate (g m^{-2} y^{-1})	121	h
Inflow Particle Concentration (mg L^{-1})	7.4	b
Water Column Particle Concentration (mg L^{-1})	1.5	b
Water Inflow / Outflow Rate (m^3 y^{-1}) (Steady State)		
1993	5.36 x 10^6	b
1994	5.32 x 10^6	b
Aerosol Concentration (μg m^{-3})	0.2	i
Aerosol Deposition Velocity (m h^{-1})	1.8	j
Organic Carbon Fractions:		
Water Column Particles	0.05	b
Inflow Suspended Particles	0.05	b
Sediment Solids	0.012	k
Densities (kg m^{-3})		
Aerosol Particles	2000	f
Suspended Particles	1500	f
Sediment Solids	2000	f

a = Environment Canada (1990). b = Semkin *et al.* (unpublished data). c = Jenson *et al.* (1997) d = Assumed value, thickness of surface slice in which concentrations have been measured in Muir *et al.* (1995; 1996). e = Calculated as in Freitas *et al.* (1997) [(sediment deposition rate) x (water area) / (sediment burial rate)]. f = Freitas *et al.* (1997). g = Settling of particles to the surface sediments, determined by sediment traps (Semkin, unpublished data). h = Removal from exchange processes with water column assumed to occur over "active" sediment area or zone of sediment focussing which is 37% of total sediment surface area. Average value from three cores taken in 1993 determined by methods listed in Muir *et al.* (1995) (Semkin, unpublished data). i = Barrie (1986). j = Barrie *et al.* (1992). k = Muir *et al.* (1995).

TABLE C.2.2 Mean OC concentrations and ranges in inflow streams, mean concentrations ± standard deviations in Amituk Lake, the outflow, and Alert air, and Amituk Lake sediment concentrations

	Inflow Streams (pg/L)				Lake (pg/L)[2]						Outflow (pg/L)[4]		Air (pg/m³)[5]		Sediment (ng/g)[6]
	1993		1994		1992	1993		1994			1993	1994	1993	1994	1992
Compound	Mean[1]	Range	Mean[1]	Range	20 m	3 m	20 m	3 m	20 m	40 m					
α-HCH	1470	n.d.–5429	602	6–2969	1963 ± 1018	1164 ± 322	1090 ± 261	677 ± 134	564 ± 86	646 ± 122	1383 ± 424	854 ± 313	49.9 ± 26.4	41.8 ± 15.7	0.35
β-HCH	39	9–109	29	n.d.–101	48 ± 18	30 ± 10	41 ± 39	32 ± 14	30 ± 12	31 ± 10	63 ± 33	27 ± 11	0.4 ± 0.2	0.1 ± 0.1	0.01
γ-HCH	376	5–1818	212	n.d.–863	351 ± 155	191 ± 58	193 ± 120	191 ± 63	153 ± 29	163 ± 45	194 ± 111	255 ± 111	6.3 ± 3.0	7.0 ± 4.2	0.55
Dieldrin	71	16–171	95	5–255	60 ± 44	51 ± 8	56 ± 22	86 ± 16	80 ± 27	81 ± 22	33 ± 33	111 ± 35	1.7 ± 1.0	1.4 ± 0.6	1.18
Endo I	137	n.d.–477	87	n.d.–408	45 ± 66	45 ± 23	3.9 ± 5.4	43 ± 36	20 ± 11	20 ± 9	69 ± 46	71 ± 62	4.4 ± 3.2	4.4 ± 2.8	–
HepEx	61	6–129	47	7–130	44 ± 25	64 ± 7	63 ± 27	41 ± 13	38 ± 19	47 ± 12	63 ± 15	51 ± 19	1.6 ± 0.8	1.5 ± 0.6	0.10
ΣDDT	44	n.d.–125	19	n.d.–58	182 ± 145	39 ± 13	40 ± 37	8.6 ± 4.5	36 ± 16	35 ± 12	40 ± 28	9 ± 4	0.4 ± 0.2	0.8 ± 0.4	1.67
ΣPCB[3]	576	n.d.–1438	399	117–1706	1380 ± 663	417 ± 155	305 ± 115	421 ± 244	400 ± 190	294 ± 108	462 ± 266	487 ± 142	39.5 ± 29.7	32.1 ± 12.0	12.2

[1]Weighted to stream contributions to total discharge. [2]n = 12 in 1992, n = 9 in 1993, n = 11 in 1994. [3]Excludes outlier on June 16, 1993. [4]n = 7 in 1993, n = 17 in 1994. [5]Summarized in Stern *et al.* (1997), Halsall *et al.* (1998). [6]Muir *et al.* (1995; 1996). n.d. – not detected.

TABLE C.2.3 Selected physicochemical properties of OCs included in model calculations

Compound	MW (g/mol)	Solubility[1], C_L (mg/L)	Vapor Pressure[2], P_L (Pa)	Henry's Law Constant[3], H (Pa m³/mol)	log K_{OW}[1]	Degradation[4], $t_{1/2}$ (y)
α-HCH	290.85	33.3	0.02188	0.1235[9]	3.81	0.90[10]
β-HCH	290.85	66.0	0.0264[7]	0.116	3.8	12.6[11]
γ-HCH	290.85	53.5	0.005864	0.07554[9]	3.7	1.26[10]
Dieldrin	380.93	6.5	0.0006127	0.036	5.2	22.0
Endosulfan I	406.95	0.68	0.0003923	0.23	3.62	0.05
Heptachlor Epoxide	389.2	7.4	0.054[7]	2.8	5.0	5.71[10]
ΣDDT[5]	330.2	0.49	0.0000779	0.02	6.05	20.0
ΣPCB[5]	302.3	0.479[6]	0.00985[8]	1.261	6.1[6]	12.6[12]

[1]From Mackay *et al.* (1997), Vol. V.

[2]Calculated from Hinckely *et al.* (1990) @ 275 K, unless noted.

[3]Calculated from P_L/C_L unless noted.

[4]From Mackay *et al.* (1997) Vol. I and V unless noted. Selected values were doubled to adjust for the "Q-10 rule".

[5]Properties for ΣDDT and ΣPCB are averages weighted according to the occurrence of DDT, DDE and DDE for ΣDDT and PCB homologs for ΣPCB in Amituk Lake and streams.

[6]Weighted average of homolog values from Mackay *et al.* (1997) Vol. I.

[7]From Mackay *et al.* (1997) Vol. V.

[8]Weighted average of homolog values from Falconer *et al.* (1994) calculated @ 275 K.

[9]Calculated from Kucklick *et al.* (1991) @ 275 K.

[10]Determined from model calibration.

[11]From Mackay *et al.* (1997) Vol. V, Class 9, doubled.

[12]From Mackay *et al.* (1997) Vol. I, Class 9, doubled.

The steady-state version of the model indicated that stream inflow contributed from 96 to > 99% of total loadings to the lake, with the remainder coming from direct atmospheric deposition to the lake surface. This being the case, it is clear that snow scavenging, over-winter aging and melting processes over the extent of the watershed exert considerable control over lake loadings. Between 57 to 98% of loadings were exported from the lake (Figure C.2.5). The higher values were for chemicals with low Henry's Law constants (HLC) and longer degradation half-lives such as Σ-HCH, dieldrin, and ΣDDT. Conversely, percentage losses via export are lower for chemicals with shorter degradation half-lives and having high HLC such as endosulphan I and heptachlor expoxide.

As a consequence of limited chemical scavenging and sedimentation, > 99% of the total inventory of HCHs and endosulphan I in the lake (water column plus the top 1 cm) was estimated to be in the water column. This value decreased to ~32% for ΣDDT and ΣPCB.

The percentage of loadings retained in the sediments was only 0.4 to 3.4% with a higher percentage retention for chemicals with higher K_{OW} such as dieldrin, ΣDDT, and ΣPCB. Air-water exchange accounted for minimal chemical movement. The fugacity ratios calculated between air and water indicate that the lake is close to equilibrium with the air and is not a significant source of chemicals to Arctic air. Degradation was the other process that accounted for chemical loss. There is very limited data on degradation rates, and none for Arctic waters. Consequently, the the unsteady-state version of the model was used to estimate these values.

An unsteady-state version of the model was used to consider temporal variations in lake hydrology, ice cover, inflowing chemical concentrations, precipitation, and year-round degradation. Average concentrations obtained in 1992 for the water column and sediments were used to initialize the model that was then run using 1993 and 1994 chemical loadings (Figure C.2.6).

Chemicals fell into two categories depending on their persistence. In the first category, ΣDDT and ΣPCB concentrations rose in spring by about ~10% due to cryoconcentration, and then fell as spring meltwaters diluted concentrations. In the second category, containing α- and γ-HCH and endosulphan I, concentrations declined over winter due to their relatively rapid degradation rates (Figure C.2.6). The unsteady-state version of the model was then used to estimate degradation times by holding constant all other parameter values obtained by modelling the first group of chemicals.

The estimated degradation half-lives were based on the assumption that only degradation was responsible for inventory changes between "ice-on" (1993) and "ice-off" (1994), which is very reasonable considering that ice prevents any loadings and loss by volatilization, and because there is minimal sediment-water exchange.

The calculated degradation half-life for α-HCH was 0.9 years (0.60–1.38 years) which is 3-8 times faster than the overall α-HCH degradation half-life of 4.7 years found for the Arctic Ocean (Harner *et al.*, 1999; 2001a). Microbial process are assumed to be responsible for this rapid degradation since hydrolysis can account for only 2% of losses and biotic activity does occur year-round although the lakes are cold, dark, and ice-covered (Kalff and Welch, 1974).

Degradation of γ-HCH was also calculated to be much faster in Amituk Lake than the Arctic Ocean, i.e., a degradation half-life of 1.25 years (0.80–1.70 y in Amituk versus 16.1 years for microbial degradation in the Arctic Ocean, Harner *et al.*, 1999; 2001a) and 42 years for hydrolysis under ice (Ngabe *et al.*, 1993).

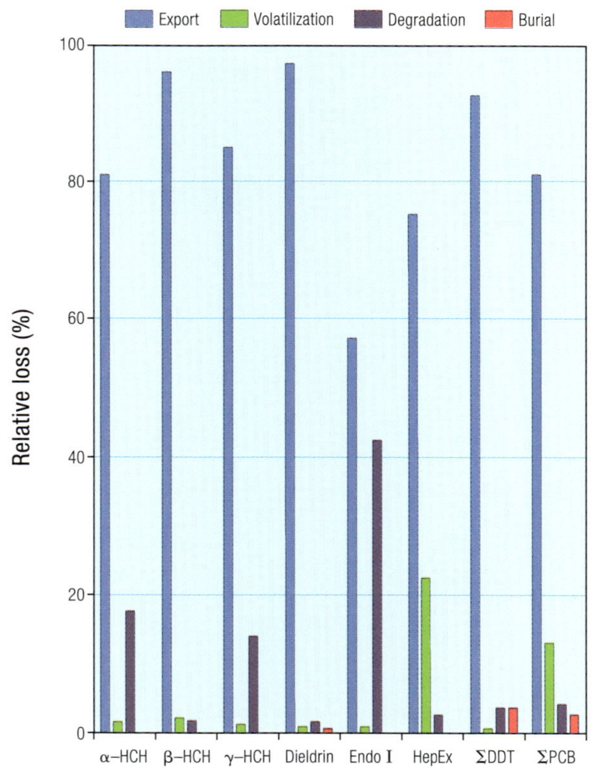

FIGURE C.2.5

Percentage losses of organochlorines from Amituk Lake, estimated by mass balance modelling.

Endosulfan I was estimated to have the most rapid degradation rate since low or non-detectable levels were observed at the end of winter before ice melt and loss to the sediments was minimal. Runoff then replenished the endosulfan I inventory annually. The estimated degradation half-life was rapid, ranging from 0.046–0.147 years. The process by which endosulfan I degraded is unclear. At pH 8.2 and 25°C, the hydrolysis half-life would be ~0.016 years (Peterson and Batley 1993). If the "Q10" rule is applied and the hydrolysis rate constant is divided in half for every drop of 10°C, then the half-life would be ~0.129 years at 3°C. This is very short time could account for the loss of endosulfan I. Peterson and Batley (1993), however, did not detect loss by hydrolysis in experiments conducted at 4°C with "creek water". Endosulfan I may also be subject to biodegradation,

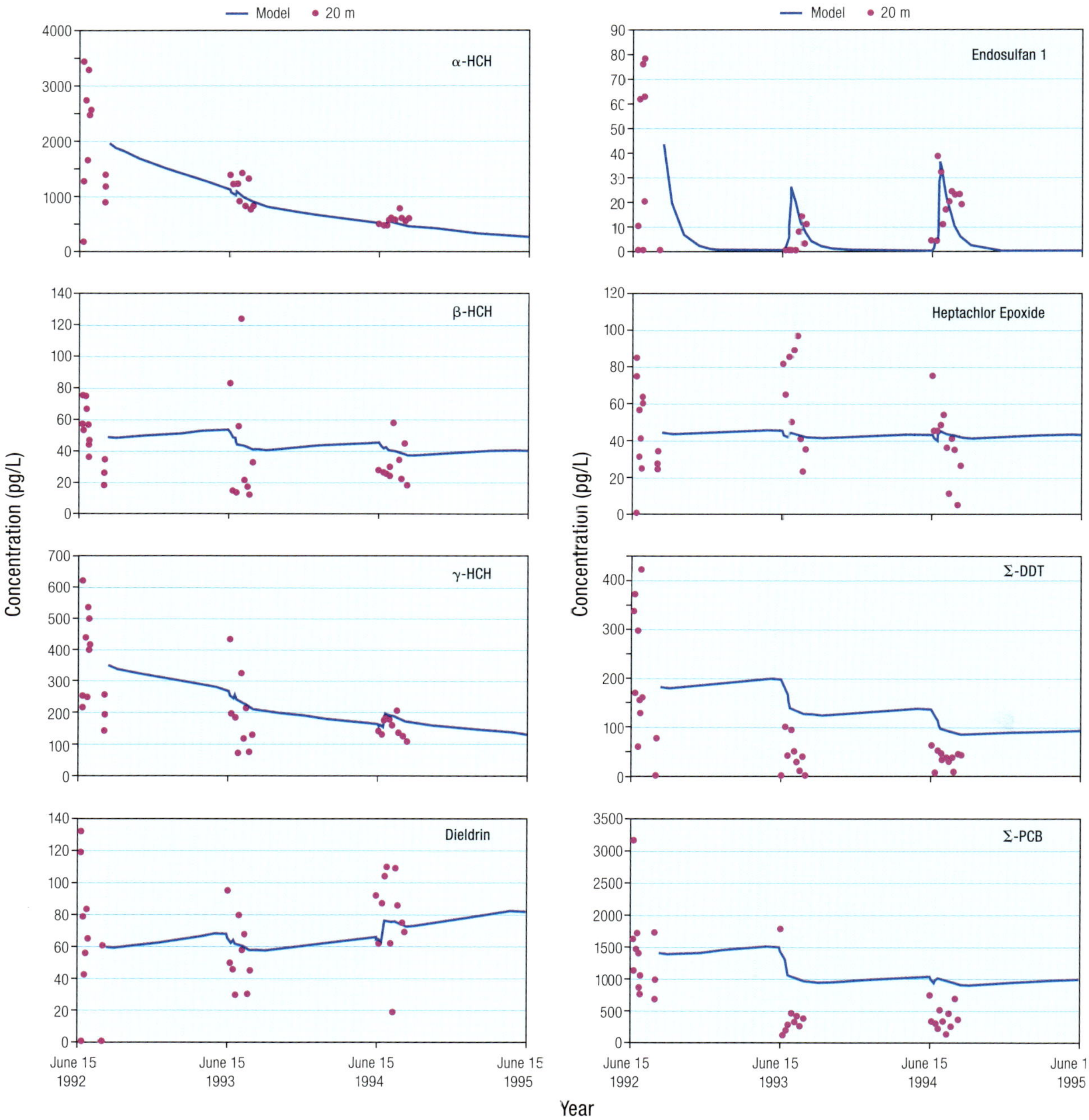

FIGURE C.2.6

Measured and modelled concentrations of organochlorines in Amituk Lake from 1992 to 1995.

with reported loss rate constants similar to that for hydrolysis at warm temperatures. It is interesting that endosulfan disappears so quickly from Amituk, because it is ubiquitous in Arctic Ocean surface water at about 2–9 pg/L (Jantunen and Bidleman, 1998).

The final chemical modelled was heptachlor epoxide for which the degradation half-life was difficult to estimate because of the large scatter in measured concentrations. Based on mean and the range of measured concentrations, the estimated mean and range of half-lives were 5.7 years and 3.4–11 years, respectively. These estimates are considerably longer than that reported for one test under temperate conditions (Mackay *et al.*, 1997).

As noted, the estimated degradation half-lives obtained from Amituk Lake are either shorter (α- and γ-HCH), similar to (endosulfan I) or longer (heptachlor epoxide) than those reported for Arctic or temperate systems. The phenomenon of shorter half-lives may be explained by the ability of microbes in this high Arctic lake to metabolize or co-metabolize this relatively "unpalatable" source of carbon, as discussed in the following text.

The relatively short half-lives of several OCs, evidence that chemical concentrations in the lake are near equilibrium (i.e., calculated air-water fugacity ratios near 1), and minimal sediment-water exchange that can buffer water-column concentrations, suggest that the lake responds relatively rapidly to changes in loadings. This conclusion was confirmed by running the model initialized to June 1994 concentrations and loadings set to zero. The model results indicate that water-column concentrations will fall by 90% within one year for endosulfan I, and three to four years for α- and γ-HCH, losses of which are largely controlled by degradation. The remaining OCs would decline by 90% within six to eight years of reducing loadings as a result of meltwater dilution and chemical export. Sediment response times are much longer and poorly known. This knowledge gap may impair our ability to estimate changes in concentrations in biota that consume benthos; however, it does not impair the ability to estimate water-column concentrations because the potential for sediment "feedback" is limited by minimal sediment-water exchange.

Overall, these results indicate that high Arctic lakes are surprisingly sensitive indicators of changes in loadings for some chemicals such as α- and γ-HCH, endosulfan I, and heptachlor epoxide. The lakes are not major repositories of chemicals because of very low sedimentary retention rates, but rather the oceans and air are the main sink for OCs deposited to snow in Arctic watersheds.

Pat Roach

C.2.2.3 Organochlorine degradation and relative age

One of the more interesting findings from the Amituk Lake field study and modelling effort was the short half-life of α- and γ-HCH that was attributed to microbial degradation. These short half-lives are consistent with previous reports of relatively low ratios of the enantiomers of α-HCH in Arctic and subarctic freshwater systems that were also attributed to enantioselective microbial degradation (Falconer *et al.*, 1995a; Alaee 1997).

Two studies were conducted to improve our understanding of factors related to these surprisingly high rates of degradation. Results that indicate higher rates of, at least, enantioselective degradation in Arctic and subarctic freshwater systems in comparison to temperate systems, which is the reverse of what would be expected based on temperature-related enzymes efficiencies.

Helm *et al.* (2000b) investigated enantiomer ratios as an indicator of enantioselective degradation in several streams and lakes on Cornwallis Island, including Amituk Lake and streams. They confirmed the results of Falconer *et al.* (1995a) that the ratio of (+)/(-)-α-HCH (ER) varies from a racemic value of 1.01 to as low as 0.36 in stream water (Table C.2.4). The lowest ratios occur when temperatures are highest and stream flow rates are lowest. At low flow rates there is maximal contact between the water, waterborne chemicals and the stream bed, which is the presumed location of the greatest microbial populations. Further evidence of the importance of a high contact rate between water and sediments came from correlations between ER and indicators of bedrock weathering such as conductivity and alkalinity. Helm *et al.* (2000b) also found that enantioselective degradation was greatest in the most biologically

TABLE C.2.4 Median and range α-HCH concentrations and ERs, stream temperatures and discharge, and water chemistry parameters measured at Amituk Lake June to August, 1994

	Creeks			Lake			
Measured Parameter	**Gorge**	**Mud**	**Cave**	**3 m**	**20 m**	**40 m**	**Outflow**
Enantiomer Ratio	0.72 (0.36–1.00)	0.71 (0.64–0.88)	0.74 (0.60–0.92)	0.76 (0.73–0.90)	0.74 (0.73–0.77)	0.72 (0.66–0.74)	0.81 (0.75–0.90)
α-HCH Conc. (pg/L)	419 (89–2330)	119 (20–307)	529 (293–958)	644 (583–944)	520 (469–598)	688 (342–727)	851 (514–1069)
Temperature (°C)	3.0 (0.0–4.0)	6.8 (2.0–10.0)	1.5 (0.0–2.0)	2.4 (1.7–3.7)	2.7 (2.4–3.7)	2.8 (2.4–3.7)	–
Total Discharge (x10^6 m^3)	33.8	8.64	10.3	–	–	–	61.6
Ca^{2+} (mg/L)	19.0 (9.8–21.1)	26.1 (19.5–31.1)	19.4 (14.4–21.8)	20.6 (15.4–26.0)	22.9 (20.2–25.2)	24.3 (21.1–25.9)	–
Mg^{2+} (mg/L)	4.21 (2.52–7.60)	3.35 (2.32–5.90)	3.11 (1.69–3.37)	3.38 (2.75–4.58)	4.19 (3.92–4.70)	4.47 (3.94–5.03)	–
Conductivity (μS/cm)	126 (70–166)	152 (117–200)	126 (86–141)	137 (104–161)	157 (137–158)	157 (136–170)	–
Alkalinity (meq/L)	1.20 (0.65–1.66)	1.51 (1.16–1.91)	1.19 (0.82–1.38)	1.30 (0.96–1.51)	1.50 (1.30–1.51)	1.51 (1.32–1.62)	–
pH	8.13 (7.81–8.27)	8.20 (8.04–8.44)	8.12 (8.02–8.28)	8.24 (8.13–8.33)	8.25 (8.20–8.34)	8.21 (8.09–8.32)	–
DOC (mg/L)	0.61 (0.19–1.17)	0.78 (0.33–1.14)	0.59 (0.34–0.89)	0.59 (0.46–0.76)	0.55 (0.38–0.61)	0.59 (0.48–0.77)	–
SS (mg/L)	0.26 (0.11–17.76)	0.51 (0.20–6.86)	5.15 (1.10–18.10)	1.14 (0.15–2.22)	1.11 (0.25–1.19)	1.11 (0.43–2.06)	–
N-NO_3 (x10^{-2} mg/L)	4.1 (1.8–8.0)	2.4 (1.2–12.8)	2.2 (1.8–3.1)	2.1 (1.4–2.8)	2.2 (1.6–2.8)	2.1 (0.8–8.4)	–

productive streams and wetlands that support communities of mosses versus streams that run as rills over the exposed bedrock.

Helm *et al.* (2000b) then assessed the importance of enantioselective degradation as a loss process for α-HCH. They found that enantioselective degradation accounted for the greatest mass loss when ratios were relatively high (around 0.95 to 0.80), which coincides with high flow rates. The low ERs measured in streams during late runoff accounted for negligible mass loss. Based on these data, Helm *et al.* (2000b) estimated that enantioselective degradation in the streams accounted for about 7% of α-HCH losses before entering Amituk Lake. Once in the lake, there was little evidence for enantioselective degradation, but rather most of the degradative losses were thought to be non-enantioselective microbial degradation, accounting for elimination of 33–61% of the α-HCH water column inventory from ice-on (1993) to ice-off (1994).

Law *et al.* (2001) further investigated the factors related to enantioselective degradation of α-HCH as well as ratios of α- and γ-HCH that are indicative of the relative age of the chemical mixture. They measured and statistically analyzed concentrations of α- and γ-HCH and the enantiomers of α-HCH in streams, wetlands, and lakes in the high Arctic (Cornwallis Island), subarctic (southern Yukon), Great Lakes, and temperate locations (Muskoka and Kawartha regions located on or nearby the Canadian Shield and in the Greater Toronto Area). These data are summarized in Figure C.2.7.

Law *et al.* (2001) reported concentrations of α-HCH measured in the late 1990s ranging from 0.64 to 1.7 ng/L in Char (Cornwallis Island) and Atlin (Yukon) Lakes, to 2.12 and 1.7 ng/L in Lakes Superior and Huron to 0.04 to 0.43 ng/L in temperate lakes in southern Ontario. Concentrations of γ-HCH varied from below detection (10 pg/L) to 0.95 ng/L in southern lakes to 0.13 to 0.41 ng/L in northern lakes.

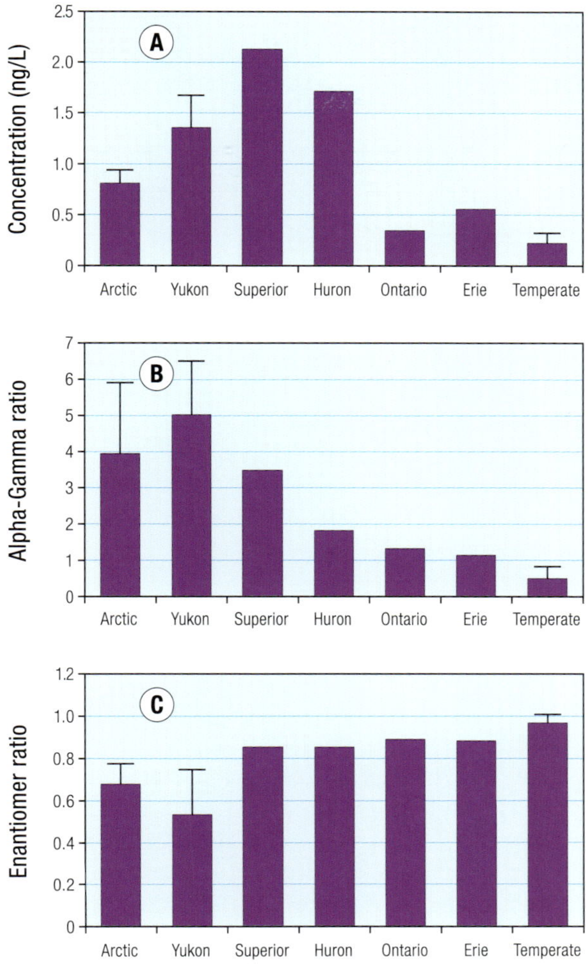

FIGURE C.2.7

Summary of concentrations and ratios of α-HCH/γ-HCH and enantiomer ratios of α-HCH in high Arctic, Yukon and temperate lakes.

The analysis of Law *et al.* (2001) confirmed the hypotheses of Helm *et al.* (2000b) that enantioselective degradation of α-HCH is promoted by contact with sediment substrates, as occurs in wetlands and streams especially at low flow rates. Counter-intuitively, they reported minimal enantioselective degradation in nutrient-rich, warm streams and wetlands in southern Ontario. For lakes, enantioselective degradation was greatest (ERs from 0.31 to 0.7) in small high Arctic lakes and in large lakes in the subarctic and the Great Lakes, all of which had high concentrations of α-HCH and, at least for the larger lakes, had long water residence times (e.g., Atlin and Tagish Lakes in Yukon). The long water residence time would allow more contact time between chemical and sediment substrates.

These results led Law *et al.* (2001) to confirm the importance of contact time between chemical and sediment substrates, noting that microbes tend to live attached to substrates as biofilms. Thus, wetlands and moss-lined streams in the high Arctic are acting as "biotrickling filters", efficiently and rapidly degrading α-HCH enantioselectively. Law *et al.* (2001) hypothesized that enantioselective degradation is greatest in the presence of oligotrophic bacteria that occur in nutrient-poor northern lakes. Oligotrophic bacteria are known to induce multiple enzymes, shift metabolic pathways, and use mixed carbon sources, whereas eutrophic bacteria are adapted to using sources of carbon that are more easily degraded than HCHs. The relationship between high α-HCH concentrations and enantioselective degradation was concluded to be ancillary to nutrient status and water residence time. Large, cold lakes have elevated concentrations of α-HCH because they act as "cold" traps by maximizing atmospheric deposition and minimizing chemical losses due to volatilization and sedimentation.

Finally, Law *et al.* (2001) analyzed ratios of α- and γ-HCH to discern the relative age of the HCH mixture. Alpha-HCH, that constitutes 60 to 70% of technical HCH, has been banned or restricted in most countries. Gamma-HCH or lindane, is still widely used in Canada, Europe, and some Asian countries. The expected ratio of of α-/γ-HCH in the technical mixture is 4 to 7, whereas lower ratios indicate recent lindane usage and higher ratios indicate long-range transport of the HCH isomers. The ratio of α-/γ-HCH increases due to the preferential scavenging of γ-HCH from air and its greater susceptibility to atmospheric hydroxylation.

Law *et al.* (2001) found that α-/γ-HCH ratios increased with latitude, which is a function of higher concentrations of α-HCH rather than systematic variations in concentrations of γ-HCH. The ratio ranged from 5.2–9.4 and 3.5-8.0 in high Arctic and Yukon lakes, respectively, ratios that are indicative of an atmospherically transported source. These ratios are higher than those reported for the Bering and Chukchi Seas (Hinckley *et al.*, 1992) and the Eastern Arctic Ocean (Harner *et al.*, 1999). In contrast, ratios in the Great Lakes and small temperate lakes varied from 1.1–3.5 and 0.3–0.4, respectively. The very low ratios for the small temperate lakes suggest recent inputs of γ-HCH since it can be lost rapidly by photolysis and hydrolysis in these warm and relatively shallow systems.

C.3 Transport and fate of hexachlorocyclohexanes in the North American Arctic Ocean

Authors:

Y. Li (1); R. Macdonald (2); L. Jantunen (1); T. Harner (1); T. Bidleman (3); W. Strachan (4).

Affiliations:

1. Meteorological Service of Canada, 4905 Dufferin Street, Toronto, ON, M3H 5T4, Canada;
2. Institute of Ocean Science, 9860 West Saanich Road, Sidney, BC, V8L 4B2, Canada;
3. Centre for Atmospheric Research Experiments, 6248 Eighth Line, R.R. #1, Essa Township, Egbert, ON, L0L 1N0, Canada;
4. National Water Research Institute, 867 Lakeshore Road, Burlington, ON, L7R 4A6, Canada.

C.3.1 Introduction

Hexachlorocyclohexane (HCH) is the most abundant organochlorine pesticide in Arctic water and air, and its transport and distribution have been the topics of several studies, summarized in the 1997 Canadian Arctic Contaminants Assessment Report (Jensen *et al.*, 1997) and subsequently updated by Macdonald *et al.* (2000). A relatively complete knowledge of α-HCH's physical and chemical properties, together with a simple representation of global compartments, has provided the basis for multimedia models that appear to reproduce the major features of the global distribution of α-HCH in the upper ocean (Wania *et al.*, 1999c; Wania and Mackay, 1999a). These studies explain a perplexing observation that the world's highest α-HCH concentrations are to be found in the surface water of the Arctic Ocean under the permanent pack — a long way from where any technical HCH has been used (Macdonald *et al.*, 1997a; 2000). The distribution of α-HCH throughout surface waters of the northern hemisphere provides the clearest example of "cold condensation", primarily because of its strong partitioning from the atmosphere into cold water.

A budget for α-HCH and γ-HCH in the Arctic Ocean was presented as a case study in CACAR-I. The budget included input through ocean-currents, atmospheric deposition and rivers, and losses by outflow of ocean water, volatilization, degradation by chemical hydrolysis, and ice export. The mass balance was subsequently updated with new information and published in Macdonald *et al.* (2000). This version includes an estimate of microbial degradation rates for γ-HCH and the two enantiomers of α-HCH in the water column of the eastern Arctic Ocean (EAO). This process appears to be considerably faster than chemical hydrolysis and may account for about a third of the annual HCH loss (Harner *et al.*, 1999; 2001a).

Case studies for chemicals released to the environment by human activities show that their transport and fate depends on their physical properties, which are often reasonably well known, and on environmental pathways and processes, which are not completely understood and often yield surprises (Macdonald *et al.*, 2001a). Here, the HCH budget is considered further based on new information gathered since publication of CACAR-I and for the first time β-HCH is included. Although isomers other than γ-HCH exhibit much reduced insecticidal activity, both α and β-HCH disrupt endocrine processes (Kendall *et al.*, 1998; Willett *et al.*, 1998). Of these two isomers, β-HCH tends to bioaccumulates more strongly and is more resistant to metabolism and microbial degradation. These characteristics are consistent with the observation that α-HCH is the most abundant isomer in surface water but β-HCH predominates in human fat, serum, and breast milk (Van Oostdam *et al.*, 1999; Willett *et al.*, 1998). Residues in marine mammals vary, some seal and whale species show preferential accumulation of β-HCH while others contain predominantly α-HCH with lesser amounts of β- and γ-HCHs (Willett *et al.*, 1998). Residues of α-HCH and γ-HCH decreased in thick-billed murres at Prince Leopold Island in the Canadian Arctic between 1975-98, whereas β-HCH increased (Braune *et al.*, 2001).

The bioaccumulating and toxic properties of β-HCH give cause for concern about the consequences of releasing nearly 1 Mt (see following) to the environment during the past 50 years. Unfortunately, the data are sparse and until recently this isomer has been neglected in studies of the Arctic physical environment. This is results from β-HCH being less abundant than α-HCH or γ-HCH, and also less responsive by electron capture detection or negative ion mass spectrometry, two techniques frequently used to determine organochlorine compounds.

The available data for HCHs in the Arctic physical environment are summarized in this section and β-HCH is contrasted with α-HCH in the context of the Arctic as a receptor. Measurements of HCHs in Arctic water and air are collected from various published and unpublished sources. These, along with the recently determined Henry's law constants (HLC) for the three major isomers are used to formulate a scenario for HCH transport into the North American Arctic Ocean (NAAO) over the last two decades.

C.3.2 Usage and emissions of hexachlorocyclohexanes

Technical HCH was manufactured and heavily used during the 1970s and 80s, largely in China, India, and the former Soviet Union. Much effort has gone into compiling the production and usage data, interpolating for years and countries where data were lacking, and synthesizing the results into a detailed picture of northern hemisphere usage and emissions on a 1° x 1° grid. These reports have been published (Li *et al.*, 1998a; 1998b; 1999a; 1999b; 2000; 2001) and results are summarized in Section A.2.2. The isomer percentages in technical HCH are (Iwata *et al.*, 1993b; Kutz, 1991): 60–70% α-HCH, 5–12% β-HCH, 10–12% γ-HCH, 6–10% δ-HCH, and 3–4% ε-HCH. The application of 9700 kt of technical HCH between 1948 and 1997 (Li, 1999a) has led to the cumulative release of 6300 kt α-HCH, 820 kt β-HCH, and 1060 kt γ-HCH to the environment. The γ-HCH released through application of technical HCH was augmented by use of lindane (pure γ-HCH), estimated at 720 kt from 1950–93 and 3 kt for the year 2000 (Li, 2001).

Global emissions of α-HCH to the atmosphere have been estimated at 4300 kt from 1948–97, 184 kt in 1980, and 59 kt in 1990. Corresponding β-HCH emissions for 1980 and 1990 were 16.6 kt and 4.1 kt, about 81–86% from current usage and 14–19% from residues in the soil from previous applications (Li, 2001) (Figure A.2.14). Global γ-HCH emission estimates are not yet available.

C.3.3 Physical and chemical properties of the hexachlorocyclohexane isomers

Properties of the HCHs that affect partitioning and degradation in the Arctic environment are summarized in Table C.3.1. Henry's law constants (HLC), liquid-phase vapour pressures (VP), and octanol-air partition coefficients (K_{OA}) are functions of temperature, and values at 0oC are given in the table. Log octanol-water (K_{OW}) and organic carbon-water (K_{OC}) partition coefficients have been reported only at 20–25°C; however, these values should not be much different from those for 0°C since Kow appears to vary only slightly with temperature for organochlorine compounds (Bahadur *et al.*, 1997). Microbial degradation rates for γ-HCH and the two enantiomers of α-HCH were estimated from in-situ measurements available only for one area of the eastern Arctic Ocean (Harner *et al.*, 1999; 2001a) and, although their general applicability remains uncertain, it is clear that these two isomers undergo microbial attack throughout Arctic surface waters (Falconer *et al.*, 1995b; Harner *et al.*, 1999; Jantunen and Bidleman, 1997; 1998; Moisey *et al.*, 2001).

Because the HCHs are fairly close to air-water equilibrium in the Arctic Ocean, estimating their gas exchange requires highly accurate HLC values as functions of temperature. The previous HCH budget (Macdonald *et al.*, 2000; 2001) used the HLCs of Kucklick *et al.* (1991) for α- and γ-HCH. New measurements have been made for α- and γ-HCHs by Jantunen and Bidleman (2000) and Sahsuvar *et al.* (2001), and the latter study included β-HCH. These HLCs for α- and γ-HCH at 0°C are about 2–3 times lower than those of Kucklick *et al.* (1991) (Table C.3.1). The Sahsuvar *et al.* (2001) set contains all three isomers and these HLCs are used in the following discussions.

TABLE C.3.1 Properties of α-, β-, and γ-HCH affecting partitioning and fate in the Arctic

Property	HLC Pa m^3/mol 0°C	Liquid VP Pa, 0°C	Log K_{ow}	Log K_{oc}	Log K_{oa}	Hydrolysis $t_{1/2}$, y, 0°C	Microbial deg. $t_{1/2}$, y, Arctic Ocean	Gas-phase OH reaction $t_{1/2}$, d
α-HCH	0.057[1] 0.064[2] 0.10[3]	0.018[4]	3.80–4.44[6]	3.25–4.10[6]	7.61[7]	64[8]	6–23[10]	120[11]
β-HCH	0.0030[1]	0.0061[5]	3.78–4.15[6]	3.36–3.98[6]	8.88[7]	β>>α-HCH[9]	unknown	unknown
γ-HCH	0.021[1] 0.032[2] 0.061[3]	0.0047[4]	3.24–5.32[6]	1.18–4.30[6]	7.85[7]	110[8]	16[10]	96[11]

(1)Sahsuvar *et al.* (2002).

(2)Jautunen and Bidleman (2000).

(3)Kucklick *et al.* (1991).

(4)Hinckley *et al.* (1990).

(5)Leone, A.D. (unpublished work, Meteorological Service of Canada, 2001).

(6)Mackay *et al.* (1997).

(7)Shoeib and Harner (2002).

(8)Using second-order rate constants of Ngabe *et al.* (1993) and seawater pH = 8.1.

(9)No reaction in 0.005 M sodium hydroxide in 72 h; Cristol (1947).

(10)Harner *et al.* (1999; 2001a). 6 y for (+) and 23 y for (-) enantiomer of α-HCH.

(11)Using second-order rate constant k = 1 x 10^{-13} cm^3/s (α-HCH) or 1.2 x 10-13 cm^3/s (γ-HCH) at 4°C and 24-h average [OH] = 9.7 x 10^5 (Brubaker and Hites, 1998c).

C.3.4 Current understanding of α-HCH transport to the Arctic

A large body of research has been synthesized to produce an Arctic Ocean budget for α- and γ-HCH and to outline the means by which they enter and leave (Macdonald *et al.*, 2000). The salient features of that synthesis are provided here for α-HCH.

The atmosphere plays a significant role in the global distribution of α-HCH as demonstrated by two key pieces of evidence: the record of α-HCH in Arctic air between 1980 and 1996 (Bidleman *et al.*, 1995b; Li *et al.*, 1998a), and the spatial distribution of α-HCH in surface waters of the North Pacific and Arctic Oceans (Jantunen and Bidleman, 1997; Harner *et al.*, 1999; Macdonald *et al.*, 1997a; 2000; Wania and Mackay, 1996). When air concentration data for α-HCH, collected by various groups at differing Arctic locations since 1979, are collated into a time series up to 1996 (Figure A.2.11), two significant decreases are seen to correspond to prominent emission reductions. The first decrease (1982–83) occurred when China abandoned technical HCH use and the second (1990–92) followed reductions in technical HCH use in India and the former Soviet Union (Li *et al.*, 1998a). The simplest explanation for the in-step response of α-HCH concentration in Arctic air to the change of global (mainly Asian) emissions is that atmospheric long-range transport provides rapid dispersion of α-HCH from its release point into the Arctic. A global-scale model of atmospheric transport shows the easy movement of α-HCH in upper-level air currents from source regions, across the North Pacific to North America (Koziol and Pudykiewicz *et al.*, 2001). Figure A.2.11 further implies that a reasonable estimate of historical air concentrations of α-HCH in the western Arctic can be inferred from the global emission data (Li *et al.*, 2000). Since the last stepwise decrease in α-HCH emissions (and Arctic air concentrations) between 1990–92, α-HCH has continued to decline slowly throughout the 1990s with a half-life at Alert of 16.9 years (Hung *et al.*, 2002a).

The second piece of evidence, the relatively high concentrations of α-HCH in northern surface waters of the Pacific and Arctic oceans (Figure C.3.1), illustrates the importance of atmospheric transport especially in the early "ocean-loading" stage. Figure C.3.1 shows the gradient in α-HCH concentration which increases from about 0.7 ng/L in the regional seas off the Asian coast (20 to 40°N) to 2 ng/L in the Bering and Chukchi seas (55 to 72°N) to 2.5 ng/L in the Canada Basin north of 72° (see also Iwata *et al.*, 1993b; Wania and Mackay, 1996; Carmack *et al.*, 1997). Even higher concentrations, 2–6 ng/L, are found in surface waters of the

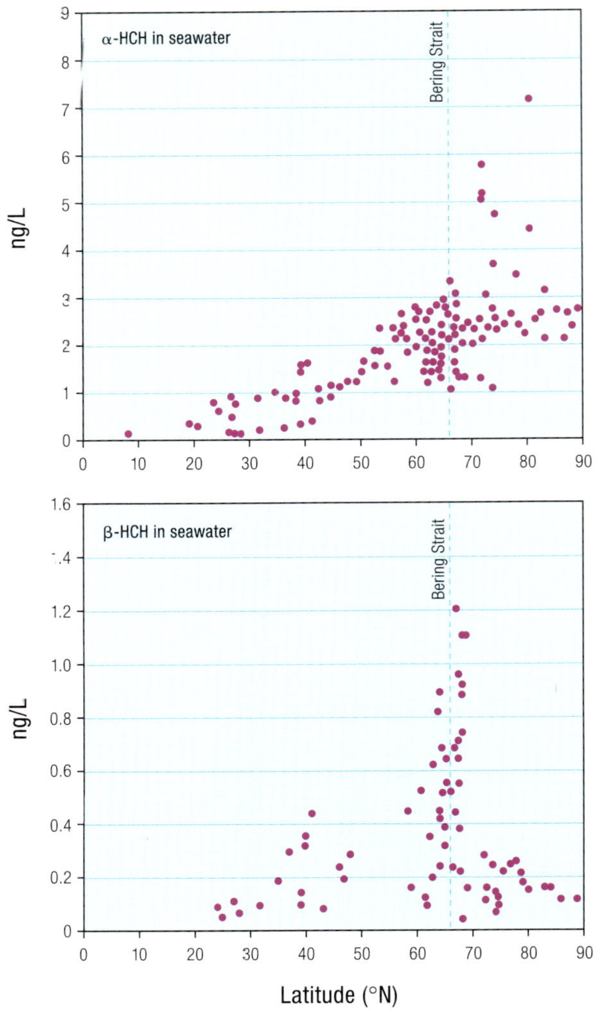

FIGURE C.3.1

Concentrations of α-HCH and β-HCH in surface water of the North Pacific and western Arctic Ocean measured between 1988 and 1999. Although these data span a 10-year period and are collected from a wide range of longitudes (Li *et al.*, 2002), they illustrate well the general increase in α-HCH in cool, northern waters with exceptionally high values under the pack ice of the Canada Basin, in contrast with β-HCH which shows the highest concentrations centred on the Bering Strait.

Beaufort Gyre and Canadian Archipelago (Table C.3.2; see also Table B.2.1 which gives ΣHCH concentrations). Concentrations of α-HCH are lower in the eastern Arctic Ocean, but nevertheless increase significantly with latitude from 0.4 ng/L in the Barents Sea at 73–79°N to 1 ng/L over the Lomonosov Ridge at 85–88°N (Harner *et al.*, 1999; Carmack *et al.*, 1997; Strachan, EC, 2002).

Arctic marine mammals probably mirror the ocean concentration of α-HCH, as suggested by the higher concentrations found in ringed seals from the Canadian Archipelago compared to those in Greenland and the

TABLE C.3.2 Mean surface water concentrations of HCHs, ng/L

	Year	α-HCH	n	β-HCH	n	γ-HCH	n	α/β Ratio	α/γ Ratio	Ref.
Western Arctic Ocean										
Northern North Pacific and Bering Sea	1981	2.65	2	0.17	2	0.64	2	16	4.1	1
Bering Sea	1979	3.00[(1)]	7	0.19[(1)]	7	0.73[(1)]	7	16	4.1	2
	1988	2.34	14			0.59	14		4.0	3
	1988	1.76	19	0.39	19	0.94	19	4.5	1.9	4,5
	1990	1.50	4			0.19	4		7.9	6
	1993	2.14	26	0.16	26	0.36	26	13	5.9	7
	1993	1.95	36			0.46	36		4.2	8
Chukchi Sea	1988	2.38	5			0.60	5		4.0	3
	1988	2.15	15	0.55	15	0.93	15	4.0	2.3	4,5
	1990	1.40	4			0.18	4		7.8	6
	1993	2.30	10	0.16	10	0.25	10	15	9.2	7
	1993	2.08	19			0.44	19		4.8	8
	1993	2.20		0.22				10		9
	1994	1.50	2	0.25	2	0.28		6.0	5.4	10
Beaufort Sea	1986	4.40	6			0.61	6		7.2	11
	1987	7.10	7			0.81	7		8.8	12
	1992	5.10		0.12				43		9
	1993	5.70		0.14				41		9
	1997	2.77	6	0.074	6	0.95	6	37	2.9	9,13
N. Canada Basin	1993	2.66		0.22				12		9
	1994	2.41	15	0.18	7	0.47	15	14	5.1	10,14
Lincoln Sea	1994	3.10		0.16				19		9
Canadian Archipelago										
West and Central	1992	4.70	15			0.44	15		11	15
	1993	3.64	34			0.52	34		7.0	16
	1997	2.70	10	0.12	10	0.35	10	22	7.7	7
	1997	3.30	5							9,13
Northwater Polynya	1998	1.10	19	0.070	16	0.20	19	19	5.5	17
Hudson Strait, Foxe Basin, Prince Regent Inlet, Gulf of Boothia	1999 1999	1.01 2.68	6 22	0.14 0.092	6 6	0.19 0.26	6 22	7.2 29	5.3 10	7 14
Eastern Arctic Ocean										
Barents Sea	1992	0.47	3	0.12	3	0.29	3	3.9	1.6	5
	1996	0.40	7			0.16	6		2.5	18
Eastern Arctic Ocean	1996	0.72	44	0.030	44	0.21	44	24	3.4	7
	1996	1.01	33			0.29	33		3.5	18

[(1)]Estimated from reported α-HCH = 3.9 ng/L and the α/β, α/γ ratios reported by Kawano *et al.* (1988).

1 = Kawano *et al.* (1988). 2 = Tanabe and Tatsukawa (1980). 3 = Hinckley *et al.* (1991). 4 = Rice and Shigaev (1997). 5 = Chernyak *et al.* (1996). 6 = Iwata *et al.* (1993a). 7 = Strachan (pers. comm.) 8 = Jantunen and Bidleman (1995). 9 = Macdonald *et al.* (1997b, 1999b,c and pers. comm.) 10 = Jantunen and Bidleman (1998). 11 = Hargrave *et al.* (1988). 12 = Patton *et al.* (1989). 13 = Stern (pers. comm.) 14 = Jantunen *et al.* (pers. comm.) 15 = Falconer *et al.* (1995b). 16 = Hargrave *et al.* (1997). 17 = Moisey *et al.* (2001). 18 = Harner *et al.* (1999).

TABLE C.3.3 Mean concentrations[1] of HCHs in air (pg/m^3) and snowpack (ng/m^2)

Location	Year	α-HCH	β-HCH	γ-HCH	α/β Ratio[2]	α/γ Ratio[2]	Reference[3]
Technical HCH					7.6	5.9	
Source Regions							
South India	1993–94	3550	1180	2070	2.6	1.7	1
India	1989	2240	1160	829	19	2.7	2
Thailand	1989–90	200	4	50	75	4.0	2
Vietnam	1990	9800	1070	4140	17	2.4	2
China					5.7	2.4	3
Arctic							
Alert, Canada	1988	156	0.9	23	206	6.8	4
	1993	60	0.75	9.7	80	6.2	5
	1994	61	0.39	11	156	5.5	5
	1995	56	0.13	8.2	430	6.8	6
	1996	55	0.24	7.6	229	7.2	6
	1997	47	0.089	6.4	528	7.3	6
Tagish, Canada	1993	79	1.08	11	73	7.2	5
	1994	78	0.47	13	166	6.0	5
Dunai, Russia	1993	40	1.02	10	39	4.0	5
Ny Ålesund, Svalbard	1993	77		14		5.5	7,8
	1994	62		16		3.9	8
	1995	63		13		4.8	8
	1996	60		13		4.6	8
Pallas, Finland	1996	24		11		2.2	8,9
	1997	28		11		2.5	8,9
Bering-Chukchi Seas	1993	40	1.5	8.8	27	4.5	10
	1993	91		23		4.0	11
	1994	125		18		6.9	12
Northern Canada Basin	1994	64		14		4.6	12
Barents Sea	1996	19		36		0.5	13
Eastern Arctic Ocean	1996	37		17		2.2	13
Snow Deposition, Canadian Arctic							
Fresh snowfall[4]	1991–94	210		96		2.2	14
Snowpack[5]	1992	44		20		2.2	14
Snowpack[6]	1992–94	150	9.8	66	15	2.3	14
Snowpack[7]	1986–87	100–300		70–100		1.4–3.0	14,15

[1]Geometric means given for source regions, since concentrations vary by orders of magnitude; arithmetic means given for the Arctic.

[2]Average of individual sample ratios, if available, otherwise ratio of average concentrations.

[3]1 = Rajendran *et al.* (1999). 2 = Iwata *et al.* (1994). 3 = Cai *et al.* (1985). 4 = Patton *et al.* (1991). 5 = Halsall *et al.* (1998). 6 = Hung *et al.* (2001). 7 = Oehme *et al.* (1996). 8 = Berg and Hjelbrekke (1999). 9 = Brorström-Lundén *et al.* (2000). 10 = Strachan (pers. comm.) 11 = Jantunen and Bidleman (1995). 12 = Jantunen and Bidleman (1996). 13 = Harner *et al.* (1999). 14 = Macdonald *et al.* (2000). 15 = Gregor and Gummer (1989).

[4]Average of 8 locations.

[5]Average of 17 locations.

[6]Amituk Lake.

[7]Average of 21 locations.

eastern Arctic (Muir *et al.*, 2000b) and by the fact that, like the water, concentrations of α-HCH in Holman seal blubber have responded little to the dramatic atmospheric turndown in the early 1980s, remaining constant between 1981 and 1991 (Addison and Smith, 1998).

The increase of α-HCH northward into the Arctic Ocean cannot be explained solely by a water transportation mechanism as that would produce dilution away from the prominent sources off India and Asia (for example, see the evidence for dilution in radionuclides transported by ocean-currents, Dahlgaard, 1995). Rather, the ocean surface concentrations and the atmospheric measurements imply that α-HCH was distributed quickly in the atmosphere from which it partitioned preferentially into cold water due to an HLC that decreases appreciably with water temperature (Iwata *et al.*, 1993a; 1993b; Macdonald *et al.*, 1997a; 2000; Wania and Mackay, 1996; Wania *et al.*, 1999c).

Another crucial component in the α-HCH story is the transient nature of the emission, and therefore the atmospheric concentration, of α-HCH (Figure A.2.11). The first report of HCHs in Arctic air was from the 1979 measurements of Tanabe and Tatsukawa (1980) in the Bering Sea, where α-HCH (estimated from their total HCH data by assuming $\alpha/\gamma = 5$) was 830 pg/m^3. In the same paper, these authors reported total HCH concentrations averaging 3600 pg/m^3 (3000 pg/m^3 α-HCH) from research cruises off China and Japan in 1975–77. In the peak years of technical HCH usage, therefore, HCH isomers were abundantly available to be carried over the North Pacific to the western Canadian Arctic. Recent work has confirmed and clarified the Pacific atmospheric route as an important pathway for rapidly delivering organochlorines from Asia to western Canada and the eastern Pacific Ocean (Bailey *et al.*, 2000; Koziol and Pudykiewicz, 2001; Wilkening *et al.*, 2000).

Because air-water partitioning for α-HCH overwhelmingly favours the water phase, especially for cold water, α-HCH moved northward by air, accumulated in the water, and slowly built a large reservoir in the frigid surface waters of the Arctic Ocean. Impediments to exchange, such as ice cover and ocean stratification, slowed the process but did not stop the relentless invasion of α-HCH from the atmosphere either in the marginal seas, which are seasonally clear of ice, or through open leads and polynyas in the permanent ice pack. Partly due to slow loading and partly due to degradation within the ocean, the surface water of the Arctic Ocean appears to have substantially lagged in coming to steady state with the atmosphere of the early 1980s. Henry's law equilibrium with the atmospheric α-HCH concentrations of the early to mid 1980s (~400–800 pg/m^3; Figure A.2.11) would have produced surface water concentrations of about 15–30 ng/L. The interior Arctic Ocean in the Beaufort Gyre and downstream in the Canadian Archipelago exhibited 3–6 ng/L α-HCH in the late 1980s and early 1990s (Table C.3.2), where these concentrations likely reflect the retention of chemical built up between 1945 and 1985 and dissipation by slow exchange mechanisms (Macdonald *et al.*, 1997a; 2001).

The dramatic downturn in atmospheric concentration of α-HCH in the 1980s (Figure A.2.11) produced the situation where previously undersaturated surface water achieved near-equilibrium in some areas and supersaturation in others during the early 1990s. The net exchange reversed, and the western Arctic Ocean became a source of α-HCH to the atmosphere (Jantunen and Bidleman, 1995; 1997). These changes in α-HCH accumulation and gas exchange reversal over time have been reproduced by a fugacity model which gives excellent agreement with measurements (Wania *et al.*, 1999c). By the mid-1990s, the α-HCH in Arctic air was responding to the ocean as demonstrated by the atmospheric enantiomeric composition over ice-free regions, which displayed depletion in the same enantiomers as in surface water (Jantunen and Bidleman, 1997). It is important to note that despite the impact of oceanic α-HCH on atmospheric composition, α-HCH evasion from the Arctic Ocean remains a relatively minor pathway due to the extensive ice cover. The large ocean reservoir will shed its accumulated HCH predominantly through water currents (Macdonald *et al.*, 2000; 2001) and degradation (Harner *et al.*, 2001a) (Section C.3.7).

C.3.5 The enigma of β-HCH

Assuming a constant composition for technical HCH, the emission function for β-HCH will be similar to that of α-HCH, but reduced in scale because of the lower abundance of β-HCH in the technical mixture (Figure A.2.10). Once released, α and β-HCH pathways diverge. β-HCH is often enriched relative to α-HCH in soils to which technical HCH has been applied (Cai *et al.*, 1985; Li *et al.*, 1998b, Suzuki *et al.*, 1977; Willett *et al.*, 1998; Wu *et al.*, 1997). This appears to mainly reflect the greater resistence to degradation of β-HCH compared to α-HCH rather than differences in volatility from soil (Li *et al.*, 2002), and in fact this leads to carry-over and emissions of β-HCH from previous years of technical HCH applications (Li *et al.*, 2001, Section A.2.2.9.1.).

Fairly high concentrations and proportions of β-HCH are found in the air of source regions (Table C.3.3), indicating that atmospheric transport of β-HCH can occur. Assuming that β-HCH is at least as resistant to atmospheric degradation as α-HCH, β-HCH ought to

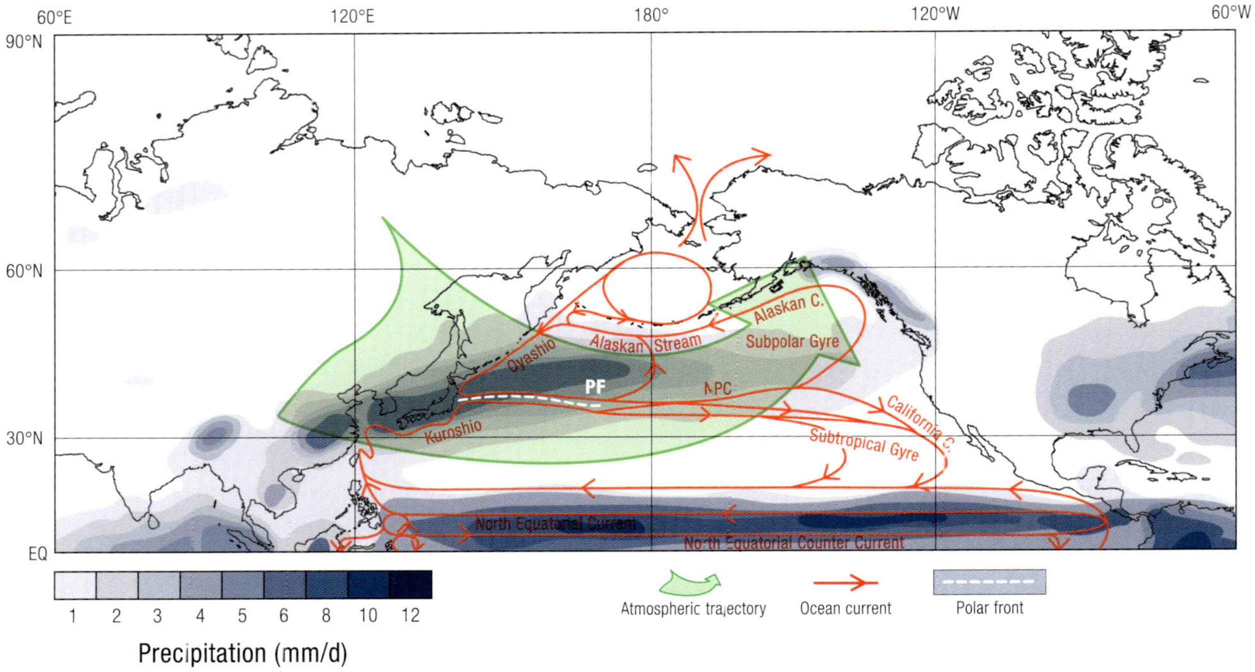

FIGURE C.3.2

Schematic diagram showing how precipitation isopleths, atmospheric trajectories (Bailey *et al.*, 2000), and ocean current structure for the North Pacific work together to load the Bering Sea with Asian β-HCH emissions.

mimic α-HCH and quickly arrive in the Arctic by the atmospheric pathway. However, α/β ratios in Arctic air are far higher than in source regions. Ratios observed in air over the Bering-Chukchi seas and the high Arctic range from 27–528 (median = 160), compared to 3-75 (median = 17) in Asian source regions (Table C.3.3). This implies selective removal of atmospheric β-HCH between the emission regions and the Arctic. Other compelling evidence of selective β-HCH loss during northward atmospheric transport is found in lake sediments, where β-HCH accounted for 45–65% of total HCH residues in four Canadian lakes between 49.6–63.6°N, but less than 2% in four other lakes between 75.0–81.75°N (Muir *et al.*, 1995).

Despite the relative inefficiency of the atmosphere to transport β-HCH northward, β-HCH has been found in the surface water of the North Pacific and Arctic oceans in samples collected from the early 1980s to the present (Table C.3.2; Figure C.3.1). The median α/β ratio in NAAO (including the Chukchi Sea) surface water is 17 with a range of 4–39 (Table C.3.2). A ratio of 8 in water would be consistent with the median ratio in Arctic air (160, see above) based on the appropriate HLCs (Table C.3.1).

A striking difference between α-HCH and β-HCH is their spatial distribution in the surface ocean. Whereas α-HCH increases steadily with latitude along a transect from southeast Asian seas and the North Pacific to the polar cap, β-HCH peaks in the Bering-Chukchi region and then decreases northward into the Arctic Ocean interior (Figure C.3.1). There appear to be processes, not completely understood, that caused β-HCH concentrations in seawater to maximize in the low Arctic regions during the early 1980s to mid 1990s, and these are discussed in the following text.

Depletion of β-HCH in air of the high Arctic can be explained by its exceptionally low HLC (Table C.3.1) which forces removal to water surfaces early during the transport through air-sea exchange and scavenging by precipitation. Fractionation of isomers due to differential air-sea exchange has previously been invoked to explain higher α/γ ratios in air at northern versus temperate latitudes (Iwata *et al.*, 1993b), and the difference in HLCs between α-HCH and β-HCH is much larger than between α-HCH and γ-HCH (Table C.3.1). The HLCs for α-HCH and β-HCH at 0°C are 0.056 and 0.0030 respectively (Table C.3.1), which yield gas-phase scavenging ratios $W_g = RT/H = 4.1 \times 10^4$ and 7.6×10^5 for these two isomers. Rain scavenging, therefore, is predicted to be more efficient for β-HCH than for α-HCH.

The high Arctic is a desert, with only 10–20 cm precipitation annually, whereas precipitation in the north Pacific and Bering Sea is 40–200 cm/y (Walsh *et al.*, 1994; Royer, 1982; Baumgartner and Reichel, 1975). The pattern of precipitation over the North Pacific together with wind trajectories that frequently move air from Asia across the North Pacific to the North American coast (Figure C.3.2) (Bailey *et al.*, 2000; Koziol and Pudykiewicz, 2001) provide a means whereby β-HCH, initially airborne over continental Asia, may be selectively deposited into the Pacific Ocean in water north of the Polar Front to the northeast of Japan while allowing more of the β-HCH to continue its atmospheric journey. Once in this cold surface water, β-HCH may be transported via ocean-currents to the Bering Sea following drift routes of the major gyres (Figure C.3.2). Heavy precipitation along the west coast of North America provides a further screen to remove β-HCH selectively and thence transport it northward in the Alaskan coastal current (Figure C.3.2). In addition to the ubiquitous air-to-sea gas exchange which would also favour selective deposition of β-HCH, the combination of wind and precipitation patterns provide a mechanism to load the surface of the Pacific Ocean with β-HCH north of about 30°N in waters that form part of the great circulation loop of the Oyashio and the North Pacific Current. According to the emission history of β-HCH, (Figure A.2.10), this loading would have maximized during the period between 1970 to 1985. The surface ocean of the North Pacific, then forming a large reservoir, would have delivered this "pulse" to the Bering Sea as a more diffuse signal, delayed by some unspecified time, perhaps a decade or more.

C.3.6 Comparative loadings of hexachlorocyclohexane isomers to the North American Arctic Ocean

To determine how HCHs have entered the NAAO (i.e., north of the Bering Strait and 50°W–180°W, about 25% of the total Arctic Ocean area) an input budget is constructed following the framework given in Macdonald *et al.* (2000). Calculations are made for 1980, 1985, and 1995, corresponding to three distinct regimes of HCH concentrations in the Arctic atmosphere (Figure A.2.11).

Input to the NAAO is assumed to take place by direct atmospheric deposition and by advection through the Bering Strait (2.6 x 10^4 km^3/y). Inflow from the Mackenzie River is neglected because previous estimates indicate this to be a minor pathway for HCHs (Macdonald *et al.*, 2000). Henry's Law Constant values for estimating precipitation scavenging and air-sea gas exchange are taken from Sahsuvar *et al.* (2001) and adjusted for temperature. Values of other parameters (e.g., gas-phase mass transfer coefficient, monthly ice cover, precipitation) are as assumed by Macdonald *et al.* (2000).

Historical measurements of HCHs in surface water of the Bering and Chukchi seas are sparse and so estimates of temporal trends are crude. The first reported measurements of HCHs in the Bering Sea were from 1979, and only total HCH (sum of α-, β-, and γ-HCHs) was given (Tanabe and Tatsukawa, 1980). Kawano *et al.* (1988) collected two samples from the northern North Pacific and Bering Sea in 1981 and reported concentrations of the three individual isomers. Using their isomer ratios, concentrations of α-, β-, and γ-HCHs in the Tanabe and Tatsukawa study can be estimated (Table C.3.2). Later measurements were made between 1988-94 by different research groups (Table C.3.2).

Regression analysis of the ln transformed data suggest declines for both α-HCH (r^2 = 0.32, p = 0.034) and γ-HCH (r^2 = 0.30, p = 0.054) (Figure C.3.3). From these regressions, concentrations for 1980, 1985, and 1995 are estimated as: α-HCH = 2.49, 2.17 and 1.66 ng/L; γ-HCH = 0.83, 0.60 and 0.31 ng/L. Measurements of (β-HCH in Arctic waters are more sparse, but suggest elevated levels in the Bering-Chukchi region around the mid-late 1980s (Table C.3.2, Figures C.3.1 and C.3.3).

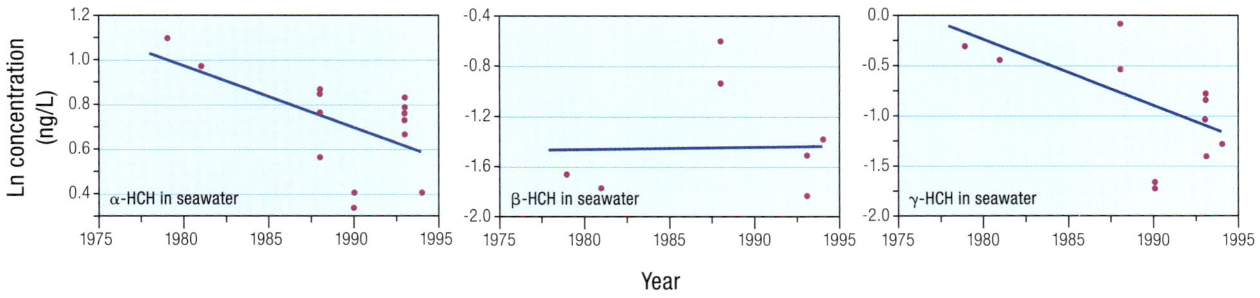

FIGURE C.3.3
Temporal trends of HCH isomers in the Bering and Chukchi seas; mean concentrations from Table C.3.2.

TABLE C.3.4 Input of HCHs to the North American Arctic Ocean, 1980–95 and losses for 1995

	1995			1985			1980		
Concentration Data	**α-HCH**	**β-HCH**	**γ-HCH**	**α-HCH**	**β-HCH**	**γ-HCH**	**α-HCH**	**β-HCH**	**γ-HCH**
Air, pg/m^3	56	0.32	8.3	420	2.6	34	750	4.7	34
Snowpack, ng/m^2	44	3.0	20	200	14	85	500	35	85
Inflow water, $\mu g/m^3$	1.66	0.20	0.31	2.17	0.47	0.60	2.49	0.17	0.83
Loadings, t/y									
Rain	0.44	0.047	0.17	3.28	0.38	0.68	5.86	0.69	0.68
Snowmelt	0.11	0.0080	0.051	0.51	0.036	0.22	1.29	0.090	0.22
Dry particle deposition	0.018	0.0003	0.011	0.13	0.0021	0.043	0.24	0.0040	0.043
Gas exchange	4.02	0.023	0.59	30.1	0.19	2.42	53.8	0.34	2.42
Bering Strait inflow	43.2	5.20	8.06	56.4	12.2	15.6	64.7	4.42	21.6
Total	47.8	5.28	8.88	90.4	12.8	19.0	125.9	5.54	24.9
Removal, t/y									
Volatilization	-5.2	-0.014	-0.31						
Advection	-274	-16.0	-46.2						
Microbial degradation	-123		-9.1						
Hydrolysis	-19		-1.6						
Total	-421	-16.0	-57.2						

Concentrations of β-HCH in the Bering-Chukchi region in 1981, 1988, and 1993–94 were 0.17, 0.47, and 0.20 ng/L, respectively, and these values were assumed to apply to the 1980, 1985, and 1995 budget years.

Atmospheric concentrations of α-HCH were taken as 750 and 420 pg/m^3 in 1980 and 1985, from the trend in Figure A.2.11 (Li *et al.*, 2002) (Table C.3.3); β-HCH in the corresponding years was estimated as 4.7 and 2.6 pg/m^3, assuming the median α/β ratio of 160 (Section C.3.5). Levels of γ-HCH show no clear trend in the earlier years and the average for the 1980s, 34 pg/m^3 (Bidleman *et al.*, 1995b), was taken for both 1980 and 1985. Air concentrations for 1995 were obtained from regressions of the means from 1993–97 in Table C.3.3, resulting in 56, 0.32 and 8.3 pg/m^3 for α-, β- and γ-HCH, respectively.

In a previous budget (Macdonald *et al.*, 2000), the estimate of precipitation scavenging included gas-phase adsorption to falling snow crystals which was thought to be more efficient than rain scavenging. Recent estimates indicate that, although snow scavenging at low temperatures is more important for hydrophobic chemicals like PCBs, rain removal dominates for the more soluble HCHs (Section C.1.4). Furthermore, much of the HCH deposited by snow probably revolatilizes over the winter, due to diagenesis of the snowpack and loss of surface area (Wania *et al.*, 1998; Section C.1.3). Measurements in the Canadian Arctic at multiple locations indicate that by the time of spring melt, the snowpack contains only 20% of the HCHs deposited by fresh snow (Gregor *et al.*, 1991; Macdonald *et al.*, 2000). Thus, the input of HCH to the ocean via snow deposition was probably overestimated in the previous budget.

Calculations of HCH loadings to the Arctic Ocean via precipitation in this revised budget are based on: a) the snowpack burden at the end of spring; b) rain scavenging over June-September, estimated from Henry's law; and c) particle scavenging. Seasonal snowpack burdens of α- and γ-HCH are derived from measurements at multiple locations across the Canadian Arctic between 1986–87 and 1992. β-HCH is not reported in these studies, but is estimated here from the ratio of β/α = 0.069 measured at one location (Amituk Lake) in 1992–94 (Table C.3.3). The median burdens for 1986–87 (α-HCH = 200, β-HCH = 14, γ-HCH = 85 ng/m^2) and 1992 (α-HCH = 44, β-HCH = 3.0, γ-HCH = 20 ng/m^2) are assumed to hold for 1985 and 1995, respectively. The estimated 1980 burdens are α-HCH = 500, β-HCH = 35, γ-HCH = 85 ng/m^2, based on the higher air concentrations of α-HCH and the mean snowpack/air ratios in other years. Note that the 1980 snowpack burden for γ-HCH is assumed to be the same as in 1985 because no change in its air concentrations is indicated (see previous text). It is assumed that all the HCH from the melted snowpack drains into the Arctic Ocean.

The estimated loadings in Table C.3.4 reveal striking differences between the HCH isomers in their mode of entry into the Canada Basin over the 15 years. For α-HCH in 1980, flow through the Bering Strait (51.4%) and direct atmospheric deposition (48.6%, gas exchange + precipitation + snowmelt + dry particle deposition) are equal contributors (ratio = 1.06). The relative loading from Bering Strait inflow (62.4%) to atmospheric deposition (37.6%) is greater in 1985 (ratio = 1.66), and even more so in 1995 (inflow = 90.4%, atmosphere = 9.6%, ratio = 9.4) (Figure C.3.4). The role of the ocean increases with time because α-HCH in the inflowing water decreases by only 33% over 15 years (Table C.3.4), whereas α-HCH in air declines by 14-fold (Figure A.2.11).

For γ-HCH, and especially β-HCH, direct atmospheric deposition into the western Arctic Ocean is always a smaller term (Table C.3.4; Figure C.3.4.). β-HCH may be deposited into the North Pacific via precipitation and gas exchange with subsequent flow through the Bering Strait (Section C.3.5). Selective deposition of γ-HCH relative to α-HCH during air transport across the Pacific is also likely (Iwata *et al.*, 1993b). The ocean dominates the transport earlier and more completely due to the lower HLCs of these two isomers (Table C.3.1).

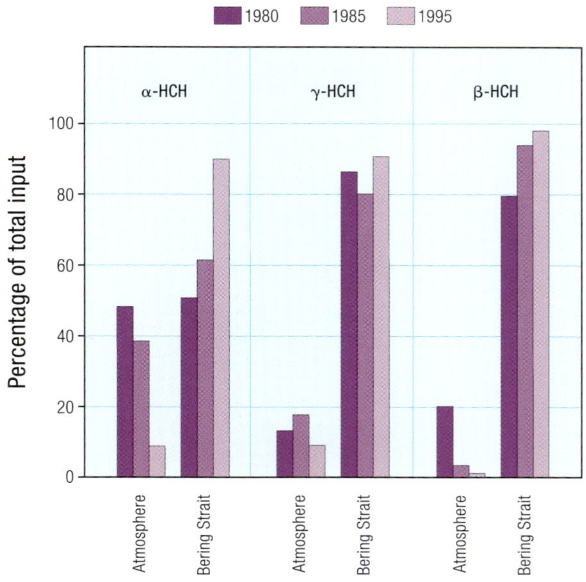

FIGURE C.3.4

Relative loadings of HCHs to the North American Arctic Ocean from atmospheric processes and ocean advection.

Based on the input budget (Table C.3.4) and a very sparse set of vertical profiles (Macdonald *et al.*, 1995; 1997b; 1999a,b; Hinckley *et al.*, 1991; Jantunen and Bidleman, 1995; 1997), a sequence of diagrams is constructed to illustrate how the loading of the NAAO might have occurred for α and β-HCH between 1980 and 1995 (Figure C.3.5). At the height of the transient emission for (α-HCH (1980), inflow from the Bering

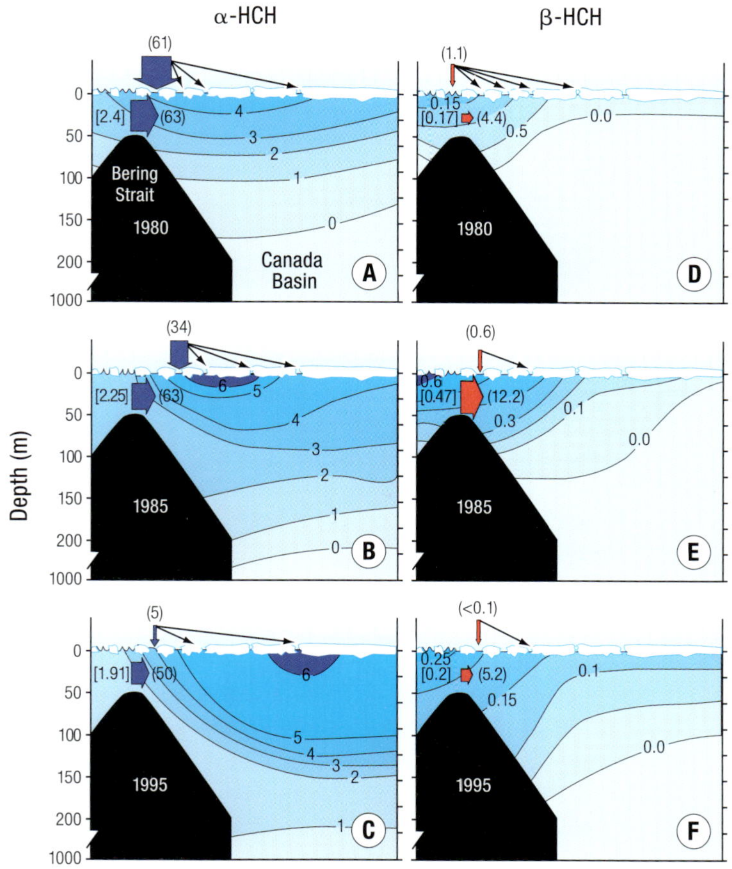

FIGURE C.3.5

A schematic diagram contrasting the evolution of α-HCH (a, b, c) and β-HCH (d, e, f) in surface waters of the Bering and Chukchi Seas. Shown are speculative sections from the northern Bering Sea, through the Bering Strait and across the Chukchi Shelf to the Canada Basin for three time periods (1980, 1985, 1995). These time periods have been chosen from different regimes of the emission history (Figure A.2.11) and the fluxes are based on the input budgets provided in Table C.3.4. For α-HCH, direct atmospheric loading to the Arctic Ocean is significant, if not dominant. Because the interior ocean is slower to respond than the Chukchi or Bering Seas, during the later stages (1995) the rapidly declining concentration in the Bering Sea produces a gradient in α-HCH that increases northward. For β-HCH, where direct atmospheric transport into the Arctic has been minimal, the buildup of the chemical in the Bering Sea produces a gradient in β-HCH that decreases northward.

Sea provides a strong foundation (2.49 ng/L) upon which atmospheric loading can further build such that the highest α-HCH concentrations are seen at the ocean surface, probably toward the outer edge of the seasonally ice-covered Chukchi Sea or under the pack ice. By 1985, the atmospheric delivery has considerably slowed and the base concentration entering through Bering Strait (2.17) produces slightly lower concentrations in the Chukchi Sea. Higher concentrations of α-HCH, however, have by this time grown into the region under the permanent pack, supported primarily by continued atmospheric exchange. Finally, in 1995 α-HCH concentrations in the margins of the Chukchi have decreased in response to decreased concentration in the Bering inflow (1.66 ng/L). The atmospheric input to the NAAO is by now too meagre to build up concentrations as the water crosses the shelf, but the inventory accumulated by the Beaufort Sea during the previous decades is maintained as a large pool trapped under the ice, now mixed deeper in the water column. The α-HCH distributions shown in Figure C.3.5 can be understood given that the residence time of water in the Chukchi is less than a year with rapid replacement from the Pacific inflow in contrast to the ten-year residence time of surface water in the interior ocean (Macdonald, 2000) and that ice forms only a seasonal cover for much of the Chukchi shelf but provides almost permanent cover for the interior ocean.

The situation for β-HCH is more speculative due to a paucity input data for the Bering Strait during the 1980s and lack of vertical profiles for Canada Basin waters. In 1980 (Figure C.3.5) inflow from the Bering Strait provides the dominant pathway, with atmospheric exchange contributing only about 20% of the loading as a broadly distributed input. As a result, concentrations of β-HCH decrease away from the point of inflow, unlike the case for α-HCH. In 1985, about five years after the emission maximum, the Bering Strait inflow completely dominates the β-HCH loading to the NAAO as a consequence of the buildup of β-HCH in the Bering Sea. The direct atmospheric input to the NAAO has faded to a negligible pathway with the consequence that a gradient is maintained between the point of inflow at Bering Strait and the large, dilute reservoir in the top 200 m of the Arctic Ocean interior. The Chukchi Sea can, therefore, be seen as a relatively small reservoir (~30 000 km^3) with short residence time sandwiched between a large (~300 000 km^3), upstream β-HCH loaded reservoir (Bering Sea surface water) and an even larger (700 000 km^3) weakly loaded downstream reservoir (Canada Basin surface water). By 1995, the concentration of β-HCH in the Bering inflow has declined, but the short residence time of the Chukchi and the relatively low concentrations of β-HCH in the interior ocean maintain the gradient between Bering Strait and the interior NAAO (Figure C.3.1).

C.3.7 Hexachlorocyclohexane removal processes in the Arctic Ocean

Loss pathways of HCHs from the water column were discussed in CACAR-I (Jensen *et al.*, 1997) and updated in Macdonald *et al.* (2000). The major improvement to our understanding between these two reports is an estimation of microbial degradation rates for α- and γ-HCH. Based on vertical profiles of HCHs and changes in the enantiomer composition of α-HCH in the eastern Arctic Ocean, microbial degradation appears to take place 3–10 times faster than basic hydrolysis (Harner *et al.*, 1999), and could account for one-third of the annual losses from the Arctic Ocean as a whole (Harner *et al.*, 2001a; Macdonald *et al.*, 2000).

Here the removal budget HCHs is revisited, specifically for the NAAO in 1995. The budget presented in Table C.3.4 uses the framework formulated by Macdonald *et al.* (2000), and their integrated burden of α-HCH and γ-HCH in the upper 200 m of the NAAO (1750 and 250 t). From the median α/β ratio of 17 in surface water of the Beaufort Sea and Canadian Archipelago (Table C.3.2), the corresponding burden of β-HCH is 103 t. Loss occurs by outflow through the Canadian Archipelago and East Greenland Current, chemical hydrolysis, microbial degradation and gas exchange. Concentrations of α-HCH and γ-HCH in outflowing water are those assumed by Macdonald *et al.* (2000), and β-HCH is estimated using the median α/β ratio of 17. Ice export is ignored, since it accounts for only about 1% of the removal budget (Macdonald *et al.*, 2000). Hydrolysis and microbial degradation of β-HCH are assigned a value of zero in the budget because there are no data applicable to the Arctic Ocean. Both processes are likely to be much slower than for α- and γ-HCHs (Cristol, 1947; Gupta *et al.*, 2001). Gas exchange (volatilization) depends on surface water concentrations, and the following concentrations were assumed (mean of regional averages for the 1990s in Table C.3.2): α-HCH = 2.9, β-HCH = 0.15, γ-HCH = 0.47 ng/L.

The estimated total losses (Table C.3.4) indicate that both α and β HCH are undergoing net removal from the NAAO, but that α-HCH is being depleted more rapidly. Loss/input ratios for the two isomers are 8.8 for α-HCH, 3.0 for β-HCH and 6.4 for γ-HCH. Due to the extensive ice cover in the NAAO, volatilization is a minor pathway relative to total losses: α-HCH = 1.2%, β-HCH < 0.1% and γ-HCH = 0.5%. The difference in loss/input among the isomers is due to more rapid degradation of α-HCH, predominantly by microbial activity.

C.3.8 Conclusions

Perhaps the most significant insight gained from our understanding of α-HCH is that a single property (low HLC) together with a rapid atmospheric pathway can provide a means to concentrate this chemical in cold aquatic reservoirs far removed from sources. If the mechanism proposed here is correct, β-HCH offers the important caveat that it is dangerous to project the environmental behaviour of one chemical based on another similar chemical without first developing a comprehensive understanding of the environmental pathways. The even lower HLC for β-HCH appears not to exacerbate the direct atmospheric transfer to the cold aquatic reservoir in the Arctic Ocean; rather, scavenging by rain and air-to-sea gas exchange intercede over the North Pacific to transfer the β-HCH from the atmosphere to the ocean before it can reach the Arctic. However, this does not ultimately protect the Arctic Ocean from β-HCH, but only delays the input and alters its spatial distribution.

Although not insecticidal, β-HCH appears to have other harmful properties and becomes concentrated in humans. The perspective presented here suggests that the Bering and Chukchi Seas are the most vulnerable locations for β-HCH loadings and that these loadings come primarily from Asia via the Pacific. Just as the Archipelago receives elevated loadings of α-HCH from the Canada Basin (Macdonald *et al.*, 1997a), the Chukchi Sea receives elevated loadings of β-HCH from the Bering Sea. The traditional atmospheric pathway across the Arctic Ocean, as implied by Arctic haze and springtime atmospheric pressure fields (e.g., Heinzenberg, 1989), together with the demonstration of rapid trans-Pacific movement of organochlorine pesticides (Bailey *et al.*, 2000) suggests that the Bering — Chukchi Seas may be particularly vulnerable to airborne contaminants, receiving them both from Eastern Europe across the pole and from Asia across the Pacific.

C.4 Modelling in support of international negotiations

Author:

F. Wania

Affiliation:

Division of Physical Sciences, University of Toronto at Scarborough,1265 Military Trail, Toronto, Ontario, M1C 1A4, Canada.

C.4.1 Introduction

An important scientific aspect of contamination in the Arctic is to understand quantitatively how the various contaminant chemicals reach the Arctic and accumulate there. Numerical simulations in the form of computer models play a key role in gaining such a quantitative understanding, because they allow to synthesize and quantify the complex interaction of a set of processes that would otherwise be difficult to integrate. Models also are the only means of developing a predictive capability in terms of identifying chemicals that may have the potential to accumulate in the Arctic and cause detrimental effects. The first Canadian Arctic Contaminants Assessment Report (CACAR) (Jensen *et al.*, 1997) described models of variable complexity that had been developed to describe the global transport behaviour of contaminants accumulating in the Arctic. In the past five years such modelling has taken two different, yet highly complementary routes of further development. On the one hand, great efforts have been directed towards very simple modelling approaches that allow the screening of large numbers of chemicals in terms of their potential to undergo long-range transport into remote regions. On the other hand, the existing more complex models have been improved, evaluated and applied to the study of transport into Arctic regions. These two developments are summarized in the following sections.

C.4.2 Model simulations of the global transport and fate of persistent organic pollutants

CACAR-I and a follow-up article (MacDonald *et al.*, 2000) discussed the initial efforts to describe the global transport behaviour of persistent organic pollutants with the help of numerical models. The progress since then and the current state of knowledge on modelling the global fate and transport of persistent organic pollutants has recently been reviewed by Scheringer and Wania (2003); therefore, only a brief account of this topic will be given here.

TABLE C.4.1 Comparison of the major characteristics of four multimedia box models that have been used to describe the global transport behaviour of persistent organic pollutants[1]

	Model Name			
	Bergen model	**Globo-POP**	**CliMoChem**	**ChemRange**
Principle design	Zonally averaged units that differ in size and environmental parameters.			Closed loop.
Zonal subdivision	Six zones of 30°latitude.	Ten zones based on climate zones defined in Troll (1966).	Flexible number *n* of zones corresponding to latitudinal range of (180°/*n*).	Boxes are not zones, but discretization for numerical solution; analytically solved version has no zones.
Compartments	Atmosphere, surface ocean, several deep ocean layers, two soil types.	Atmosphere, surface ocean, fresh water, fresh water sediment, two soil types.	Atmosphere, surface ocean, soil.	Atmosphere, surface ocean, soil.
Atmosphere				
Vertical atmospheric resolution	Four atmospheric layers (stratosphere, middle/upper troposphere, lower/middle troposphere, atmospheric boundary layer.		None	None
Treatment of atmospheric transport	Zonally averaged advection and macro-diffusion coefficients interpolated from GCM (Strand and Hov, 1993).		Zonally averaged macro-diffusion using diffusion coefficients from Keeling and Heimann (1986).	One macro-diffusion coefficient based on average of values by Keeling and Heimann (1986).
Treatment of gas-particle partitioning	None	K_{OA}-based as described by Finizio *et al.* (1997a)	Sorbed fraction can be chosen as a user-defined chemical input parameter or is derived from K_{OA} (Finizio *et al.*, 1997a).	
Treatment of atmospheric degradation	User-specified gas phase reaction rate.	Gas phase reaction rate calculated as a function of spatially and temporally variable OH concentration and temperature; no reaction of sorbed fraction.	Gas phase reaction rate calculated as a function of spatially and temporally variable temperature; reaction rate of sorbed fraction can be specified by user; default: no reaction of sorbed fraction.	User-specified gas phase reaction rate; reaction rate of sorbed fraction can be specified by user; default: no reaction of sorbed fraction.
Treatment of atmosphere-surface exchange	Diffusive gas exchange and rain dissolution; no particle bound deposition.	Diffusive gas exchange, rain dissolution and wet and dry particle deposition.		
Marine Environment				
Treatment of oceanic transport	Based on vertically resolved HILDA model by Siegenthaler and Joos (1992).	Zonally averaged macro-diffusion coefficients based on average water residence time in various marine areas from various literature sources.	One macro-diffusion coefficient based on data from Okubo (1971).	
Treatment of sorption to suspended solids and vertical particle settling	None	Sorption based on empirical relationship between K_{OC} and K_{OW} by Seth *et al.* (1999); settling based on carbon export rates by Falkowski *et al.* (1998).	Sorption to particles and settling was not included in the versions described in the references. In the present versions of the models, sorption and settling have been included.	
Treatment of oceanic degradation	User specified half-life.	Reaction rate calculated as function of spatially and temporally variable ocean water temperature.		User-specified reaction rate.
Terrestrial Environment				
Soil types	Cultivated (receives emissions) and uncultivated soil (receives no emissions).		One soil type (receives emissions).	
Composition of the soil types	Air, water, and organic carbon with fractions of 0.2, 0.3, and 0.0125.	Air, water, and organic carbon; relative size of these constitutents is zonally variable and can be changed by user.	Air, water, and organic carbon with fractions of 0.2, 0.3, and 0.02.	
Depth/vertical resolution	Constant soil depth of 0.15 m.	Zonally variable soil depth that can be changed by user.	Constant soil depth of 0.1 m.	
Other				
Treatment of innerannual variability	Four seasonal values for air temperature, precipitation and atmospheric transport parameters.	Linear interpolation of monthly values for temperature, OH radical concentration, sea ice cover and atmospheric transport parameters.	User-defined number (max. 12) of "seasons" per year; temperatures based on Legates and Wilmott (1990).	No interannual variation.

C.4.1 continued

	Model Name			
	Bergen model	**Globo-POP**	**CliMoChem**	**ChemRange**
Mathematical solution	Step-wise solution of zonally averaged continuity equation using finite difference approximation.	90 linear mass balance equations in fugacity notation solved step-wise using finite difference approximation.	3·*n* linear mass balance equations solved analytically for each time period with constant conditions.	System of 3·*n* linear mass balance equations solved by matrix inversion (Scheringer, 1996) or analytically derived continuous functions c(x) for the three media (Held, 2001).
Application using realistic emission scenarios	α- and γ-HCH.	α-HCH, seven PCB congeners.	α-HCH, p,p'-DDT.	Not applicable.
Software	Turbo Pascal	MS Visual Basic	Mathematica	MS Excel + Visual Basic
Availability	Not available.	www.utsc.utoronto.ca/~wania	Not yet available.	http://ltcmail.ethz.ch/hungerb/research/product/product.html
Reference	Strand and Hov (1996).	Wania *et al.* (1999c); Wania and Mackay (2000a).	Wegmann (1999); Scheringer *et al.* (2000).	Scheringer (1996; 1997); Held (2001; 2002).

C.4.1 continued

[1]Modified from Scheringer and Wania (2003), pages 252–254.

One of the modelling approaches that emerged at the time of CACAR-I was based on box models that divide the global environment into a series of zonal bands, each of which is described by a series of well-mixed environmental compartments. This type of global model has seen further development and use during the past five years. In particular, the global distribution model by Wania and Mackay (1993b; 1995), has seen further improvement, evaluation, and use, and a version of that model, named Globo-POP, is now freely available on the internet (Wania and Mackay, 2000a). Also, two new models of this type were introduced by Scheringer and co-workers (Wegmann, 1999; Scheringer *et al.*, 2000; Scheringer, 1996; 1997; Held, 2002). Table C.4.1, adapted from the review by Scheringer and Wania (2003), compares technical details of these models by expanding on a similar compilation presented in CACAR-I (MacDonald *et al.*, 2000). For comparison, this table still includes details of the early model by Strand and Hov (1996). Figures C.4.1 to C.4.4 show the general structure of these four models.

These global-scale box models can be used in two different ways, either to describe the real fate of persistent organic pollutants using historical emission estimates, or in an evaluative fashion to understand global transport and accumulation processes without the availability of realistic emission scenarios. The first of these uses is primarily limited by the availability of suitable emission estimates, which need to be global in spatial scale and span decades in time. Such estimates presently only exist for α-HCH and PCBs and the Globo-POP model has been used to simulate the real environmental fate of these two POPs. The model has also been used to identify the chemical property combinations that make a chemical susceptible to accumulation in remote polar regions. Each of these studies is briefly described in the following paragraphs.

C.4.2.1 Simulation of the global fate of α-HCH

Alpha-HCH is a component of a technical pesticide formulation that has been used in massive quantities in Asia in the 1970s, however, has seen a steady decline in use since then and is now virtually out of use worldwide (Li *et al.*, 1996). Compared to most other POPs, α-HCH is relatively water soluble, relatively volatile, and less persistent. It is of particular interest from an Arctic perspective because by the 1990s the concentration levels in the Canadian Arctic Ocean were higher than anywhere else in the global marine environment, notably an order of magnitude higher than in the Indian Ocean and the Western Pacific which are much closer to major source regions (Iwata *et al.*, 1993a; Wania and Mackay, 1996). Calculations using global distribution model suggested that this inverted concentration profile developed during the two decades of declining emissions and resulted from α-HCH being trapped and preserved in the cold Arctic Ocean (Wania, 1999a). The HCH levels in source regions declined as a result of degradation and volatilization. The Arctic Ocean thus constitutes the last global refuge of α-HCH (MacDonald *et al.*, 1997). Interestingly, inverted concentration profiles can be established even if only a small percentage of the globally emitted amount is transferred to Arctic latitudes. In fact, the model predicted that the bulk of the α-HCH never left the agricultural systems in which it had been applied. Relatively small amounts can result in high concentrations in the Arctic because of the relatively small size of the Arctic, and the environmental phases with a high capacity for hydrophobic organic chemicals (organic soils, vegetation, organic sediments) within the Arctic. By binding substances such as HCH these phases have the ability to reduce the amount present in the atmosphere and in aquatic systems. The model also

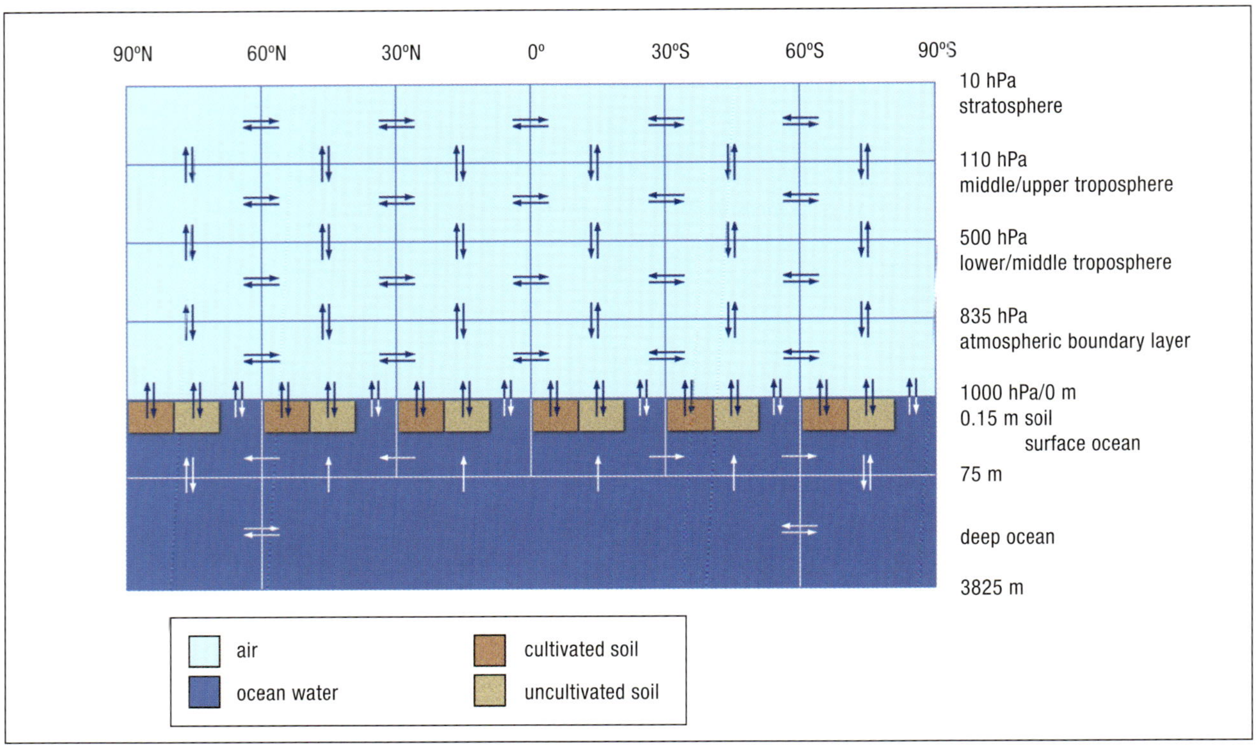

FIGURE C.4.1
Structure of the global model by Strand and Hov (1996).

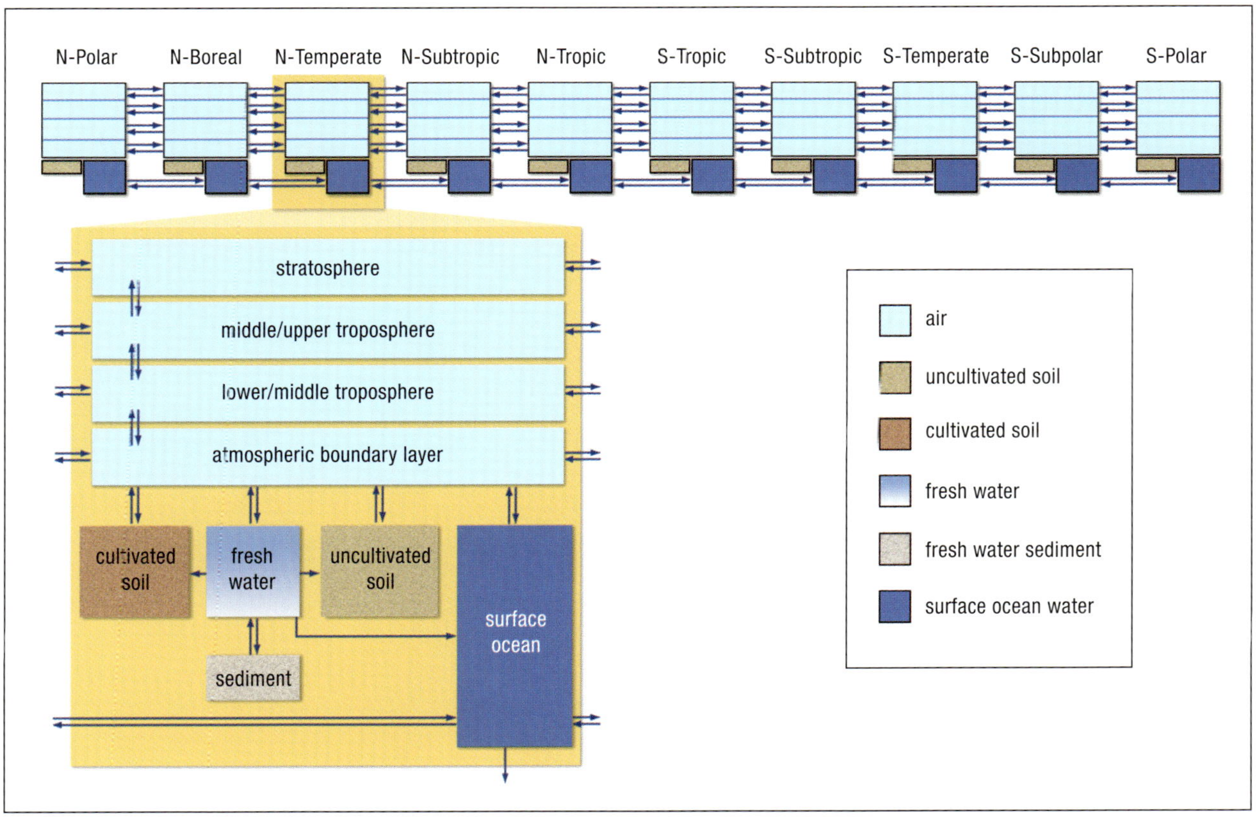

FIGURE C.4.2
Structure of the Globo-POP model by Wania and Mackay (2000a).

suggested that oceanic currents are as important in the northbound transport of α-HCH as the atmosphere. Most of the α-HCH in the Arctic originated in the Northern temperate zone which also was a net exporter of α-HCH to low latitudes. Whereas α-HCH accumulates in the North, it is rapidly degraded under tropical conditions. Comparisons show that calculated and measured α-HCH concentrations in the atmosphere and seawater generally agree within one order of magnitude (Wania *et al.*, 1999c; Lakaschus *et al.*, 2002). Deviations are due to the zonal averaging characteristics of the model and uncertainties associated with the environmental degradation rates of α-HCH.

C.4.2.2 Simulation of the global fate of PCBs

The Globo-POP model has also been used to describe the global fate of PCBs over a time scale of several decades (Wania, 1999b; Wania *et al.*, 1999b; 2000; Wania and Daly, 2002). Polychlorinated biphenyls have been used as mixtures consisting of individual substances which differ substantially in their physicochemical characteristics and persistence. It is also likely that the temporal and spatial pattern of release into the environment has been different for different congeners (Breivik *et al.*, 2002a; b). Large differences in the simulated fate of the various PCB-congeners reaffirm the need to perform calculations for individual chemicals rather than chemical mixtures. Calculations were thus performed for a selection of congeners that vary in the number of chlorine substitutions.

One of the motivations for modelling PCBs globally is to identify the major global loss processes to assess the likely rate of future concentration decline in the Arctic environment and elsewhere (Wania and Daly, 2002). The model calculations show, that historically atmospheric degradation and transfer to the deep sea contributed the most to the loss of PCBs from the global environment, whereas burial in freshwater sediments is of little significance on a global scale. Reaction of the gaseous compound with OH radicals is the loss process of primary importance to the lighter congeners, whereas deep-sea transfer increases in relevance with the degree of chlorination. The model further predicts that the relative importance of the various loss processes has been changing in time, with degradation in soils becoming a major loss process in the past twenty years. As primary emissions decreased, the concentrations (and thus the loss rates) of PCBs in the atmosphere and in ocean water have declined quickly, whereas soils retained a high pollutant load due to their slow response time (large capacity, but slow evaporation and degradation). Unfortunately, this implies that the future rate of purification of the global environment will be determined by the slow and poorly quantified degradation rate in the soil environment.

The global distribution model can also be used to investigate compositional shifts among the PCB-congeners between compartments, zones, and different time periods. Due to their wide range of physicochemical properties, PCBs have played an important role in the derivation of the concept of global fractionation, which results in

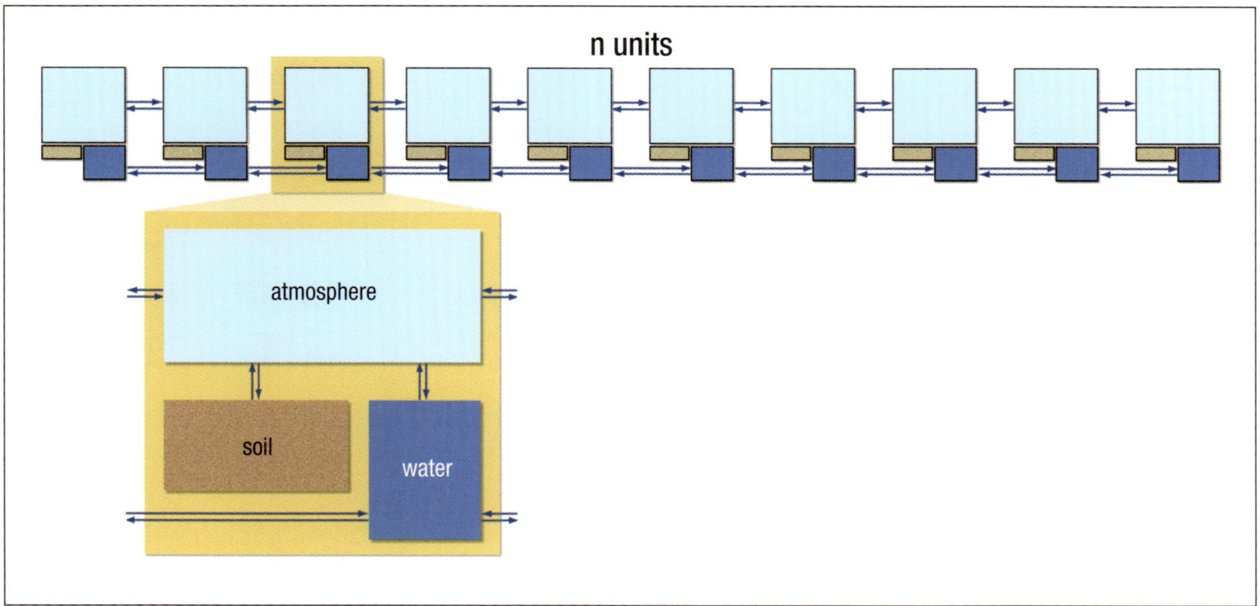

FIGURE C.4.3
Structure of the Chemrange model by Scheringer (1996).

compositional shifts of compound mixtures with latitude (Wania and Mackay, 1993b; 1996). The model succeeded in reproducing shifts towards lighter PCB-congeners with increasing latitude, as observed in various measurement campaigns (Muir *et al.*, 1996b; Ockenden *et al.*, 1998). The model further suggests that these shifts are rather complex, namely differ between various environmental media, have sometimes surprising anomalies, and change in time. This is illustrated in Figure C.4.5, which shows how the simulated homologue composition of PCBs in soil shifts to increasingly more volatile homologues with increasing latitude. In the time period between 1970 and 2000, the less volatile congeners increased in relative importance as the lighter ones are degraded more readily. It should be stressed that these global fractionation patterns have been established with only relatively minor fractions of the global inventory of PCBs being transferred northward (Figure C.4.6). Compositional shifts consistent with the global fractionation hypothesis, as well as inverted concentration profiles with higher concentrations in the Arctic than at lower latitudes, do not require that the bulk of a chemical's global inventory is transferred to Arctic latitudes. It is sufficient if differences exist in the extent of re-distribution of the various PCB homologues, even if that re-distribution may be small relative to the total global amount.

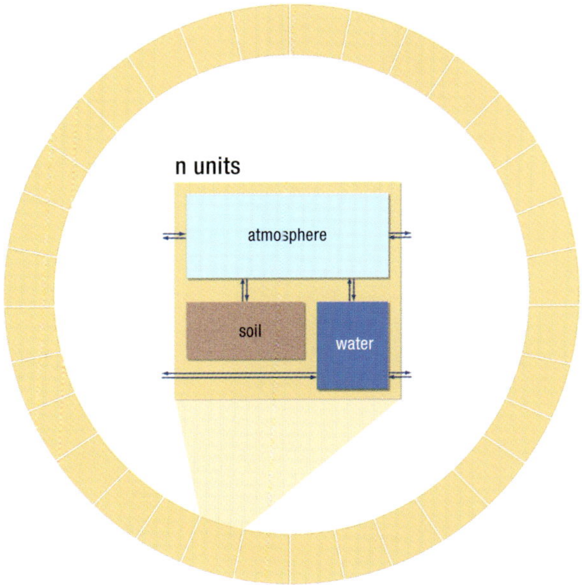

FIGURE C.4.4

Structure of the CliMoChem model by Scheringer *et al.* (2000)

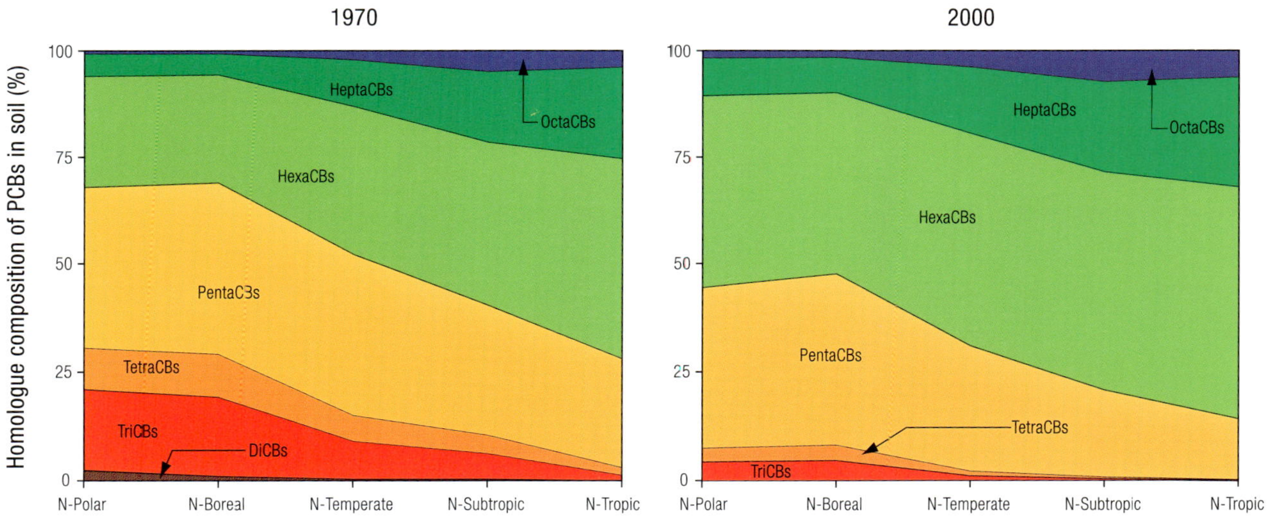

FIGURE C.4.5

Simulated homologue composition of PCBs in soil in 1970 and 2000. The composition shifts to increasingly more volatile homologues with increasing latitude. In the past thirty years of steadily declining emissions, the less volatile congeners increased in relative importance as the lighter ones are degraded more readily.

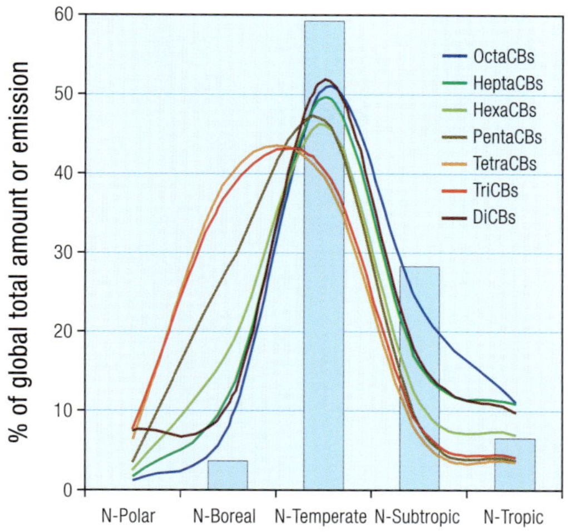

FIGURE C.4.6

Simulated zonal distribution of the total global amount of various PCB homologues in 2000. The zonal distribution of the cumulative emissions from 1930 until 2000 are depicted as bars. The bulk of the distribution of the lighter PCBs, in particular those with three and four chlorines, is shifted towards the higher latitudes, whereas the distribution of the octachlorobiphenyls is similar to that of the emissions. The highly reactive dichlorobiphenyls are also restricted largely to the zones of emission.

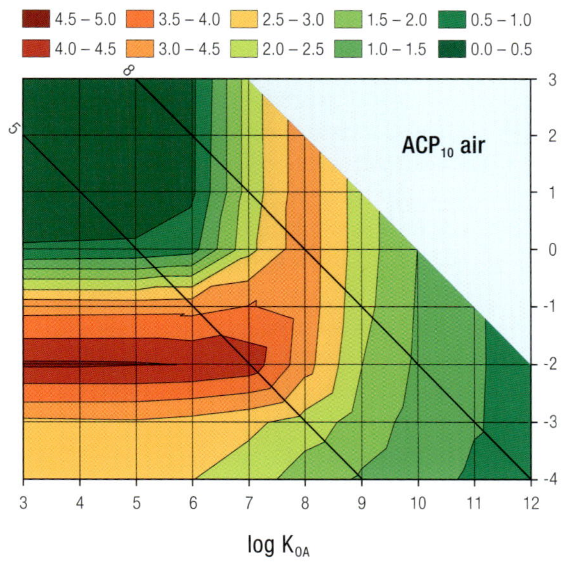

FIGURE C.4.7

The Arctic Contamination Potential of a perfectly persistent organic chemical with a particular log K_{OA} / log K_{AW} property combination after one year of steady emissions into the atmosphere as obtained with the Globo-POP model (Wania *et al.* 2001). Property combinations shown in red indicate a large potential for accumulation in the Arctic, while those in green indicate a very low potential.

C.4.2.3 Identification of the potential for accumulation in the Arctic

Finally, the Globo-POP model has been employed to identify chemical properties and emission situations that favour accumulation in Arctic ecosystems. The criterion defined to quantify this Arctic Contamination Potential (ACP) is the fraction of the total amount in the global environment that is in the surface media of the Arctic zone after 1 and 10 years of steady emissions (Wania, 1998a; Wania *et al.*, 2001; Wania, submitted), reflecting an immediate and a long-term potential for Arctic enrichment, respectively. The simulations use hypothetical emission estimates with a generic spatial and temporal distribution. Simulations for a two-dimensional "space" of hypothetical chemical property combinations (Figure C.4.7) indicate that the potential of a perfectly persistent organic chemical to accumulate in the Arctic is determined by a complex set of processes, but tends to be higher for two combinations of partitioning properties: volatile substances (log $K_{OA} < 9$) that are relatively water soluble ($4 > \log K_{AW} < -0.5$) and substances which are semi-volatile (log K_{OA} 6.5 to 10) and relatively hydrophobic (log $K_{AW} > -3$). Very volatile chemicals with log $K_{OA} < 6.5$ and log $K_{AW} > -0.5$ remain in the atmosphere even under Arctic temperature conditions, whereas very involatile chemical with a log $K_{OA} > 10$ are efficiently and irreversibly deposited with atmospheric particles before reaching the Arctic. The two sets of partitioning characteristics with elevated ACP overlap in the range $6.5 < \log K_{OA} < 10$ and $-0.5 > \log K_{AW} > -3$, which also corresponds to a log K_{OW} range of 5 to 8 (i.e., comprises substances with a potential for bioaccumulation). Organic contaminants known to accumulate in the Arctic, such as hexachlorobenzene and the PCBs indeed have such partitioning properties. Marine currents contribute significantly to the long range transport of chemicals with log $K_{AW} < -2$. Emissions to surface media greatly reduce the ACP, except for chemicals with octanol/water partition coefficients log $K_{OW} < 5$. The ACP of chemicals with different partitioning properties is sensitive to different sets of environmental parameters, reflecting the different pathways which determine their global transport behavior. The ACP of most chemicals is sensitive to the temperature dependence of the partition coefficients, temperature, atmospheric mixing coefficients, and sea ice cover. Finally, variable persistence of a chemical in air and surface media can either enhance or decrease its potential to become relatively enriched in Arctic ecosystems (Wania *et al.*, 2001).

ITK/Eric Loring

C.4.2.4 *Spatially highly resolved models of global fate and transport*

At the time CACAR-I was published, three-dimensional global atmospheric transport models with high spatial resolution were beginning to be adapted to describe persistent organic pollutants. This requires taking into account the multimedia distribution of POPs (Wania and Mackay, 1999b). Substantial progress has been made in this regard, and recent models of this type include description of the contaminant fate processes occurring in surface media such as soil. Specifically, Koziol and Pudykiewicz (2001) modelled the global transport of α- and γ-HCHs in 1993 and 1994 on a 2°x 2°grid and compared the measured and modelled air concentrations for several Arctic locations. The assumption of constant ocean water concentrations limits the model to fairly short time periods. The model also does not include particle-associated atmospheric transport, which may be important at the low temperatures of high altitudes and latitudes.

A different modelling approach to quantifying the transport of POPs to the Arctic was used by Commoner *et al.* (2000). Using emission estimates for PCDDs for North America in combination with air mass trajectories calculated with the HYSPLIT (Hybrid Single-particle Lagrangian Integrated Trajectory) air transport model, they derived source-receptor relationship for several communities in the Canadian High Arctic. The model estimated the amount of PCDD emitted by each of 44 000 sources that is deposited at each of eight receptor sites in Nunavut over a one-year period, 1 July 1996–30 June 1997. A fairly limited number of North American sources outside of Nunavut were found to be responsible for almost all of the PCDDs deposited on that territory.

C.4.3 Models in support of defining the persistent organic pollutants attributes of persistence and long-range transport potential

In the context of the recent UNEP and UN-ECE protocols on persistent organic pollutants the need arises for identifying and classifying chemicals as "persistent" in the environment and as "being subject to long-range transport" (LRT) (Rodan *et al.*, 1999; Mackay *et al.*, 2001). The requirement for predicting LRT potential is clearly prompted by the occurrence of many POPs in the Arctic environment far from any sources, and the desire to identify chemicals that may have the potential to experience a similar transport to remote regions. Early on multimedia mass balance models have been identified as playing an important role in that context. The purpose of the models is to take into account the influence of a chemical's environmental phase distribution on its ability to persist and be transported over long distances. The environmental phase distribution is influenced by a large number of factors related to both environmental and chemical characteristics, and is not necessarily intuitive or easily comprehended. A multimedia model provides a tool to take most of these factors into account in a transparent, objective, and reproducible manner (van de Meent *et al.*, 2001). Accordingly, there has been intense scientific activity within the last few years aimed at developing model-based approaches to persistence and LRT potential assessment (Scheringer *et al.*, 2001a).

C.4.3.1 Models to estimate persistence

The UN protocols on POPs rely on the comparison of a chemical's degradation rates or half-lives in individual environmental media with threshold values when evaluating whether that chemical should be considered persistent or not. By accounting for the multimedia partitioning of a substance, a model allows the relative importance of degradation rates in the various environmental compartments to be weighed. The case can be made that the half-life of a chemical in a medium into which it does not partition to a significant extent is irrelevant. It is further possible to calculate a single persistence indicator for the environment as a whole, variably referred to as global life-time (Müller-Herold, 1996; Müller-Herold *et al.*, 1997) overall persistence (Scheringer, 1996; 1997; Webster *et al.*, 1998; Gouin *et al.*, 2000), or characteristic time (Bennett *et al.*, 1999). Such a parameter allows an easier comparison between, and thus ranking of, different chemicals with respect to persistence.

An overall persistence indicator τ, may be defined as the ratio of the total amount of a chemical in a system M_{tot} and its total loss rate from that system by means of degradation N_{Rtot}. If a multimedia mass balance model has "n" compartments, "M_i" is the amount of chemical in compartment "i", and "k_i" is the degradation rate in that compartment, "τ" can be calculated using:

$$\tau = \frac{M_{tot}}{N_{Rtot}} = \frac{\sum_{i}^{n} M_i}{\sum_{i}^{n} (M_i \cdot k_i)}$$

Inverting this to obtain the overall rate constant gives:

$$k_{tot} = \frac{1}{\tau} \sum_{i}^{n} \frac{M_i}{M_{tot}} \cdot k_i$$

which implies that the overall rate constant is the sum of the rate constants in the individual compartments, weighted by the mass fraction present in each of these media. If all rate constants are assumed to be first order or pseudo-first order, the overall persistence becomes independent of the release rate. Essentially, all existing approaches for calculating an overall persistence indicator using multimedia models adopt this relationship. Differences relate to the way the weights or mass fractions are derived, to the number and nature of compartments that make up the multimedia model, to whether transport processes that result in a permanent and irreversible removal from the sphere accessible to organisms should be included, and to whether or how environmental variability is treated.

The simplest method of estimating the distribution between the model compartments is to assume equilibrium partitioning among the model compartments (Müller-Herold, 1996; Gouin *et al.*, 2000). A persistence estimation based on equilibrium partitioning, however, may severely underestimate the real environmental half-life of a chemical by assuming partitioning into a phase with fast degradation, which the chemical cannot reach because of significant intermedia transfer resistances (Wania, 1998b). Most approaches assume a steady-state situation, which allows for deviation from equilibrium, yet assumes that the rates of chemical input and loss are equal at all points in time. Bennett *et al.* (1999) calculated and compared an overall persistence for several chemicals using equilibrium, steady-state, and transient or dynamic distributions. They concluded that a persistence "based on the steady-state distribution in the environment closely approximates the dynamic characteristic time for a range of chemicals and can be used in decisions regarding chemical use in the environment". Fenner *et al.* (2000) showed that scenarios involving a steady-state assumption and a pulse release gave the same media distribution, and thus yielded equivalent estimates for persistence.

In principle, even fairly complex models such as those described in Table C.4.1 can be used to estimate an overall persistence indicator. Notably, zonally averaged global models have been used to derive an overall global persistence or half-life, using both steady-state (Klein, 1999) and non-steady-state model versions (Wania and Mackay, 1999b). The latter study using Globo-POP revealed that such a global life-time estimate is time variant (seasonally and over a longer time scale), and strongly depends on where a chemical has been discharged into the global environment.

C.4.3.2 Models to estimate potential for long-range transport

Multimedia model-based approaches for assessing the potential of a chemical to undergo long-range transport, usually take the form of calculating a chemical's characteristic distance or spatial range. A chemical with a large spatial range has high long-range transport potential (LRTP). Often it is assumed that long-range transport occurs only in the atmosphere. It is increasingly obvious, however, that water can be a medium of long-range transport for some chemicals, both in the form of rivers and ocean-currents.

Van Pul *et al.* (1998) suggested the calculation of a spatial range in air using a one-compartmental modelling approach for the atmosphere, avoiding the need to describe surface compartments by using net deposition velocities. Using a multimedia approach, Bennett *et al.* (1998) calculated a travel distance in air using a model involving a Lagrangian cell of air passing over a stationary soil/plant surface in a one dimensional system, such that the stationary terrestrial phase in contact with the air cell is at steady-state with respect to the air concentration. Beyer *et al.* (2000) extended these studies and showed that existing multimedia models, such as the EQC-model (Mackay *et al.*, 1996), can be used directly to calculate travel distances in air and water. If emission is assumed to occur into a mobile phase, the travel distance in that mobile phase "M" can be calculated using:

$$L_M = \frac{u \cdot M_M}{N_{Rtot}} = \frac{u}{\sum_i^n \frac{M_i}{M_M} \cdot k_i} = \frac{u}{k_{eff}}$$

where "u" is the advective velocity of the mobile phase, "M_M" is the amount in the mobile phase and "N_{Rtot}" and "k_i" as defined previously. This equation does not apply if emission occurs to a medium other than the mobile phase or to more than just the mobile phase. The denominator in above equation has been termed an effective loss rate constant, "k_{eff}" (Bennett *et al.*, 1998). In the case of emissions into the mobile phase only travel distance and overall persistence are then related with this simple relationship:

$$L_M = u \cdot \frac{M_M}{M_{tot}} \cdot \tau$$

A more thorough discussion of how persistence and long-range transport potential are related can be found in Scheringer *et al.* (2000). The travel distance is an expression of how far the chemical can be transported within the average life-time "τ" it has available for transport. This is controlled by the speed of phase movement u and the fraction of the chemical in the transport medium "M_M/M_{tot}". As is the case in the overall persistence calculation, a multimedia model can be used to derive these compartmental mass fractions. Again, various approaches to derive mass proportions, based either on equilibrium (e.g., Müller-Herold and Nickel, 2000) or steady-state distributions (e.g., Beyer *et al.*, 2000), have been used.

Whereas most simple LRTP assessment models assume steady concentrations of 25°C, Beyer *et al.* (submitted) investigated the effect of temperature on the estimated LRTP of persistent organic chemicals. The effect was significant (LRTP can change by a factor of six in response to a 25 K temperature change), sometimes counter-intuitive, and highly dependent on chemical characteristics. The LRTP can monotonically increase or decrease with increasing temperature, or it can have a maximum in the environmental temperature range, but it cannot have a minimum. The results suggest that as a chemical is transported to a cooler region its transport potential may increase or decrease and further atmospheric transport might be inhibited or accelerated (Beyer *et al.*, submitted).

Most approaches to LRTP assessment have focussed on the transport in the atmosphere, although Beyer *et al.* (2000) showed that the very same principles can be applied to assess LRTP in oceans. Beyer and Matthies (2001) presented an approach that is capable of assessing the LRTP in coupled air-ocean systems, suggesting that the LRTP in such systems is higher than if transport in each of the two phases is considered separately.

Even more complex approaches to LRTP assessment rely on the global models characterised in Table C.4.1. The spatial range as defined by Scheringer (1996; 1997) measures the spatial extent of a concentration distribution by quantifying the distance (or area) that contains 95% of the weight of this concentration distribution. It can, for example, be calculated with the ChemRange model described in Table C.4.1. For each of the three types of compartments in that model, a different spatial range can be calculated from the spatial distribution among the various units. The rationale is that on a global scale, chemical dispersion is more appropriately described with a macro-diffusive approach than the one-dimensional advective approach described previously. A similar macro-diffusive approach is adopted in the Globo-POP model by Wania and Mackay (1995; 2000a) or the steady-state version of that model by Klein (1999). The one-dimensional advective approach is more suited for describing atmospheric transport on a smaller scale. A more thorough discussion of this issue and a comparison of advective and diffusive model types can be found in Beyer *et al.* (2001) and Bennett *et al.* (2001b).

The calculation of an Arctic Contamination Potential (Wania *et al.*, 2001) briefly discussed previously, is really an approach to assessing an LRTP that explicitly takes into account the potential to deposit and accumulate in Arctic regions. The cold condensation potential by Scheringer *et al.* (2000) calculated with the CliMoChem Model (Table C.4.1) also intends to describe a chemical's tendency to move to and accumulate in polar regions. It indicates whether a chemical's concentration profile increases towards the poles, and is calculated as the ratio of concentration in the polar zone and the minimum concentration between equator and polar zone.

C.4.3.3 Approaches to reduce input data requirements for the model based evaluation.

The approaches discussed so far require usually a minimum of chemical-specific data, namely: (1) degradation rates in each of the various model compartments, and (2) physicochemical partitioning data (vapour pressure, solubility in water and solubility in organic matter or octanol; alternatively, the air-water and the octanol-water partition coefficient). If the mass proportions are to be calculated using a non-equilibrium distribution, information is also needed on the mode of entry into the environment for the assessment of overall persistence. For some chemicals, this information, in particular the media-specific degradation half-lives and the mode-of-entry may not be readily obtained. For this reason several suggestions have been made to reduce the input data requirement in the assessment.

One such method suggests identifying those chemicals among the large population of substances to be screened which, based on their physicochemical properties, will partition almost exclusively into a single environmental medium. This identification could be based on compartmental distributions calculated using multimedia models. No actual modelling is required if tables or graphs are used to display which combination of physicochemical properties results in single-media distributions (Gouin *et al.*, 2000). If, for such chemicals, the degradation half-life in the medium of predominance is lower than the critical threshold for the overall persistence, the chemical could be classified as non-persistent without information on the degradation rate in the remaining compartments. It is possible to devise a simple graphical method by which a chemical is located on one or more two-dimensional surfaces, and a decision made based on location (Gouin *et al.*, 2000).

An intriguing approach suggested by Bennett *et al.* (2000; 2001a) is to run a number of chemicals through a multimedia model and determine their persistence. A Classification and Regression Tree (CART) analysis is then done to identify the key input properties which can control persistence in the form of a decision tree. The tree can be used for other chemicals not in the "training set" and, although there may be errors in the evaluation, they are predictable in magnitude.

Another approach which reduces the number of required input parameters has been suggested by Pennington (2001a; b). It consists of several guidelines that determine the pertinence of a degradation half-life in the calculation of overall persistence and characteristic travel distance. These guidelines are based on either mass fractions estimated using an equilibrium model, thus resembling the approach by Gouin *et al.* (2000), or directly on the underlying partition coefficients, most notably the Henry's Law Constant and the octanol-water partition coefficient. The error associated with eliminating the use of a particular degradation half-life compared to using all degradation data can be estimated.

C.4.3.4 Other aspects of long-range transport potential and persistence assessment

Some of the compounds that are found as contaminants in the Arctic, such as DDE or heptachlor epoxide, have not been emitted to the environment directly, but are formed from precursors such as DDT and heptachlor. When evaluating the persistence or LRTP of such precursors it becomes imperative to take into account the persistence and LRTP of the transformation products. Fenner *et al.* (2000) showed how the persistence of transformation products can be included in the overall persistence calculation of the parent compound, whereas Quartier and Müller-Herold (2000) showed this for the spatial range of a transformation product.

Considering the multitude of different models and approaches to persistence and LRTP assessment it becomes imperative to compare predictions from various multimedia fate models. In one such study by Wania and Mackay (2000b) several research groups were asked to calculate parameters characterizing overall persistence and long-range transport potential for a set of 26 diverse chemicals using their respective multimedia modelling approaches (12 for persistence and 8 for LRT). Absolute values for persistence and LRTP differed substantially among models, suggesting that these characteristics cannot be defined independently of the model used to generate the estimate. However, all persistence models based on non-equilibrium steady-state multimedia calculations, which include a sediment compartment and all long-range transport potential models, produced similar relative persistence and LRTP rankings. This suggests that the use of specific chemicals as benchmarks for persistence and long-range transport appears to be a viable option in assessing LRTP and P. Other studies compared two or more modelling approaches in greater detail (Bennett *et al.*, 2001b; Beyer *et al.*, 2001; Pennington, 2001a; b; Wania and Dugani, 2003).

Chapter D

Special Topic

D.1 The interaction of climate change with contaminant pathways to and within the Canadian Arctic

Authors:

R. Macdonald (1); T. Harner (2); and J. Fyfe (3).

Affiliations:

1. Institute of Ocean Science, 9860 West Saanich Road, Sidney, BC, V8L 4B2, Canada;
2. Meteorological Service of Canada, 4905 Dufferin Street, Downsview, ON, M3H 5T4, Canada;
3. Meteorological Service of Canada, Centre for Climate Modelling and Analysis, 3964 Gordon Head Road, Victoria, BC, V8N 2Y2, Canada.

D.1.1 Introduction

In the previous Northern Contaminants Program assessment (CACAR I), physical pathways (Barrie *et al.*, 1997), ecosystem uptake (Muir *et al.*, 1996b), and human health (Gilman *et al.*, 1997) were discussed as separate topics. Because these three aspects of Arctic systems are strongly interrelated, this arbitrary compartmentalization makes it difficult to produce a holistic discussion of why contaminants end up where they do and how the environment conspires to put certain ecosystem components at risk far from any contaminating sources.

The chapter on physical pathways, in particular, was organized around a logical structure of compartments and fluxes (see Figure 1.2 in Barrie *et al.*, 1997). This allowed a coherent approach to discussing contaminant transfers from one compartment to another (e.g., atmosphere to ocean), an identification of where our most serious knowledge gaps lay, and a quantitative assessment of progress made in determining rates for the various arrows and inventories for the different boxes.

Although there were hints in the chapter on pathways that the environment contained certain features that might be very sensitive to change (e.g., the role of snow in contaminant scavenging, the sensitivity of gas — particle partitioning to temperature, the importance of ice cover in air-sea exchange), the 'box and arrows' approach implied that the Arctic could be understood as a rather rigid set of processes into which contaminants entered, and it was physical and chemical properties that set the rules for where a contaminant might end up. Nothing was said about how the Arctic's processes (e.g., hydrology, winds, currents, ice cover, organic carbon cycle, precipitation) might themselves be subject to change. We have learned, however, that system components can not only change (e.g., ice cover), but the pathways can also change (diversion of Russian rivers into the Canada Basin). This shows that some of the conclusions of CACAR I are inaccurate and some may be wrong.

This chapter, therefore, is a departure from the previous assessment in that it 1) identifies and discusses change in the Arctic in the context of contaminant pathways and 2) includes interactions that cross between physical and biological components of Arctic systems. This chapter draws on valuable insights developed under NCP Phase II work, such as the role of snow in scavenging contaminants, the mechanism by which Arctic lakes interact with their basins, the detailed chemistry behind mercury depletion events and the trophic distribution of contaminants in aquatic foodwebs. Most of the material in this chapter has been drawn from other literature, particularly concerning recent change in the Arctic. Much of the material remains speculative as to how such change will affect contaminants simply because the right kinds of measurements have yet to be made.

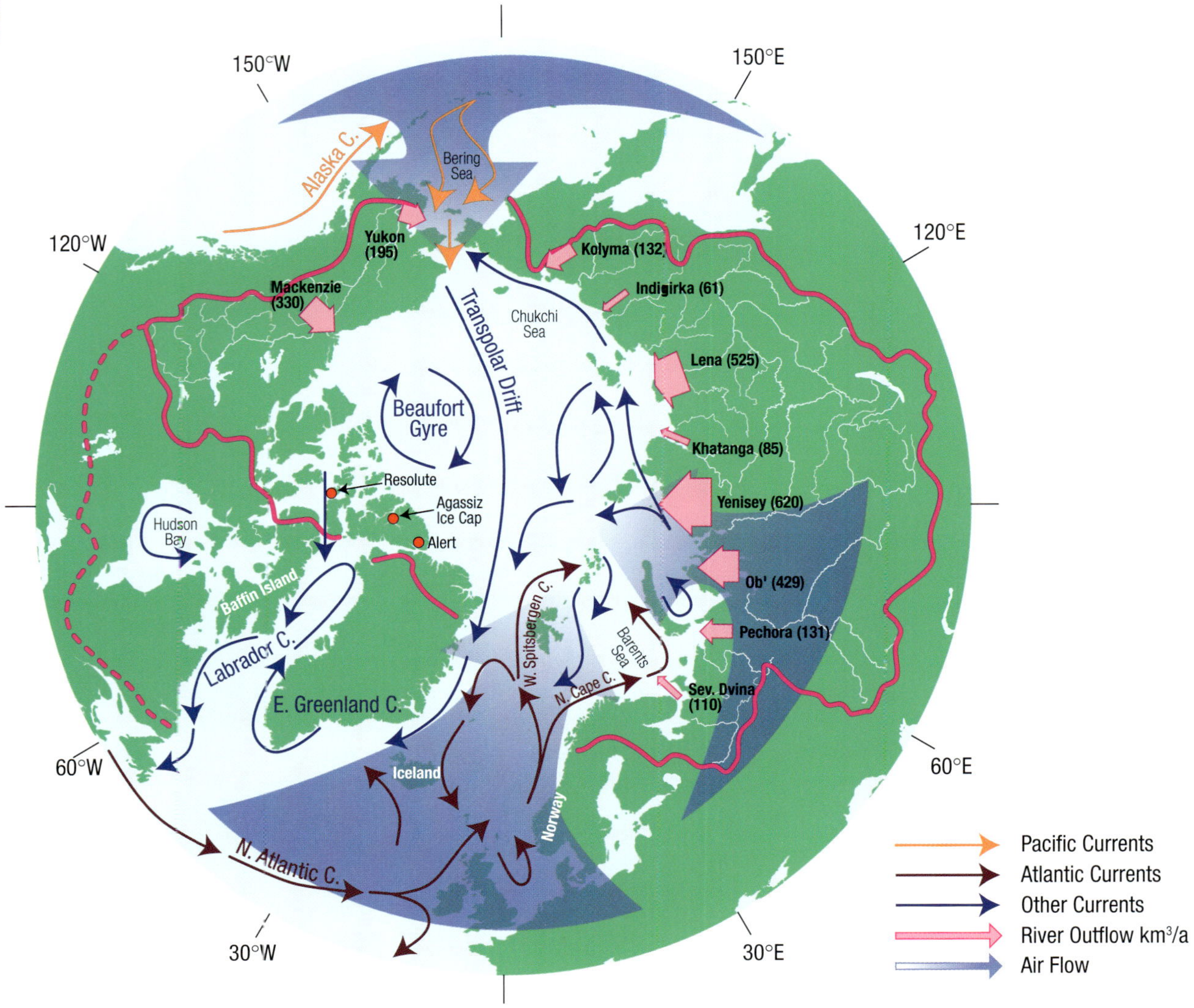

FIGURE D.1.1

The major physical pathways (wind, rivers, and ocean currents) that transport contaminants to the Arctic.

D.1.1.1 Contaminant transport in a changing world

Predicting how climate change will alter contaminant transport in the global environment poses an exceptional challenge. It requires a detailed knowledge of the physical and chemical properties of contaminants, an area in which much progress has been made during the past decade. It also requires a profound understanding of environmental pathways and how they might respond to change forced by, for example, alteration of the atmosphere's greenhouse gas composition. We presently lack this depth of understanding. It is clear that greenhouse gases such as CO_2, CH_4, N_2O and CFCs (chlorofluorocarbons) and aerosols are being released to the Earth's atmosphere by human activities (IPCC, 1990; 1995; 2002), that the world's hydrological cycle is being massively altered by damming for power and irrigation (Dynesius and Nilsson, 1994), and that humans are assaulting the global marine food web (Pauly *et al.*, 1998). Sufficient evidence has been assembled from paleorecords and much shorter instrumental observations to convince most — but not all — climate scientists that these kinds of disturbances will contribute to global change if they have not already done so (e.g., see IPCC, 1990; 2002; Parsons, 1996; Showstack, 2001).

As a component of the cryosphere with a large seasonal amplitude in ice and snow cover, the Arctic will be pivotal both as a region sensitive to change (sentinel) and as an exporter of change to other parts of the world (amplifier) (Aagaard and Carmack, 1989; Vörösmarty *et al.*, 2001; Walsh, 1995; Walsh and Crane, 1992). The 0°C isotherm is an especially important threshold of

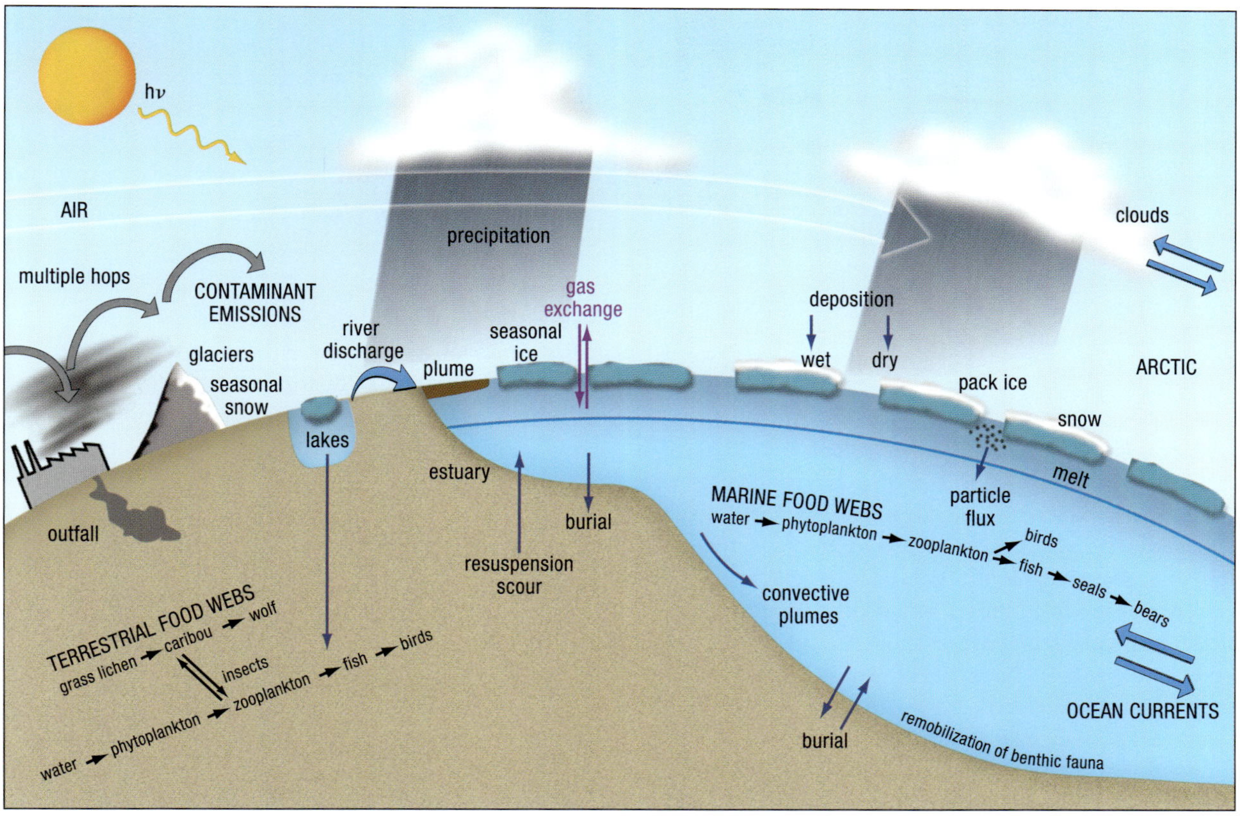

FIGURE D.1.2

A simplified schematic diagram showing how physical pathways deliver contaminants emitted from northern industrial regions to the Arctic where they may be concentrated in biota or removed through degradation and burial.

change because shifts between liquid and solid water have immense consequences for physical and biological systems and for humans.

Two major difficulties face us in understanding change in the Arctic and projecting its future. First, recent trends are difficult to detect and comprehend due to short, sparse instrumental records. This is especially true of the Arctic where any climate trend must be discriminated from an enormous seasonal amplitude in weather, and natural variation at time scales from annual to 5 years to centuries and longer (Fischer *et al.*, 1998; McGhee, 1996; Polyakov and Johnson, 2000; Proshutinsky and Johnson, 1997; Stirling *et al.*, 1999; Tremblay *et al.*, 1997; Vanegas and Mysak, 2000; Wang and Ikeda, 2001).

Second, our understanding of environmental processes in the Arctic is not sufficiently complete to allow a confident linking of primary changes (sea-level air pressure, air temperature, ice cover) with those of higher complexity but of much greater significance (e.g., thermohaline circulation, ecological structure and function, the hydrological cycle). Presently, these difficulties form an insurmountable hurdle to making confident projections of exactly how exposure of Arctic biota to contaminants will be affected by global change. Nevertheless, sufficient evidence has accumulated during the past decade to assure us that contaminant pathways can change and will continue to change.

Experience strongly warns us to expect the environment to deliver surprises (Macdonald *et al.*, 2001). One of those surprises is likely to be the abruptness of change — something for which the Arctic has recently shown unexpected predisposition (Alley *et al.*, 2002; Dickson, 1999; Johnson and Marshall, 2002; Macdonald, 1996; Mysak, 2001; Rothrock *et al.*, 1999). Other surprises lurk in the sometimes subtle and non-intuitive connections between global and regional pathways that put the Arctic at risk from contaminants in the first place (Figure D.1.1, AMAP, 1998; Macdonald *et al.*, 2000; Macdonald *et al.*, 2001).

To project how global change may alter contaminant exposure in the Arctic, we build on the foundation developed in previous assessments (AMAP, 1998; Jensen *et al.*, 1997), particularly those components addressing pathways (Barrie *et al.*, 1998; Macdonald *et al.*, 2000). We might then consider model predictions that the globe will experience a mean global air temperature rise of 3 to 5°C during the coming century (IPCC, 2002). Increased temperature will have direct effects on contaminants (enhanced volatility, more rapid degradation, altered partitioning between phases) and on the environment (loss of permafrost, change in the seasonal cycle of snow or ice).

However, a general air-temperature projection like this prepares us very poorly for addressing the more important changes that will inevitably occur in the connections between the Arctic and regions to the south and in the detail of pathways within the Arctic. In the context of Arctic ecosystems and humans who depend on them, it will not be temperature rise that counts but, rather the consequent change in the timing of seasons. During a contaminant's voyage to the Arctic, it may spend varying proportions of time in air, soil, water, ice, and in food webs or it may become degraded (Figure D.1.2). Each step along the path and every point of transfer can be altered by global change, which for a contaminant may mean dilution, concentration, bifurcation, shortcut, or delay.

D.1.1.2 Objectives of this chapter

This synthesis focuses on the question of how global change might alter contaminant pathways to and within the Canadian Arctic. As such, this is neither a review of change nor is it a review of contaminant pathways or of newly emerging contaminants. These topics have been, or are being, thoroughly reviewed elsewhere (AMAP, 1998; IPCC, 1995; 2002; Macdonald *et al.*, 2000; Ruddiman, 2000). Rather, we seek here observations and projections of global change that seem most likely to play a significant role in the life history of contaminants headed for a destination in Canada's Arctic.

D.1.1.3 Organization of this chapter

We first discuss change in *physical pathways*, emphasizing especially observations during the past decade. Recent dramatic changes, many of which can be related directly to variation in atmospheric pressure fields [popularly referred to as the Arctic Oscillation (AO) (Wallace and Thompson, 2002)], include winds, sea-ice drift and cover, ocean-currents, precipitation, and other environmental pathway components (Morison *et al.*, 2000; Serreze *et al.*, 2000). The contrast between high and low AO indices allows us to discuss with authority some of the ways in which the Arctic actually has changed.

Second, we discuss the likely consequences that these physical changes will have for *biological pathways*, noting that the organic carbon cycle and the food web are crucial pathway components for many contaminants (e.g., mercury, cadmium and most organochlorine compounds).

Third, we briefly review how *human responses* to global change are likely to alter contaminant pathways in the Arctic. The detailed review of physical and biological pathway changes that have occurred or are likely to occur then puts us in a position to discuss consequences for each category of contaminant of concern to the Northern Contaminants Program (metals, organochlorines, radionuclides, and hydrocarbons).

Finally, we discuss the difficulty of interpreting time-series data given that contaminant trends observed in sediment, water, air, and biota collected from the Arctic may include aspects related to contaminant emissions as well as aspects related to change in pathways.

D.1.2 Long-term change in the Arctic

D.1.2.1 The distant past

The Arctic is not static. Over the last 400 000 years, the earth has experienced four ice ages which are recorded or partially recorded in glacial ice accumulating in Antarctica (Petit *et al.*, 1999) and in Greenland (Dansgaard *et al.*, 1993; Sowers and Bender, 1995). The overall surface air-temperature change between glacial and interglacial periods is thought to have been about 12°C. Perhaps more significant than temperature, however, were the accompanying changes in continental ice masses, sea-ice climate, and global ecosystems.

Sea-ice cover in particular has proven to be a master variable in the equation of change. During the last glacial maximum, sea ice was locked within the Arctic and seasonal or perennial sea ice extended well south into the North Atlantic Ocean (Darby *et al.*, 1997; de Vernal *et al.*, 1993). The change from glacial to de-glacial to interglacial can be seen widely in Arctic sediments both in sedimentation rate and in the amounts and sources of organic material deriving from primary production or land (see for example, Darby *et al.*, 2001; Nørgaard-Petersen *et al.*, 1998; Phillips and Grantz, 1997; Stein *et al.*, 2001; 1994).

During the last glacial maximum, sea level dropped by about 120 m (Fairbanks, 1989). This exposed much of the Arctic Ocean's enormous continental shelves, forcing rivers to cut channels across them to enter the interior sea directly. Furthermore, the lower sea level cut the connection between the Arctic and Pacific Oceans and reduced or eliminated flow through most of the channels in the Canadian Archipelago. With sea-level rise, about 15,000 years ago the Bering land bridge was flooded (Hopkins, 1979) and then gradually submerged (Dyke *et al.*, 1996a) allowing the Pacific Ocean access to the Arctic Ocean. This sequence together with inundation of the continental shelves and the reflooding of the Archipelago channels must have had enormous consequences for the oceanography and regional biogeography of the western Arctic and the Archipelago (Dunton, 1992; Dyke *et al.*, 1996a; b; Héquette *et al.*, 1995).

Although climate during the past 10 000 years has been described as 'exceptionally stable' (Dansgaard *et al.*, 1993), it has actually continued to undergo substantial fluctuations. Indeed, it seems that very small shifts in temperature, perhaps only a degree or two, account for the so-called Medieval Warm Period (1100–1400 AD) and Little Ice Age (1450–1850 AD) that followed it. (For the relevance of these terms see Bradley and Jones, 1993; Crowley and Lowery, 2000.) Both of these minor and sporadic deviations in the temperature record had dramatic consequences for humans — especially those living on the margins of Northern oceans (Alley *et al.*, 2002; McGhee, 1996; Ogilvie and Junsson, 2000). During the past two centuries, small changes in the distribution of ice and watermass distribution have continued to affect humans and ecosystems, sometimes leading to migration or abandonment of locations, but certainly requiring adaptation (McGhee, 1996; Miller *et al.*, 2001; Vibe, 1967).

For most of the past 10,000 years (the Holocene), climate change was not accompanied by the added complexity of anthropogenic pollutants. Over the past two millennia, however, and especially during the past two centuries, Arctic glacial ice has recorded the transient rise of virtually every contaminant emitted by human activities (Boutron *et al.*, 1995; Boutron *et al.*, 1998; Gregor *et al.*, 1995; Hong *et al.*, 1994; Masclet and Hoyau, 1994; Rosman *et al.*, 1997). Included in these contaminants are the green house gases (GHGs) that force atmospheric temperature change (Petit *et al.*, 1999). The dramatic rise of these GHGs during the past several decades make future projections based on past climates subject to uncertainty.

D.1.2.2 The present and future

The 20th century was the warmest in the Arctic for the past 400 years (Figure D.1.3 Overpeck *et al.*, 1997). The Intergovernmental Panel on Climate Change (IPCC) suggests that over the past century, the global mean surface temperature has increased by about 0.3 to 0.6°C, attributable mostly to human activities, and will probably further increase by 1.4 to 5.8°C between 1990 and 2100

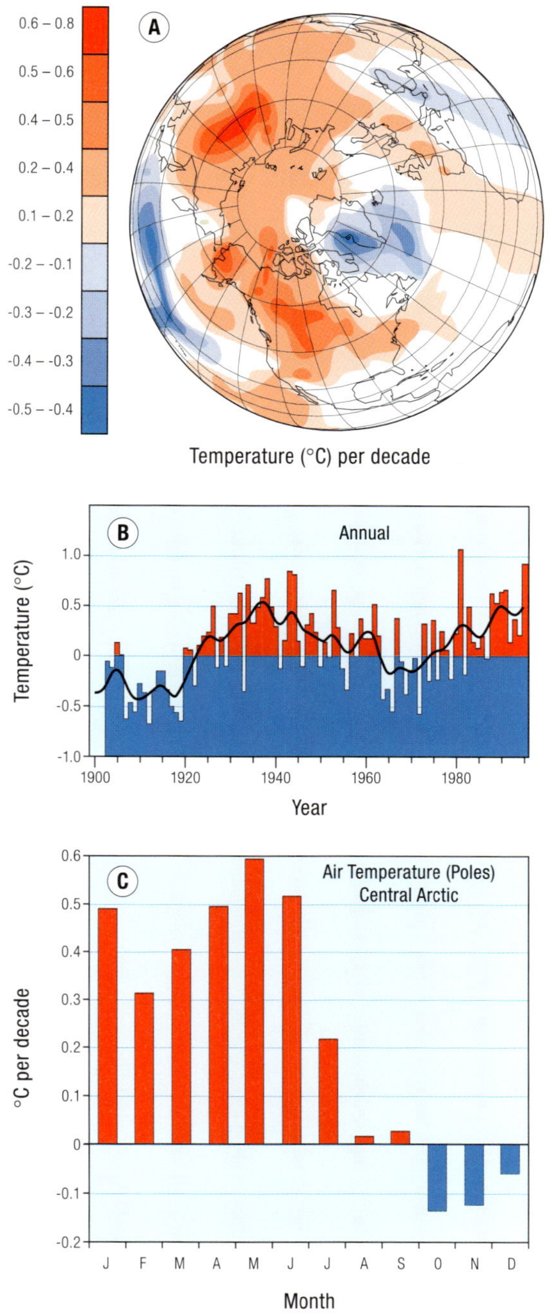

FIGURE D.1.3

Temperature trends for the Arctic showing a) The annual surface temperature trends over the Northern Hemisphere expressed as rates of change for the period 1961–90 [courtesy of the Climate Monitoring and Data Interpretation Division of the Atmospheric Environment Service (Stewart *et al.* 1998)]; b) temperature anomalies (55–85°N) for 1900–1995 evaluated against the average for 1951–1980 showing the high temperatures of the late 1980s and 1990s are matched by equally high temperatures during the 1930s and 40s (adapted from Serreze *et al.* 2000); and c) the trend by month in surface air temperature of the central Arctic Ocean for the period 1979–1995 illustrating that recent warming is mainly a winter-spring phenomenon (adapted from Serreze *et al.* 2000).

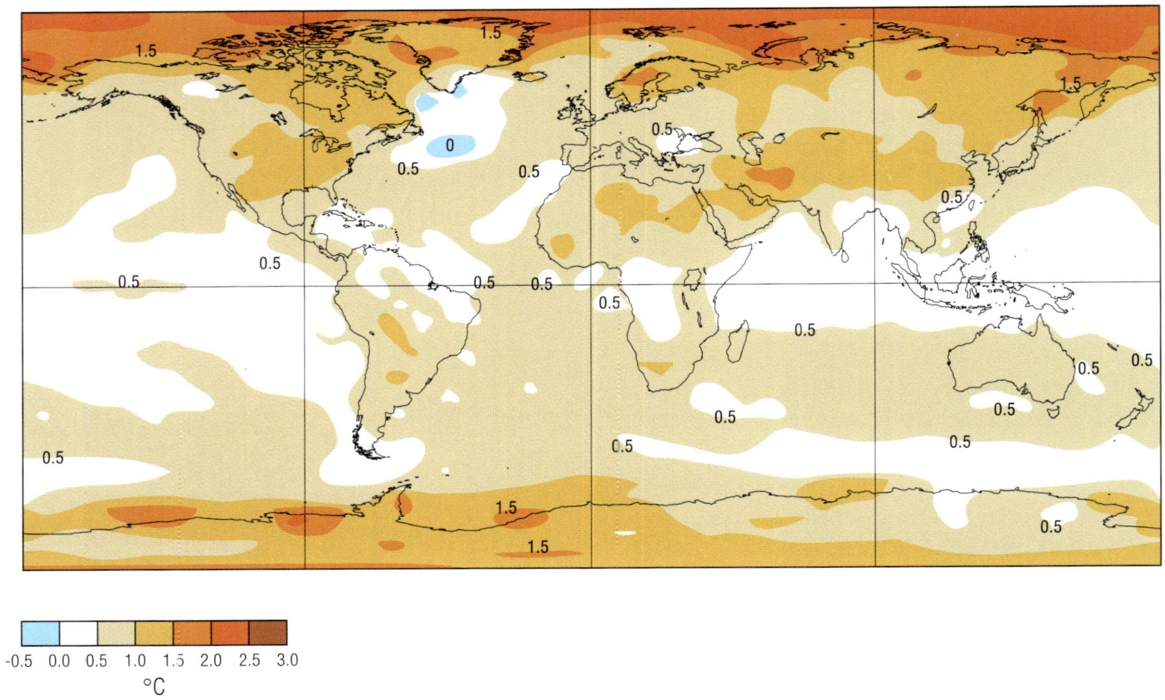

FIGURE D.1.4

Change in surface air temperature for 2020–30 relative to 1990–2000 as projected by the Canadian Centre for Climate Modeling and Analysis (CCCMa). Global warming is expected to have an uneven geographic distribution with the Arctic experiencing the highest projected warming (courtesy of CCCMa and see Zwiers 2002).

(Houghton *et al.*, 1995; IPCC, 1995; 2002; Showstack, 2001). According to models, warming will be more pronounced in polar regions (Figure D.1.4), perhaps 5°C or more near the pole and 2 to 3°C around the margins of the Arctic Ocean, with a decreasing temperature contrast between poles and the equator (Manabe *et al.*, 1992; Mitchell *et al.*, 1995; Zwiers, 2002).

For gases emitted to the atmosphere, climate warming will increase inter-hemispheric exchange times, mixing times, and mean transit times by perhaps 10% (Holzer and Boer, 2001). Furthermore, the greatest warming will occur in autumn-winter due to delay in the onset of sea-ice cover (Manabe *et al.*, 1992; Serreze *et al.*, 2000). Continental interiors will become dryer and sea level will continue to rise, perhaps by a further 50 cm in addition to the estimated rise of 10–25 cm during the past century (Proshutinsky *et al.*, 2001; Serreze *et al.*, 2000).

Models predict that, after about 80 years of atmospheric CO_2 increase at 1% per year, precipitation will increase within the Arctic and subpolar regions to 0.5–1 m per year (Manabe *et al.*, 1992), making the Arctic a considerably 'wetter' place. Over the past four decades, the extent of sea-ice in the Arctic Ocean has decreased in summer by as much as 25% (Figure D.1.5a Vinnikov *et al.*, 1999). By the end of the 21st century, GHG forcing might produce an Arctic Ocean seasonally clear of ice (Figure D.1.5b, Flato and Boer, 2001).

Simulations based on GHG forcing predict that mean annual river discharge will increase by about 20% for the Yenisei, Lena, and Mackenzie Rivers, but decrease by 12% for the Ob (Arora and Boer, 2001; Miller and Russell, 1992). Furthermore, the projection that high-latitude rivers will undergo marked changes in amplitude and seasonality of flow due to decreased snowfall and earlier spring melt (Arora and Boer, 2001) may already have some support in observations (Lammers *et al.*, 2001).

The coupling of the runoff cycle with the hydrology of northern lakes is probably one of the points most sensitive to climate change (see for example, Vörösmarty *et al.*, 2001) although the processes involved are not yet understood well enough to make confident projections. If Arctic lakes become more 'temperate' in character, productivity will likely be enhanced due to less ice cover and more mixing and there will be greater opportunity for runoff to mix into the lake during freshet, further supporting a more vigorous aquatic food web.

With these primary changes, an acceleration of permafrost melting can be expected, which disrupts vegetation and enhances nutrient, organic carbon, and sediment loading of rivers and lakes (Vörösmarty *et al.*, 2001). The loss of sea ice in the marginal seas, particularly during the period of autumn storms, together with sea-level rise will promote further erosion of poorly bonded, low-gradient coasts.

D.1.3 Recent change in the Arctic and the Arctic oscillation

D.1.3.1 The Arctic oscillation

During the 1990s, a quiet revolution took place in our perception of the Arctic (Carmack *et al.*, 1997; Dickson, 1999; Johannessen *et al.*, 1995; Johnson and Polyakov, 2001; Kerr, 1999; Levi, 2000; Macdonald, 1996; Macdonald *et al.*, 1999a; Maslanik *et al.*, 1996; Maslowski *et al.*, 2000; McPhee *et al.*, 1998; Morison *et al.*, 2000; Morison *et al.*, 1998; Moritz *et al.*, 2002; Parkinson *et al.*, 1999; Polyakov and Johnson, 2000; Quadfasel *et al.*, 1991; Smith, 1998; Steele and Boyd, 1998; Vanegas and Mysak, 2000; Vörösmarty *et al.*, 2001; Walsh, 1991; Welch, 1998; Weller and Lange, 1999). Despite early evidence of cyclical change in northern biological populations and ice conditions (see for example, Bockstoce, 1986; Gudkovich, 1961; Vibe, 1967), the general view among many western physical scientists throughout the 1960s, 70s and 80s, was that the Arctic was a relatively stable place (Macdonald, 1996). This view has been replaced by one of an Arctic where major shifts can occur in a very short time, forced primarily by natural variation in the atmospheric pressure field associated with the Northern-hemisphere Annual Mode (NAM).

The NAM, popularly referred to as the Arctic Oscillation (AO) (Wallace and Thompson, 2002), is a robust pattern of the surface manifestation of the strength of the polar vortex (for a very readable description, see Hodges, 2000). The AO correlates strongly (85–95%) with the more commonly used indicator of large-scale wind forcing, the North Atlantic Oscillation (NAO) (the NAO is the normalized gradient in sea-level air pressure between Iceland and the Azores — see for example, Deser, 2000; Dickson *et al.*, 2000; Hurrell, 1995; Serreze *et al.*, 2000).

We will use the AO and NAO more or less interchangeably because they carry much the same information. We recognize, however, that in both cases the term 'oscillation' is a bit misleading because neither index exhibits quasi-periodic behaviour (Wallace and Thompson, 2002). The AO captures more of the hemispheric variability than does the NAO which is important because many of the recent changes associated with the AO have occurred in the Laptev, East Siberian, Chukchi, and Beaufort Seas — a long way from the NAO's centre of action (Thompson and Wallace, 1998). Furthermore, the Bering Sea and the Mackenzie Basin are both influenced to some degree by atmospheric processes in the North Pacific [e.g., the Pacific Decadal Oscillation (PDO), and see Bjornsson *et al.*, 1995; Niebauer and Day, 1989; Stabeno and Overland, 2001] whereas Baffin Bay ice climate appears to have an association with the Southern Oscillation (Newell, 1996) and the Canadian Archipelago and Hudson Bay probably respond to varied atmospheric forcings in ways not yet fully understood.

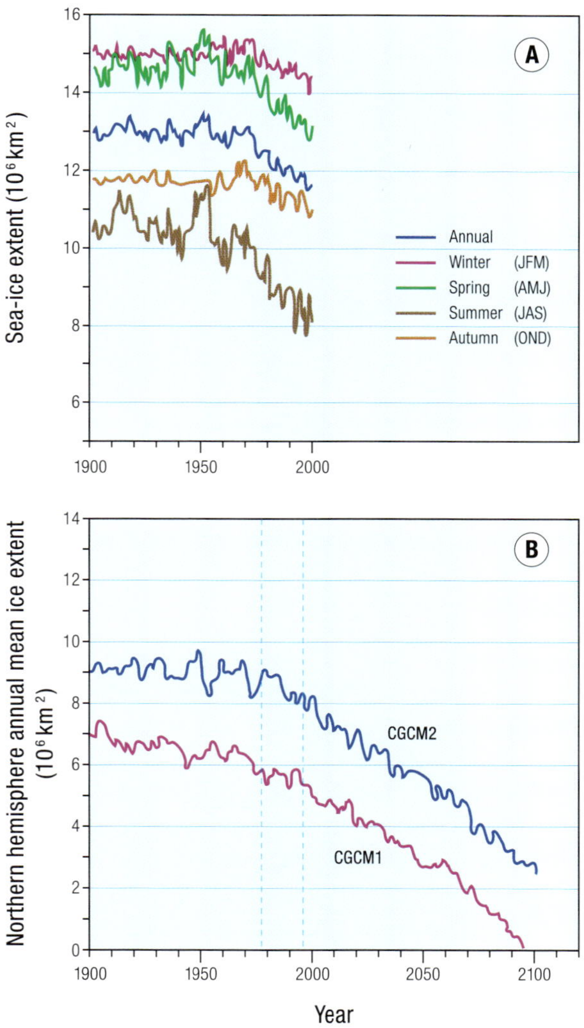

FIGURE D.1.5

a) Time series from 1900 to 2000 of annual and seasonal sea-ice extent in the Northern Hemisphere (Walsh and Chapman, 2000). b) Model projections of annual mean sea-ice extent for the Northern Hemisphere as simulated by CGCM1 and CGCM2 where the latter model differs from the former in mixing parameratization (figure from Flato and Boer 2001).

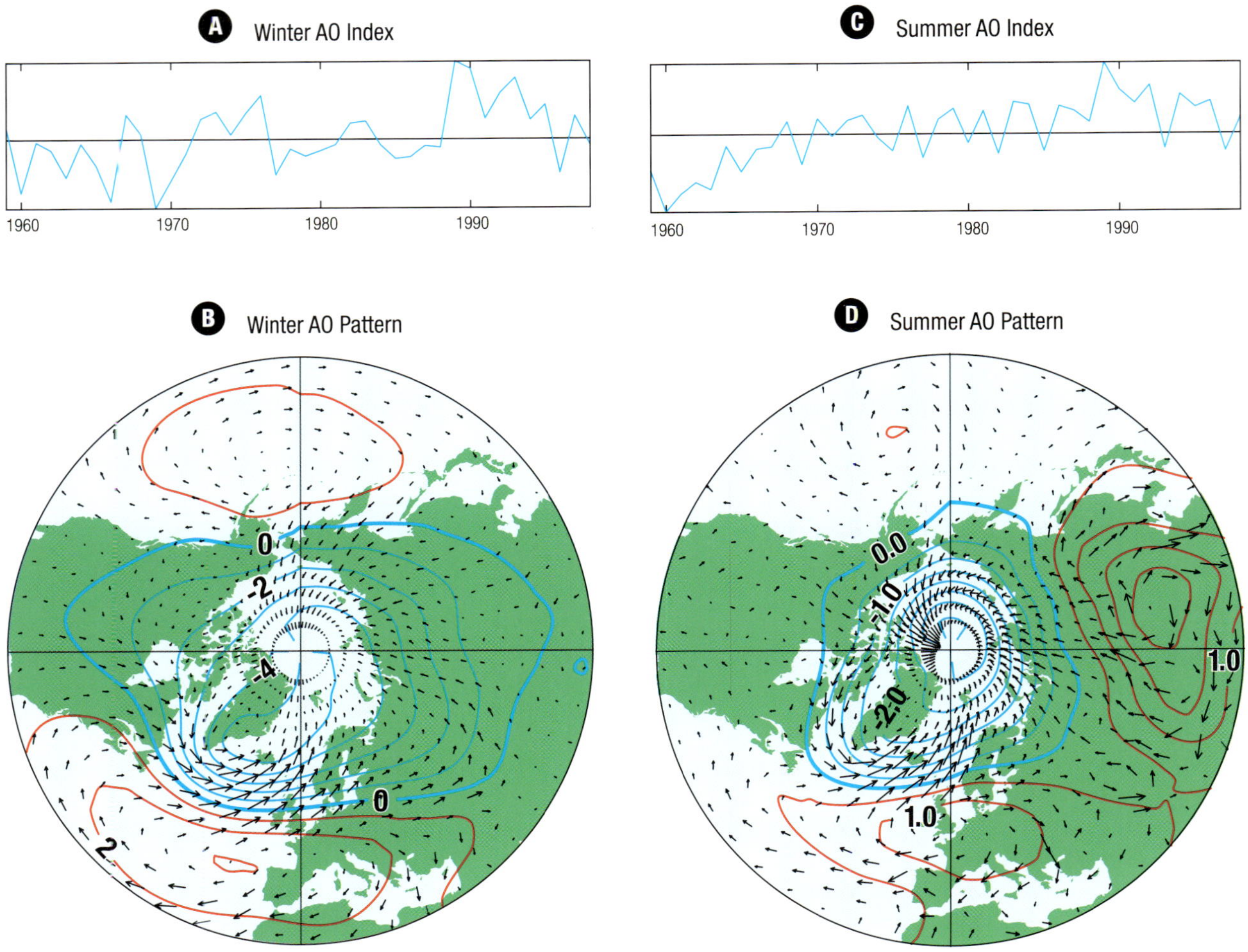

FIGURE D.1.6

a) The Arctic Oscillation index between 1958 and 1998 in winter with b) the winter pressure pattern associated with it and, c) the Arctic Oscillation index between 1958 and 1998 in summer with d) the summer AO pressure pattern associated with it.

Although the AO is an important component of change in Arctic climate, it contains only 20% of the variance in the atmospheric pressure field and much else can change in the atmospheric forcing. Because the AO is usually 'smoothed', it does not well represent events and short-term variations that are known to be important in the delivery of contaminants to the Arctic and that may also be important locally in forcing ice and surface water motion (see for example, Sherrell *et al.*, 2000; Welch *et al.*, 1991). It is important to note, therefore, that one of the projections of climate change is an increase in cyclone activity; extreme events may become a prominent component in atmospheric transport in the coming century [an example may have been the earliest significant rain event on record observed widely throughout the Canadian Archipelago on May 26, 1994 and see Graham and Diaz, 2001; Lambert, 1995].

In about 1988–89, the AO entered a positive phase of unprecedented strength (Figure D.1.6a, c). The pressure distribution pattern of the AO for winter and summer (Figure D.1.6b, d) shows that a positive shift is manifest by lower than average sea-level pressure (SLP) distributed somewhat symmetrically over the pole (the blue region on Figure D.1.6b, d) and higher pressures over the North Atlantic and North Pacific in winter and over Siberia and Europe in summer. As might be expected by examining the AO SLP pattern (Figures D.1.6b, d), when the AO index is strongly positive, conditions become more "cyclonic" — i.e., atmospheric circulation becomes more strongly counter-clockwise (Proshutinsky and Johnson, 1997; Serreze *et al.*, 2000).

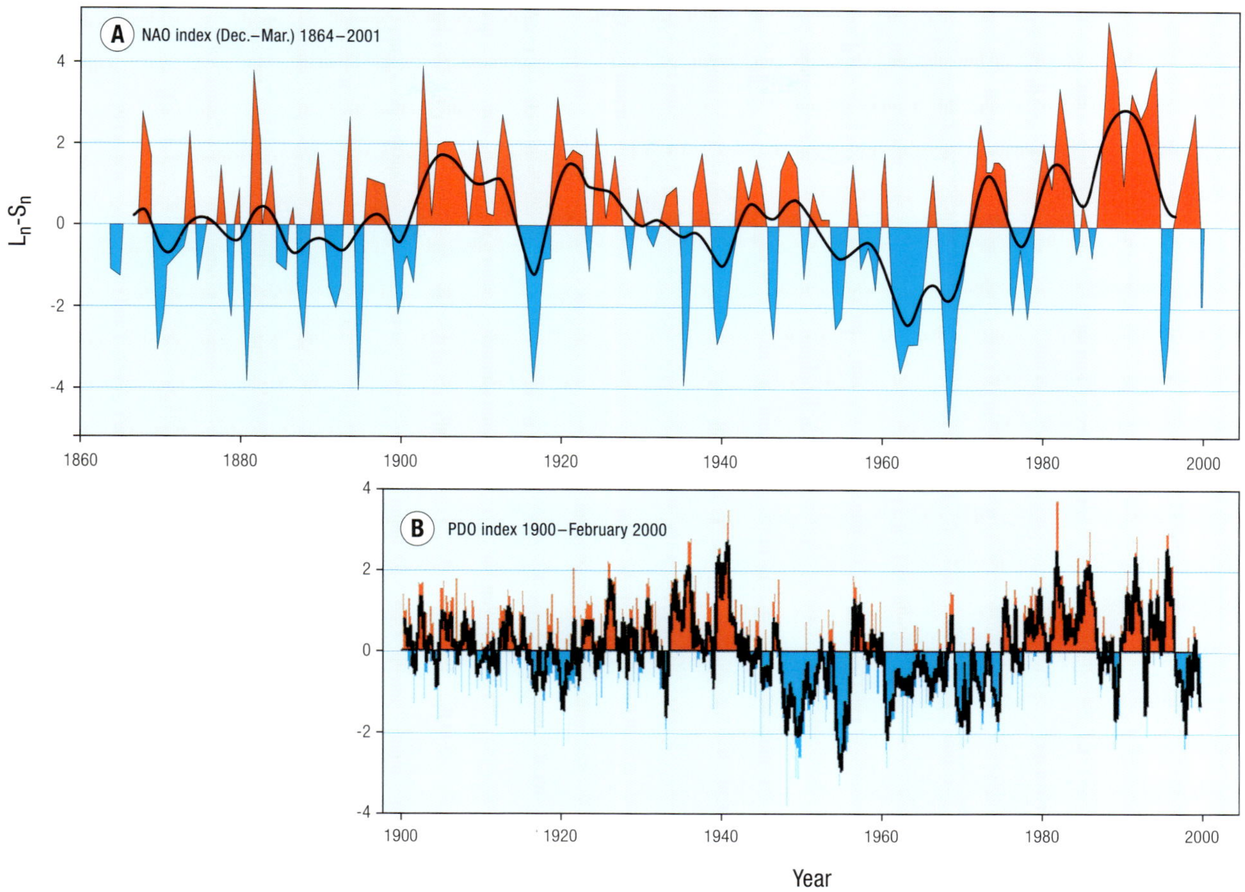

FIGURE D.1.7

The North Atlantic Oscillation (NAO) index from 1860 to 2000 (source Hurrell 2002).

In discussing change, it is important to distinguish between variability, which can occur at a variety of time scales (Fischer *et al.*, 1998; Polyakov and Johnson, 2000), and trends caused by, for example, GHG warming. It has been argued that locking the AO into a positive position might actually be one way that a trend forced by GHGs can manifest itself in the Arctic (Shindell *et al.*, 1999). Others suggest, however, that the extraordinary conditions of the 1990s were produced naturally by a reinforcing of short (5–7 year) and long (50–80 year) time-scale components of SLP variation (Polyakov and Johnson, 2000; Wang and Ikeda, 2001), and that GHG forcing will affect the mean property fields rather than alter the AO itself (Fyfe, 2002; Fyfe *et al.*, 1999). Longer records of the NAO index (Figure D.1.7a) indeed suggest that there have been other periods of high AO index during the past 150 years (e.g., 1900–1914) but none so strong as that experienced during the early 1990s. Recent data suggest that the AO index has decreased and that the Arctic system has to some degree begun to return to AO^- conditions (Björk *et al.*, 2002; Boyd *et al.*, 2002).

The contrast in conditions between the Arctic as we knew it in the 1960s, '70s, and '80s (generally low AO index) and the Arctic as it became in the early 1990s (exceptionally high AO index) provides an extraordinary opportunity to investigate how the Arctic might respond to climate change. The similarity between climate-change projections and AO^- induced change suggests that examining the differences between AO^- and AO^+ states should provide insight into the likely effects of climate change forced by GHG emissions. Variation in SLP as recorded by the AO index demonstrates that the Arctic exhibits at least two modes of behaviour (Morison *et al.*, 2000; Proshutinsky and Johnson, 1997) and that these modes cascade from SLP into wind fields, ice drift patterns, watermass distributions, ice cover, and probably many other environmental parameters.

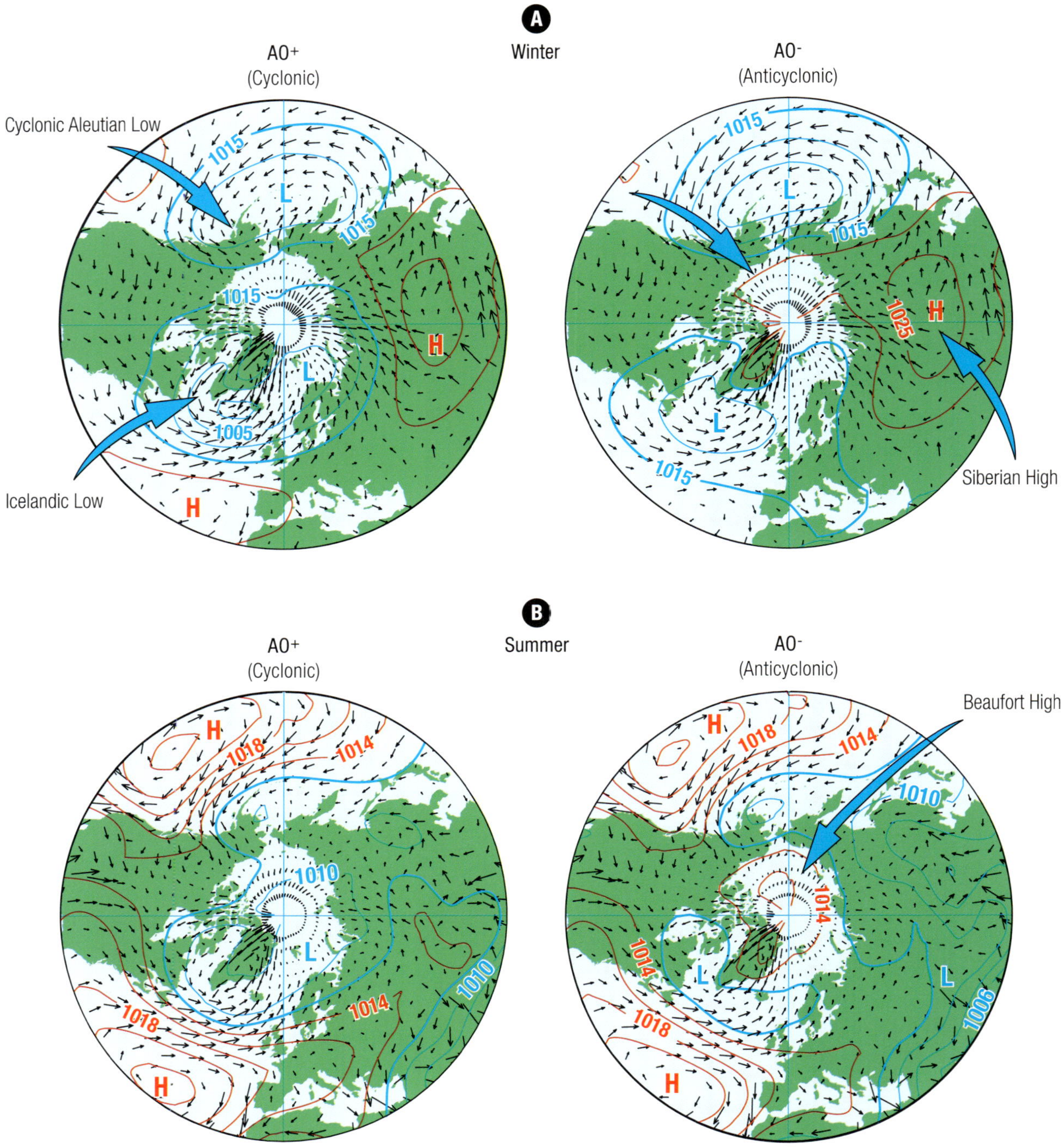

FIGURE D.1.8

Atmospheric pressure fields and wind stream lines for a) winter with the Arctic Oscillation index strongly positive (left) and strongly negative (right) and b) summer with the Arctic Oscillation index strongly positive (left) and strongly negative (right).

The Arctic is to a large degree constrained by overarching structures and processes in how it can respond to change. As illustrated in previous assessments, the Arctic Ocean is, and will remain, a 'mediterranean' sea much influenced by land-ocean interaction and with restricted exchange with other oceans (Figure D.1.1). Topography, bathymetry, and global distribution of salinity in the ocean, require that water from the Pacific Ocean will predominantly flow *in* to the Arctic and the shallow sill at Bering Strait (50 m) guarantees only surface water will be involved in the exchange. Pacific water will remain above Atlantic Layer water which is denser. Deep basin water communicates predominantly with the Atlantic Ocean through the deep connection at Fram Strait.

Ocean circulation within the Arctic is tightly tied to bathymetry through topographic steering of currents (Rudels *et al.*, 1994). Considering these kinds of constraints, rapid change can occur in ocean-current pathways or in the source or properties of the water carried by currents when, for example, fronts shift from one bathymetric feature to another (McLaughlin *et al.*, 1996; Morison *et al.*, 2000), when a given current strengthens or weakens (Dickson *et al.*, 2000), when source-water composition alters (Smith *et al.*, 1998; Swift *et al.*, 1997), when relative strength of outflow varies between the Canadian Archipelago and Fram Strait (Macdonald, 1996), but not by reversal of flow in boundary currents or reversal of mean flow in at Bering Strait or out through the Archipelago.

Change associated with the NAM forces us to consider large-scale variability in the Arctic, to recognize in greater detail that physical pathways can change rapidly, and to assess potential effects of GHG emissions against this naturally variable background.

D.1.3.2 Winds

Winds transport contaminants directly to the Arctic by delivering volatile and semi-volatile chemicals from the south within a few days (Bailey *et al.*, 2000; Barrie *et al.*, 1998; Halsall *et al.*, 1998; Hung *et al.*, 2001; Stern *et al.*, 1997). Winds also provide the primary forcing for ice and ocean (Mysak, 2001; Proshutinsky and Johnson, 1997) thereby indirectly affecting transport by these two media as well.

To understand how swings in the AO affect atmospheric circulation, we have constructed AO^+ and AO^- wind field/SLP maps for winter (Figure D.1.8a) and summer (Figure D.1.8b) by adding (AO^+) or subtracting (AO^-) patterns in Figures D.1.6b, d to the mean pattern for the period of record in the time series (Figures D.1.6a, c; 1958–1998). The changes discussed below can be considered generally as the difference between conditions during the 1960–70s (low AO index) and during the early 1990s (high AO index) (see for example, Proshutinsky and Johnson, 1997).

In winter (Figure D.1.8a), the lower tropospheric circulation is dominated by high pressures over the continents and low pressures over the northern Pacific (Aleutian Low) and Atlantic Oceans (Icelandic Low). The Siberian High tends to force air on its western side into the Arctic, acting as an effective atmospheric conduit from industrialized regions of Siberia and Eastern Europe into the high Arctic. The high pressure ridge over North America then forces air southward giving a net transport out of Eurasia into the Arctic, across the Arctic, and south over North America. The Icelandic Low produces westerly winds over the eastern North Atlantic and southerly winds over the Norwegian Sea, providing a conduit for airborne contaminants from eastern North America and Europe to reach the Arctic rapidly. Finally, the Aleutian Low tends to steer air that has crossed the Pacific from Asia up into Alaska, Yukon, and the Bering Sea (Bailey *et al.*, 2000; Li *et al.*, 2002; Wilkening *et al.*, 2000). During winter, these three routes into the Arctic — southerlies in the Norwegian Sea (40%), Eastern Europe/Siberia (15%), and Bering Sea (25%) account for about 80% of the annual south-to-north air transport (Iversen, 1996).

With a higher AO index (Figure D.1.8a, left side), the Icelandic Low intensifies and extends farther into the Arctic across the Barents Sea and into the Kara and Laptev Seas (Johnson *et al.*, 1999). This has the effect of increasing the wind transport east across the North Atlantic, across Southern Europe, and up into the Norwegian Sea. During high NAO winters, westerlies onto Europe may be as much as 8 m/s (~700 km/day) stronger (Hurrell, 1995). At the same time, strong northerly winds are to be found over the Labrador Sea (Mysak, 2001).

The extension of the Icelandic Low into the Arctic also implies an effect of the AO on the storm track. During the strong positive AO conditions of the early 1990s, there was a remarkable increase in the incidence of deep storms to around 15 per winter and such storms penetrated farther into the Arctic (Dickson *et al.*, 2000; Maslanik *et al.*, 1996; Semiletov *et al.*, 2000). Increased cyclone activity increases poleward transport of heat and other properties carried by the air masses involved. Anomalous southerly airflow over the Nordic Seas enhances the connection between industrial regions of North America and Europe and the Arctic. At the same time, increased cyclones provide the means to transfer contaminants from atmosphere to surface through precipitation. Deep within the Arctic, the high SLP ridge that extends across the Canada Basin during AO^- conditions (the Beaufort High), disappears and withdraws into Russia (Johnson *et al.*, 1999; Morison *et al.*, 2000). Of note, the Pacific mean atmospheric pressure field and wind patterns appear to change little between strong positive and strong negative phases of the AO in winter. Penetration of air from the Pacific into the Arctic is hindered by the mountain barrier along the west coast of North America, where intensive precipitation also provides a mechanism to remove contaminants and aerosols to the surface (Figure D.1.8a).

Summer pressure fields and air-flow patterns are markedly different from those of winter (compare Figures D.1.8a with Figure D.1.8b). In summer, the continental high-pressure cells disappear and the oceanic low pressure cells weaken, with the result that northward transport from low latitudes weakens (Figure D.1.8b). According to Iversen (1996) summer accounts for only 20% of the annual south-to-north air transport [southerlies in the Norwegian Sea (10%), Eastern Europe/Siberia (5%), and Bering Sea (5%)]. The streamlines show that winds provide a means to transport contaminants from industrialized North America and Europe to the North Atlantic but penetration into the Arctic weakens. In the North Pacific, there remain atmospheric pathways to move air masses into the Gulf of Alaska from the east coast of Asia (Figure D.1.8b). In particular, during AO^+ conditions, the Beaufort High weakens or disappears (Figure D.1.8a, b), altering mean wind fields.

D.1.3.3 Surface air temperature

A strong trend of warming has been observed in the Arctic for the period from 1961–1990 (Figure D.1.3a). This warming, which has been especially evident over northwestern North America and Siberia, has been accompanied by cooling in northeastern Canada and Baffin Bay. An almost identical pattern of warming to that shown in Figure D.1.3a is produced by taking the difference between mean surface air temperatures on high AO index days and low AO index days (Wallace and Thompson, 2002).

Due to an extensive temperature record collected from drifting buoys, manned drifting stations, and land stations, direct relationships can be drawn between air surface temperature over sea and land in the Arctic and the changes in pressure field discussed above. Over the period 1979–97, a trend of +1°C per decade was found for winter surface air temperature (SAT) in the eastern Arctic Ocean, offset by a trend of -1°C per decade in the western Arctic Ocean (Rigor *et al.*, 2000). However, in spring almost the entire Arctic Ocean shows significant warming — as much as 2°C per decade in the eastern Arctic, where a trend toward lengthened melt season was also observed.

The trend of increasing SAT over the ocean is matched by temperature increases over Arctic land masses of 2°C per decade during winter and spring (Figure D.1.3a). Long records of temperature anomalies since 1900 (Figure D.1.3b) clearly show the warming trend since the 1970s but note also the similar episode of warming that occurred in the 1930–1940s.

Taken together, the trends in surface air temperature over the central Arctic Ocean suggest that warming has occurred predominantly during January to July (Figure D.1.3c; (Serreze *et al.*, 2000)). Over half of the changes in SAT for Alaska, Eurasia and the eastern Arctic Ocean have been ascribed to the AO but less than half in the western Arctic (see Dickson *et al.*, 2000; Rigor *et al.*, 2000; Serreze *et al.*, 2000). The temperature changes associated with the AO are considered large enough to have an immediate effect on polar circulation (Morison *et al.*, 2000).

D.1.3.4 Precipitation and runoff

Precipitation is a key pathway for contaminant transport (Figure D.1.2). Rain or snow scavenge aerosols and gases from the atmosphere to deposit them at the surface (Li *et al.*, 2002; Macdonald *et al.*, 2000; Mackay and Wania, 1995; Wania and Mackay, 1999b). Scavenging by precipitation may presently be relatively weak due to the desert-like conditions of the Arctic. For example, mean precipitation for the Arctic Ocean is estimated at about 25.2 cm yr^{-1} and evaporation about 13.6 cm yr^{-1} for a net moisture flux to ground of 11.9 cm yr^{-1} (Barry and Serreze, 2000). The net precipitation over land in the Arctic drainage basins is apparently greater as implied by runoff yield [precipitation minus evaporation (P-E)] being estimated at 21.2 cm yr^{-1} from the network of gauged discharge by rivers (Lammers *et al.*, 2001).

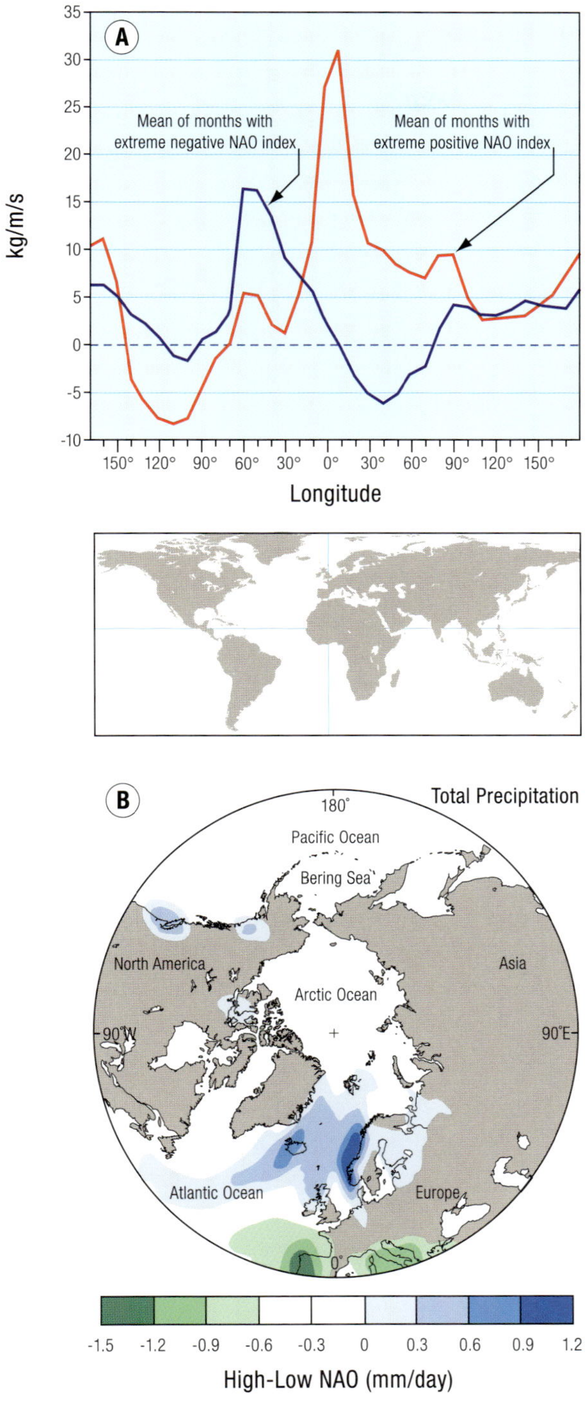

FIGURE D.1.9

The effect of the Arctic Oscillation on precipitation in the Arctic showing a) vertically integrated meridional flux crossing 70°N in winter for low NAO conditions (blue) and high NAO conditions (red) and b) the change in winter precipitation between low-index NAO and high-index NAO in mm per day (modified from Dickson *et al.*, 2000).

Given the changes in winds (Figure D.1.8a, b) and temperature that occur with shifts in the AO, it is to be expected that precipitation and evaporation within the Arctic will also be affected either in amount or seasonality (Serreze *et al.*, 2000). Due to sparse networks and short time series, it is difficult to assess with confidence the spatial or temporal variation of precipitation within the Arctic. Nevertheless, records suggest that precipitation has been increasing over Canada's North by about 20% during the past 40 years (Serreze *et al.*, 2000).

The increase in southerly winds in the Norwegian Sea in winter and penetration of cyclones from the Atlantic into the Barents, Kara, and Laptev Seas, when the AO (or NAO) index is high is reflected in increased moisture flux and precipitation during autumn and winter especially near 10°W–50°E (Figure D.1.9a, b Dickson *et al.*, 2000; Serreze *et al.*, 2000; Walsh, 2000). The composite difference in precipitation (Figure D.1.9b), which may actually underestimate the change between index extremes (Dickson *et al.*, 2000), shows an increase of up to 15 cm/yr precipitation in the Norwegian-Greenland Sea atmosphere-ocean conduit to the Arctic when the NAO is strongly positive. The response over the central Arctic to changes in the AO/NAO index is clearly much less but it is likely that conditions there become wetter during index highs (Serreze *et al.*, 2000). Overall, it is estimated that precipitation minus evaporation (P-E) north of 70°N is 36% higher during periods of high index compared to low index (Serreze *et al.*, 1995). Over central and northern Canada, flux of moisture out of the Arctic increases when the AO/NAO is high, but toward the western Beaufort moisture flux into the Arctic again increases (Figure D.1.9a).

Whether precipitation falls as snow or as rain, and how long snow covers surfaces are crucially important components of climate that control the interaction of contaminants with the hydrological cycle (Helm *et al.*, 2002b; Macdonald *et al.*, 2002a; Wania, 1997). Snow cover varies from a maximum of about 46 x 10^6 km^2 to as little as 4 x 10^6 km^2 (Serreze *et al.*, 2000). As might be predicted from recent warming trends over Arctic land masses (Figure D.1.3a), there is evidence that the average area covered by snow has been decreasing by about 2% (450 000 km^2) per decade between 1979 and 1999 (Armstrong and Brodzik, 2001).

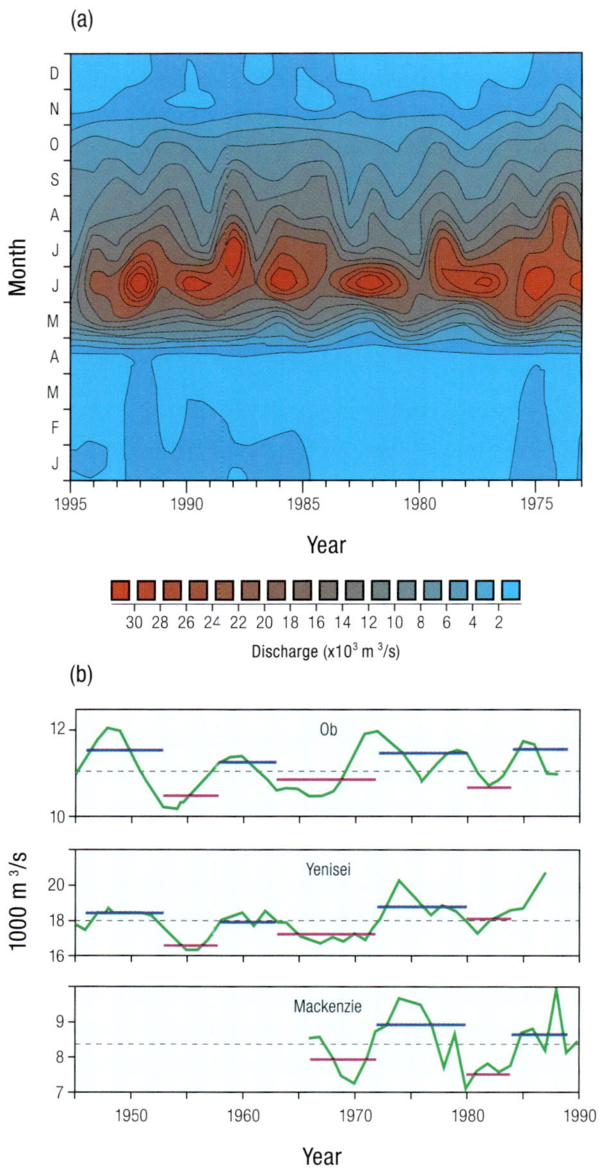

FIGURE D.1.10

a) The monthly river discharge for the Mackenzie River from 1973 to 1995 inferred from observations at Arctic Red River and the Peel River (after Stewart, 2000) and b) the relationship between the cyclonic (AO^+)/anticyclonic (AO^-) states and the three-year smoothed monthly river discharge for the Ob, Yenisei, and Mackenzie Rivers (after Johnson *et al.*, 1999).

A correlation between the AO and snow cover in Eurasia from 1972 to 1997 suggests a change from minimum to maximum AO index is accompanied by a loss of about 4×10^6 km^2 of snow cover which could account for much of the trend described above (Vörösmarty *et al.*, 2001). The snow-cover anomalies plotted by Armstrong and Brodzik (2001) show a downward step in about 1989 when the AO index sharply increased. The late 1980s up to at least 1998 has been identified as a period of low snow cover for both Eurasia and North America, with the largest changes occurring in spring-summer (Serreze *et al.*, 2000). For Canada, there has been a decrease in snow depth, especially in spring, since 1946 (Brown and Goodison, 1996).

Precipitation minus evaporation (P-E) integrated over a drainage basin should be equivalent to river discharge for the basin. Arctic rivers exhibit large interannual variation (Semiletov *et al.*, 2000; Shiklomanov *et al.*, 2000; Stewart, 2000) making it difficult to link river flow to precipitation or temperature trends or to climate variables like the AO. For example, Shiklomanov *et al.* (2000) suggested little change in mean annual discharge for Arctic rivers between the 1920s and 1990s, whereas Semiletov *et al.* (2000) found recent increases for several Eurasian rivers and Lammers *et al.* (2001) found evidence of increased winter discharge from rivers in Siberia and Alaska in the 1980s compared to the 1960s–70s.

Within Canada, the Mackenzie Basin has undergone an exceptional warming between 1961 and 1990 (Figure D.1.3a, Stewart *et al.*, 1998). Nevertheless, increased basin temperatures are not obviously recorded in this river's hydrology (Figure D.1.10a, Stewart, 2000) or in other Arctic rivers (Shiklomanov *et al.*, 2000). Instead, one sees evidence of 3- to 4-year periodicity in peak flow and alterations in the seasonal shape of the hydrograph with higher flows delayed well into August suggesting changes in the total annual discharge and its seasonality and possibly changes in the relative importance of the river's sub-drainage-basins.

Such patterns appear to be only partially related to the AO as evidenced by significant correlations between runoff and precipitation for the Mackenzie Basin and variation in North Pacific storm tracks (Bjornsson *et al.*, 1995). These correlations hint that trans-Pacific transport of airborne contaminants may be the dominant component of contaminant loading for northwestern Canada, something that is borne out by time-series data collected at Tagish in Yukon (Bailey *et al.*, 2000). Hence, change related to atmospheric contaminant pathways for this region is more likely to come from the North Pacific, and it is possible that such change might be manifested as an alteration in the domains of influence for Pacific air masses versus Eurasian air masses.

The discharges for the Ob, Yenisei, and Mackenzie Rivers appear to show a positive relationship with the North Pole pressure anomaly with a lag in discharge of about 0.5–0.7 years (Figure D.1.10b, Johnson *et al.*, 1999) but such a relationship runs counter to the enhanced precipitation observed during AO+ (cyclonic) conditions (Figure D.1.9). Even if all variation in Arctic river discharge at the 4–5 year time scale is assigned to shifts in AO/NAO index, the maximum effect on annual flow would be about 5–15% which is within the range of interannual variability (for example see Johnson *et al.*, 1999; Semiletov *et al.*, 2000).

D.1.3.5 The ocean

D.1.3.5.1 Sea ice

D.1.3.5.1.1 Sea-ice cover

Sea ice controls the exchange of heat and other properties between atmosphere and ocean and, together with snow cover, determines the penetration of light into the sea. Ice also provides a surface for particle and snow deposition, a biological habitat above, beneath, and within the ice and, when it melts in summer, stratification of the upper ocean.

During the 1990s, the science community recognized with some alarm that Arctic sea ice had been undergoing retreat over the past three decades. Observed changes include: a reduction in area covered by sea ice (Johannessen *et al.*, 1999; Levi, 2000; Maslanik *et al.*, 1996; Parkinson *et al.*, 1999; Vinnikov *et al.*, 1999), an increase in the length of the ice melt season (Rigor *et al.*, 2002; Smith, 1998), a loss of multiyear ice (Johannessen and Miles, 2001), a general decrease in the thickness of ice over the central Arctic Ocean (Rothrock *et al.*, 1999), and an increase of ice melt in the Beaufort Sea (Macdonald *et al.*, 1999a; McPhee *et al.*, 1998).

Analyses of satellite data from 1978–87 revealed a decrease in Arctic sea-ice area of about 2.4% per decade (Gloersen and Campbell, 1991). Subsequent analyses have revised that figure upward to 4% per decade for the period from 1987–94, with an estimated average loss during the entire period (1978–97) of 3% per decade. This amounts to the disappearance of 0.3 x 10^6 km^2 per decade of sea ice (Cavalieri *et al.*, 1997; Parkinson *et al.*, 1999). Shelves of the Eastern Arctic contribute significantly to the estimated ice losses. Multi-year ice is apparently being lost at an even greater rate, estimated at 7% per decade, partly replaced by first-year ice (Johannessen and Miles, 2001).

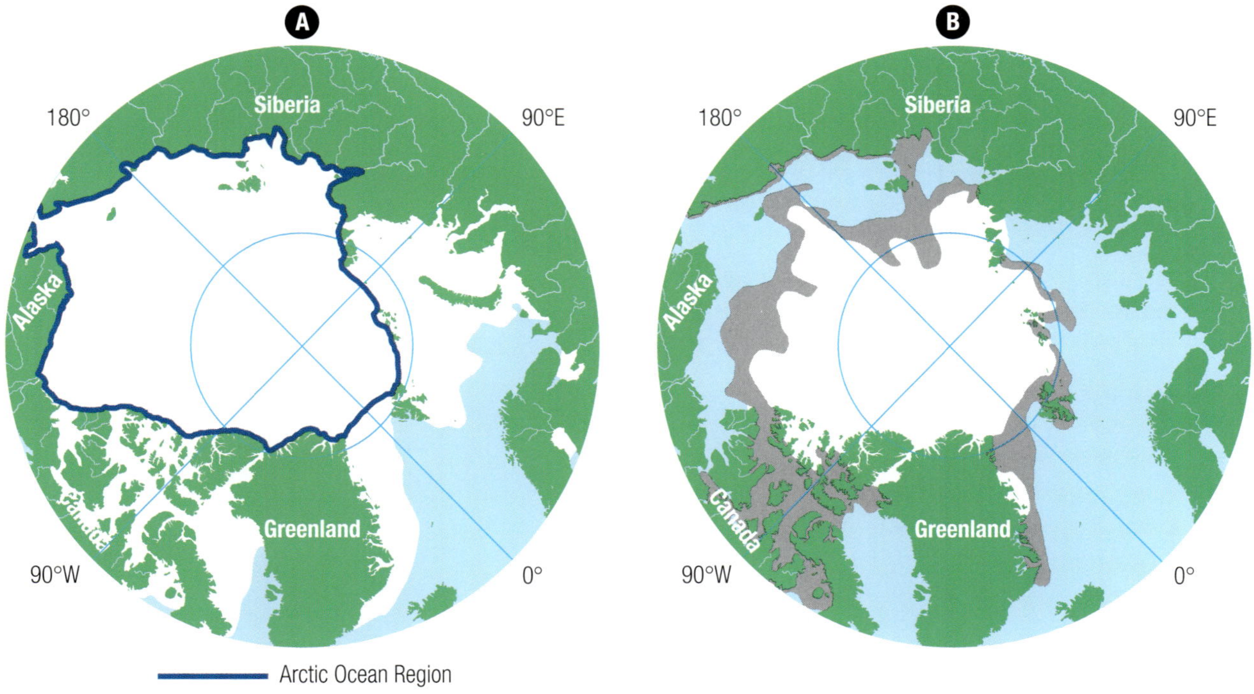

FIGURE D.1.11

Arctic sea ice cover in a) Winter maximum and b) Summer minimum as derived from satellite imagery (Johannessen and Miles, 2001). Grey areas indicate partial ice cover (greater than 15%). The blue line in a) outlines the area defined as Arctic Ocean for the ice area trends given in Figure D.1.11a (Parkinson *et al.*, 1999).

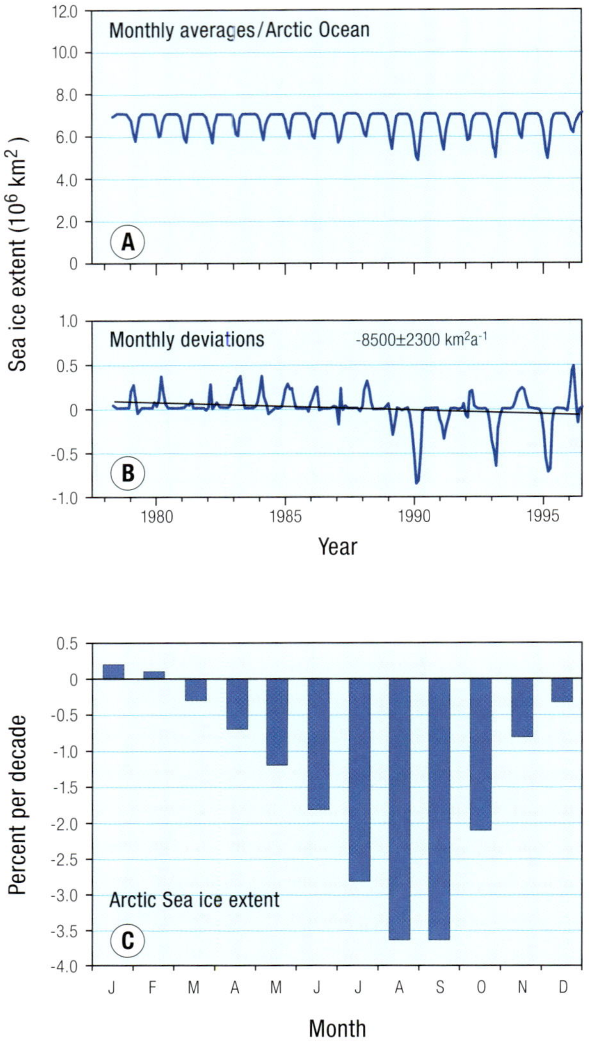

FIGURE D.1.12

Ice cover in the Arctic Ocean showing a) monthly averages in ice cover for the period from 1979 to 1996 for the Arctic Ocean (as defined by blue line in Figure D.1.11a, [after Parkinson *et al.* (1999)], and b) the monthly deviations in ice cover for the same area showing the transition in 1990 to seasonally clear shelves, (after Parkinson *et al.*, 1999) and c) The change in sea ice extent in % per decade (1979–1995) showing the ice loss to be predominantly a late winter–summer phenomenon (after Serreze *et al.*, 2000).

The large seasonal amplitude in area covered by ice (Figure D.1.11) makes it difficult to assess trends. Furthermore, various authors have partitioned the Arctic differently to assess changes in ice cover or have compared different years and/or different seasons (Dickson *et al.*, 2000; Johannessen and Miles, 2001; Maslanik *et al.*, 1999; Maslanik *et al.*, 1996; Parkinson et *al.*, 1999). Despite these difficulties, the satellite data available since the late 1970s clearly imply a reduction of 2% per decade of total ice area in winter (Johannessen *et al.*, 1999) and a significant shift in the marginal seas toward first-year ice, which is easier to melt than multiyear ice because it is thinner and saltier. The total area of Arctic sea ice, including the marginal seas, varies from about 13 x 10^6 km^2 in winter to 5 x 10^6 km^2 in summer, and has shrunk by about 0.6 x 10^6 km^2 between 1978 and 1997 (Johannessen and Miles, 2001).

The Arctic Ocean component as defined by Parkinson *et al.* (1999) (Figure D.1.11), which is about 7 x 10^6 km^2 in area, began to exhibit a much stronger seasonal modulation in ice cover in about 1989 (Figure D.1.12a, b) with the East Siberian and Beaufort Seas experiencing anomalous areas of open water in late summer at various times during the 1990s (Maslanik *et al.*, 1999; Parkinson *et al.*, 1999; Rigor *et al.*, 2002; Serreze *et al.*, 1995). That the loss of sea-ice cover is predominantly a late winter — summer phenomenon is clearly shown by seasonal monthly trends for which June to October show the greatest change (Figure D.1.12c, Serreze *et al.*, 2000).

What part does the AO play in the variation of Arctic sea-ice distribution? The trends in ice cover with time (Figure D.1.12a, b) suggest that the wholesale clearing of ice from shelves is a phenomenon of the 1990s, timed with (Russian Shelves) or slightly delayed from (Beaufort Shelves) the shift to strong positive AO index in 1989. In the Beaufort Sea, Macdonald *et al.* (1999) used stable isotope data ($\delta^{18}O$) collected from 1987 to 1997 to show that amounts of ice melt contained in the water column increased substantially at the same time the AO index increased in 1989. During such conditions, the cyclonic circulation leads to greater ice divergence, more new ice in leads, enhanced heat flux, and reduced ridging, all of which imply thinning (Flato and Boer, 2001; Macdonald *et al.*, 1999a; Rigor *et al.*, 2002).

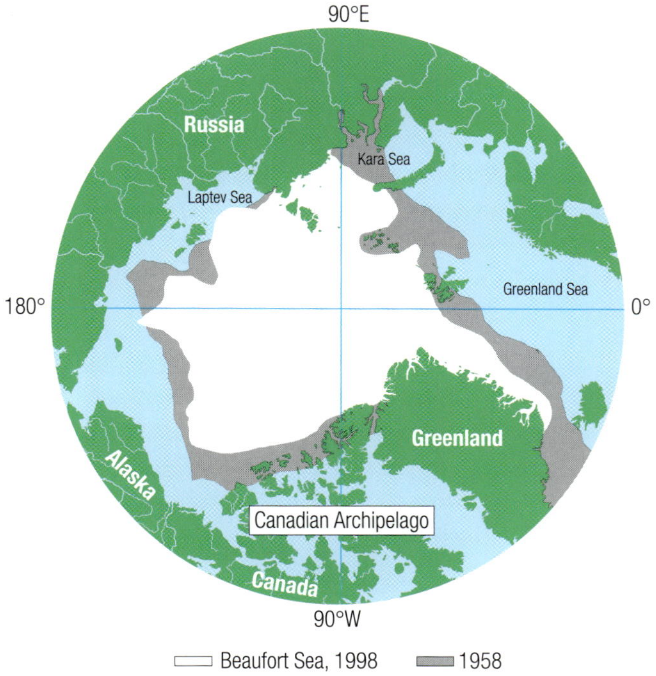

D.1.13

The contrast in ice cover between low AO⁻ index (1958) and high AO⁺ index (1990s) (Dickson *et al.*, 2000, Maslanik *et al.*, 1999, Serreze *et al.*, 1995).

Maslanik *et al.* (1996) draw the connection between increased cyclones and increased poleward transport of heat, which are observed during AO^+ conditions, and the absence of ice in late summer over the Siberian shelves. Based on results of a coupled sea/ice/ocean model, Zhang *et al.* (2000) suggest that there is a strong correlation between sea-ice thinning and the AO (~80%) due to dynamical effects, and that the Eurasian and Canada Basins respond differently to the AO forcing. The removal of the supply of ice from the Beaufort to the East Siberian Sea when the index goes strongly positive (discussed in section D.1.3.5.1.2) results in depletion of thick ice in the Eastern Arctic Ocean but may enhance thick ice buildup to the north of the Archipelago.

This point is important in light of the findings from repeated submarine surveys that ice thickness decreased over the central Arctic by about 1.3 m between 1958–1976 and the 1990s (Rothrock *et al.*, 1999; Wadhams, 1997; 2000). According to several models (Holloway and Sou, 2001; Polyakov and Johnson, 2000; Zhang *et al.*, 2000), the submarine observations may have been conducted primarily in that part of the ocean that underwent thinning in response to a shift to high AO index.

The conclusion of reduction of ice thickness, while valid for the domain of submarine measurements, is not necessarily true for the whole Arctic Ocean and an alternative hypothesis that ice thickness *distribution* changed but ice *volume* may not have changed in response to the AO needs to be carefully evaluated. The loss of ice cover between NAO^- and NAO^+ conditions is estimated at 590,000 km^2 in the Barents and Greenland Seas (Dickson *et al.*, 2000) and, if the remarkably open ice in the East Siberian Sea in 1990 and the Beaufort Sea in 1998 (Figure D.1.13) is a product of the high AO index of the early 1990s, then perhaps half as much again ice loss occurred over the Russian and North American Shelves due to AO forcing.

In light of the changes observed in ice cover during the 1990s, it is worth noting that over a century ago the Pacific whaling fleet experienced similar dramatic changes in ice conditions in the western Arctic. Extraordinarily open water from 1861 to 1867 perhaps led to a complacency that resulted in the loss of 32 ships, which were crushed in the ice along the Alaska coast in 1871 (Bockstoce, 1986). One is reminded of the caution given by Polyakov and Johnson (2000) that both short (decadal) and long (60–80 yr) time-scale variations are associated with the AO.

From data collected between 1979 and 1997, Rigor *et al.* (2000) determined that sea-ice melt begins in the marginal seas by the first week of June and advances rapidly to the pole in two weeks. Freezing begins at the pole on 16 August, returning to the marginal seas by late September, for a total melt season of about 58 days at the pole and 100 days toward the margin. Based on satellite data (SSMR and SSM/I) predominantly from the Beaufort Sea, Smith (1998) estimated that the length of the melting season has been increasing by about 5.3 days per decade during 1979–1996. In contrast, Rigor *et al.* (2000) found a shortening of the melt season in the western Arctic of 0.4 days per decade and an increase of about 2.6 days per decade in the Eastern Arctic. These trends in length of melt season parallel the general observations of a 1°C per decade decrease in temperature for the Beaufort compared to a 1°C per decade increase in the Eastern Arctic for the same time period.

Change in ice cover and its seasonality are especially important for contaminants like HCH, toxaphene and PCBs where air-sea exchange is a significant component of regional budgets (Macdonald *et al.*, 2000; Macdonald *et al.*, 2001). Furthermore, change in sea-ice cover, which alters light penetration and mixing, may also alter primary production and carbon flux (Gobeil *et al.*, 2001), which then alters the vertical flux of particle reactive and bio-active contaminants from the ocean surface to depth.

D.1.3.5.1.2 *Sea-ice drift*

General ice motion in the Arctic Ocean is organized into the Transpolar Drift (TPD) on the Eurasian Side of the ocean and the Beaufort Gyre in the Canada Basin (Barrie *et al.*, 1998). Although it has long been recognized that large-scale ice-drift patterns in the Arctic undergo change (Gudkovich, 1961), it was not until the International Arctic Buoy Programme (IABP) that sufficient data was available to map the ice drift in detail and thereby directly evaluate the role of the AO in changing ice drift trajectories.

The IABP data from 1979-98 suggest two characteristic modes of Arctic ice motion, one during low index (AO$^-$) and the other during high index (AO$^+$) [Figure D.1.14a, b, (Proshutinsky and Johnson, 1997; Rigor *et al.*, 2002)]. The ice-motion scheme shown by drifting buoys is reasonably well corroborated by models investigating the influence of the atmospheric variability inherent in the AO (Maslowski *et al.*, 2000; Polyakov and Johnson, 2000). There are two overarching differences between the two ice circulation modes; 1) during AO$^-$ conditions (Figure D.1.14a), ice in the trans-polar drift (TPD) tends to move directly from the Laptev Sea across the Eurasian Basin and out into the Greenland Sea whereas during strong AO$^+$ conditions (Figure D.1.14b), ice in the TPD takes a strong cyclonic diversion across the Lomonosov Ridge and into the Canada Basin (Mysak, 2001); and 2) during AO$^+$ conditions (Figure D.1.14b), the Beaufort Gyre shrinks back into the Beaufort Sea and becomes more disconnected from the rest of the Arctic Ocean exporting less ice to the East Siberian Sea and importing little ice from the region to the north of the Archipelago — a region known to contain the Arctic's thickest multi-year ice (Bourke and Garrett, 1987).

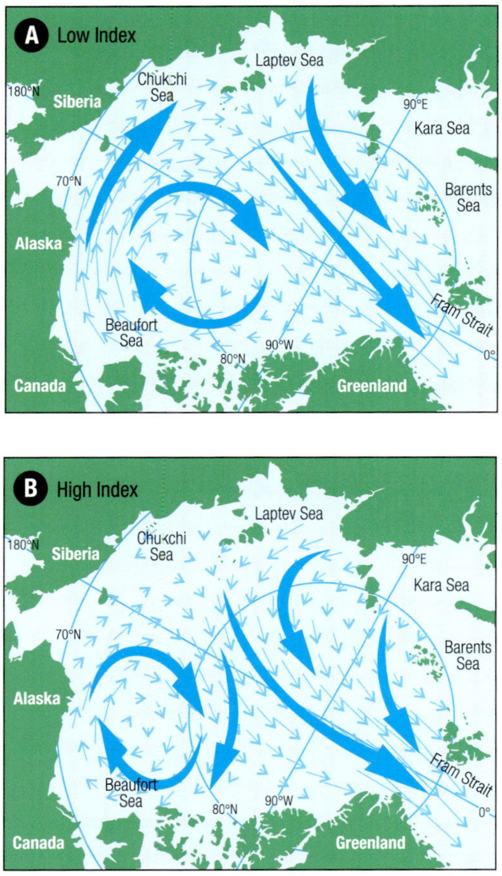

FIGURE D.1.14

Ice drift patterns for a) years with low AO$^-$ index (anticyclonic conditions) and b) high AO$^+$ index (cyclonic conditions) (after Maslowski *et al.*, 2000; Polyakov and Johnson, 2000; Rigor *et al.*, 2001).

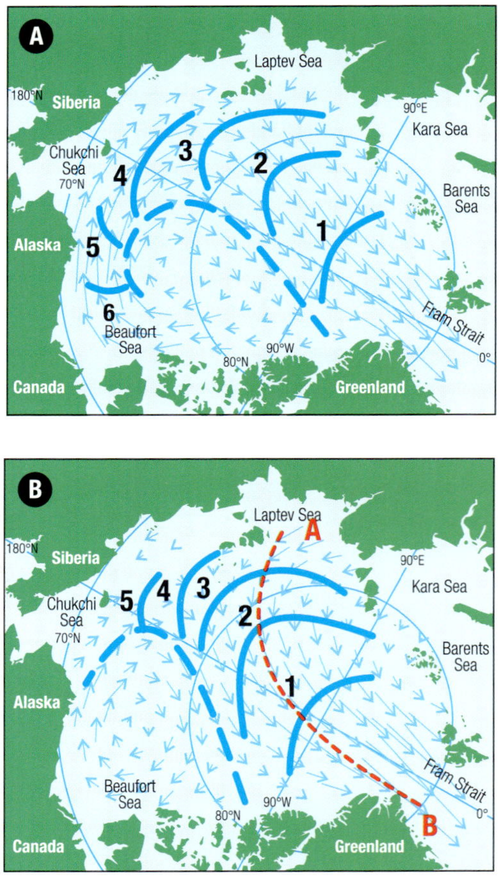

FIGURE D.1.15

Time taken in years for sea ice to reach Fram Strait during a) low AO$^-$ index conditions and b) high AO$^+$ index conditions (after Rigor *et al.*, 2001). Line A-B represents the transect used to describe change in sea ice during drift shown in Figure D.1.17.

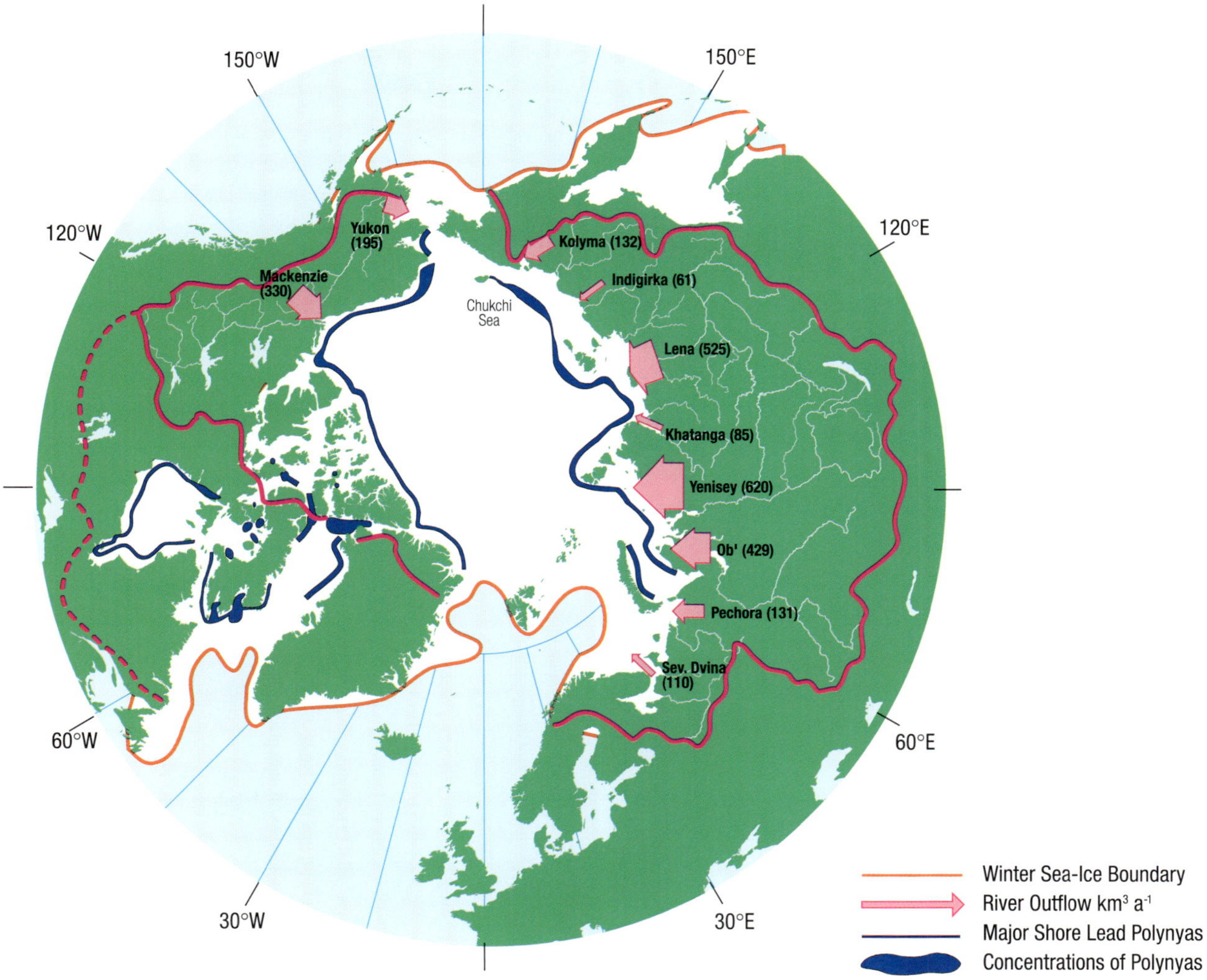

FIGURE D.1.16

The Arctic Ocean showing its predominant connections with the Atlantic and Pacific Oceans, the drainage basin, the runoff, the distribution of polynyas and the sea ice production.

There are also changes in the time required for ice to transit the ocean (Figure D.1.15a, b, (Rigor *et al.*, 2002)) and in the destinations of ice exported from shelves. During winter under AO^+ conditions, there is an increase in ice advection away from the East Siberian and Laptev Sea coasts. This leads to the production of more new, thin ice in the coastal flaw leads (Figure D.1.16, Polyakov and Johnson, 2000; Rigor *et al.*, 2002), a decrease in the advection of ice from the western Arctic into the eastern Arctic, possibly an increased advection of ice from the Arctic Ocean to the Barents Sea through the Spitzbergen-Franz Josef Land passage (Polyakov and Johnson, 2000), and an increase in the 900,000 km^2 of ice advection out of the Arctic at Fram Strait (Morison *et al.*, 2000; Rigor *et al.*, 2002). Interestingly, increased ice export through Fram Strait can be produced by shifts to both negative and positive AO phases (Dickson *et al.*, 2000).

Comparing the two modes of ice drift (Figure D.1.14a, b), it is apparent that, during AO^- conditions, the East Siberian Sea imports much of its ice from the Beaufort Sea and that there is an efficient route to carry ice clockwise around the Arctic margin of the East Siberian Sea and out toward Fram Strait. Under the strong AO^+ conditions of the early 1990s, the Beaufort Sea ice became more isolated, whereas ice from the Kara, Laptev, and East Siberian Seas was displaced into the central Arctic and toward the Archipelago. It is not clear from the IABP data how much ice from the Russian Shelves might transport into the Canadian archipelago or the Beaufort Gyre under AO^+ conditions, but models (Maslowski *et al.*, 2000; Polyakov and Johnson, 2000), palaeo-studies of Eurasian wood (Dyke *et al.*, 1997; Tremblay *et al.*, 1997), and sediment records (Darby *et al.*, 2001) all suggest that such transport is likely and may at times be important.

Sea ice provides a rapid means for accumulating and transporting contaminants long distances without dilution (Pfirman *et al.*, 1995; Wadhams, 2000). The response in ice-drift trajectories to change in AO index (Rigor *et al.*, 2002) therefore carries immense implication for the connectivity between contaminant source and sink regions for ice pathways within the Arctic Ocean.

D.1.3.5.1.3 Sea-ice transport of material

Sea ice is an important mechanism for transporting sediments from coastal and continental shelf to the interior ocean and out into the Greenland Sea (Barrie *et al.*, 1998; Dethleff *et al.*, 2000; Nürnberg *et al.*, 1994). Although all the shelves of the Arctic are implicated in this process, the Laptev Sea has so far proven so far to be the most efficient exporter of sediment-laden ice (Eicken *et al.*, 2000; Eicken *et al.*, 1997; Reimnitz *et al.*, 1994; Reimnitz *et al.*, 1992; Reimnitz *et al.*, 1993).

This transport process involves several steps including: 1) the delivery of sediment to the shelf from rivers or from coastal erosion where much of it may become trapped; 2) the incorporation of sediment into the ice, either through ice grounding or through suspension freezing in mid-shelf flaw polynyas; 3) the export of ice from the shelf to the interior ocean; 4) the transport of ice across Arctic basins potentially with some loss of sediment during transport; and 5) the release of sediment at the location where the ice melts (Figure D.1.17).

During transport, the ice 'weathers', ablating at the surface during summer and incorporating more ice on the bottom during winter, with the result that some of the sediment entrained over the shelf migrates to the surface of the ice. Additionally, atmospheric particulates accumulate on the ice along its transport route and, therefore, an increase or decrease in the time taken for ice to cross the Arctic Ocean (Figure D.1.15) respectively increases or decreases the time it takes for atmospheric aerosols and sediments to accumulate at the ice surface. Each step in the ice pathway can be altered by climate change. For example, fine river sediments (known to carry contaminants) become trapped in estuaries by the so-called 'marginal filter' (Lisitzin, 1995). Sea level rise, change in the ice climate, or change in the river's hydrology alter the location of this filter. The process of suspension freezing might be enhanced by larger amounts of open water over shelves in the fall, whereas more sediment might be lost from the ice during transport due to predominance of thin, first-year ice and augmented melting. Finally, the location at which ice melts and drops its particulate and dissolved loads can be changed. There are no direct data on how the above components of the ice-transport pathway respond either individually or collectively to the AO; however, long-term sediment records (Darby *et al.*, 2001), disequilibria in sediments (Gobeil *et al.*, 2001) and the distribution of sediments within the Arctic Ocean (Stein, 2000) suggest that climate forcing akin to the AO likely occurs.

D.1.3.5.2 Ocean-currents and water properties

D.1.3.5.2.1 Surface water

For ocean-currents that deliver contaminants to Arctic ecosystems, surface water is most important because it interacts more directly with biota and ecosystems. Surface water pathways will to some extent reflect ice-drift trajectories (Morison *et al.*, 2000), responding in like manner to the state of the AO (Figure D.1.14a, b). With a high AO index, water in the TPD makes a diversion into the Makarov Basin and the Beaufort Gyre contracts and retreats into the Canada Basin. However, the AO produces other crucial changes in surface water not represented by ice drift. With the enhanced inflow and spreading of water in the Atlantic layer, a retreat of the cold halocline in the Eurasian Basin was also noted (Steele and Boyd, 1998). The halocline (Figure D.1.18) provides stratification between the Atlantic layer and surface water thereby preventing or reducing the transfer of properties like heat or contaminants between deep and surface layers. The salinification of surface water in the Eurasian Basin noted by Steele and Boyd (1998), however, was not due to enhanced inflow from the Atlantic, which actually freshened slightly with the high AO/NAO index of the late 1980s, but rather to the diversion of river inflow at the margins of the Arctic Ocean.

Models (Figure D.1.19a Dickson, 1999; Johnson and Polyakov, 2001; Maslowski *et al.*, 1998) and geochemical measurements (Ekwurzel *et al.*, 2001; Guay *et al.*, 2001; Macdonald *et al.*, 1999a; 2002; Schlosser *et al.*, 2002) clearly show that, with the high AO index of the late 1980s, river water entering the Laptev and Kara shelves was forced to the east rather than directly off the shelf and into the TPD. Under strong AO^+ conditions, perhaps 1000 $km^3\ yr^{-1}$ or more of runoff from the Lena,

Ob, and Yenesei rivers stopped entering the Eurasian Basin and entered, instead, the East Siberian Shelf and thence the Canadian Basin, possibly to exit the Arctic Ocean via the Archipelago (Figure D.1.19b) (Morison *et al.*, 2000). This diversion resulted in a reduction of stratification in the Eurasian Basin (Steele and Boyd, 1998) and an increase in stratification in the Canadian Basin (Macdonald *et al.*, 1999a; Macdonald *et al.*, 2002a). The drop in the AO index toward the end of the 1990s (Figure D.1.6) appears to have initiated a return to the former pathways for river water in the Eurasian Basin (Björk *et al.*, 2002; Boyd *et al.*, 2002).

At the same time, Atlantic surface water invaded the Makarov Basin, displacing water of Pacific origin from the top 200 m of the water column (McLaughlin *et al.*, 1996). This represents a rapid change of water source and properties for about 20% of the Arctic Ocean's area (Figures D.1.18, D.1.19). Although there does not appear to be a strong AO signal in the Pacific inflow through Bering Strait (~0.8 Sv), there has been a general decline by about 15% since the early 1940s (Coachman and Aagaard, 1988; Roach *et al.*, 1995) and that flow may also have freshened due to runoff and precipitation in the Bering Sea (Weingartner, pers. comm.).

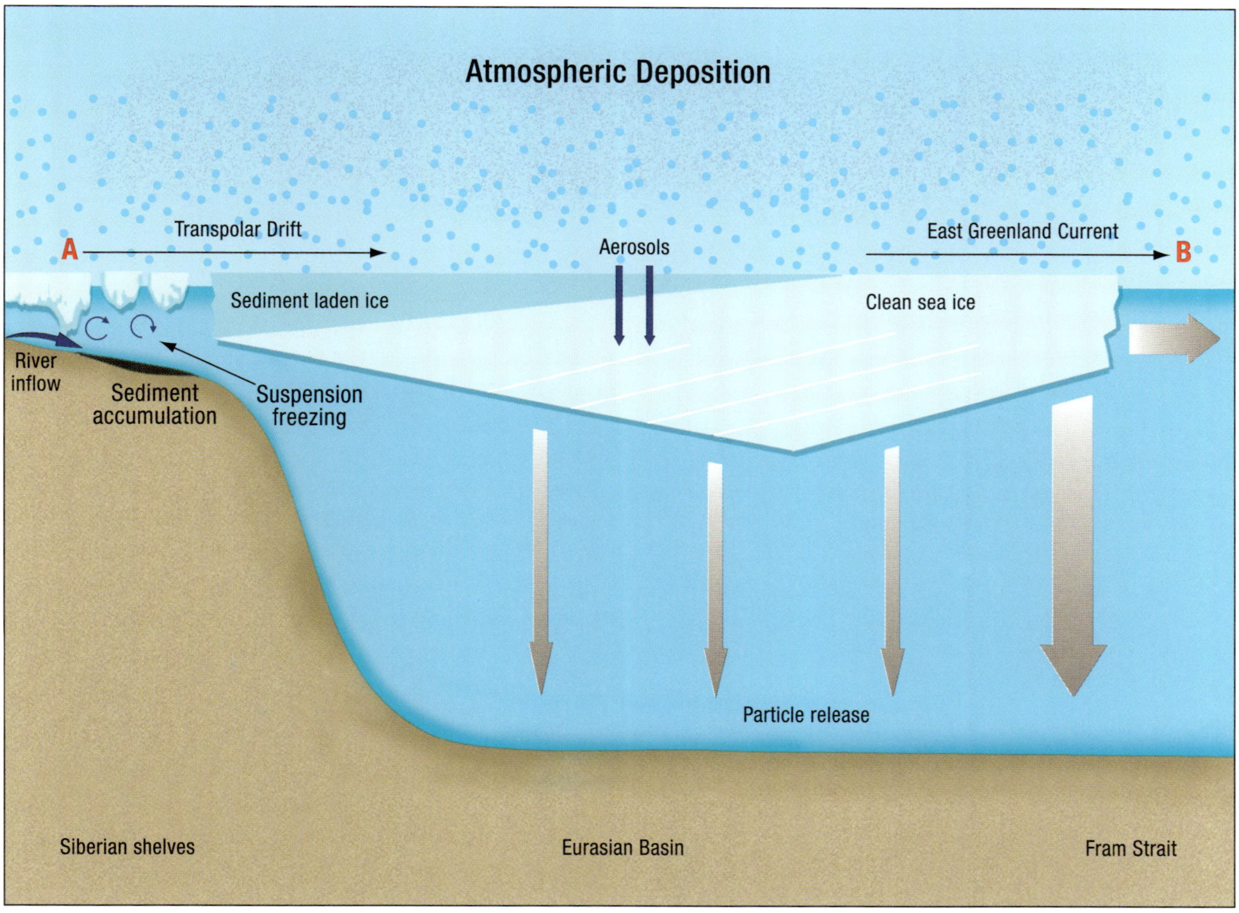

FIGURE D.1.17

A schematic diagram showing the accumulation and transport of sediments and contaminants by ice following the path marked A-B on Figure D.1.15 (modified from Lange and Pfirman, 1998). Sediments become incorporated in ice formed over shelves. Once exported to the interior ocean, the ice joins general ice circulation patterns (Figures D.1.14 and Figure D.1.15) which deliver much of the ice to the Greenland Sea in a time scale of two to six years. During its drift, the ice accumulates contaminants through surface deposition and it also weathers, ablating in summer and accumulating more ice on the bottom of floes in winter. Change in the location where ice melts or in the time taken to arrive there alters the contaminant pathway associated with ice transport.

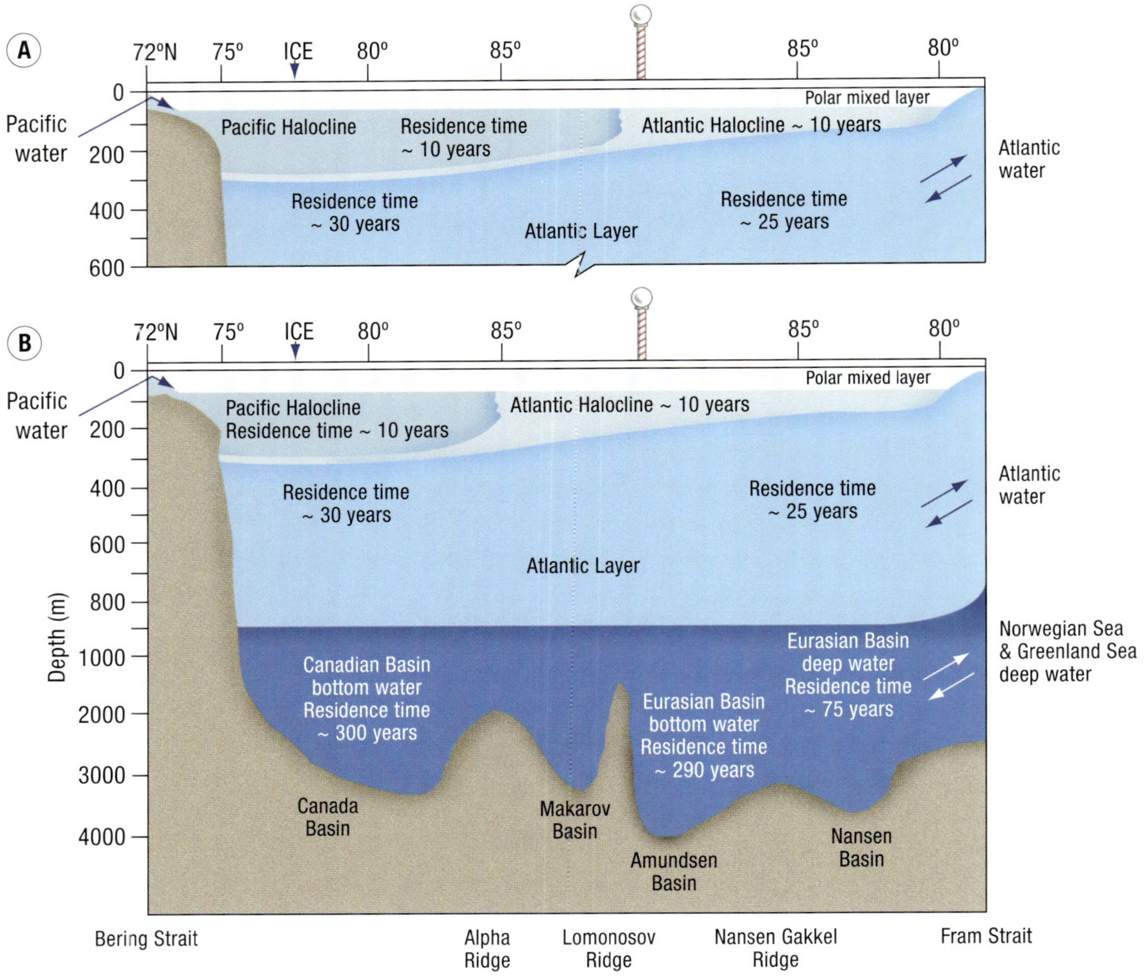

FIGURE D.1.18

The stratification of the Arctic Ocean showing the polar mixed layer, the Pacific and Atlantic domains of influence, and the haloclines. The graphic at top centre shows the displacement of the Atlantic Pacific front during the high Arctic Oscillation index of the early 1990s.

D.1.3.5.2.2 The Atlantic Layer

Repeat hydrographic surveys of Arctic basins, starting in 1987 (Aagaard *et al.*, 1996; Anderson *et al.*, 1989; Carmack *et al.*, 1995; McLaughlin *et al.*, 1996; Morison *et al.*, 1998; Quadfasel *et al.*, 1991; Swift *et al.*, 1997) have revealed an Arctic Ocean in transition. The timing of that transition in the late 1980s implicates the Arctic Oscillation (or North Atlantic Oscillation) as a major source of forcing that has altered connections between the Atlantic and the Arctic Oceans and changed the distribution of Atlantic water within the Arctic, both in the surface layer, discussed above, and in the deeper Atlantic Layer water (Dickson, 1999; Macdonald, 1996). Ironically, some of the clearest evidence of these changes has come from contaminant time series — in particular, the artificial radionuclides released from European sources to the eastern North Atlantic (Carmack *et al.*, 1997; Smith *et al.*, 1998).

Starting in about 1989, a major change was a possible intensification of flow from the Atlantic into the Arctic through Fram Strait and the Barents Sea in response to the shift to strong AO^{+} or NAO^{+} conditions (Figure D.1.20 — for detailed reviews, see Dickson *et al.*, 2000; Morison *et al.*, 2000; Serreze *et al.*, 2000). The winds associated with AO^{+} conditions (Figure D.1.8a, b) increased the rate of northward transport of surface water in the Norwegian Sea and produced warmer air temperatures which, together with the shorter transit times, contributed to warming of the Atlantic water entering the Arctic by about 2.3°C (Swift *et al.*, 1997). The Atlantic water also exhibited slightly decreased salinity (by 0.03–0.05), probably reflecting increased precipitation in the Nordic Seas during NAO^{+} conditions (Figure D.1.9b).

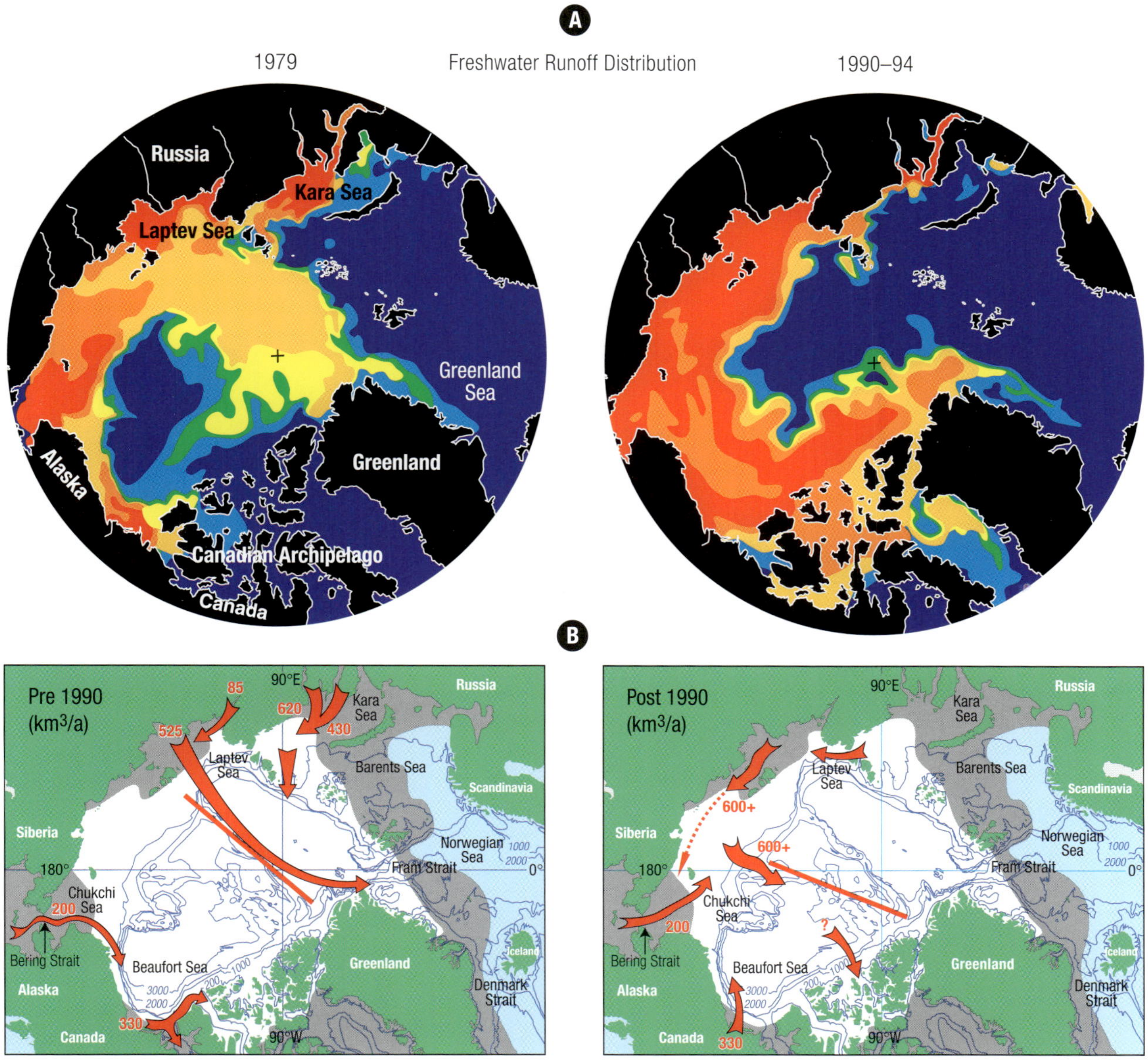

FIGURE D.1.19

a) The change in river inflow pathways between low AO^- index (1979) and high AO^+ index year(1990-94) (based on model results from W. Maslowski, reproduced in Dickson, 1999) and b) a schematic diagram showing the amounts and changes in pathways for river inflow to the Arctic Ocean under AO^- conditions (left) and AO^+ conditions (right).

Within the Arctic Ocean, the changes in the distribution and composition of the Atlantic layer water were spectacular when set against the traditional perception of a quiet, steady-state ocean (see Figures D.1.18 and D.1.20 for this layer's disposition in the water column). The front between Atlantic water and Pacific water, traditionally located over the Lomonsov Ridge, was forced over to the Alpha-Mendeleev Ridge [Figure D.1.18 and see McLaughlin *et al.* (1996); Morison *et al.* (2000)]. At the same time, the inflowing water could be detected in the Atlantic layer by an approximately 1.5°C temperature rise above the climatological norm (Carmack *et al.*, 1995). The changes in volume and composition of Atlantic water entering the Arctic Ocean through Fram Strait continue to cascade through the Arctic basins, first as change in properties along the boundaries (McLaughlin *et al.*, 2002; Newton and Sotirin, 1997), then as change propagated into the basin interiors along surfaces of constant density (Carmack *et al.*, 1997) (Figure D.1.18). Woodgate *et al.* (2001) estimated that in 1995–96, the boundary flow over the southern margin of the Eurasian Basin was transporting 5 ± 1 Sv (1 Sv = 32,000 $km^3\ yr^{-1}$)

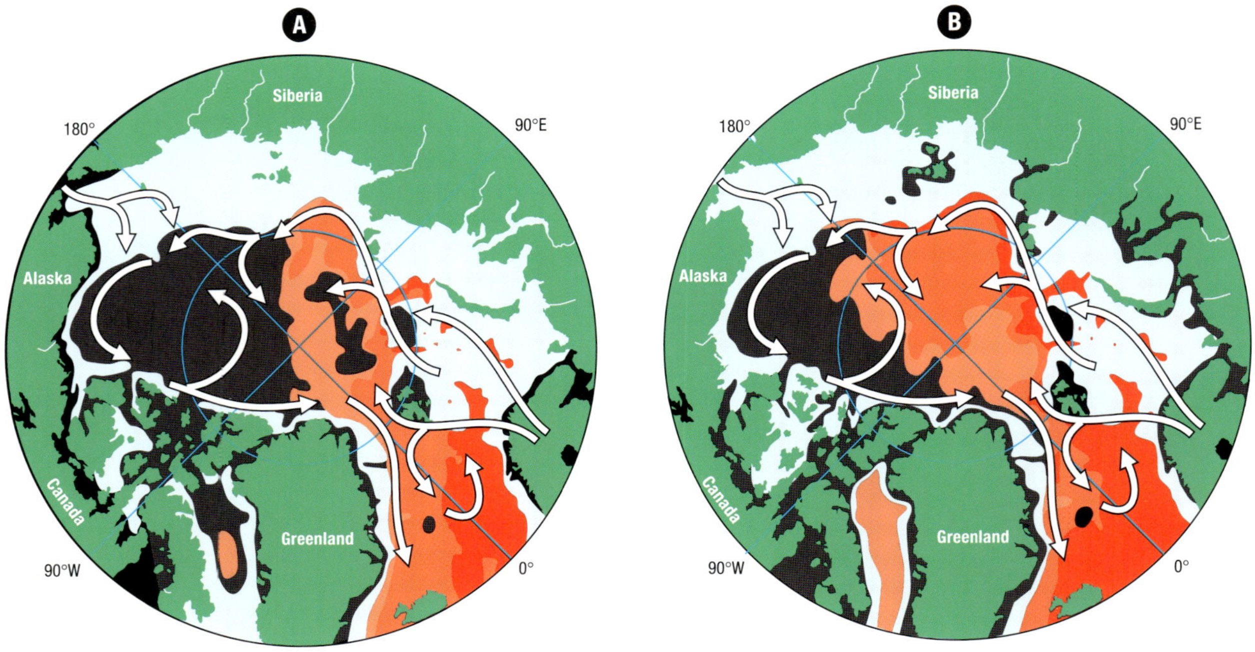

FIGURE D.1.20

The change in Atlantic water inflow to the Arctic and distribution within the Arctic produced by the exceptionally strong shift to AO+/NAO+ in about 1989. The distribution of the Atlantic layer (see Figure D.1.18) is based on Hodges (2000), Maslowski *et al.* (2000), McLaughlin *et al.* (1996), and Morison *et al.* (2000; 1998). The Atlantic layer boundary currents, which are relatively fast, transport properties along basin margins at about 1–5 cm/s (300–1600 km/y) (Woodgate *et al.*, 2001).

at about 1–5 cm s^{-1} (300–1600 km yr^{-1}). When water in the boundary current reached the Lomonosov Ridge, the flow split, with about half entering the Canada Basin along its margin and half returning towards Fram Strait along the Lomonosov Ridge. The high NAO index of the late 1980s (Figure D.1.19) also strengthened and warmed the inflowing Barents Sea Branch of Atlantic water, perhaps by as much as 25% since 1970 (Dickson *et al.*, 2000). This probably led to a parallel warming and salinification of the Barents Sea (Zhang *et al.*, 2000).

D.1.3.6 Adjacent polar seas and regions

D.1.3.6.1 The Canadian Archipelago

The Canadian Archipelago provides one of the important outlets for Arctic Ocean surface water (Figure D.1.1). Therefore, changes in Arctic Ocean surface water contaminant burdens or changes in the source of water flowing out through the Archipelago both have the potential to alter contaminant concentrations within the Archipelago's channels. There are few data with which to evaluate how seawater within the Archipelago channels responds to the Arctic Oscillation. However, changes in distribution of surface water properties (Figures D.1.18, D.1.19) and ice drift trajectories (Figure D.1.14) in the Arctic Ocean itself, together with non-uniform aerial distribution of properties including river water and contaminants (e.g., see Carmack *et al.*, 1997; Guay and Falkner, 1997; Li *et al.*, 2002; Macdonald *et al.*, 1997a) should alert us to the potential for upstream basin changes to be recorded as variable contaminant loadings in water flowing through the Archipelago. Furthermore, the remains of bowhead whale and driftwood on Archipelago shores suggest that ice-drift trajectories and ice cover have both varied greatly over time (Dyke *et al.*, 1996b; 1997; Dyke and Savelle, 2000; 2001) implying that the Archipelago is sensitive to rapid and dramatic change.

D.1.3.6.2 Hudson Bay

Hudson Bay is a large, shallow, semi-enclosed sea strongly influenced by seasonal runoff. The annual discharge (710 km^3 yr^{-1}) is equivalent to a freshwater yield of about 65 cm (Prinsenberg, 1991). Presently, this sea exhibits a complete cryogenic cycle, with summer (August-September) being ice free and winter fully ice covered. Climate models suggest that a doubling of CO_2 may lead to the virtual disappearance of ice from Hudson Bay, thereby raising winter air temperatures leading to the thawing of permafrost in adjacent land areas (Gough and Wolfe, 2001). These same models predict that the complete loss of ice will be preceded by years exhibiting earlier breakup and later freezeup. According to Stirling and co-workers (1999), some of these projected changes may already be occurring (Figure D.1.21), putting considerable stress on the western Hudson Bay polar bear population.

The hydrological cycle of Hudson Bay has been strongly altered through immense damming projects in the drainage basin which will lead to an increase in winter runoff to Hudson Bay of over 50% (Prinsenberg, 1991). Not only do such changes affect stratification and hence nutrient-cycling in this sea, but newly flooded reservoirs are well known for their secondary effect of releasing mercury to downstream aquatic environments (Bodaly and Johnston, 1992).

Due to its southern location, Hudson Bay is clearly in the vanguard of Arctic change and is, therefore, a vital region to collect time series. According to Figure D.1.3a, Hudson Bay lies on a divide between warming and cooling. Regional temperature maps and other evidence (Gilchrist and Roberston, 2000; Skinner *et al.*, 1998) confirm that between 1950 and 1990, the western side has warmed at about the same rate as the eastern side has cooled. In agreement with this observation, bears on the eastern side of Hudson Bay do not show the same pattern of weight loss as the bears on the western side (I. Stirling, EC. pers. comm.), further emphasizing the importance of this region as a laboratory to study detailed consequences of change by contrast.

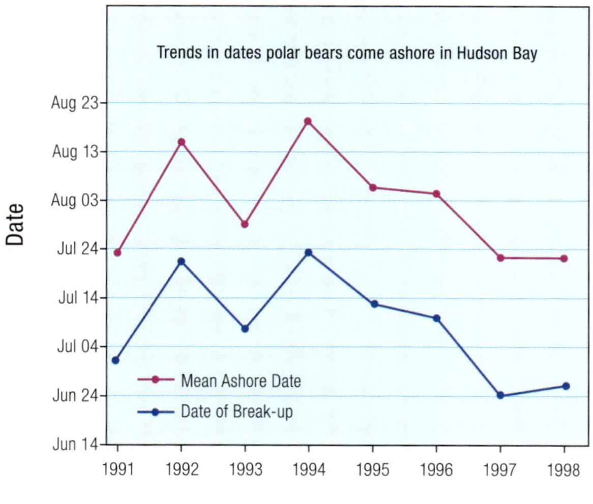

FIGURE D.1.21

Trends in the date of ice breakup in Hudson Bay (blue) and the mean date at which polar bears return to shore in Hudson Bay (pink). The breakup dates refer to the region in Hudson Bay where satellite-tagged female polar bears spent at least 90% of their time (1991–98). Breakup during the 1990s has steadily come earlier by as much as 2–3 weeks giving bears less time to hunt seals in spring (modified from Stirling *et al.*, 1999).

D.1.3.6.3 Baffin Bay

Baffin Bay, Davis Strait, and the Labrador Sea occupy a unique position in that they may receive contaminants from ice and water that exit the Arctic Ocean in the East Greenland Current but they may also receive contaminants from water and ice passing through the Archipelago (Figure D.1.1). Change, therefore, can be produced by variation of contaminant composition within either of these sources or by altering their relative strength and the strength of direct exchange with the atmosphere. Furthermore, decadal-scale modulation very likely differs for the various sources, with the Arctic Oscillation perhaps influencing the Archipelago through flow or Fram Strait outflow, whereas the ice cover in Baffin Bay is more closely associated with the Southern Oscillation (Newell, 1996). In agreement with spatial temperature patterns (Figure D.1.3a), whereas ice season has been getting shorter within the Arctic Ocean and its marginal seas, Davis Strait and the Labrador Sea have recently exhibited an increase in the length of ice season (Parkinson, 1992). Long-period cycles in the ice climate of this region (50 years or more) appear to have had dire consequences for both terrestrial and marine biological populations — including humans (Vibe, 1967). Like Hudson Bay, this region appears to be an important one to study in the context of contaminants (Fisk *et al.*, 2001a), biogeographical variation (Johns *et al.*, 2001), and the impact of change on humans (Woollett *et al.*, 2000).

D.1.3.7 Lake and river ice

Arctic lakes and rivers are likely to provide sensitive sentinels of climate change in their freeze, melt, and hydrological cycles (Vörösmarty *et al.*, 2001). Whereas there appear to be no studies showing a relationship between freshwater ice cover and the AO, significant trends have been demonstrated in these properties over the past 150 to 286 years (Magnuson *et al.*, 2000; Semiletov *et al.*, 2000). There has been a mean delay of 5.8 days per century for freezeup and a corresponding 6.5 days per century advance in breakup between 1846 and 1995. This change in the freeze/melt cycle implies increasing temperatures of about 1.2°C per century.

Most Arctic lakes receive their contaminant burdens atmospherically with the catchment area acting as a receptor through snow fall in winter and a conveyor through runoff in spring. From a very limited set of studies, it appears that Arctic lakes presently retain only a small fraction of contaminant inputs because the main runoff pulse, which precedes lake turnover and peak primary production, simply traverses the lake surface under the ice (Macdonald *et al.*, 2000). With lakes exhibiting more temperate characteristics, the coupling

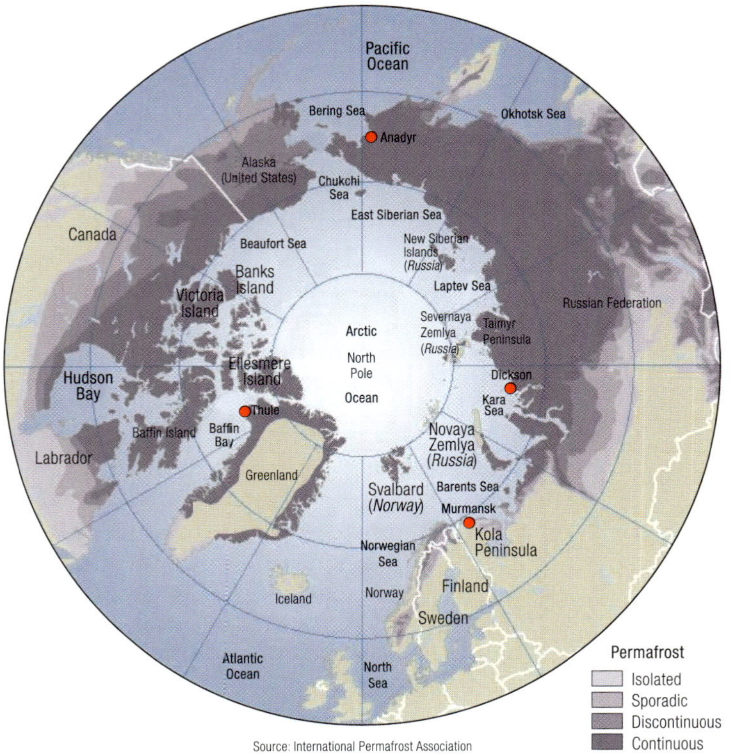

FIGURE D.1.22

The distribution of permafrost in northern landmasses. Discontinuous permafrost, found in large areas of Canada, Alaska, and Russia, are particularly vulnerable to destruction with projected warming (source AMAP, 2002).

of runoff with lake mixing and primary production will change, probably allowing lakes to capture more of the inflowing contaminant burden. In particular, the potential for snow surfaces to enhance contaminant fugacity in lake settings is extremely large (Macdonald *et al.*, 2002c). Quantitative measurements of contaminant-snow interactions, however, are required because the significance of snow in contaminant cycling cannot be projected simply from hydrological measurements.

D.1.3.8 Permafrost

Permafrost underlies about 25% of the land in the Northern Hemisphere including large areas of Canada, Russia, China, and Alaska (Figure D.1.22, Zhang *et al.*, 1999a). Permafrost can also be found in sediments of the continental shelves (not shown). Especially vulnerable to change are regions of discontinuous permafrost, which include large parts of northern Canada. The IPCC (2002) suggests that permafrost area could be reduced by 12–22% by 2100, with perhaps as much as half of the present-day Canadian permafrost disappearing.

In regions of permafrost, the active layer of soil, typically limited to the top 1 m, supports almost all of the biological processes. The loss of permafrost, therefore, alters these biological processes, including the kind of vegetation that can grow, and changes the way soil interacts with the hydrological cycle (Osterkamp *et al.*, 2000; Vörösmarty *et al.*, 2001), both of which have consequences for contaminant transport. In particular, thawing frozen ground releases sediment, nutrients, and organic carbon which then enter ground water, rivers, and lakes to impact biological cycles [see, for example, the studies done in the Mackenzie Basin (Cohen, 1997a)]. The observed thawing trends in Alaska and Russia, but not in northeastern Canada, appear to match the observed trends in SAT (see Figure D.1.3a and Rigor *et al.*, 2000). Accelerated permafrost degradation during the 1990s can probably be ascribed at least partly to the AO (Morison *et al.*, 2000) with, for example, the advection of warm air into the Russian Arctic during high AO index contributing to thawing in that region.

In addition to the changes in biogeochemical pathways that will accompany permafrost degradation, there will also be the widespread problem of re-mobilization of contaminants [see, for example, conference proceedings dedicated to the issue of contaminants in frozen ground (Anonymous, 2001a; b)]. Historical disposal of waste substances has occurred in the form of sewage lagoons, dump sites at DEW-line installations, solid waste dumps in small Arctic communities, mine tailings, and oil drilling sumps. A large component of the containment strategy for these sites is the permafrost. When permafrost degrades, landfills can be washed directly into rivers or the ocean or runoff can be leached into local groundwater. In locations like river deltas and coastal plains, low relief may provide a shortcut between such waste sites and drinking water.

D.1.3.9 Glacial ice

Most Arctic glaciers have experienced net loss in ice mass over the past few decades (Dowdeswell *et al.*, 1997). The Greenland ice mass appears presently (1994–1999) to be decreasing, predominantly at lower elevations, at a rate of about 51 km^3/yr (Krabill *et al.*, 2000). Data also point clearly to loss of ice mass for small glaciers in the Arctic from 1961 to 1993 (Arendt *et al.*, 2002;

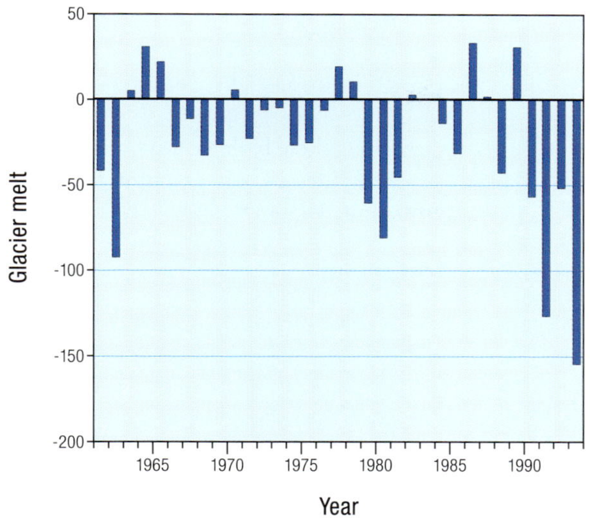

FIGURE D.1.23

The loss of glacial ice mass between 1961 and 1993 based on data compiled by Serreze *et al.* (2000) and Dyurgerov and Meier (1997).

Dyurgerov and Meier, 1997; Serreze *et al.*, 2000). Since about 1960, glacial melt back in the Canadian Arctic alone (Figure D.1.23) is estimated at over 800 km^3 — about half of the melt estimated for the whole Arctic (Dyurgerov and Meier, 1997). In conformance with the high AO index of the early 1990s, the loss of glacial ice mass in the Archipelago was exceptionally strong in the early 1990s, amounting to 390 km^3 (Figure D.1.23).

Glaciers may act as long-term archives, sequestering and preserving airborne contaminants during peak emission years (1950–1970) to release them later during periods of melt back (Blais *et al.*, 1998).

D.1.4 Biological responses to climate change

It is not feasible at this time to give an exhaustive projection of the biological consequences of the kinds of physical change projected for the Arctic, nor is it warranted for the purpose of identifying how biological changes might effect major change in contaminant pathways. Here, we highlight ecosystem changes that appear to have a strong potential to alter the exposure of Arctic biota to contaminants or to alter their resilience to that exposure. We are confident that the kinds of changes discussed in the following text have taken place, or will take place in the Arctic, but we are much less confident of their probable scope and timing. Our primary intent, therefore, is to provide examples of processes that ought to be included explicitly in models and to help focus future attention on biological connections of significance to contaminants.

Whereas aquatic foodwebs in the Arctic exhibit endemic contamination from biomagnifying chemicals, terrestrial foodwebs are found to be among the world's cleanest (de March *et al.*, 1998; de Wit *et al.*, 2003). Therefore, apex feeders that adapt to change by switching between land-based and aquatic foodwebs have a particularly large potential to change their exposure to contaminants like organochlorines and mercury. Humans probably provide the best example of such flexibility, but other animals (e.g., Arctic foxes and grizzly bears) can also adjust diet to opportunity.

D.1.4.1 Terrestrial systems

In this report, terrestrial systems include forests, grasslands, tundra, agricultural crops, and soils. Surface-air exchange between airborne contaminants and terrestrial systems is important in the overall fate and long-range transport of chemicals, especially for the semi-volatile chemicals which are split between the gaseous and condensed states. As a result of their high organic content, terrestrial phases (e.g., soils, forests, grasslands) act as reservoirs for many POPs (Simonich and Hites, 1994) with PCBs, DDT, HCH and chlorobenzenes figuring prominently (de Wit *et al.*, 2003).

Air-surface exchange of POPs into these phases is a dynamic process that controls air burdens of chemicals. Thus any change in the extent of vegetation cover associated with global warming will have implications for contaminant fate and transport. Wania and McLachlan (2001) have shown that forests have a unique ability to mitigate atmospheric concentrations of organochlorines by "pumping" chemicals from the atmosphere into foliage and from there to a long-term reservoir in forest soil. This process is likely to be most important for OC compounds with log K_{OA} of ~9–10 and log K_{AW} of ~ -2 to -3 (where K_{OA} and K_{AW} are octanol-air and air-water partition coefficients — see Wania, 2001). Because these key properties are strongly temperature-dependent (discussed in greater detail in Section D.1.6.3.4), even a small change in climate may alter the dynamics of this process and the cycling of contaminants. Terrestrial vegetation also indirectly affects contaminants by altering snow accumulation and soil temperature (Sturm *et al.*, 2001).

Arctic terrestrial populations have provided some of the clearest examples of large temporal cycles (Krebs *et al.*, 2001; Predavec *et al.*, 2001) and it is against this natural background that the effects of global change will have to be evaluated. Warmer winter temperatures promote the growth of woody shrubs and encourage the northward migration of the tree line (MacDonald *et al.*, 1993; Serreze *et al.*, 2000; Vörösmarty *et al.*, 2001). Although the advance of the tree line might be expected to occur slowly over time scales measured in centuries [estimated at 100 km per°C warming (IPCC, 2002)], the particular sensitivity of tundra to water-table fluctuations and permafrost melt could produce widespread alteration in ground cover more rapidly with, for example, the replacement of tundra by vascular plants (Gorham, 1991; Rouse *et al.*, 1997; Weller and Lange, 1999). Gradual climate change can affect species distribution, population diet, abundance, morphology, behaviour, and community structure (Easterling *et al.*, 2000; Predavec *et al.*, 2001). Although there appears to be no compelling evidence of recent large change in the Arctic's tundra, ecosystem models suggest that tundra may decrease to one third its present size (Everett and Fitzharris, 1998).

Warmer summer temperatures are likely to promote forest fires which will be accompanied by direct emissions of PAHs, PCDD/Fs, and other POPs produced by combustion (see for example, Gribble, 1994; Yunker *et al.*, 2002). Forest fires will also damage terrestrial soils, leading to erosion and an increased release of organic carbon, which will impinge aquatic systems.

D.1.4.2 Aquatic systems

D.1.4.2.1 Lakes, rivers, and estuaries

The changes in snow and ice cover and in the hydrological cycle will alter the light and nutrient climate of freshwater systems. These changes, together with loss of permafrost, which will enhance the supply of nutrients and particulates to lakes, will increase aquatic productivity and particle flux (Douglas *et al.*, 1994; McDonald *et al.*, 1996; Schindler, 1997). Spring bloom will probably advance with early loss of ice cover, but hydrological processes in a lake's drainage basin will probably also advance. Increased summer temperatures will disadvantage fish like trout and grayling, whereas an increase in winter temperatures may enhance microbial decomposition. Shifts in the seasonal light/temperature cycle may also advantage or disadvantage species lower in the food web including insects, phyto-, and zooplankton. Change in water level will have obvious effects on important fish stocks, especially species dependent on small refugia to over-winter (Hammar, 1989) or species dependent on freshwater coastal corridors for their life cycle. The Arctic cisco (*Coregonus autumnalis*) provides a relevant Canadian example of the latter (see, Gallaway *et al.*, 1983). Warming and loss of nearshore or estuarine ice may accordingly eliminate indigenous fish, to be replaced by anadromous fish from the Pacific Ocean (see Babaluk *et al.*, 2000). With warming, we can expect widely distributed shifts in zoogeographic distributions with the potential to affect every step in the freshwater food chain. Prediction, however, will probably founder on "counter-intuitive" surprises (Schindler, 1997).

D.1.4.2.2 The ocean

The effect of ice on Arctic marine ecosystems has long been understood by those who harvest the sea (Bockstoce, 1986; McGhee, 1996; Scoresby, 1969; Vibe, 1967). Change in ice climate, therefore, has a large potential to modify marine ecosystems, either through a *bottom-up* reorganization of the food web by altering the nutrient or light cycle, or a *top-down* reorganization by altering critical habitat for higher trophic levels (see for example Parsons, 1992). Any reorganization that changes the number of trophic levels in the food web or alters the flow of carbon between, for example, pelagic and benthic food webs would have particular significance for contaminants that biomagnify like mercury and the organochlorines: the complexity of the interaction between ice and aquatic ecosystems provides much scope for surprises in contaminant pathways (Figure D.1.24). Arctic and sub-arctic marine ecosystems are also altered by ocean climate changes such as regime shifts involving the displacement of water masses and associated populations or temperature change (Dippner and Ottersen, 2001; Hare and Mantua, 2000; Helland-Hansen and Nansen, 1909; Hunt *et al.*, 1999; Loeng, 2001; Saar, 2000; Sakshaug *et al.*, 1994; 1991).

There are far too many examples of how variations in ice climate can affect ecosystem structure to list them all (see for example Sakshaug and Slagstad, 1992; Tynan and DeMaster, 1997). It is not even likely that we have observed all changes that have occurred in Arctic systems. The thickness and distribution of ice can influence the amount of organic carbon produced, the types of algae that produce it, and connections between the algal production and communities in the water column or sediments (Niebauer and Alexander, 1985). Ice controls wind mixing and light penetration, especially when it is covered with snow, and it may support upwelling at the ice edge but suppress it beneath the ice. Through its annual cycle, ice formation decreases stratification in winter but increases it when the ice melts in spring.

These physical factors impact the nutrient supply to surface water, the light climate, and the water stability which together control primary production. Furthermore, mats of algae that grow on the bottom of the ice support an epontic food web that feeds polar cod

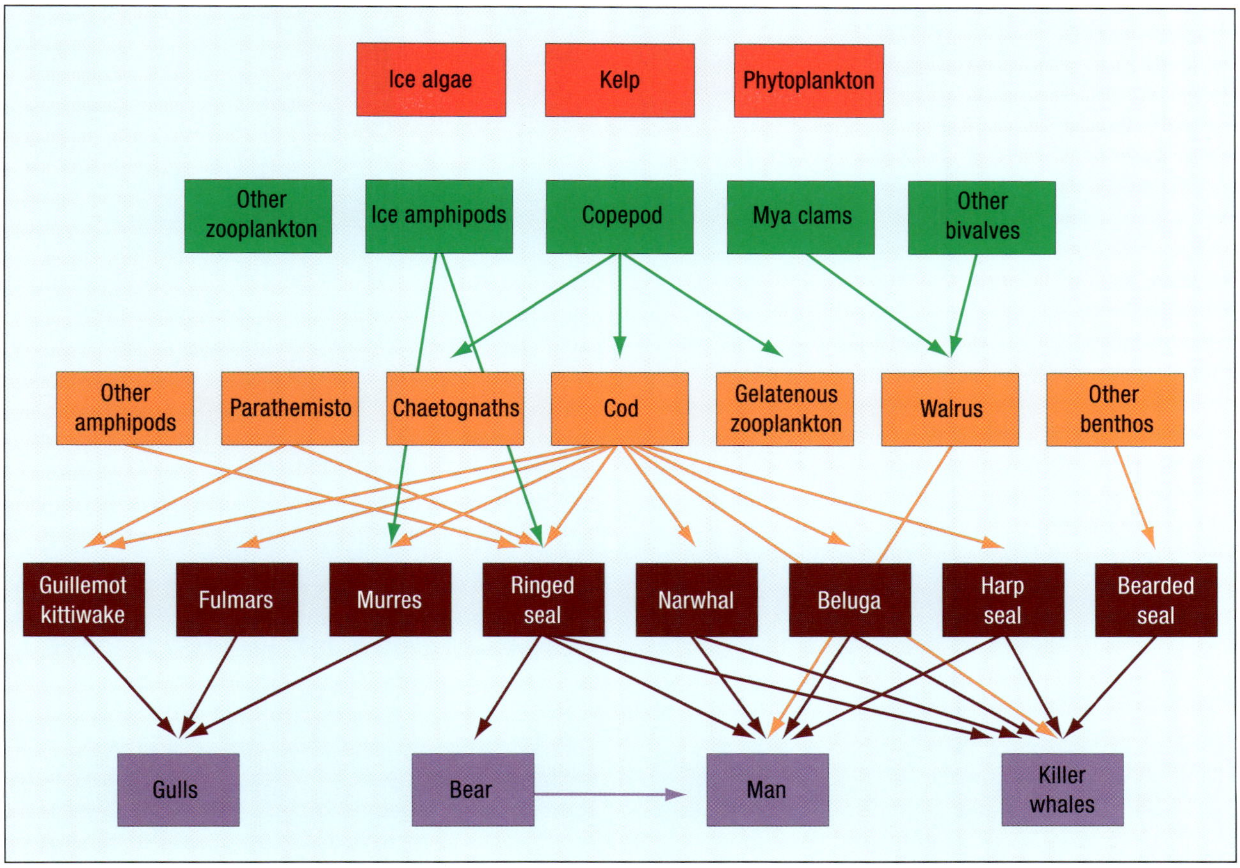

FIGURE D.1.24

A simplified schematic diagram showing the marine food web (based on Welch *et al.*, 1992).

(*Boreogadus saida*), ringed seals (*Phoca hispida*), and polar bears (*Ursus maritimus*). Alternatively, by being shed from melting ice in spring, these mats of algae support a benthic food web that feeds molluscs, walrus (*Odobenus rosmarus*), bearded seals (*Erignathus barbatus*), and king eiders (*Somateria spectabilis*). Similarly, primary production in the water column may be partially grazed to support a pelagic food web, or descend ungrazed and, together with fecal pellets and zooplankton carcasses, feed the benthos (Grebmeier and Dunton, 2000). The bifurcation between pelagic and benthic food webs is very much influenced by the distribution of ice and its impact on nutrient and light climate.

Shifts in the distribution of benthic species due to temperature, carbon flux, or other climate-related change have the potential to alter completely the coupling between sediments and bottom water. In one well-documented example from a temperate region, the invasion of Echiura (*Listroiolobus pelodes*) into coastal benthic communities off California, for as yet unknown reasons, resulted in aerated and biomixed sediments that reduced the evidence of wastewater impacts regionally (Stull *et al.*, 1986).

D.1.4.2.2.1 Bottom-up trophic change

The projected loss of ice for the Arctic Ocean, particularly over the shelves, intuitively should increase primary production in the marginal seas through enhanced mixing, penetration by light, and upwelling. In other words, Arctic shelves would begin to look more 'temperate'. Greater new production implies greater particle flux and greater secondary production, but the complexity of marine ecosystems should alert us to possible surprises. Massive blooms of jellyfish were observed in the Bering Sea during the 1990s (Brodeur *et al.*, 1999; Hunt *et al.*, 1999). Their emergence was ascribed to an increase in the temperature of the sea surface and loss of ice cover — the same two key changes poised over the Arctic Ocean.

Parsons (1979) has drawn attention to the fundamental ecological differences between western seaboards, where coastal water exhibits divergence and upwelling, and eastern seaboards, which are convergent. The former have been of greater commercial interest to humans but they are also characterized by jellyfish (Parsons, 1979). The AO does not cause reversal of large-scale wind circulation but it does produce more divergent Arctic Ocean margins under AO^-/anti-cyclonic conditions and less divergent margins under AO^+/cyclonic conditions. The inherently noisy events of coastal upwelling and downwelling could then act together with the AO in a form of 'stochastic resonance' (Rahmstorf and Alley, 2002) to enhance upwelling during AO^- conditions which might then have the capacity to produce large-scale modal shifts in shelf ecosystems and their commercial significance. Changes in ocean climate like those associated with the AO/NAO have long been known to impact fisheries in sub-polar seas either directly through water property changes (temperature, salinity) or indirectly through changes in the community structure (Hare and Mantua, 2000; Klyashtorin, 1998; Marteinsdottir and Thorarinsson, 1998).

A dramatic example of large-scale, bottom-up biological change was witnessed during the SHEBA drift across the Beaufort and Chukchi Seas in 1997-98 (Melnikov *et al.*, 2002). Compared to Soviet observations from drifting stations that passed over the same region 20 years earlier, there was a marked decrease in large diatoms in the water column and microfauna within the ice. The freshening and strong stratification of the surface water, due to river discharge diverted into the basin under the high AO conditions of the early 1990s (described above), reduced the supply of nutrients from below and produced species more typical of fresh water. As a result, there was a high proportion of recycled production and less new production. The loss of relatively large diatoms could reduce the size of herbivores, potentially inserting an extra 'small-carnivore' step at the bottom of the food web which would increase the number of trophic levels. Because biomagnification of organochlorines is often exponential (Fisk *et al.*, 2001a), slightly higher concentrations at low trophic levels (e.g., copepods) can have a large impact on apex feeders. Stratification, which is altered at the basin scale under AO/NAO shifts, affects plankton composition and vertical flux dramatically as evident from studies in the Barents Sea (Wassmann, 2001). For example, Wassman *et al.* (1990) showed that algal blooms by *Phaeocycstis* sp. along the Greenland coast and in the Barents Sea tend not to get grazed resulting in a large transfer of organic carbon to the benthos. Climate change, in the form either of loss of ice cover or increase in stratification, has the potential to alter the quantity of available food and redistribute its flow between epontic, pelagic, and benthic habitats.

The Bering Sea provides another outstanding example of recent change from the bottom up permeating an entire ecosystem. In view of the Bering Sea's vulnerability to airborne contaminants from Asia (Bailey *et al.*, 2000; Li *et al.*, 2002), it is particularly regrettable that the observations of ecosystem change since the 1970s have not been matched by studies of contaminant pathways. Evidence from stable isotope records in bowhead whale (*Balaena mysticetus*) baleen suggests that the carrying capacity of the Bering Sea ecosystem began to decline in the mid-1970s (Schell, 2000). This change may relate to a larger picture of change throughout the North Pacific (Hare and Mantua, 2000) and, in particular, to the switch in the Pacific Decadal Oscillation from cold to warm phase in the mid 1970s. The change in regime rapidly permeated the entire ecosystem of the Bering Sea, altering fish community structure and seabird and mammal populations (Springer, 1998).

More recently, blooms of small phytoplankton (*Emiliana huxlei*) were observed in 1997 and 1998 (Saar, 2000). Because these phytoplankton are smaller than the diatoms that typically bloom in the Bering Sea, they were grazed by smaller copepods instead of larger euphausids which in turn probably led to die-offs of the short-tailed shearwaters (*Puffinus tenuirostris*) that feed on the latter (Stockwell *et al.*, 2000). Similarly, the alteration of primary production both in quantity and distribution probably decreased food availability for fish, whale, seal, and walrus populations forcing die-offs, migration, or redistribution throughout the foodweb (Botsford *et al.*, 1997; Grebmeier and Cooper, 1995; Grebmeier and Dunton, 2000; Hare and Mantua, 2000; Rugh *et al.*, 1999; Stabeno and Overland, 2001).

Large as these ecosystem changes appear to have been, they may pale in comparison to the natural fluctuations that have occurred during the past two millennia (Finney *et al.*, 2002). Clearly, the dramatic changes in the Bering Sea system could spill over into the Chukchi Sea, and the decline of Bering inflow by ~15% since the 1940s suggests a matched decline in new and advected production in the Chukchi Sea simply due to reduced nutrient and organic carbon supply.

In the Barents and Nordic Seas, it has long been recognized that fish populations respond to climate variability (Helland-Hansen and Nansen, 1909). Indeed, the distribution of capelin (*Mallotus villosus*), the single most important food species for Arcto-Norwegian cod, is known to vary from year to year depending on the inflow of Atlantic water (Sakshaug *et al.*, 1994). Fluctuations in large- and regional-scale atmospheric pressure conditions affect winds and upper ocean-currents, modify water temperature, alter drift patterns of fish larvae, and change availability of prey items. Mixing during summer alters the nutrient cycle and the coupling between primary production and benthos (Peinert *et al.*, 2001; Wassmann, 2001).

D.1.4.2.2.2 Top-down trophic change

Ice-covered seas have a unique capacity for top-down trophic change. To understand and predict how the partial or complete loss of ice will impact the trophic structure requires a detailed understanding of how top predators take advantage of ice (Carmack and Macdonald, 2002; Lowry, 2000; Vibe, 1967). In an incisive review, Tynan and deMaster (1997) discuss how whales, walrus, seals, bears, and cod, are likely to be impacted by change in ice climate, showing that their response to change depends on how 'plastic' their dependence on ice might be.

Change in the landfast ice may give an advantage to seals or to bears with the result that Arctic cod (*Boreogadus saida*) would be subject to more, or less, predation, respectively. Walrus use drifting ice to haul out in winter because it provides better access to benthos but they also use terrestrial haul outs in ice-free periods, perhaps with detrimental energy costs (Lowry, 2000; Tynan and deMaster, 1997). In contrast, eiders and other benthic-feeding birds prefer open water with a relatively shallow bottom (< 50 m) (Dickson and Gilchrist, 2002; Grebmeier *et al.*, 1988; Suydam *et al.*, 2000).

Loss of ice (landfast or drifting) in critical regions or at critical times of the year, or movement of the ice edge to deeper water where benthos can no longer be accessed, therefore, can mean a substantial rearrangement of the top of the food web advantaging some animals, disadvantaging others and possibly causing wholesale migration (Dyke *et al.*, 1999; Dyke *et al.*, 1996a; Dyke and Savelle, 2001; Fay, 1982; Lowry, 2000; Moore and Clarke, 1986; Tynan and deMaster, 1997; Woollett *et al.*, 2000). With benthos not readily available, walrus might turn to predation on seals, thereby raising their trophic position considerably (Muir *et al.*, 1999a), or with absence of ice, killer-whale predation on bowhead whales might decimate their population leaving their prey (zooplankton) to feed something else. Early breakup in the Bering and Beaufort Seas during 1995–98 probably led to the observed abandonment of seal pups in 1998 and the decline or starvation of walrus.

The polar bear population in Hudson Bay provides perhaps one of the clearest warnings to us of the consequence of change. Polar bears rely on ringed seals for food and ringed seals prefer landfast or stable first-year ice for pupping (Finley *et al.*, 1983; Stirling, 2002; Wiig *et al.*, 1999). The loss of landfast ice in spring, the loss of food supply for seals, or the inability of bears to access seals during the few critical weeks in spring when pupping occurs, means life or death and can produce large population shifts (Harwood *et al.*, 2000; Smith and Harwood, 2001; Stirling *et al.*, 1999). In Hudson Bay, bears probably accumulate most of their annual energy requirements during the few months of late spring before breakup when they can access older pre-weaning ringed-seal (*Phoca hispida*) pups or naïve post-weaning pups — exactly the period of time that has seen recent dramatic change (Figure D.1.21). Furthermore, permafrost is a critical habitat for bears because they dig maternity dens in frozen peat. This habitat is threatened by warming or increased incidence of fire initiated by more frequent lightning strikes. In Hudson Bay, at the southern limit of their population, polar bears presently appear in a very precarious position (Stirling and Derocher, 1993; Stirling *et al.*, 1999).

Arctic cod is the most important forage fish in the Arctic Ocean's food web (Figure D.1.24, Bradstreet *et al.*, 1986; Tynan and DeMaster, 1997; Welch, 1995). The loss of ice, either in the marginal seas or, as projected by models, in the entire ocean (Figure D.1.5, Flato and Boer, 2001), would have a massive impact on the distribution and life history of Arctic cod and, therefore, on seals, beluga and birds who depend heavily on them. One thing is clear: the ice edge, which is most vulnerable to change, is an especially critical habitat for cod and marine mammals.

Finally, climate change can alter the routes and destinations of migratory species. For example, under the AO$^+$ conditions of the early 1990s, Pacific salmon began to enter Arctic rivers (Babaluk *et al.*, 2000). Similarly, bowhead whales and belugas range widely in search of food and their range varies enormously in time and space with changes in ice climate (Dyke *et al.*, 1996b; Dyke and Savelle, 2001; McGhee, 1996). Nor are long migrations limited to whales. Harp seals of the Northwest Atlantic undergo 8000 km round trips to feed on Arctic cod in Baffin Bay (Finley *et al.*, 1990) and bird species migrate inordinately long distances, often depending on critical areas along their migration pathways where they may ingest contaminants (see, for example, Braune *et al.*, 1999; Savinova *et al.*, 1995; Springer, 1998).

D.1.5 The effects of climate change on human activities

There are at least six very different ways in which climate change may lead to an alteration of contaminant pathways through a modification of human activities. First, people on the margins of the Arctic Ocean will make dietary choices, as they have always done, based on the availability of traditional/country foods including terrestrial and marine animals (Krupnik, 2000; McGhee, 1996; Vibe, 1967). Second, a marginal sea that clears of ice for large portions of the year will encourage shipping, tourism, oil exploration, and other industrial activities, each of which brings with it associated contaminants. Furthermore, enhanced shipping increases the risk of introducing exotic species or diseases, which then affect indigenous species. Third, the encroachment of commercial fisheries (fish, shellfish and marine mammals) into the Arctic could alter the structure of the food web in Arctic seas (Bockstoce, 1986; Parsons, 1992; Pauly *et al.*, 1998; Pauly *et al.*, 2001) and lakes (de Graff and Mychasiw, 1994; de Wit *et al.*, 2003). Fourth, climate change may promote the spread of insect pests globally and some countries may choose to re-introduce or increase the use of banned or restricted pesticides, especially those that are proven cheap and effective. Fifth, climate change toward conditions suitable for domestic crops may encourage further expansion of agriculture or silviculture within the Arctic's drainage basin. Sixth, the various changes listed above will likely contribute to demographic shifts and population increases in northern regions, which will then lead to increased contaminant release locally (for example from burning, power and fuel consumption, use of industrial or agricultural products).

It is well known that the dietary composition (e.g., marine vs. terrestrial, fat vs. protein, old fish vs. young fish) of the human 'food basket' controls the amounts and kinds of contaminants ingested (AMAP, 1998; Kinloch *et al.*, 1992; Van Oostdam *et al.*, 1999). These dietary changes can be brought about by fluctuations in the populations of target species as previously discussed (e.g., beluga, bowhead whales, walrus, seals, bears, birds, fish, caribou, muskox) or by changes in access to the species caused by early melt, permafrost degradation, open water, loss of multiyear ice, or late freeze-up (see, for example, Fast and Berkes, 1998; Riedlinger, 2001; Vibe, 1967). Because the food webs are affected so differently by contaminants, the balance between terrestrial and aquatic food items in the food basket will be a pivotal point of change in exposure to biomagnifying contaminants.

With marginal seas clear of ice for long periods of the year, it is inevitable that the Arctic will become a favoured shipping route between Europe, Asia, and North America either via the Northern Sea Route (Russia) or the Canadian Archipelago. Shipping brings with it specific, well-known contaminants like hydrocarbons and marine antifoulants (e.g., tributyl tin) and potentially exotic species in ballast water. The major concern, however, is likely to come from accelerated oil exploration and development on the Arctic continental shelves of North America and Eurasia (Bakke *et al.*, 1998).

Outside the Arctic, global warming and alteration of hydrological cycles will probably cause insects and other pests to flourish in some locations. Many of the pesticides that are problematic in the Arctic (toxaphene, DDT, HCH) continue to be used in central America, Africa, and Asia, particularly by developing countries. These countries will be forced to rely more heavily on pesticides in coming decades (Harner, 1997).

Within Arctic drainage basins, warming may expand the area suitable for agriculture. Much of the southern portion of the Mackenzie Basin in North America is presently cultivated. In a global warming scenario this region is projected to contribute an additional 10 million hectares of land suitable for small grain crops (Cohen, 1997b), an area that might be further expanded with the development of new 'climate'-resistant crops. In Russia, most of the major Arctic river basins contain agricultural land particularly within the river valleys and as far north as 65°N (including the Sv. Dvina, Ob, Yenesey, and Lena River basins). Agriculture brings with it industrial pesticides and other chemicals. If pests also thrive in a warmer climate, farmers may resort to increased reliance on pesticides to protect crops.

D.1.6 The effects of climate change on contaminant pathways

We now consider how these changes will affect specific contaminant groupings — heavy metals, radionuclides, organochlorines, and hydrocarbons. For each contaminant grouping, we will generally discuss the significance of the recent shift to high AO index and then consider the more general, long-term changes. To avoid repetition, direct, unsupported statements will be made for which arguments and citations have already been provided in preceding sections.

We will emphasize connections in the contaminant pathways (Figure D.1.1) such as:

1) mobilization from global primary or secondary contaminant sources and/or a change in delivery pathways to Arctic ecosystems;
2) entry into the base of the food web from water, snowpack, ice, soil, and runoff;
3) shifts in the relative importance of source of primary productivity in aquatic systems (ice vs. aquatic or coastal vs. deep ocean);
4) change in food web structure affecting the degree of biomagnification (bottom up effects);
5) change in the feeding ecology of key higher order consumers (top down effects); and,
6) change in the age structure of higher trophic order populations where contaminant concentrations in tissue are age-dependent.

D.1.6.1 Heavy metals

D.1.6.1.1 Lead, cadmium, zinc

Lead, cadmium and zinc are commonly released to the atmosphere through high-temperature processes. They are also released, in even greater quantities, to water through runoff, municipal discharges and dumping (Pacyna *et al.*, 1995). Contaminant lead has had a unique, predominant source in leaded gasoline combustion. Due to the strong atmospheric connection in winter between Eurasia and the high Arctic (Figure D.1.8a), long recognized in events like Arctic haze (Hileman, 1983) and brown snow (Welch *et al.*, 1991), much attention has been focused on the air as the means of transporting contaminant metals (Akeredolu *et al.*, 1994; AMAP, 1998; Boutron *et al.*, 1995; Pacyna, 1995; Rosman *et al.*, 1993; Sirois and Barrie, 1999; Sturges and Barrie, 1989).

Based on back trajectories, models, and stable lead isotope composition, the sources of atmospheric metals have been established primarily as Eurasia (Norilsk), then western and eastern Europe and finally, North America — each of which has the opportunity at particular times during the year to supply air masses to the Arctic (Figure D.1.8a, b). The shift between AO^- and AO^+ conditions alters mean wind fields thereby effecting change on the balance and timing of air movement from the three sources, but the connections remain intact. Air transport from eastern North America and Western Europe — both of which no long use lead in gasoline — strengthens under high AO index, especially in winter, due to the intensification and extension northward of the Icelandic Low. Changes in air mass trajectories, while probably shifting the pathway and rate of transport between temperate sources and Arctic sinks, will not likely substantially change the net delivery of airborne contaminants. This hypothesis could be tested by running transport models (e.g., see Akeredolu *et al.*, 1994) under AO^+ and AO^- conditions.

The greatest leverage for change with aerosol metals resides in the wet and dry removal processes within the Arctic which are not completely understood. Because the Arctic is a poor trap for atmospheric contaminants, sequestering < 10% of the emissions that pass through it (Akeredolu *et al.*, 1994; Pacyna, 1995), there is a great deal of scope for enhancing the deposition of airborne contaminants to surfaces by altering the location and intensity of precipitation (Figure D.1.9a, b) and/or by changing the balance between snow and rain (see, for example, Sherrell *et al.*, 2000). Under AO^+ conditions, the atmospheric corridor from eastern North America and Western Europe will become a much more efficient trap for particulates, raining them out in the Nordic Seas and in the southern portion of the Eurasian Basin. Particle scavenging will generally increase wherever higher precipitation prevails, such as over northern Europe and the Eurasian Basin in general. Contaminants deposited on the eastern side of the Nordic Seas will then enter the Barents Sea and the Eurasian Basin via ocean-currents. Contaminants deposited to the west will be delivered back into the North Atlantic via the East Greenland Current.

Given the above scenario, it seems likely that, under AO^+ conditions, metal-contaminated aerosols entering the Arctic near the prime meridian will be subject to enhanced scavenging *en route*. Larger areas of open water (Figure D.1.13) mean that scavenging will tend to place a greater proportion of these airborne contaminants directly into the surface ocean rather than on the sea ice. The decline of aerosol metal concentrations at Alert after about 1991, ascribed by Sirois and Barrie (1999) to the collapse of industry in the Soviet Union following its collapse and fragmentation, could be explained partly by changes in wind and precipitation patterns at the end of the 1980s (Figures D.1.8 and D.1.9). Enhanced loadings to sea-ice surfaces under AO^+ conditions are most likely to occur in the southern Eurasian Basin and this ice would then be exported back into the Greenland Sea.

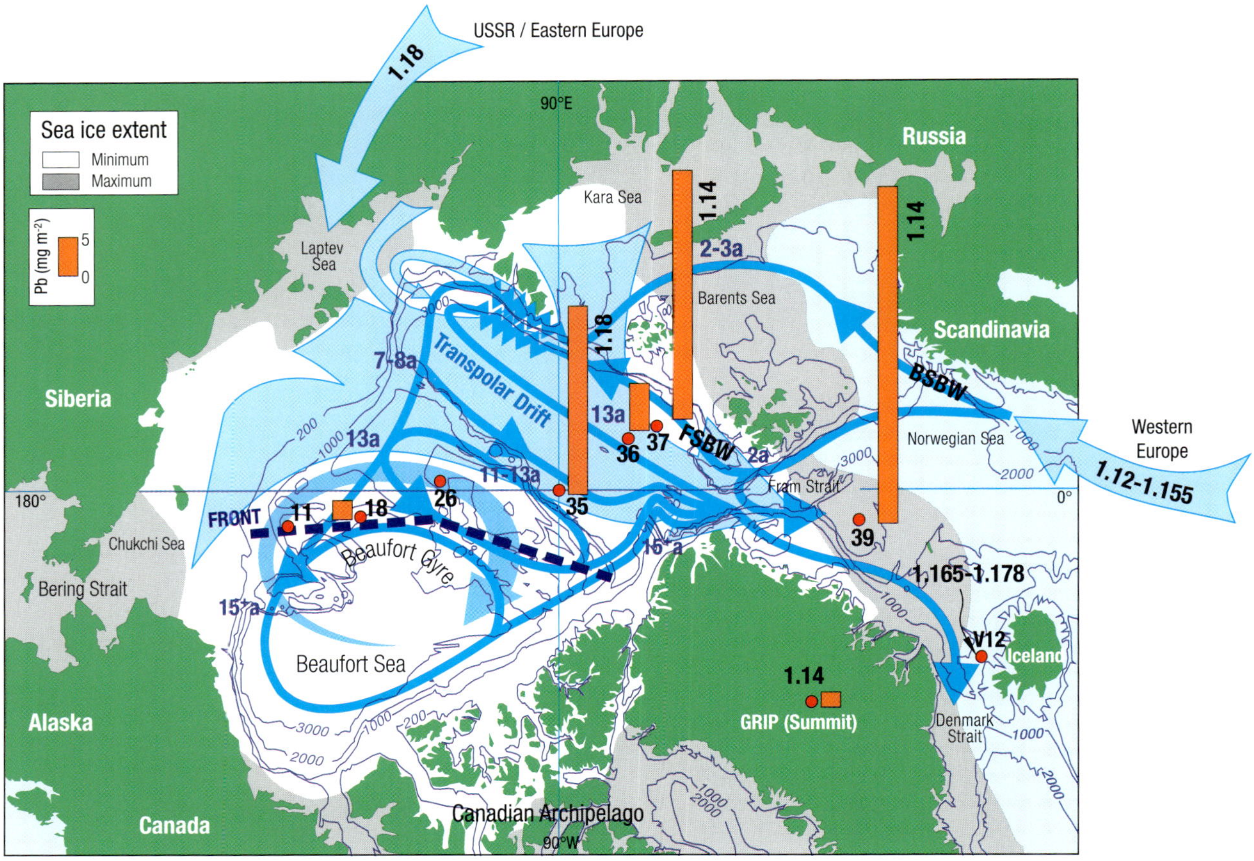

FIGURE D.1.25

The transport of lead into the Arctic Ocean following boundary currents shown by the contaminant lead inventory in sediment cores. Source of the lead (western Europe or Eurasia) is shown by the stable lead isotope composition (figure adapted from Gobeil *et al.*, 2001a).

The focus on the atmosphere as a pathway for contaminant metals to the Arctic has to some degree diverted attention from the ocean. Sediment cores collected along the margins of the Eurasian and Canadian Basins (Figure D.1.25) suggest that a major route for contaminant lead to the Arctic Ocean has been the same ocean-current that transports radionuclides northward from the European reprocessing plants (Gobeil *et al.*, 2001). The residence time of lead in surface water, which is relatively short (< 5 years), is still long enough to permit contaminant lead to be transferred from the North Atlantic and Nordic Seas into the Arctic (Gobeil *et al.*, 2001). Under AO^+ conditions an even more efficient transfer of metals can be expected from western Europe to the Arctic either via rainout in the air transport corridor to the northwest of Europe (Figures D.1.8 and D.1.9) or via coastal discharges to the North or Baltic Seas.

Lead comprises four stable isotopes, ^{204}Pb (1.48%), ^{206}Pb (23.6%), ^{207}Pb (22.6%), and ^{208}Pb (52.3%), with the composition varying between geological reservoirs (Sangster *et al.*, 2000). That variation has provided an incisive way to determine the sources of contaminant lead in global environmental media, including Arctic aerosols and ice (Rosman *et al.*, 1993; Sturges and Barrie, 1989; Sturges *et al.*, 1993). Accordingly, Gobeil *et al.* (2001) were able to relate the contaminant lead accumulating in sediments along the Barents Sea margin to a western European source ($^{206}Pb/^{207}Pb$ ~1.14), with ocean-currents acting as the major transporting mechanism (Figures D.1.20 and D.1.25). Contaminant lead in sediments near the North Pole, however, had an eastern Europe or Russian composition ($^{206}Pb/^{207}Pb$ ~1.18) (Figure D.1.25). Based on the distribution of contaminant lead in the Arctic sediments, these authors proposed a transport scheme wherein lead enters the Arctic Ocean via the Laptev Sea, either in ice or, perhaps more likely, in water of the trans-polar drift (Gobeil *et al.*, 2001). The observation that contaminant lead was conspicuous in the margins of the Eurasian Basin, but not in the Canadian Basin, led these authors to conclude that ocean and ice transport pathways during peak lead emission years (~ last 60 years) must have been predominantly those associated with the AO- conditions. Under AO+ conditions, pathways for ice, ocean-currents, and runoff change dramatically (Figures D.1.8, D.1.14, D.1.19) so that contaminant metals entering the Russian Shelves (atmospherically or by runoff) would be diverted to the east into the Canada Basin and toward the Archipelago (Gobeil *et al.*, 2001; Mysak, 2001). The pathways illuminated by lead contamination are probably relevent for other contaminants that attach to particles — for example the highly chlorinated PCBs.

Atmospheric aerosols of cadmium and zinc will to some degree behave like lead, with the exception that the predominant source for the latter in leaded gasoline. Time series of aerosol composition at Alert (Sirois and Barrie, 1999) and records from ice cores (Boutron *et al.*, 1995; 1991) and glacial snow (Sherrell *et al.*, 2000) reveal contamination due to industrial activity in Asia, Europe, and North America. Like lead, these elements are poorly captured within the Arctic (< 15%, Akeredolu *et al.*, 1994; Pacyna, 1995) and changes in precipitation patterns probably have the greatest potential to change metal delivery to surfaces.

Of these three heavy metals, cadmium provides the greatest risk through bioaccumulation and biomagnification (Figure D.1.26) especially into the liver and kidneys of marine and terrestrial mammals (Braune *et al.*, 1999; Muir *et al.*, 1999a). Observed high concentrations of cadmium in Arctic biota, however, appear to be natural and not obviously related to human activities except, possibly, at locations close to sources (< 100 km). As a result, significant changes in cadmium exposure are likely to be delivered by changes in the natural cycle of cadmium and not by changes in contaminant cadmium pathways. An exception to this rule may occur locally when cadmium contamination is accompanied by or followed by system changes that alter cadmium biogeochemistry. Croteau *et al.* (2002) provide a clear example in which reductions in cadmium loadings to a contaminated lake were accompanied by increases in pH with the consequence that organisms actually exhibited increased cadmium uptake.

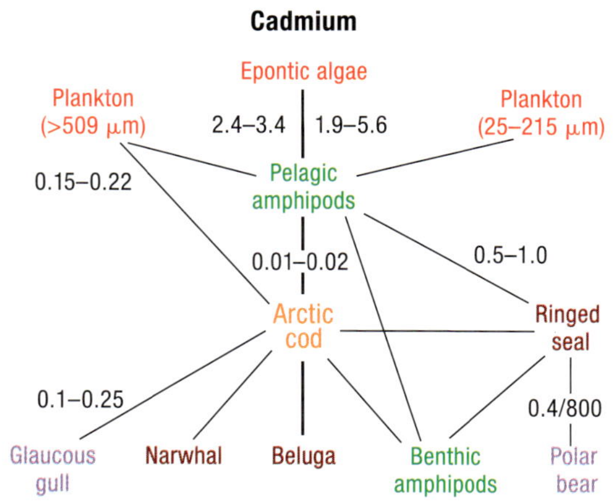

FIGURE D.1.26

Biomagnification factors (BMFs) and pathways for cadmium in the Arctic marine food web. BMFs are based on dry weight concentrations in whole organisms (invertebrates and fish) and on wet weight concentrations for fish to muscle to top predators (for polar bear, liver was used instead of muscle). (Figure from p. 116, Muir *et al.*, 1999).

In the Arctic Ocean, natural cycles completely dominate cadmium fluxes and budgets (Macdonald *et al.*, 2000). Because cadmium follows soft body parts in the marine biogeochemical cycle, it exhibits a strong correlation with phosphate (Boyle, 1988; de Baar *et al.*, 1994). The interaction between the biogeochemical cycle and circulation of the world ocean results in sub-surface water of the North Pacific containing naturally higher cadmium concentrations than those of the North Atlantic (by a factor of about 5 — see Bruland and Franks, 1983). This in turn makes the Pacific inflow through Bering Strait by far the dominant source of cadmium to surface waters of the Arctic Ocean (Macdonald *et al.*, 2000). Reduced Bering inflow since 1940 probably entails a similar, 15% reduction of the supply of cadmium to the Arctic Ocean from that source. The encroachment of Atlantic water into the surface of the Makarov Basin, seen under recent AO+ conditions, will further reduce the domain of cadmium-rich water within the Arctic (Figures D.1.18, D.1.20). The accompanying increased stratification and recycled production in the smaller Pacific domain of the Canada Basin will, however, tend to maintain cadmium from runoff or atmospheric deposition at the surface. It is noteworthy that the Archipelago is the downstream recipient of water from the Pacific Ocean and, therefore, the recipient of cadmium and nutrients from that source.

Recent work on metal-impacted lakes near a copper-smelting centre in Quebec, Canada, shows that metal loadings (cadmium, copper, zinc) can alter the ecosystem structure, causing the demise of medium to large benthic invertebrates and producing fish populations shifted to smaller sizes (Sherwood *et al.*, 2001; Sherwood *et al.*, 2000). This suggests a strong possibility that contamination by these heavy metals could interact with the accumulation and biomagnification of other contaminants like mercury and the organochlorines, actually reducing concentration of the latter in apex feeders.

D.1.6.1.2 Mercury

Due to its volatility and tendency to undergo biogeochemical transformation, mercury must be considered separately from other heavy metals (see, for example, Fitzgerald *et al.*, 1998; Mason and Fitzgerald, 1996; Mason *et al.*, 1994). Aquatic environments are particularly important because it is there that mercury poses its greatest threat through biomagnification (Atwell *et al.*, 1998; Evans and Lockhart, 1999; Muir *et al.*, 1999a). To a degree, processes leading to enhanced mercury concentration in the environment can be considered as either "solvent switching" or "solvent reduction" (see section D.2.6.3 and Macdonald *et al.*, 2002c). In the former, mercury moves between phases such as air, water, and particles based simply on partition coefficients, whereas in the latter, mercury may achieve high fugacity through the loss of surfaces or through chemical reactions mediated by photons or microbes (see for example processes described in Lindberg *et al.*, 2002; Malcolm and Keeler, 2002).

The natural mercury cycle has been enhanced by human activities such that two to three times as much mercury is presently cycling through the atmosphere and upper ocean than before the rise of industry (Lamborg *et al.*, 2002; Mason and Fitzgerald, 1996; Mason *et al.*, 1994; Pacyna and Keeler, 1995). Because atmospheric mercury is almost entirely gaseous Hg^0, it is tempting to assume that the polar regions might be global sinks simply due to low temperature. Accordingly, projections could be made of the effects of climate change based on this alone. In fact, thermal forcing for Hg^0 to Arctic aquatic systems either by rain or by air-water exchange is weak, simply due to a relatively high Henry's Law constant (Macdonald *et al.*, 2000; Mason *et al.*, 1994).

Nevertheless, the Arctic may be especially vulnerable to mercury because of an extraordinary set of circumstances that occur at polar sunrise whereby reactive (and bioavailable) mercury is deposited to the surface (Figure D.1.27, Lindberg *et al.*, 2002; Lu *et al.*, 2001; Schroeder *et al.*, 1998; Scott, 2001). A relatively long residence time for Hg^0 of 1 — 2 years in the atmosphere (Lamborg *et al.*, 2002) ensures that winds can transport mercury to the Arctic. With polar sunrise, Hg^0 is converted to reactive gaseous mercury (RGM) through reaction sequences in which bromine, chlorine, and compounds like BrO and ClO have prominent roles (Figure D.1.27 and see Lindberg *et al.*, 2002; Lu *et al.*, 2001). Reactive gaseous mercury is then very effectively removed from the atmosphere by particles/snow. This process is estimated to account for the deposition of ~50 tonnes (> 90% of the annual total) on the Arctic Ocean and Hudson Bay during spring (Lu *et al.*, 2001) and perhaps as much as 150–300 tonnes/year (Lindberg *et al.*, 2002). Based on snow samples and the distribution of BrO from satellite measurements, reactive or particulate mercury is probably deposited on most of the land in northern Canada and in the Archipelago, and on much of the Arctic's marginal seas.

With spring warming, it is estimated that about two-thirds of the mercury deposited in snow re-volatilizes (Brooks, pers.comm.) with the remaining third entering aquatic environments through meltwater. Before projecting the impact of global change on the mercury cycle in the Arctic, one more fact needs to be grasped. The invasion of mercury to global aquatic reservoirs via wet and dry deposition of reactive forms, is balanced by gaseous evasion of reduced forms of mercury such as Hg^0 or methylmercury (Mason *et al.*, 1994). The Arctic's vulnerability to global mercury emissions probably lies in the mismatch between invasion and evasion processes. Meltwater and runoff can drain into surface water below the ice cover, whereas the evasion of gaseous mercury is partly or completely blocked by ice cover. Indeed, this set of circumstances may provide the foundation for elevated mercury levels which have been observed in Arctic biota (Evans and Lockhart, 1999; Macdonald *et al.*, 2000; Muir *et al.*, 1999a; Wagemann *et al.*, 1995; 1996). Unfortunately, appropriate geochemical studies investigating the cycling of mercury in Arctic aquatic systems still need to be conducted and we are forced at this time to the above speculations.

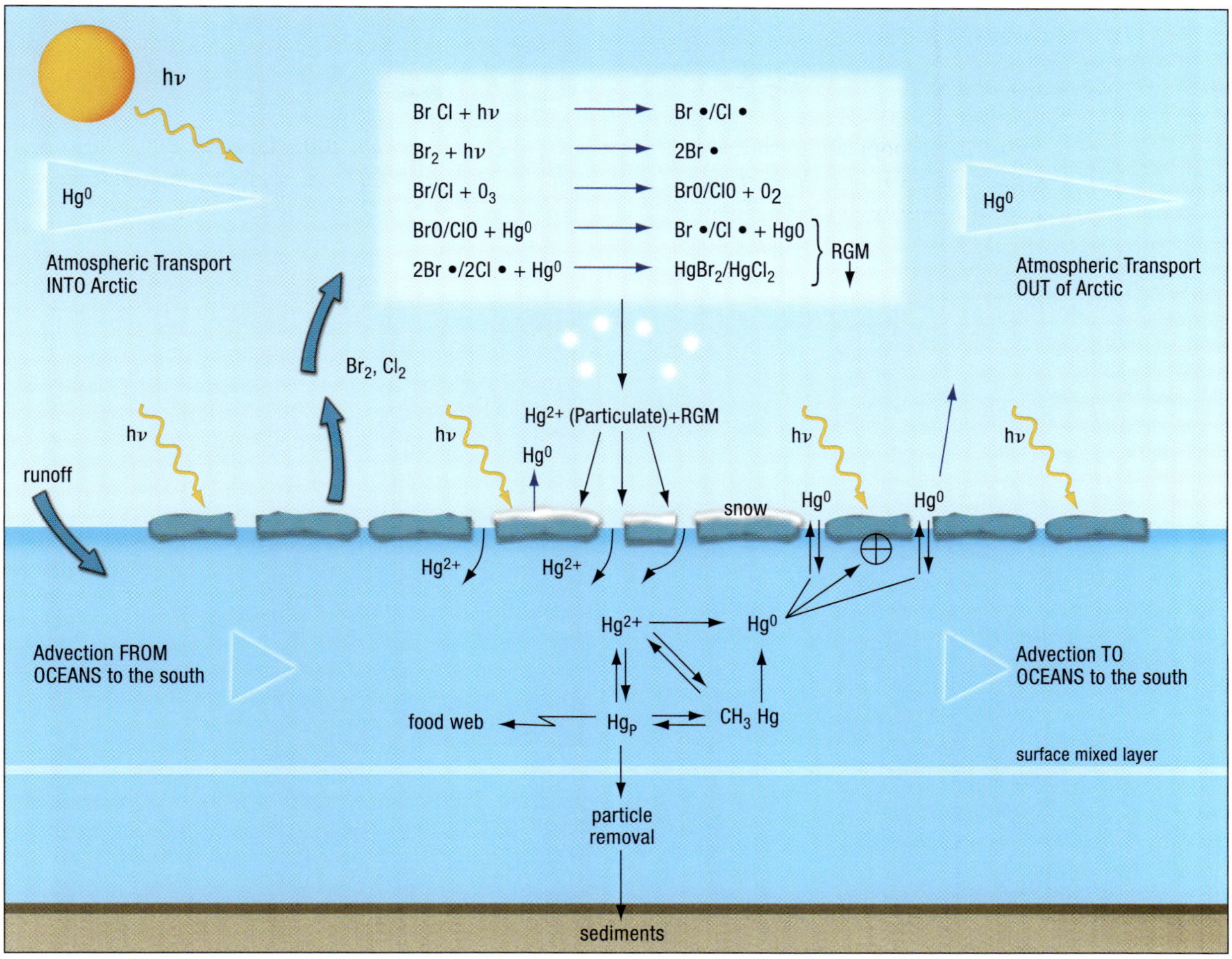

FIGURE D.1.27

A schematic diagram illustrating how atmospheric mercury gets scavenged after polar sunrise and thence enters surface waters (for mercury reactions schemes in atmosphere and ocean, see Lamborg *et al.*, 2002, Lindberg *et al.*, 2002).

Climate change can affect both the invasion and evasion routes for mercury. Springtime depletion of atmospheric mercury depends on the availability of sea salt, calm weather, a temperature inversion, the presence of sunlight and sub-zero temperatures (Lindberg *et al.*, 2002; Lu *et al.*, 2001). With climate change, it is likely that increased amounts of first-year ice around the polar margins will contribute initially to generally saltier ice and snow in spring, which will enhance the production of BrO and ClO. Depending on what controls the rate of supply of mercury to the Arctic, increased BrO and ClO will either enhance scavenging or maintain it at present levels, possibly extending the area of springtime mercury depletion beyond that implied by recent satellite measurements of the distribution of BrO/ClO (see Lu *et al.*, 2001).

Considering that global emissions of mercury have actually been decreasing recently, Lindberg *et al.* (2002) proposed that the recent increases in mercury levels observed in Arctic biota are, in fact, evidence that MDEs may be a recent phenomenon due to change in sea-ice climate over the past decade or two. However, the entire mercury cycle must be considered before linking MDEs with mercury concentrations in apex aquatic feeders. Larger areas of open water in spring, either in ocean or lakes, will enhance exchange, allowing gaseous forms of mercury to escape back to the atmosphere (Figure D.1.28). With further warming, parts of the Arctic will become more temperate in character, implying that atmospheric mercury depletion would decrease, and evasion from the water increase leading eventually to lower mercury concentrations in water.

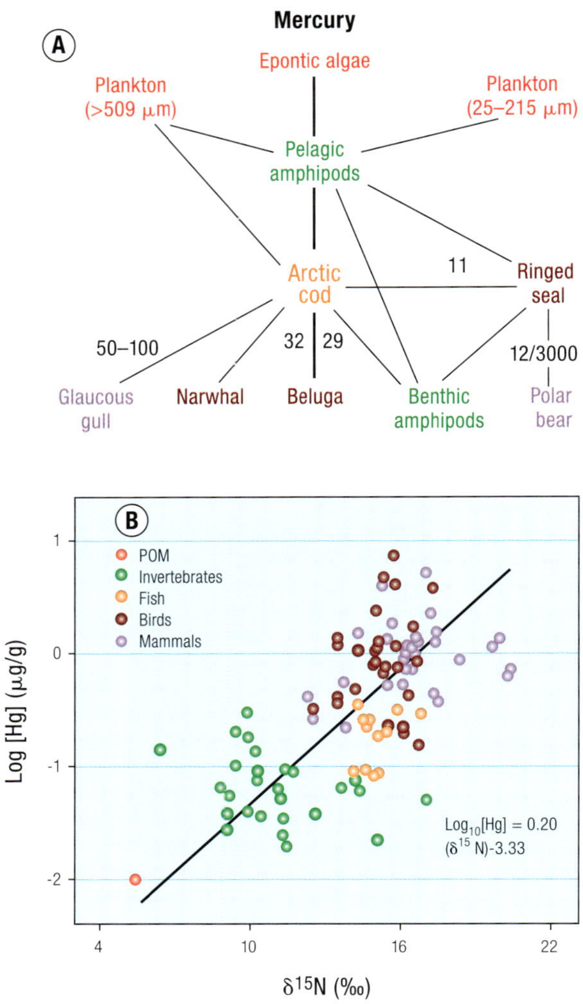

FIGURE D.1.28

a) Biomagnification factors for a simplified food web [based on dry weight concentrations in whole organisms (invertebrates and fish) and on a wet weight concentration for fish to muscle to top predators (for polar bear, liver was used instead of muscle) (Muir *et al.*,1999)] and b) biomagnification as a function of trophic level based on $\delta^{15}N$ measurements (adapted from Atwell *et al.*, 1998).

Aquatic food webs have strong leverage for change in mercury exposure in the Arctic because methylmercury biomagnifies, exhibiting a concentration increase of about 1000–3000 from particulate organic matter to apex predators (Figure D.1.28a, b, Atwell *et al.*, 1998; Kidd *et al.*, 1995; Muir *et al.*, 1999a). Mercury concentration also increases with age/size in predatory fish. Large, old fish often exceed thresholds considered safe for unrestricted human consumption (Lockhart and Evans, 2000), containing anywhere from two to five times the mercury concentration in smaller fish. Therefore, adding an extra step in the food web could enhance mercury concentrations in higher trophic levels by a factor of five. Likewise, altering the population distribution of fish in a lake could produce a change rivaling or exceeding that caused by alteration of physical pathways.

With warming (Figure D.1.3) will come the loss of permafrost (Figure D.1.22), which then leads to altered hydrology, potentially more wetland, and enhanced fluxes of soil and organic carbon to rivers, lakes, and estuaries. Warming of drainage basins in the Arctic, therefore, would appear to provide a widespread mechanism to increase mercury fluxes to northern aquatic environments and to the atmosphere. A recent study of the pathway of mercury from snow-covered land to streams in Vermont (Stanley *et al.*, 2002) showed that mercury export from soils correlated with particulate organic carbon, and that mercury concentrations in runoff increased with flow — unlike most solutes (see also Bishop *et al.*, 1995). These two factors together suggest that episodic, large releases of organically-bound mercury (contaminant and natural) may accompany permafrost degradation. Clearly, Arctic lakes would be most vulnerable to this process, but enhanced input of terrestrial carbon is projected to occur to Arctic seas as well (Kabat *et al.*, 2001) suggesting that mercury loadings there may, similarly, be increased. In the ocean, Hudson Bay would seem especially vulnerable partly due to its large drainage basin, already affected by reservoir flooding (Bodaly and Johnston, 1992), and partly due to the likelihood of permafrost melting within that drainage basin (Gough and Wolfe, 2001). It seems noteworthy that enhanced mercury concentrations in snow are observed generally in that region (Lu *et al.*, 2001) and that a recent increase in mercury flux to Hudson Bay sediments has also been observed (Lockhart *et al.*, 1995).

Historical records from dated sediment cores have been used to infer mercury fluxes increasing by factors of 3–7 in some Arctic lakes during the past two centuries (Bindler *et al.*, 2001; Landers *et al.*, 1995; Lockhart *et al.*, 1998). What is not so clear is whether such increases are due to increased atmospheric deposition or to alteration of processes that transfer mercury from wetlands to atmosphere, or from the drainage basin into lakes, or from the water to the sediments. In regard to the latter, Gajewski *et al.* (1997) have shown major increases in diatom fluxes to varved sediments from a lake on Devon Island, which they attribute to climate change (i.e., longer ice-free summers). Not only could such a mechanism explain enhanced mercury fluxes to Arctic lake sediments but it could also have the non-intuitive result of *reducing* the exposure of higher trophic levels to mercury through bloom dilution at the algal stage (Pickhardt *et al.*, 2002).

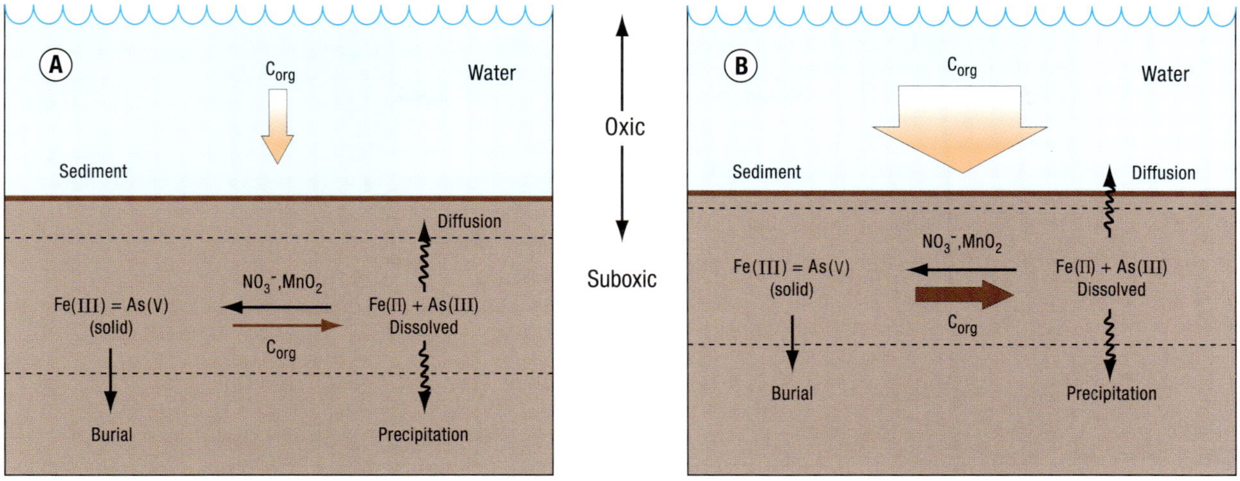

FIGURE D.1.29

A schematic diagram showing how arsenic cycles in sediments (modified from Sullivan and Aller, 1996). The solid-phase form of arsenic (As(V)) is released to pore water through reduction to As(III) which then may diffuse back out of sediments. Enhanced fluxes of organic carbon to sediments may therefore lead to a release of dissolved arsenic to bottom waters.

Mercury depletion events leading to the production of bio-available mercury have been confirmed at Barrow, Alaska (Lindberg *et al.*, 2002). Given that this region receives airborne contaminants from Asia (Li *et al.*, 2002; Wilkening *et al.*, 2000), that China is increasing its reliance on coal for energy, and that coal is a leading source of anthropogenic mercury (Nriagu and Pacyna, 1988), it seems likely that the Bering/Chukchi Sea region may be especially vulnerable to further increased mercury loadings.

D.1.6.1.3 Arsenic

The sediment geochemistry of arsenic can be altered by changing the carbon cycle. Mining and smelting have been major sources of arsenic to the environment (Nriagu, 1989) where it may accumulate in aquatic sediments (Martin and Pedersen, 2002). In the Arctic, it has also been suggested that underwater testing of nuclear weapons may have provided a significant source of contaminant arsenic to the Pechora Sea (Loring *et al.*, 1995). Diagenesis often produces large natural surface sediment arsenic enrichments by remobilizing solid-phase As(V) at depth in sediments through reduction to As(III) which then diffuses upward to re-precipitate through reactions with nitrate, oxides of manganese or oxygen (Figure D.1.29). This natural redistribution of arsenic in sediments makes it exceptionally precarious to infer contamination from sediment surface arsenic enrichments (Loring *et al.*, 1995; 1998; Siegel *et al.*, 2001). However, strong sediment-surface enrichments serve as a warning that alteration of the organic geochemistry of aquatic sediments by enhanced organic carbon fluxes may have unanticipated consequences for arsenic mobility.

A geochemical study of a temperate, seasonally ice-covered lake impacted by copper, zinc, nickel, and arsenic is especially instructive (Martin and Pedersen, 2002). When action was taken to reduce metal loadings, the lake's response was enhanced phytoplankton production which then invigorated carbon fluxes to sediments. Through metabolism in sediments, the enhanced carbon fluxes then reduced oxygen, thus mediating the conversion of solid-phase As(V) to dissolved-phase As(III), which then diffused back into the lake's bottom water. The unanticipated result of decreasing metal loadings to the lake was to produce higher arsenic concentrations. If one of the responses to change in Arctic aquatic environments is enhanced aquatic productivity or enhanced organic carbon loadings, release of solid-phase arsenic, especially from sediments with high natural or contaminant burdens (e.g., see Loring *et al.*, 1995; 1998; Siegel *et al.*, 2001) is a likely response.

D.1.6.2 Radionuclides

Previous assessments have outlined the atmospheric and oceanic pathways that transport artificial radionuclides to the Arctic (AMAP, 1998; Macdonald *et al.*, 2000). Atmospheric sources of artificial radionuclides derive mainly from atmospheric weapons testing predating the mid-1960s and Chernobyl in 1986. Accordingly, predominant food web contamination from the atmosphere has occurred to terrestrial systems through fallout (AMAP, 1998). There appears little opportunity for present or future climate changes to have much impact on historical atmospheric sources, which are decaying away with half-lives of about 30 years (^{137}Cs and ^{90}Sr, predominantly).

The distribution of fallout from Chernobyl was very much controlled by wind and rainfall patterns (see Figures 8.4 and 8.5 in AMAP, 1998). These pathways are clearly subject to climate change. An important lesson to be learned from the Chernobyl accident is that climate patterns can predispose the Arctic as to how it will receive contaminants from such accidents. For example, wetter conditions in the Nordic Seas and Northern Europe during AO$^+$ conditions (Figure D.1.9) would favour deposition of ^{137}Cs fallout in that region.

In the ocean, the predominant artificial radionuclides (e.g., ^{137}Cs, ^{90}Sr, $^{239+240}$Pu) do not biomagnify sufficiently (excepting perhaps Pu (and ^{99}Tc) in brown macroalgae and see Berrow *et al.*, 1998; Brown *et al.*, 1999; Fisher *et al.*, 1999) to contribute significantly to the radiation dose for humans (Layton *et al.*, 1997; Macdonald and Bewers, 1996). This suggest that we must look to the ice and surface water motion of the Arctic Ocean, both of which undergo alteration with the AO/NAO (Figures D.1.14, D.1.19), to find pathway changes that might be cause for concern.

The enhanced northward transport of water in the Nordic Seas under the AO$^+$ conditions of the 1990s clearly increased the delivery to the Arctic of radionuclides discharged by the European nuclear reprocessing plants (Figure D.1.20). From there they continued to spread into the surface waters of the Makarov Basin (Figures D.1.18 and D.1.20 Smith *et al.*, 1998).

A shift in the climate regime toward increased NAO and stronger wind fields will probably lead to radionuclides undergoing a faster transport closer to the Norwegian coast, and a larger proportion of them will enter the Barents Sea. It is expected that other European contaminants entering the North Sea and southern Norwegian Sea will encounter similar change in their oceanic transport route as that projected for the radionuclides.

The enhanced coupling between release points for European reprocessing plant nuclear wastes and the Arctic Ocean will be more than offset by reduction in releases since the 1970s (Macdonald *et al.*, 2000), and input of these tracers to the Arctic Ocean should continue to decline. Due to extensive data collections under the Arctic Nuclear Waste Assessment Program (ANWAP, Layton *et al.*, 1997) and from icebreakers (Smith *et al.*, 1998), we have a reasonable idea of the distribution of artificial radionuclides in Arctic surface waters and are confident that they pose little risk to the health of humans or ecosystems. The ANWAP assessment concluded that the largest radiation doses to individuals living on the Alaskan coast and consuming subsistence seafoods were, in order of importance, ^{210}Po (a natural radionuclide), followed by ^{137}Cs and ^{90}Sr from atmospheric fallout. It seems that the nuclear reprocessing radionuclides have made elegant tracers of water motion and, although they will reflect recent changes in that motion, they will continue to provide almost no risk to Arctic marine biota (Macdonald and Bewers, 1996).

The changes in ice and surface water pathways forced by the AO (Figures D.1.15 and D.1.19) strongly suggest that radionuclides discharged to the Russian shelves may, under AO$^+$ conditions, enter the Canadian Basin and subsequently find their way into the Canadian Archipelago. The estimated total release of radionuclides to the Kara Sea via river water (Ob, Yenisey) is about 1.1 x 10^{15} Bq (^{90}Sr and ^{137}Cs) (Paluszkiewicz *et al.*, 2001), or about an order of magnitude less than the fallout and reprocessing plant sources to the Arctic Ocean (Aarkrog, 1994). Therefore, diversion of Russian river runoff from the Eurasian Basin to the Canadian Basin under AO$^+$ conditions would probably be matched by a diversion of associated radionuclides (see Cooper *et al.*, 1999) which, nevertheless, will provide little threat to ecosystems there.

Ice drift, the remaining transport pathway to consider, provides a distinct, but difficult to quantify risk. Sediments from the Russian shelves, known to have been contaminated by weapons testing and accidental and deliberate discharges of radiowaste, have been found to be heavily contaminated in several locations (Josefsson, 1998; Matishov *et al.*, 1999; Smith *et al.*, 2000). Suspension freezing provides an efficient mechanism for entraining fine sediments into newly-formed ice in the Russian Seas (Eicken *et al.*, 2000) and ice has been shown to carry radioactive sediments (Dethleff *et al.*, 2000; Landa *et al.*, 1998; Meese *et al.*, 1997). High radioactivity has been found in ice-entrained sediments in the Canada Basin (> 70 Bq/kg, Cooper *et al.*, 1998) and in the Archipelago (Darby, pers.comm.) but the origin of the sediment in the ice, based on mineralogy, has not been assigned to Russian shelves. Given the very few samples together with their uncertain provenance, it is impossible to quantify risks to biota in the Canada Basin and Canadian Archipelago from contaminated ice except to say that the AO^+ conditions of the early 1990s appear to produce ice transport pathways conducive to carrying sediment and surface water from the Russian Shelves into the Archipelago.

Perhaps the most significant increase in radioactivity exposure to northerners will come from the natural ^{226}Ra decay series that supports ^{222}Rn, ^{210}Pb and ^{210}Po. ^{210}Pb in aquatic systems derives partly from *in-situ* production supported by ^{226}Ra and partly from ^{222}Rn which has diffused out of soils and, with a short 3.8 day half-life, decays to ^{210}Pb which is scavenged by particles. This latter component, called excess ^{210}Pb, often exceeds the 'supported' ^{210}Pb in aquatic sediments.

Presently, excess ^{210}Pb tends to be very low in the Arctic because ^{222}Rn remains trapped in the soil by permafrost and snow/ice cover. With warming, ^{222}Rn evasion will increase, as will excess ^{210}Pb activity matched by the activity of ^{210}Po, its granddaughter. Since ^{210}Po and ^{222}Rn together account for about 75% of the radiation dose to northern residents (Layton *et al.*, 1997; Macdonald and Bewers, 1996), any substantive increase in ^{222}Rn evasion due to warming/permafrost melting would have a widespread and substantial (doubling or tripling) effect on the radioactive dose.

D.1.6.3 Organochlorines

Of all the contaminants, the organochlorines (OCs) provide the greatest challenge to predict consequences of change because they have been so widely released, comprise so many compounds and exhibit such a wide range of physicochemical properties. Furthermore, the important chemical properties, volatility, phase partitioning, and degradation kinetics, are all sensitive to temperature and hydrological change. Efforts to determine where in the environment these compounds end up has improved our understanding of global pathways enormously, but surprises alert us that intuition often fails because we do not have a complete grasp of environmental processes (Macdonald *et al.*, 2001; Oreskes *et al.*, 1994; Schindler, 1997).

Recent OC budgets underscore the importance of the atmosphere — ocean partnership in the transfer of OCs to the Arctic from their temperate and tropical release points (Li *et al.*, 2002; Macdonald *et al.*, 2000; Macdonald *et al.*, 2001). These same budgets show that the relative importance of atmosphere versus ocean in transporting contaminants will vary widely between OC compounds and over time. Therefore, change forced by the AO or by general global change will also have a varied impact depending on the particular OC and the time period in question. All of the OCs of concern (DDT, toxaphene, chlordane, PCBs, and HCHs) have transient emissions: that is, they were first released in the 1930s to 1940s, peak sometime in the 1970s to 1990s, and emissions will continue to decline or cease, or possibly increase, depending on bans, further controls, or invigorated assaults by pests provoked by climate change. As a general rule, particularly with transient releases of chemicals that partition strongly into water, the atmosphere is initially the dominant transporting medium. As aquatic reservoirs (lakes, rivers, upper ocean) become loaded, however, these then contain the important, if not dominant, budget and flux terms (Li *et al.*, 2002; Macdonald *et al.*, 2000; Macdonald *et al.*, 2001) and it is dominant budget terms that posses the greatest leverage for change.

Before discussing how climate change is likely to impact organochlorines, it is important to understand the manner in which organochlorines become concentrated in the environment. Building on earlier perspectives developed by Wania (1999a), Macdonald *et al.* (2002) suggest that there are only two fundamental concentrating processes which they termed "solvent switching" and "solvent depletion". The distinction between these two processes is particularly important in the context of change.

Solvent switching is a natural process whereby a particular contaminant distributes itself between various phases (solid, liquid, gas) according to well described thermodynamic rules. This process can lead to elevated concentrations. For example, the partitioning of HCH into cold water will produce concentrations in the water that far exceed those in the air, and indeed this process alone can cause non-intuitive divergence of chemicals over large scales (Li *et al.*, 2002). But this process cannot cause concentrations to exceed thermodynamic equilibrium (that is, solvent switching does not increase fugacity). Another very important solvent-switching process is the transfer of a POP from water to lipid at the bottom of the food web. Again, this process enhances POPs concentrations in phytoplankton or small zooplankton strictly according to thermodynamic forcing.

Solvent depletion, however, is a very different process than solvent switching in that it can lead to concentrations in selected media that exceed thermodynamic equilibrium (i.e., fugacity is increased), which requires a source of energy. Perhaps the clearest example of solvent depletion occurs in food webs where there are inefficient lipid transfers from lower to higher trophic levels. Metabolism effectively burns much of the lipid, leaving the contaminant to accumulate in a decreasing volume of stored fat. This process can lead to organochlorine concentrations in aquatic apex feeders that are well above thermodynamic equilibrium with the water, and the situation can be exacerbated by starvation cycles during which individuals reduce their fat content even further.

Arctic and other cold environments offer several solvent depletion processes by which contaminants can be ramped up above thermodynamic equilibrium (Macdonald *et al.*, 2002c). In particular, a POP can be scavenged by absorption to snow and thence transported to ground. The initial large snow surface ($0.1\ m^2\ g^{-1}$) can be reduced by a factor of over 100 and, during melting, the snow surface can actually disappear entirely, effectively removing the solvent (the snow's surface) from under the contaminant. The fugacity can be increased enormously during snow sintering/melt with the consequence that the organochlorine is forced to diffuse back to the atmosphere, enter the meltwater, or adsorb to other particles. Similarly, fogwater can provide a temporarily large surface area which is thermodynamically attractive — upon coalescence, most of the surface is lost and contaminant fugacity increases in the remaining large water drops. Finally, cryo-concentration may occur in shallow lakes or seas that form a thick ice cover. The withdrawal of water into the ice, leaving behind most of the POPs, can easily increase concentrations in the water beneath the ice to values that exceed thermodynamic equilibrium with the atmosphere by factors of 2 to 5. Due to the ice cover, diffusion back into the air is not an option.

Under change scenarios, the solvent switching processes can be modeled by taking into account changes in partition coefficients and vapour pressures with temperature. However, changes in the solvent reduction processes are much harder to project, especially since the elevated fugacities imply transports that will be sensitive to time. For example, the effect of snow on contaminants entering an Arctic lake will depend on when the snow accumulates, how long it sinters, how porous and how deep the snow is, how and when snow melt occurs, and how quickly snow melt enters the lake. None of these transfer processes has been investigated in sufficient detail to provide guidance to modelling.

D.1.6.3.1 The influence of the Arctic Oscillation on organochlorines

There appear to be several consequential ways that physical pathways could change in response to the AO. In the Nordic Seas, the atmospheric coupling of eastern North America and western Europe with the Arctic becomes more intense, especially during winter/spring (Figure D.1.8a). This suggests that spraying pesticides in these regions will, if anything, intensify 'events' like those seen at Alert for α- and γ-HCH (see, for example, Figure 22 in Macdonald *et al.*, 2000).

Furthermore, re-emission of old OC residues in soils and aquatic reservoirs of Europe or eastern North America will enter these same air pathways to be transported north. In this regard it is worth noting that the highest cumulative use of PCBs was in western Europe and Eastern United States (see Figure 2.3.1. in de Wit *et al.*, 2003). Both are source regions for air masses entering the Arctic between Greenland and Europe (Figure D.1.8a). The higher precipitation in the Nordic Seas and southern Eurasian Basin (Figure D.1.9) will provide more effective scavenging of particle-associated OCs (high molecular weight PCBs, for example) and for OCs with low Henry's Law constants (HLCs) that partition strongly into water (HCHs, toxaphene).

In the Bering Sea, gas exchange with the ocean or washout by rain can provide a mechanism to remove β-HCH selectively from the air as it heads northward simply due to its exceptionally low HLC (Li *et al.*, 2002). This process does not prevent the entry of the contaminant to the Arctic Ocean; rather, it switches the mode of delivery from winds to ocean-currents, which consequently slows the rate of transport from $m\ s^{-1}$ to $cm\ s^{-1}$. Under AO^+ conditions, therefore, we predict a more rapid atmospheric transport of OCs into the Arctic from western Europe, with the delivery shifting toward the ocean pathway for OCs that partition strongly onto particles or have low HLC. On the Pacific side, OCs will continue to enter the Arctic via the atmosphere and ocean-currents (see for example Bailey *et al.*, 2000) but the 15% reduction in Bering Sea inflow over the past several decades would cause a proportional reduction in this pathway. Variation in precipitation over the North Pacific and Bering Sea will alter the balance between atmosphere and ocean as transport pathways to the Arctic.

Larger areas of open water under AO^+ conditions or from general climate change (Figures D.1.11, D.1.12, D.1.13) will accelerate equilibrium between air and sea by an amount equivalent to the expanded open areas. Furthermore, increased numbers of polynyas in winter will enhance the production of fog over sea ice, acting to scavenge and deposit contaminants to surface (Chernyak *et al.*, 1996) at locations known to be important for biota (Stirling, 1997). Due to the drastic reduction in atmospheric concentration through emission controls, α-HCH has become oversaturated in ice-covered areas of the Arctic Ocean (Jantunen and Bidleman, 1995; Macdonald *et al.*, 2001). The opening of the pack and the clearance of shelves seasonally will, in this case, result in evasion and drawdown of HCH from the upper ocean. In contrast, PCBs and toxaphene are still loading into the Arctic Ocean via the atmosphere (Macdonald *et al.*, 2001) and, therefore, the same loss of ice cover will lead to increased loading of the Arctic's surface water with these two chemical groups. A PCB budget for the Arctic Ocean (Macdonald *et al.*, 2001) estimated a net gas exchange into the ocean of about 20 tonnes/year. Reduced ice cover evident in Figure D.1.13 might lead to as much as a doubling of the area of open water, which would similarly double net exchange.

The diversion of Russian river inflow toward the east under AO^+ conditions (Figure D.1.19) will have a significant effect on OC pathways within the Arctic Ocean. Inherent with this diversion is a shift of all the OC loadings from these Russian rivers out of the Eurasian Basin and into the Canada Basin (Table D.1.1 and see, Macdonald *et al.*, 2000; Sericano *et al.*, 2001). As the Canada Basin has a longer residence time (10 years compared to 2 years in the TPD), there would be an added consequence of increased contaminant inventories for the Arctic Ocean in general and the Canada Basin in particular. Instead of tracking across the Eurasian Basin to exit into the East Greenland current, OCs discharged by the Russian rivers might now exit via the Archipelago (see Figure D.1.19). The changes here are consequential to budgets (Table D.1.1) and to distribution within the water column, keeping in mind that the same river water that delivers contaminants also stratifies the ocean and potentially reduces new production and vertical particle flux, which together will act to maintain river-borne contaminants near the surface where they can partition into algae. Although there are few data from which to evaluate the relative importance of organochlorine pathways in the ocean, the findings of Andersen *et al.* (2001) provide a strong warning that there are sources of PCB in the region around the Kara Sea and Franz Jozef Land, possibly as a result of river inputs.

Water in the channels of the Canadian Archipelago, which is supplied from surface water in the Arctic Ocean (~0-200 m), has the potential to undergo change in its organochlorine content due to alterations in the distribution of water masses in the Canada or Eurasian Basins. As shown in several studies (Carmack *et al.*, 1997; Li *et al.*, 2002; Macdonald *et al.*, 2000; Macdonald *et al.*, 1999c), HCHs are not distributed uniformly within the Arctic Ocean and it is likely that other organochlorine compounds are not uniformly distributed either. For example, α-HCH is highest near the surface, decreasing to very low values in water deeper than several hundred meters, and the Canada Basin, under the permanent pack, exhibits much higher HCH concentrations than are observed in surface waters of the Chukchi Sea or the Eurasian Basin.

The redistribution of Pacific and Atlantic water masses in surface water of the Arctic Ocean (Figures D.1.18 and D.1.20) may therefore have been accompanied by change in the composition of water flowing into the Archipelago. Such change could occur in two ways, either by horizontal displacement of water mass domains or by vertical displacement of water properties. Although we have few data from which to evaluate how the water composition in the Archipelago channels might respond to the Arctic Oscillation, an extraordinary set of data collected by Hargrave *et al.* (1997) illustrate clearly that change in upstream water composition can have important consequences for contaminant concentrations in water flowing through the Archipelago.

TABLE D.1.1 Russian river loadings for selected organochlorine compounds

Organochlorine	Loading (tonnes/year)	% of Arctic Ocean input budget	Reference
α-HCH	25	13	(Alexeeva *et al.* 2001, Macdonald *et al.* 2000)
γ-HCH	44	51	(Alexeeva *et al.* 2001, Macdonald *et al.* 2000)
ΣPCB	15	23	(Macdonald *et al.* 2001)
ΣDDT	18		(Alexeeva *et al.* 2001)

Seasonal measurements of organochlorine compounds made at Resolute in 1993 (Figure D.1.30) reveal a coherence between the surface and 50 m water depth in contaminant trends. This coherence, together with the fact that the patterns differ among organochlorine compounds, argues strongly that the observed time trends at Resolute are produced by variation in the composition of upstream water drawn into the Archipelago from the Canada Basin. This single season's data contain concentration variations exceeding a factor of two.

D.1.6.3.2 The effect of glacial meltback

Glacier ice-mass loss and snow meltback due to warming (cyclical or trend) can release archived contaminants accumulated during years of higher fluxes (Blais *et al.*, 1998). Based on the total amount of glacial meltback (Figure D.1.23) and the range of concentration of OC contaminants measured in ice and snow from the 1960s to 1990s for the Agassiz Ice Cap (Table D.1.2, Franz *et al.*, 1997; Gregor *et al.*, 1995), we would expect the maximum total input to be about 3 kg for PCBs to perhaps as much as 400 kg for DDT. In the year exhibiting the most substantial meltback on record (1993), we estimate a maximum release of about 0.5 kg PCBs and 74 kg DDT. The amounts for HCH and PCB are miniscule compared to Arctic Ocean budgets (Macdonald *et al.*, 2001) and are also relatively small compared to the flux of these contaminants through the Archipelago (Table D.1.2, right hand column) with the exception of DDT for which glacial melt appears potentially to provide an important, climate-modulated source. For Arctic glaciers, most of the melt occurs in zones where old ice of pre-industrial age is emerging. This ice would contain little or no contaminant and, indeed, would act to dilute any released contaminants. For smaller ice caps, more recent layers of snow and ice might be involved. It is therefore concluded that glacial meltback could only be of significance for DDT and, even there, it is likely to have only local and short-lived significance.

D.1.6.3.3 The effect of warming on organochlorine cycling in lakes

Arctic lakes presently tend to retain only a small fraction of the contaminants delivered to them, something that should alert us to the potential for change. Studies and models (Helm *et al.*, 2002b; Macdonald *et al.*, 2000) show that the snowmelt and runoff cycle connects with the lake's hydrological cycle so that most of the contaminants deposited in the drainage basin throughout winter transport across the lake surface in a low density layer under the ice to exit in out-flowing water. The lack of a strong particle flux due to oligotrophic conditions further decouples deep lake water from contaminants at the surface. Reduced ice cover and loss of permafrost, leading to greater mixing and stronger primary production, will enhance the ability of Arctic lakes to retain OCs. It is less certain how this added retention will play out in the food web.

Enhanced primary production and settling of ungrazed phytoplankton in early spring might draw down contaminant burdens in lake surface water and thereby

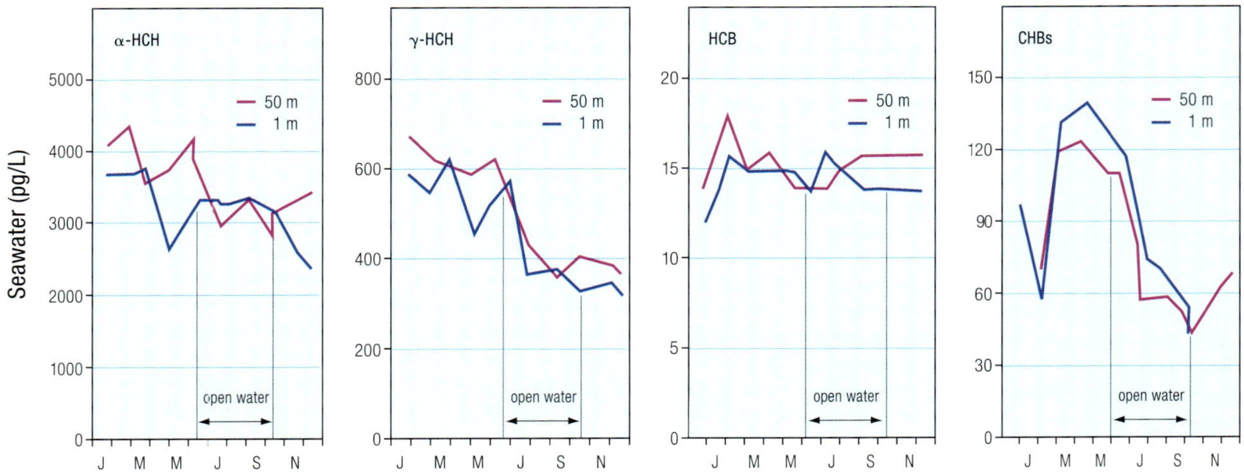

FIGURE D.1.30

Organochlorine concentrations measured at Resolute (Canadian Archipelago) during 1993. Shown are time series at the surface and 50m water depth for α- and γ-HCH, hexachlorobenzene (HCB), and toxaphene (CHB). The coherence in trend between the surface and 50 m water depth for all four contaminants strongly suggests the mechanism to be change in upstream (Canada Basin) water properties (source, Hargrave *et al.* 1997).

TABLE D.1.2 Potential input of selected organochlorines from glacial melt

Compound	Concentration (pg/L)	Total glacial input (kg)	Glacial input for 1993 (kg/yr)	Flux through the Archipelago[(3)] kg/yr
α–HCH[(1)]	256	205	39	195,000
γ-HCH[(1)]	115	92	18	27,900
ΣDDT[(1)]	480	384	74	161
CHLOR[(1)]	35	28	5	96
HCB[(1)]	65	52	10	810
PCB[(2)]	3.5	2.8	0.5	2,700

[(1)]Concentration data from Franz *et al.* (1997); samples collected for 1987, 1990 and 1992 in snow layers after first year loss.

[(2)]Concentration data from Gregor *et al.* (1995); average concentration over 30-year period from 1964/65 to 1992/93 (n = 34).

[(3)]Flux through Archipelago was estimated assuming a mean flow of 54,000 km^3/yr (Macdonald *et al.* 2000a) and using concentration data collected in the Archipelago in 1993 (HCH, HCB, CHLOR, Hargrave *et al.* 1997) or for the Canada Basin in 1997–98 (PCB, DDT, Macdonald *et al.*, 2001b).

reduce entry of contaminants into the food web or act to dilute the organochlorine concentration in algae as has been shown for mercury (Pickhardt *et al.*, 2002). However, if a given lake has a very low sedimentation rate and most of the organic carbon depositing to sediments becomes metabolized, it is likely that OCs associated with the particle flux will be released to the bottom water and re-mixed within the lake as has been observed in Lake Superior (Jeremiason *et al.*, 1994). This process of drawdown and remineralization could slowly ramp up water column concentrations below the lake's thermocline, especially if ice cover is sufficient to hinder exchange during lake turnover. It appears that climate change has the potential to cause substantive physical and biological changes in northern lakes that would alter OC pathways. We are presently woefully ill-prepared to predict what those changes might be.

D.1.6.3.4 The effect of warming on chemical partitioning and degradation

Physicochemical properties that are sensitive to temperature include vapour pressure (p^o) and Henry's Law Constant (HLC, alternatively expressed as the air-water partition coefficient, K_{AW}), octanol-air partition coefficient (K_{OA}), and the octanol-water partition coefficient (K_{OW}) (Table D.1.3). Particle-gas partition coefficient (K_p), which depends both on particle composition and chemical composition, also varies with temperature.

The extent to which chemicals are associated with aerosols is key to their atmospheric transport to the Arctic. Association with particles may on one hand slow or reduce transport to the Arctic through temporary or permanent deposition to surfaces. On the other hand, association with particles may protect a chemical from oxidation during transit to the Arctic. The potential for temperature to alter partitioning between gas and aerosol phases appears greatest for chemicals that exhibit log K_{OA}s in the range of 11 to 14 (Figure D.1.31). For example, with a temperature rise from winter (-30°C) to summer (0°C), DDT goes from being over 70% on particles to being almost entirely in the gas phase. Similar changes in partitioning are evident for many of the other organic compounds illustrated in the figure. Chemicals with K_{OA}s at -30°C of about 11-12, therefore, are most sensitive to change in atmospheric transport during temperature rise.

TABLE D.1.3 Physical parameters sensitive to temperature change

Parameter	Abbreviation	Unit	Description/ Application
Henry's Law Constant	HLC	Pa m^3/mol	Partitioning between air and water.
Air-water partitioning coefficient	K_{AW}	Unitless	K_{AW} = HLC/RT
Octanol air partition coefficient	K_{OA}	Unitless	Used as a proxy to model partitioning between air and organic phases such as vegetation, soil, sediment organic carbon, and particles in air and water.
Vapour pressure	p^0	Pa	Describes tendency of a chemical to volatilize.
Particle-gas partition coefficient	K_p	m^3/μg	The ratio of chemical concentration on atmospheric particles (ng/μg) to concentration in the gas phase (ng/m^3).

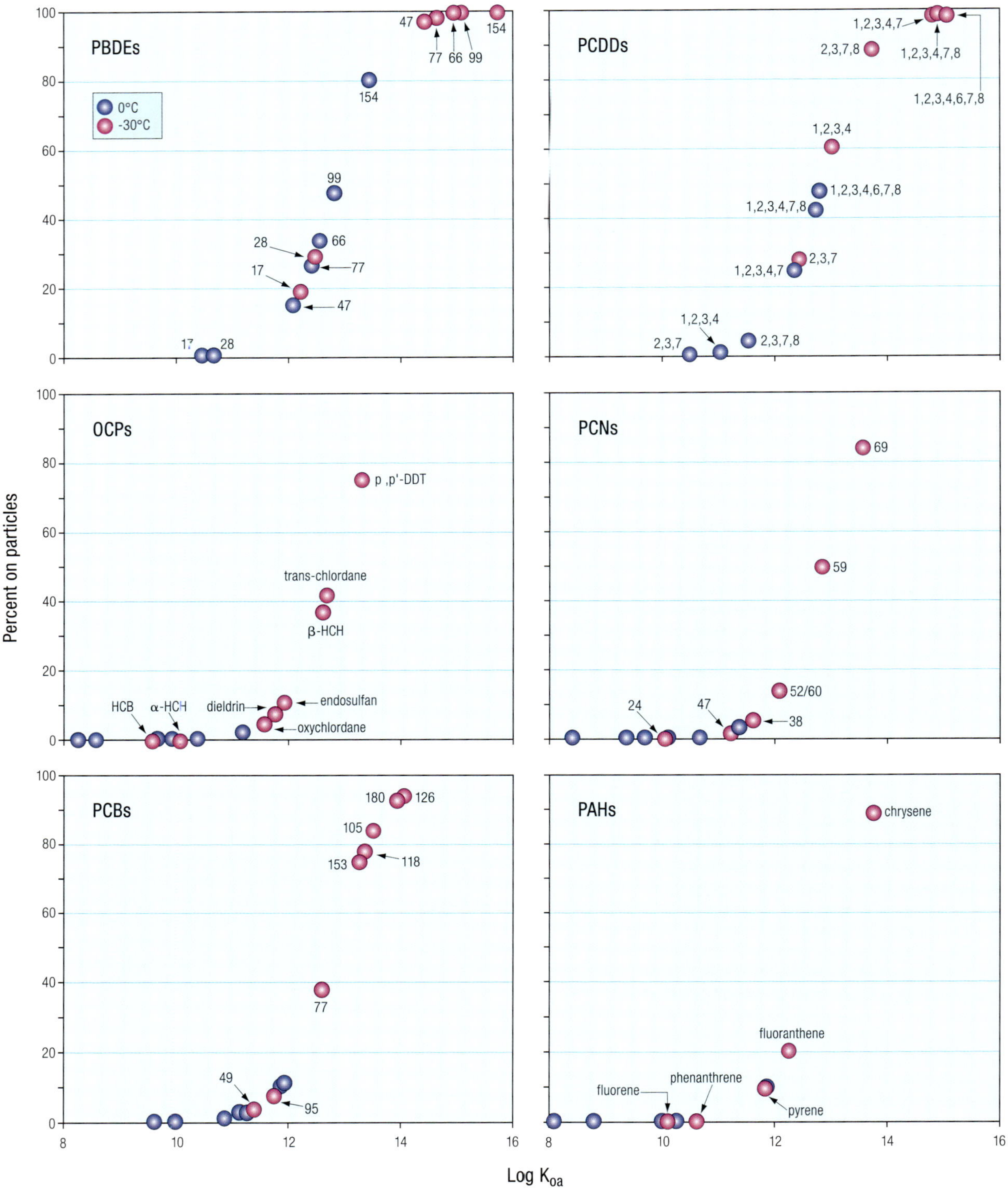

FIGURE D.1.31

The partitioning of selected POPs as a function of K_{OA}. Shown are the changes in partitioning that occur between winter (-30°C; pink) and summer (0°C; blue). Particulate fractions were calculated based on equations developed by Finizio *et al.* (1997a) from field measurements for organochlorines. Calculations are based on measurements of particulate organic carbon in Alert air during winter and summer 1998–1999 (Sharma *et al.*, 2002). Temperature-adjusted K_{OA} values were taken from the literature (Harner and Bidleman, 1996; Harner and Bidleman, 1998; Harner *et al.*, 2000; Harner and Shoeib, 2002; Shoeib and Harner, 2002).

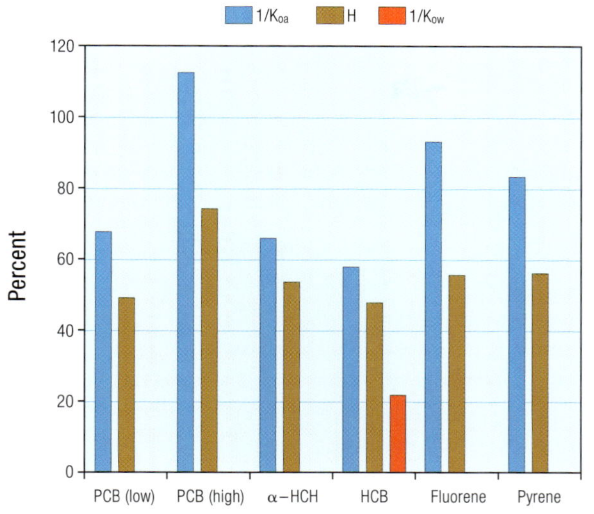

FIGURE D.1.32

Predicted percent changes in H and $1/K_{OA}$ associated with a 5°C increase in temperature for selected chemicals. "High" and "low" represent the upper and lower values reported for PCBs. Values are based on ΔH_H (enthalpy of phase change associated with transfer from water to air, kJ/mol), ΔH_{OA} (enthalpy of phase change from octanol to air), and ΔH_{OW} (enthalpy of phase change from octanol to water, Figure A.1.1, Table A.1.1).

At 0°C most of the chemicals shown in Figure D.1.32 are in the gas phase, implying that they are easily advected by air but will be prone to photolytic degradation. Timing is important: the transport processes that produce Arctic haze in spring may alter substantially if warming occurs earlier in the season with consequent change to the particle-gas partitioning. Several higher molecular weight PBDEs and PCDDs are appreciably associated with particles even at 0°C. Depending on their susceptibility to photolysis, particle-association may actually protect these compounds during transport. This might explain why the increase in PBDEs in the Arctic closely follows the production and usage to the south (Ikonomou *et al.*, 2002) — something that was not true for the PCBs and other organochlorine compounds which are mostly in the gas-phase during summer. Despite cold winter temperatures, many PCB, OCP, and PCN compounds remain in the gas phase. Winter warming projected by climate models may therefore facilitate the transport of these compounds when reduced sunlight makes them less vulnerable to degradation.

Chemical partitioning between environmental media (air, water, soil, biota) can be described using only three partition coefficients — K_{OA}, K_{AW}, and K_{OW} (Gouin *et al.*, 2000). K_{OW}, which is a ratio of two solubilities that both increase with temperature, tends to vary weakly with temperature as shown by the modest 20% increase for HCB in response to a 5°C increase (Figure D.1.32, Bahadur *et al.*, 1997). A 5°C temperature rise produces a more substantive 60–100% increase in $1/K_{OA}$, which will be manifest as an increase in volatility and greater potential for atmospheric transport. More chemicals will partition out of surface soils and aerosols to enter the gas-phase. If global warming occurs, cycling of chemicals through the atmosphere will increase.

On the other hand, models suggest that the temperature contrast between the equator and the pole will decrease (Zwiers, 2002). Accordingly, kinetic processes will increase with temperature rise but the overall thermodynamic forcing toward polar regions will decrease with reduced global thermal contrast. K_{AW} increases significantly (40–70%), which would favour evasion from surface waters. This would be particularly important for OCs like the hexachlorocyclohexanes (HCHs) which are at, or over, saturation in Arctic waters. The predicted increase in K_{AW} associated with a 5°C increase in water temperature corresponds to tonnes of HCHs that could be forced back into the atmosphere each year. We note that temperature increases of 5°C will not likely apply to partially ice-covered regions where temperatures will be buffered by melting ice. Nevertheless, under recent changes in the AO index (Figure D.1.6a, b) and with change projected by models (Figures D.1.4 and D.1.5), many lakes and major areas of Arctic shelves could be so affected.

Loss of chemicals occurs during transport in the atmosphere through reactions with hydroxyl (OH) radical, nitrate radical (NO_3), or ozone (O_3), through photolytic oxidation, and through sorptive partitioning to other phases (e.g., aerosols, precipitation, and vegetation) with subsequent deposition.

Although photolytic reactions do not strongly dependend on temperature, they will be affected by cloud cover which is predicted to increase with global warming (IPCC, 1995). Increased cloud cover will also result in lower concentrations of OH radical and less chemical removed by this and other photolytic pathways.

The dominant removal processes in soil and water include hydrolysis, photolysis, redox reactions, microbial degradation, and removal through soil-surface — air partitioning. Of these, only photolysis is not strongly dependent on temperature. The influence of temperature on the rate constant, "K", is usually described using the Arrhenius expression,

$$K = Ae^{-E_a/RT} \qquad \text{Eq. D.1.1}$$

where "A" is a constant and "E_a" is the activation energy.

Based on hydrolysis activation energies of 78–85 kJ/mol (Ngabe *et al.*, 1993) for α-HCH and γ-HCH respectively, a 5°C increase would increase removal rates by ~85-95%. The increase would be even greater if we consider that the dissociation constant for water (K_w) (e.g., pH = 8) increases with temperature, resulting in more OH^- ions. Activation energies associated with redox reactions are not reported for OCs but are usually assumed to be about 50 kJ/mol (Tratnyek and Macalady, 2000) which imply about a 50% increase in reaction rate with a 5°C rise.

Microbial degradation also follows the Arrhenius equation, but few studies report E_a. As a general rule, the biological activity in the mesophilic range (5–35°C) doubles for every 10–15°C temperature rise, which implies an E_a of 30–45 kJ/mol (Viessman and Hammer, 1985). Arctic microbial populations exhibit a large diversity and abundance (Ravenschlag *et al.*, 2001; Sahm and Berninger, 1998) and are typically cold-adapted and able to maintain efficient rates of organic degradation and mineralization down to the freezing point of seawater (-2°C) (Arnosti, 1998; Sagemann *et al.*, 1998). These facts suggest that warming may be accompanied by adaptation or population change but not necessarily that microbial degradation rates will increase. For example, reduced degradation of methyl dichlorprop was observed in experiments where field plots were warmed 5°C above normal for several years (Peterjohn, 1994).

Harner *et al.* (1999) found *in situ* microbial removal rates for α- and γ-HCH in cold Arctic Ocean waters to be surprisingly fast [$t_{1/2}$ for (+)α-HCH (5.9 yr); (-)α-HCH (22.8 yr); γ-HCH (18.8 yr)]. Assuming an E_a of 50 kJ/mol, a 2–5°C temperature rise in the upper Arctic Ocean would imply a reduction of these half-lives by 20-50%. Microbial degradation was estimated to account for over 30% of the removal of HCH suggesting that a small temperature rise could push this proportion to over 50%.

Aside from temperature, alterations in other environmental characteristics (e.g., soil moisture, soil and water pH, nutrient levels, vegetation cover and type) will be tied to global warming and affect the composition and density of microbial populations. For instance Lewis *et al.* (1999) found differences in microbial preference for microbes inhabiting forested as opposed to pasture soils. They also showed that the enantioselectivity [preference of the microbial population for a + or - enantiomer of a chiral compound (compound exhibiting mirror-image forms)] shifted with organic nutrient enrichments.

Other studies have observed high rates of microbial degradation of HCHs in Arctic lakes (Law *et al.*, 2001) and watersheds (Helm *et al.*, 2000b). Law *et al.* (2001) found that enantioselective degradation of α-HCH was greater in small, high Arctic lakes and streams than in temperate lakes and wetlands. They concluded that low nutrient levels in the Arctic systems resulted in an adapted microbial population that was more capable of degrading organic contaminants. Thus biodegradation of chemical residues in soil and water will be altered as microbial populations adapt to the changing climate. The complexity and uncertainty associated with these changes, however, does not allow us to predict whether global warming will enhance or diminish chemical removal by this pathway.

Putting many of the concepts outlined above into a numerical model, McKone *et al.* (1996) asked specifically what would be the effect of a 5°C temperature rise on the health risk from hexachlorobenzene (HCB). They found surprisingly little consequence from the projected 5°C temperature increase. Indeed, their results showed that warming would actually reduce exposure to this compound because it would enhance degradation and tend to force HCB out of water and into air. The critical step, controlled by the sensitivity of air-water partitioning to temperature, acts at the bottom of aquatic food webs which, through biomagnification, provides most of the health risk.

D.1.6.3.5 The effect of altering food web structure on organochlorines

Biomagnification can concentrate fat-soluble compounds, like the organochlorines, by factors as high as 10^5 to 10^9 from water to apex predator (Fisk *et al.*, 2001a; Kidd *et al.*, 1995; Muir *et al.*, 1999a; Muir and Norstrom, 1994). The distribution of contaminants in air, water, and for the first step in the food web (phytoplankton, particulate organic carbon), can be predicted simply by applying appropriate partition coefficients like K_{AW} and K_{OW} (Figure D.1.32). Chemical partitioning, which is based solely on thermodynamics, provides a crucial platform upon which biomagnification can then operate (Figure D.1.33). A chemical at equilibrium will have identical fugacity in the media in question (e.g., air, water, oil). This makes it relatively simple to predict how temperature will alter the distribution as discussed above.

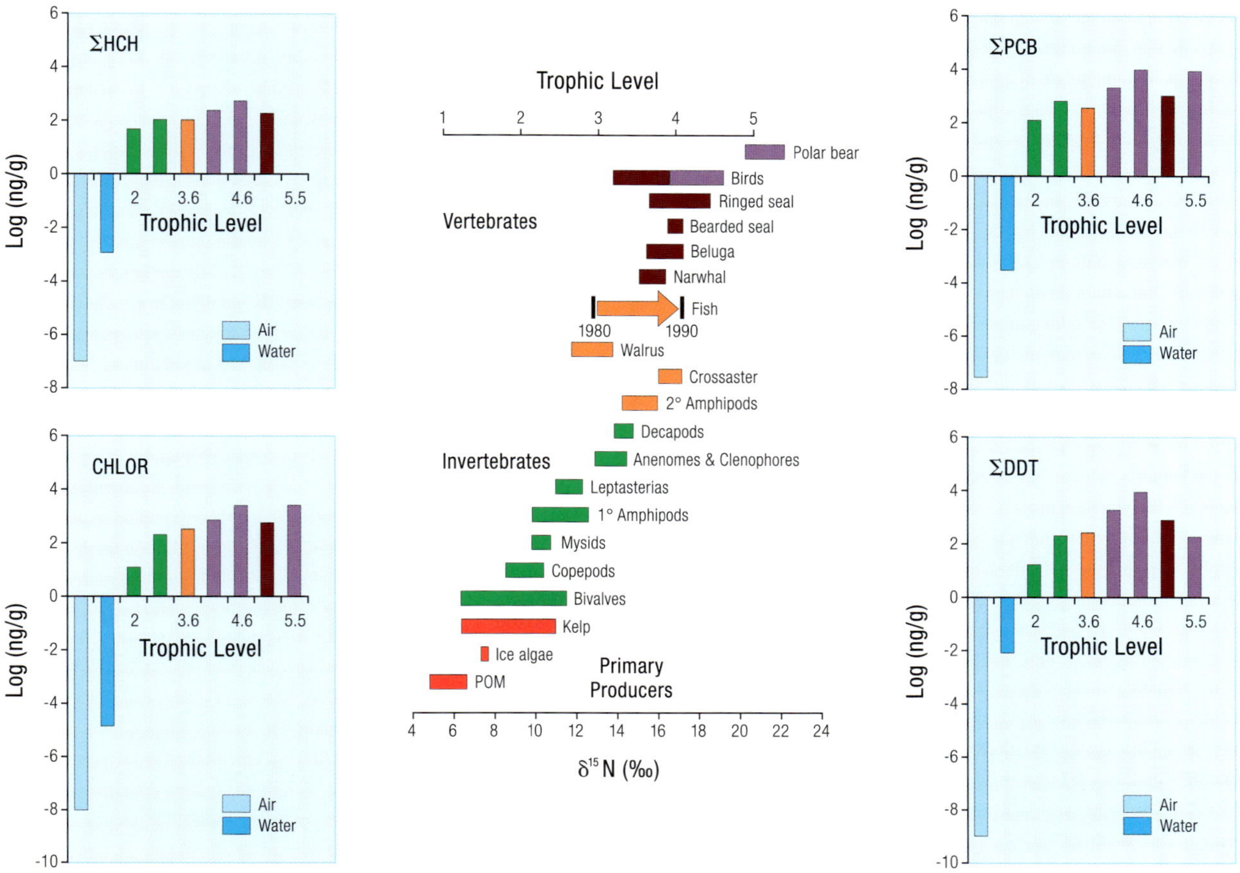

FIGURE D.1.33

The marine food web structure observed during the Northwater Project. Shown in the panels are average concentrations of selected organochlorine compounds in air, water, and various trophic levels as represented by *Calanus hyperboreus* (2), *Themisto libellula* (2.6), Arctic Cod (3.6), Black guillemot (4.3), glaucous gull (4.6), Ringed Seal (4.5), Polar Bear (male, 5.5) (see Fisk *et al.*, 2001a, b; Hobson and Welch, 1992).

Biomagnification, however, cannot be explained solely by thermodynamics and it requires energetic processes to produce the elevated concentrations in the top predators. These processes can be considered in part as a reduction of the solvent containing organochlorines (fat) through metabolism. The processes are complex and can lead to variability simply due to bio-energetics, for which the offloading of contaminants by nursing mammals is a classic example (Addison and Smith, 1998). It is not our intent to conduct a thorough review of bioaccumulation and biomagnification except to admit that the complexity of the process offers the opportunity for climate change to act in subtle ways. The data shown in the panels of Figure D.1.33 show an almost linear relationship between the log (contaminant burden) and trophic level. The slope of this relationship indicates the multiplication factor involved for each step in trophic level so that, for example, one step would multiply DDT concentration by perhaps as much as 6 (not including polar bears), whereas the other contaminants experience factors of 2 (HCH), 3 (PCB), and 4 (chlordane). The removal or addition of trophic levels in the food web mediated by climate change, therefore, will not have the same effect for all contaminants. For example, Figure D.1.33 suggests that DDT will be the most sensitive to this kind of change.

Another way in which food web structure can effect change is by bifurcation. For instance, altering the coupling between pelagic and ice production and the benthos can change the relative proportions of organic carbon (and contaminant) that enter pelagic or benthic food webs. However, these changes in pathway do not alter the relationship between contaminant concentration and trophic level. In this context, it can be seen that walrus feeding on a benthos enriched by strong coupling with primary production will be exposed to lower organochlorine concentrations by factors of 10 or more than if they switch to predation on seals.

D.1.6.3.6 The epontic food web and changes in ice climate

The entry of contaminants to a stratifying surface layer from ice melt in spring offers a mechanism vulnerable to climate change. However, studies seem not to show higher concentrations in epontic fauna than in zooplankton (Borgå *et al.*, 2002). Organochlorine concentrations in epontic amphipods (*Apherusa glacialis, Gammarus wilkitzkii, Onisimus* spp.) and zooplankton (*Calanus hyperboreus, Thysanoessa inermis, Parathemisto libellula, Chaetognatha*) from the marginal ice zone near Svalbard could be explained mostly by diet, with habitat (sea-ice underside versus the pelagic zone) accounting for a smaller part of the variance. Epontic amphipods had higher concentrations of HCB, γ- and α-HCH while DDTs, PCBs, and chlordanes did not differ between epontic and pelagic habitats.

This pattern of uptake can be explained by the vertical distribution of OCs in the water column: higher concentrations of HCHs and HCB are found near the sea surface (Harner *et al.*, 1999; Jantunen and Bidleman, 1998; Tanabe and Tatsukawa, 1983), whereas particle reactive compounds like DDTs and highly-chlorinated PCBs adsorb to sinking particles to produce a more homogenous vertical distribution (Tanabe and Tatsukawa, 1983). Furthermore, most particles transported by sea ice are not available to epontic biota because they are released in the marginal ice zone when the ice melts (Ramseier *et al.*, 1999) and descend rapidly, carrying adsorbed contaminants with them. For OCs that exhibit strong gradients in the upper ocean, the loss of ice and hence of epontic fauna can alter the dietary exposure of higher trophic levels like seabirds and seals.

D.1.6.3.7 Food deprivation or shifts in diet

Many of the Arctic's top predators undergo periods of fasting forced by lack of food, seasonality of food, or inability to access food. Perhaps the best documented example is the stress to the Hudson Bay polar bear population unable to hunt seals during spring due to change in the spring ice climate (see Figure D.1.21 and Stirling, 2002; Stirling and Lunn, 1997; Stirling *et al.*, 1999). The burning of stored fat through metabolism results in release of archived fat-soluble contaminants and, potentially, an increase of contaminant burden in the remaining fat reservoir. Longer periods of starvation due to change in ice or change in prey populations could lead to higher doses of OCs sequestered in fat — usually at a time when the animal is most vulnerable. Although there has been much concern over nourishment-deprived polar bears, similar circumstances probably apply to other species, examples including common eider, *Somateria mollissima*, (Olafsdottir *et al.*, 1998) and Arctic char (de Wit *et al.*, 2003)

Species that have dietary flexibility may respond to ecosystem change by switching to alternate prey, which changes their organochlorine intake. For example, the large variation in organochlorine concentration in the livers of glaucous gulls from the western Barents Sea reflects the wide range in the trophic level of the gull's prey as reflected by parallel variations in $\delta^{15}N$, as a proxy for trophism (de Wit *et al.*, 2003). Recent decreases in organochlorine concentrations for Svalbard Minke whales (*Balaenoptera acutorostrata*) might superficially be ascribed to banning the manufacture of PCBs during the 1970s. It seems more likely, however, that this decline reflects a dietary switch from capelin, whose stocks collapsed in 1992–93, to krill which are further down the food web (de Wit *et al.*, 2003). Polar bears also display a range in prey that can explain regional variation in organochlorine burdens. For example, Chukchi and Bering Seas bears feed more heavily on Pacific walrus which are less contaminated than ringed seals because they are at a lower trophic level, whereas bears from Svalbard feed on more heavily contaminated harp seals. These latter results indicate that climate variables expressed through prey availability and biological condition have much influence on apex feeders exposure to organochlorines.

D.1.6.3.8 Altered migration pathways and invading species

Migratory species (e.g., whales, fish, birds) can obtain contaminant loadings in one location and release them in another. Migrating animals can also be subject to varied exposure as they feed along their migration path. For one Alaskan lake, the loadings of OC contaminants returned by anadromous fish exceeded those entering the drainage basin from the atmosphere (Ewald *et al.*, 1998). The recent migration of Pacific salmon into Arctic rivers (Babaluk *et al.*, 2000) could, likewise, have an impact on contaminant budgets for rivers or lakes they enter. Certainly, these fish provide a new 'vector' for delivering contaminants to species that predate on them or depend on a food web supplied by their carcasses.

If the spatial distribution of contaminants is controlled by processes subject to climate change, then exposure during population migrations can also alter through climate change. An intriguing example of such a process has recently been described for the migration of Bering/Beaufort bowhead whales. The whales reflect in their body burdens the change in α- and β-HCH composition between the Bering and Beaufort Seas (Hoekstra *et al.*, 2002b). In turn, the ocean composition for the HCHs is likely controlled by large-scale physical processes (e.g., rainfall and air-sea partitioning — see Li *et al.*, 2002) as are migratory routes (Dyke *et al.*, 1996b; Moore *et al.*, 1995), each of which is sensitive to climate change.

Finally, invasions of new species fostered by climate change or by introduced exotic species both have the potential to re-structure food webs. An elegant example of how dramatic such change can be, both in trophic organization and contaminant pathways, was provided by the invasion of the zebra mussel into the Great Lakes (Morrison *et al.*, 1998; 2000; Whittle *et al.*, 2000). Clearly, the migration northward even of the humble filter feeder, a potential product of changing water properties or organic carbon supply, could have unexpected impact on contaminant cycling in coastal seas.

D.1.6.3.9 Organochlorines, disease and epidemics

During the past decade or so, there has emerged much evidence that mass mortality in marine mammals may occur due to a combination of factors including disease vectors, population stress, and contaminants, each of which may be affected by climate change (cf. Lavigne and Schmitz, 1990; Ross, 2002). The complexity of this interaction provides fertile ground for surprises. Some outbreaks of disease have been observed following migrations associated with large-scale ecological change and some have derived from the introduction of viruses from domestic animals. But the addition of immunotoxic chemicals, like many of the POPS, may provide the trigger for disease to emerge (Ross *et al.*, 2000; Vos and Luster, 1989). The widespread distribution of canine distemper virus or a closely related morbillivirus in seals from Greenland, led Dietz *et al.* (1989) to speculate on the possibility that large-scale migration of harp seals from the Barents Sea to north Europe in 1986-87 might have provided a disease vector. The co-factors of a naïve marine mammal (seal) population in coastal Europe manifesting suppressed immune systems through high contaminant PCB burdens would then have provided the foundation for an epidemic (Heide-Jørgensen *et al.*, 1992).

Within the Arctic, top predators would be at greatest risk due to their high exposure to contaminants, and marine mammals would likely face the added stress of change in ice climate. Accordingly, indications of immunosuppression have been found in polar bears, northern fur seals, and glaucous gulls (de Wit *et al.*, 2003). In particular, the polar bears of the Kara Sea, Franz Jozef Land, East Greenland, and Svalbard would seem especially vulnerable. First, they exhibit inordinately high contaminant burdens (Andersen *et al.*, 2001; Norstrom *et al.*, 1998), which may derive partly from the enhanced connectivity between this region and Europe/North America under the high AO/NAO indices of the 1990s (Figure D.1.8). Second, as discussed in previous sections, change in ice climate and in marine ecosystems may have provided the added stress of malnourishment. Finally, it seems these bears already have sufficient contaminant burdens to exhibit health effects (Bernhoft *et al.*, 2000; Skaare *et al.*, 2001).

D.1.6.4 Hydrocarbons

Contaminant hydrocarbons in the Arctic derive from combustion and petrogenic sources (Yunker *et al.*, 1995). The pathways of these two sources of hydrocarbon differ substantially as will their sensitivity to climate change. Contaminant hydrocarbons pose two kinds of problems; PAHs (polynuclear aromatic hydrocarbons) and their oxidation products are toxic (Zedeck, 1980) and spilled oil has direct, well known impacts on biota, especially those that inhabit surfaces (Patin, 1999; Wolfe *et al.*, 1994).

D.1.6.4.1 Combustion polycyclic aromatic hydrocarbons

Combustion PAHs are well known products of natural fires and human-related combustion processes (automobiles, burning of liquid and solid fuels, waste incineration, metallurgy). In the atmosphere, PAHs partition between the vapour phase and particulates (see for example Figures 14 and 15 in Macdonald *et al.*, 2000) and are transported long distances to be detected at remote Arctic locations (Halsall *et al.*, 1997; Macdonald *et al.*, 2000; Patton *et al.*, 1991). There is a strong seasonality in PAH concentration in air at Alert, with colder months (October to April) displaying concentrations about 10 times higher than warmer months (May to September). This suggests that the winter haze phenomenon that transports heavy metals across the pole from Eurasia similarly transports industrial combustion products. Therefore, much of what has been said about aerosol metals and climate change relates directly to PAHs. Altered wind patterns (Figure D.1.8) and enhanced precipitation (Figure D.1.9) have the potential to change pathways and to deposit PAH aerosols over parts of the Arctic Ocean, especially toward the southern Eurasian Basin. Furthermore, temperature increases may shift the equilibrium from particulate to vapour phase for PAHs like pyrene, fluoranthene, phenathrene, and anthracene, which at Arctic temperatures are partially distributed between air and solid phase (see Figure D.1.31 and Figures 14 and 14 in Macdonald *et al.*, 2000).

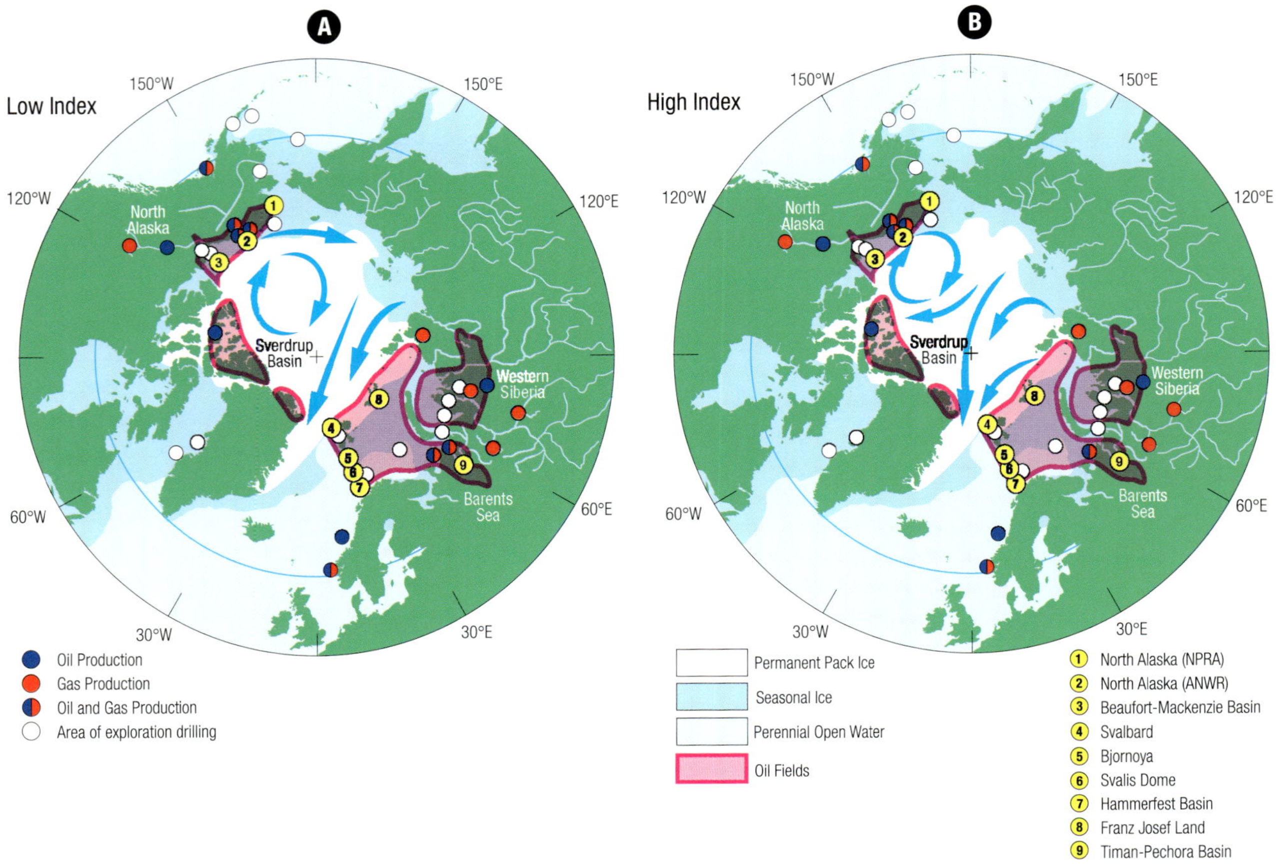

FIGURE D.1.34

A map of the Arctic showing oil-bearing regions and locations of oil production and pipelines together with the ice-drift field under a) Low Arctic Oscillation index and b) High Arctic Oscillation index [ice motion based on Rigor *et al.* (2001) and oil data are from Bakke *et al.* (1998)].

In addition to the strong seasonal signal of industrial PAHs observed in the Arctic, outliers (samples with abnormally high PAH concentrations) are also observed during summer months, particularly at Tagish, and these have been assigned to forest fires (Macdonald *et al.*, 2000). Forest fires are projected to increase through climate change as a result of warmer continental temperatures (Figure D.1.3a) and less precipitation in continental interiors. We may expect, therefore, a general increase in atmospheric PAH in the Arctic deriving from such biomass burning, and this increase will likely have a greater impact on small rivers in northern Canada which already receive almost all of their PAH from combustion sources (Yunker *et al.*, 2002). Loss of permafrost and enhanced erosion of peat may further contribute enhanced amounts of previously sequestered PAH to lakes and rivers (Yunker *et al.*, 1993).

D.1.6.4.2 Petrogenic hydrocarbons and oil

The risk of oil spills from offshore exploration and production is a major concern in the Arctic (Bakke *et al.*, 1998; Patin, 1999). Climate change that produces an ocean margin substantially clear of ice will undoubtedly encourage further exploration, perhaps in more remote locations. Producing oil from shelves, especially remote shelves, has the associated problem of transporting the produced oil south either by ship or pipeline.

Changes in the ice drift associated with changes in the AO index will clearly have a dramatic influence on where spilled oil will go if it gets into the ice pack (Figure D.1.34). In the Canadian and Alaskan sector of the Arctic Ocean during AO$^-$ conditions, spilled oil will follow ice into the East Siberian Sea to traverse the Russian shelves and exit to the Greenland Sea. During AO$^+$ conditions, oil spilled from the same location would tend to remain within the Beaufort Gyre to potentially return within a few years to where it was spilled.

Oil spilled over the Russian shelves, or entering their coastal seas from spills into rivers, would tend to track directly across the Arctic under AO^- conditions (Figure D.1.34a). However, oil spilled under AO^+ conditions could move more to the east, with a slight chance of getting into the Canada Basin and eventually reaching the Canadian Archipelago through which it would then have to pass. It is uncertain how viable this latter route might be for oil spilled in the Kara or Laptev Seas, but evidence from tree dendrology suggests that there have been periods since the Holocene when communication from Siberia to the Archipelago has been mediated by ice drift (Dyke and Savelle, 2000; Dyke *et al.*, 1997).

A final connection between spilled oil and climate change derives from the projected increase in incident UV radiation (Weatherhead and Morseth, 1998) which could lead to an increase in photo-enhanced toxicity of spilled oil (see, for example Barron and Ka'aihue, 2001; Pelletier *et al.*, 1997). Toxicological assessments of oil made in the presence of UV light reveal a toxicity up to 1000 times greater than that measured under the traditional fluorescent light. Furthermore, photoenhanced toxicity of oil can occur at the intensities and wavelengths measured for UV in aquatic water columns suggesting that increased incident UV radiation projected for polar regions would, in addition to many other effects on ecosystems (Weatherhead and Morseth, 1998), enhance damage done by spilled oil.

D.1.7 Time series

The value of time series is undisputed both for climate-related variables (Hare and Mantua, 2000; McGowan, 1990) and for contaminants (AMAP, 1998). However, recognizing the potential of climate variables to produce variance in contaminant time-series has all but been neglected (Macdonald *et al.*, 2002c). We have discussed numerous examples of how global change can alter delivery of contaminants to and within the Arctic, with alteration in wind fields and precipitation forced by the Arctic Oscillation being but one simple example. The fact that the leading 'global distillate,' water, provides one of the clearest detectors of global temperature change in its isotopic composition (Fischer *et al.*, 1998) should generate considerable anxiety about time series of volatile and semi-volatile compounds, even if the pathway between emission and point of measurement is reasonably direct (e.g., atmospheric concentrations, ice cores, sediment cores). Examples closer to contaminant time series can be found. For instance, the mercury record in ancient Antarctic ice has been suggested as a proxy for ocean productivity (Vandal *et al.*, 1993) and cadmium has been applied to paleo ocean-productivity histories (Saager and deBaar, 1993; Shen *et al.*, 1987).

The environment can be monitored at many points (Figure D.1.35), each of which will tell a separate story. For example, we might measure PCB concentration: in air at Alert every 2 weeks; in a dated sediment or ice core; in biota or surface sediments collected every 5 years; or in an ocean profile collected annually. What these PCB time-series data will tell us depends on how many environmental processes had an opportunity to operate on the original signal (the emission), and how the recorder (the medium being monitored) itself actually works. Much research has been conducted to understand and account for the latter (e.g., organochlorine cycles in female mammals; increase in mercury with age of fish, diagenesis in sediment cores) but the difficulty of climate variability creeping into the record remains relatively ignored.

Climate variability by itself may cause aliasing, which arises when sampling intervals are chosen close to natural variation frequencies. For example, the Arctic Oscillation's time scale of 5–7 years (Proshutinsky and Johnson, 1997) makes it very difficult to assess the role of such sub-decadal variation in data collected at five-year intervals. Furthermore, a trend from a time series collected from say 1980 to 2000 will potentially carry a large bias, either positive or negative, produced by a switch in the middle of the record from generally AO^- to strong AO^+ conditions. Climate change may often provide an alternate hypothesis for time-series contaminant data. A few examples are described in the following text.

D.1.7.1 Sediment-core records and surface sediments

Well-dated cores from lakes and oceans are a traditional way to estimate historical fluxes of contaminants (Lockhart *et al.*, 1995; 1998; Muir *et al.*, 1996a). Fluxes to sediments estimated from a sediment core are sometimes corrected for focusing to produce an estimate of flux to the surface of the water in order to compare it with fluxes at other sites or with emissions. Increases or decreases with depth in a sediment core are then often used to infer historical changes in emissions. However, between the emission and the sedimentation at the bottom of the lake or ocean lie atmospheric transport, deposition to the water's surface or to the drainage basin, attachment to particles, and settling to accumulate at the bottom (Figure D.1.35). Climate change may operate within this system to 1) change wind fields (important if there are atmospheric gradients); 2) change the efficiency of the air-to-ground transport (e.g., by altering precipitation or temperature); or 3) change the efficiency of capture to sediments (by processes outlined in section D.1.6.3.3).

Possibly the best example of how climate-related variables could interact with contaminant records in sediments has been unearthed from lake sediments on Bjørnøya Island (Bear Island). Lake Ellasjøen sediments contain anomalously high organochlorine concentrations attributed to large inputs of bird guano (de Wit *et al.*, 2003). What would a dated sediment core record from such a lake? It is clear that the record would contain a component deriving from the emission histories of various organochlorine chemicals, but much of that signal would be mediated by birds. The population dynamics of birds, then, will alter the contaminant delivery over time, depending on nesting locations and populations and on the source and trophic level of food which derives from the Barents and Nordic Seas (de Wit *et al.*, 2003) all of which are subject to climate variation. This example, which is particularly compelling in the way local and perhaps regional PCB patterns have been impacted by biological sources (guano — see Enge *et al.*, 1998; de Wit *et al.*, 2003) is likely not the only factor. One is reminded of the importance of fish as transporting agents for

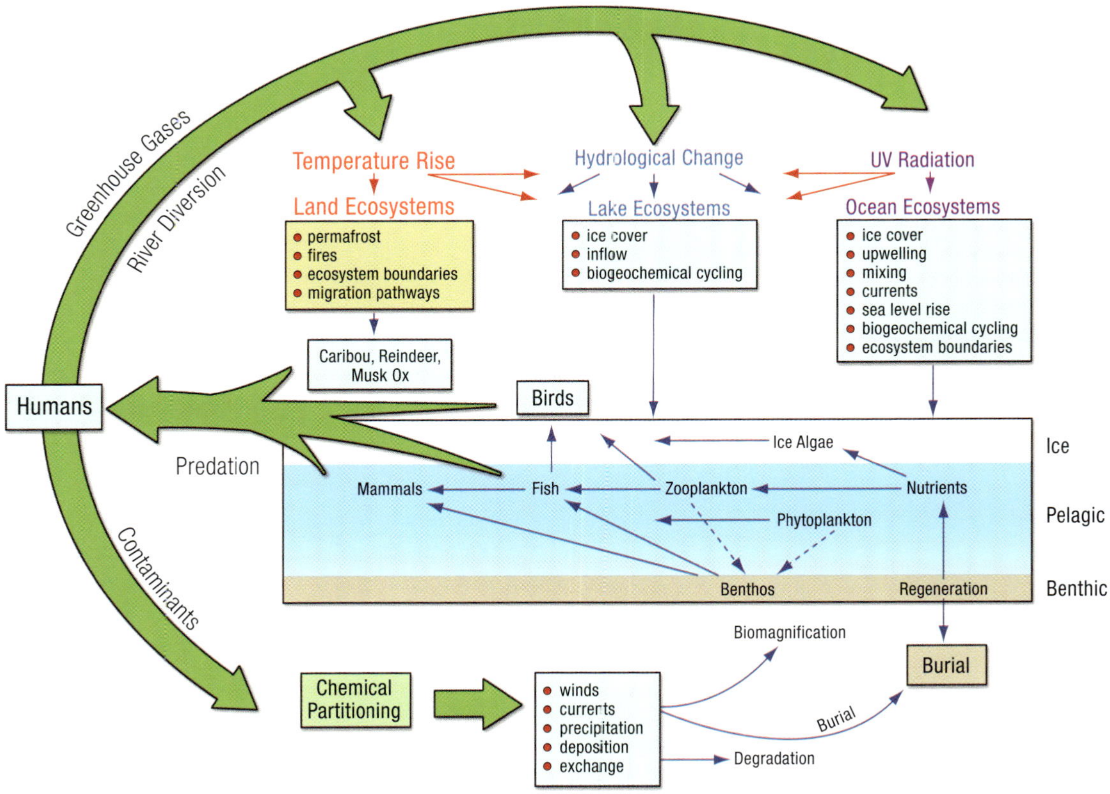

FIGURE D.1.35

A schematic diagram showing components of the pathways contaminants must traverse between emission and points at which environmental time series are collected for the Arctic.

organochlorines and mercury (Ewald *et al.*, 1998; Zhang *et al.*, 2001) which together with high and variable anadromous fish escapements (Finney *et al.*, 2000; 2002) very likely imprints as yet uninvestigated contaminant records in coastal lakes of Alaska.

Finally, natural variability in organic carbon flux to sediments can enhance sediment foraging (including small infauna and large animals such as diving birds, walrus, seals and belugas). This, together with the potential for episodic colonization of sediments by new species under changing ocean climate (see for example Stull *et al.*, 1986), provides a strong caution on using surface sediment contaminant distributions to infer spatial or temporal trends in contaminants, even where normalizing factors such as aluminum or organic carbon have been applied.

D.1.7.2 Atmospheric time series

Data collections from ground stations in the Arctic are of relatively short duration, extending back a couple of decades for metals (Sirois and Barrie, 1999) and only a few years for OCs (see for example Bailey *et al.*, 2000; Halsall *et al.*, 1998; Hung *et al.*, 2001; Macdonald *et al.*, 2000; Stern *et al.*, 1997). There are fewer processes between the emission and the recorder for these measurements (Figure D.1.35), but as discussed above, changes inherent in the Arctic Oscillation (winds, ice cover, precipitation) are sufficient to imprint themselves on the emission before it reaches this recorder.

D.1.7.3 Biological tissue time series

Marine and terrestrial biota have been collected, usually at very sparse intervals (typically 3–6 periods over a couple of decades), to monitor bioaccumulating substances, especially the organochlorines and mercury (Addison and Smith, 1998; AMAP, 1998; Braune *et al.*, 1999; Wagemann *et al.*, 1995). α-HCH concentration in seals collected at Holman (Figure D.1.36) clearly do not follow atmospheric emissions (line in Figure D.1.36) nor do they follow Arctic air concentrations which mimic emissions quite closely (see Figure 9 in Macdonald *et al.*, 2000). Instead, they approximately follow Canada Basin surface water concentrations, which makes sense given that seals obtain their HCH burden from a marine diet. The burden of α-HCH is large in surface waters of the Canada Basin (estimated at 1750 tonnes in the early 1990s, Macdonald *et al.*, 2000), and not quickly changed. However, the burden can be altered by diversion of Russian rivers or by removal of ice cover, both of which have occurred in the 1990s.

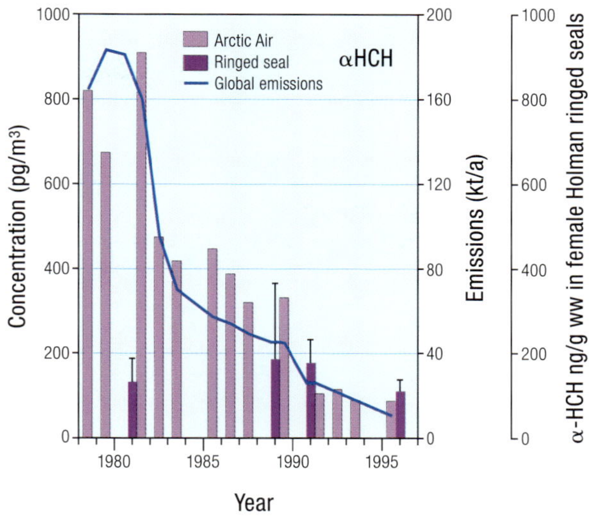

FIGURE D.1.36

A comparison of various HCH time series. Shown are 1) global emissions of α-HCH (the curve), 2) mean concentrations of α-HCH in Arctic air (the light purple bars) from 1979 to 1996 and 3) α-HCH concentration in Holman Island seals (dark purple bars). Air concentration data of α-HCH in the Arctic have been measured at different stations by several research groups [for data sources see Addison and Smith (1998), Li *et al.* (2002)]

Furthermore, change in the seal burdens can be caused by a regional change in the food web — something that has likely occurred as a result of changing ice cover and stratification (Melnikov *et al.*, 2002). The food web factor might be 'controlled for' by monitoring trophic level through isotopic measurements ($\delta^{15}N$) or by monitoring other components of the food web (e.g., Arctic cod) over the same time interval. The advantage of monitoring species at high trophic levels — that they are very sensitive to OCs and are important to human diets — is somewhat offset by their sensitivity to variance from many factors other than emission strength (Figure D.1.35). Given the 5–7 year time scale inherent in change forced by the Arctic Oscillation, time series data for biological tissue will be vulnerable to an unaccounted aliasing.

ITK/Eric Loring

References

Aagaard, K. and E.C. Carmack, 1989. "The role of sea ice and other fresh water in the Arctic circulation." *Journal of Geophysical Research*, 94: 14,485–14,498.

Aagaard, K., L. A. Barrie, E.C. Carmack, K. Garrity, E.P. Jones, D. Lubin, R.W. Macdonald, W. B. Tucker, P.A. Wheeler, and R.H. Whritner, 1996. "US, Canadian researchers explore the Arctic Ocean." *Eos, Transactions , American Geophysical Union*, 177: 209–213.

Aarkrog, A. 1994. "Radioactivity in polar regions, main sources." *Journal of Environmental Radioactivity*, 25: 21–35.

Abad, E., J. Caixach, and J. Rivera, 1999. "Dioxin-like compounds from municipal waste incinerator emissions: assessment of the presence of polychlorinated naphthalenes;" *Chemosphere,* 38: 109–120.

Addison, R.F., and T.G. Smith, 1998. "Trends in organochlorine residue concentrations in ringed seal (*Phoca hispida*) from Holman, Northwest Territories, 1972–1991." *Arctic,* 51: 253–261.

Agriculture Canada — Environment Canada. 1988. "Pesticide Registrant Survey Report Data," prepared by Commercial Chemicals Branch, Conservation and Protection, Environment Canada, Ottawa, ON.

Aigner, E.J., A.D. Leone, and R.L. Falconer, 1998. "Concentrations and enantiomeric ratios of organochlorine pesticides in soils from the US Cornbelt." *Environmental Science and Technology,* 32: 1162–1168.

Akeredolu, F.A., L.A. Barrie, M.P. Olson, K.K. Oikawa, J.M. Pacyna, and G.J. Keeler, 1994. "The flux of anthropogenic trace metals into the Arctic from the mid-latitudes in 1979/80." *Atmospheric Environment,* 128: 1557–1572.

Alaee, M, 1997. "A study of the sources and fate of organochlorine contaminants in Yukon River basin."In: *Synopsis of Research Conducted Under the 1995–97 Northern Contaminants Program.* J. Jensen (e.d.), Environmental Studies 74, QS-8573-0000-EF=A1. Indian and Northern Affairs, Ottawa, Canada.

Alaee, M., D.B.Sergeant, M.G. Ikonomou, and J.M. Luross, 2001. "A gas chromatography/ high resolution mass spectrometry method for determination of polybrominated diphenyl ethers in fish." *Chemosphere,* 44: 1489–1495.

Alegria, H.A., T.F. Bidleman, and T.J. Shaw, 2000. "Organochlorine pesticides in ambient air of Belize, Central America." *Environmental Science and Technology,* 34: 1953–1958.

Alegria, HA, T.F. Bidleman, and M. Salvador, 2001. "Pesticide levels in ambient air and soil of Chiapas, Mexico." Society for Environmental Toxicology and Chemistry (SETAC) National Meeting, Baltimore, MD, November.

Alexeeva, L.B., V.A. Surnin, E.I. Babkina. V.V. Shlychkova, L.G. Korotova, G.I. Ganin, and W.M.J. Strachan, 1997. In: *The AMAP International Symposium on Environmental Pollution in the Arctic: Extended Abstracts. Arctic Monitoring and Assessment Programme*, Oslo, Norway, Tromso, Norway, June 1–5, pp. 68–70.

Alexeeva, L.B., W.M.J. Strachan, V.V. Shlychkova, A.A. Nazarova, A.M. Nikanorov, L. G. Korotova, and V.I. Koreneva, 2001. "Organochlorine pesticide and trace metal monitoring of Russian rivers flowing to the Arctic Ocean: 1990–1996." *Marine Pollution Bulletin,* 43: 71–85.

Alho, C.J.R. and L.M. Vieira, 1997. "Fish and wildlife resources in the pantanal wetlands of Brazil and potential disturbances from the release of environmental contaminants." *Environmental Toxicology and Chemistry,* 16: 71–74.

Allen, J.O., N.M. Dookeran, K.A. Smith, A.F. Sarofim, K. Taghizadeh, and A.L. Lafleur, 1996. "Measurement of polycyclic aromatic hydrocarbons associated with size-segregated atmospheric aerosols in Massachusett.". *Environmental Science and Technology*, 30: 1023–1031.

Alley, R.B., J. Marotzke, W. Nordhaus, J. Overpeck, D. Peteet, R. Pielke, R. Pierrehumbert, P. Rhines, T. Stocker, L. Talley, and J.M. Wallace, 2002. "Abrupt Climate Change; Inevitable Surprises." National Academy Press, Washington, DC, 182 pp.

Altschuh, J., R. Bruggemann, H. Santl, G. Eichinger, and O. G. Pirringer, 1999. "Henry's Law constants for a diverse set of organic chemicals: experimental determination and comparison of estimation methods." *Chemosphere*, 39: 1871–1887.

AMAP. 1998. "AMAP Assessment Report: Arctic Pollution Issues." Arctic Monitoring and Assessment Programme (AMAP), Oslo, Norway. xii+ 859 pp.

AMAP. 1999. "AMAP Data Report: Atmospheric Subprogramme." NILU, Norway. pp. 29–32.

AMAP. 2002. Peramfrost distribution. http://www.grida.no/db/maps/pages/Arcticpermafrost.htm.

Andersen, M., E. Lie, A.E. Derocher, S.E. Belikov, A. Bernhoft, A. Boltunov, G.W. Garner, J.U. Skaare, and Ø. Wiig, 2001. "Geographic variation of PCB-congeners in polar bears (*Ursus maritimus*) from Svalbard east to the Chukchi Sea." *Polar Biology,* 24: 231–238.

Anderson, O., and G. Blomkist, 1981. "Polybrominated aromatic pollutants found in fish in Sweden." *Chemosphere,* 10: 1051–1060.

Anderson, P.N., and R.A. Hites, 1996. "OH radical reactions: The major removal pathway for polychlorinated biphenyls from the atmosphere." *Environmental Science and Technology*, 30: 1756–1763.

Anderson, L.G., E.P Jones, K.P Koltermann, P. Schlosser, J.H. Swift, and D.W.R. Wallace, 1989. "The first oceanographic section across the Nansen Basin in the Arctic Ocean." *Deep-Sea Research*, 36: 475–482.

Andrews, P., and W. Vetter, 1995. "A systematic nomenclature system for toxaphene congeners. Part 1: Chlorinated bornanes." *Chemosphere,* 31: 3879–3886.

Andrilinas, A., 1974. "Farmers' Use of Pesticides," in 1971 — *Quantities*, US Department of Agriculture, Economic Research Service, Agricultural Economic Report No. 252. 56 pp.

Anonymous, 2001a. "Contaminants in freezing ground." *Cold Regions Science and Technology* 32: 85–205.

Anonymous, 2001b. "The Second International Conference on Contaminants in freezing ground." *Polar Record* 37: 194–288.

Arendt, A, K.A. Echelmeyer, W.D. Harrison, C.S. Lingle, and V.B. Valentine, 2002. "Rapid wastage of Alaska glaciers and their contribution to rising sea level." *Science,* 297: 382–386.

Armstrong, R.L., and M.J. Brodzik, 2001. "Recent northern hemisphere snow extent: A comparison of data derived from visible and microwave satellite sentsors." *Geophysical Research Letters,* 28: 3673–3676.

Arnosti, C, 1998. "Rapid potential rates of extracellular enzymatic hydrolysis in Arctic sediments." *Limnology and Oceanography*, 43: 315–324.

Arora, V.K., and G.J. Boer, 2001. "Effects of simulated climate change on the hydrology of major river basins." *Journal of Geophysical Research,* 106: 3335–3348.

ASU (Analytical Services Unit, Queen's University). 1998. "Resolution Island 1997 Scientific Investigations," prepared for Indian and Northern Affairs Canada.

Atkins Baker, S., 1979. "A survey of environmental mercury concentrations in Yukon Territory, 1977." *Report 79–21*, Environmental Protection Service, Environment Canada, Yukon Branch, 74 pp.

Atkinson, R., R. Guicherit, R.A. Hites, W.-U. Palm, J.N. Seiber, and P. de Voogt,1999. "Transformations of pesticides in the atmosphere: a state of the art." *Water, Air, and Soil Pollution,* 115: 219–243.

AECB (Atomic Energy Control Board) and Canadian Nuclear Services, 1995. "Canada: *Living with Radiaion..*" ISBN 0-660-16036-6

Atwell, L., K.A. Hobson, and H.E. Welch, 1995. "Biomagnification and bioaccumulation of mercury in an Arctic marine food web: insights from stable nitrogen analysis." *Canadian Journal of Fisheries and Aquatic Science,* 55: 1114–1121.

Avendano, C, R. Ayres, C. Barril, S. Calumpang, G. Forget, P. Guy, S.-L. Han, N. Cortes-Maramba, D. Herod, S. Meakin, D. MacDonald, A. Ordas, R. Quijano, D. Riedel, D. Stone, J. Thomson, and H. Versteeg. 1995. "Annex II: Demonstration Substance Profiles, Meeting Background Report for the International Experties Meeting on Persistent Organic Pollutants: Towards Global Action." Joint Canadian–Philippines Planning Committee, Vancouver, Canada, June 4–8 pp. Annex II 1–171.

Babaluk, J.A., J.D. Reist, J.D. Johnson, and L. Johnson, 2000. "First records of sockeye (*Onchorhynchus nerka*) and pink salmon (*O. gorbuscha*) from Banks Island and other records of Pacfic salmon in Northwest Territories, Canada." *Arctic,* 53: 161–164.

Babkina, E.I., 1999. "Annual Report of the Ob-Irtysh UGMS Region: Soil Pollution 1992–1996." Ob–Irtysh UGMS, omsk.

Bahadur, N., W-Y Shiu, D.G.B. Boocock, and D. Mackay, 1997. "Temperature dependence of octanol-water partition coefficient for selected chlorobenzenes." *Journal of Chemical and Engineering Data.*, 42: 685–688.

Bailey, R.E., 2001. "Global hexachlorobenzene emissions." *Chemosphere,* 43: 167–182.

Bailey, R., L.A. Barrie, C.J. Halsall, P. Fellin, and D.C.G. Muir, 2000. "Atmospheric organochlorine pesticides in the western Canadian Arctic: Evidence of transpacific transport." *Journal of Geophysical Research*, 105: 11805–11811.

Baker, J.I. and R. A. Hites, 1999. "Polychlorinated dibenzo-*p*-dioxins and dibenzofurans in the remote north Atlantic marine atmosphere." *Environmental Science and Technology*, 33: 14–20.

Bakke, T., J. Hameedi, V. Kimstach, R. Macdonald, S. Melnikov, A. Robertson, R. Shearer, and D. Thomas, 1998. "Chapter 10: Petroleum hydrocarbons," pp. 661–716 in: *AMAP Assessment Report Arctic Pollution Issues*. A. Robertson, ed. AMAP, Oslo.

Bamford, H.A., D.L. Poster and J. Baker, 1999. "Temperature dependence of Henry's Law constants for thirteen polycyclic aromatic hydrocarbons between 4°C and 31°C." *Environmental Toxicology and Chemistry,* 18: 1905–1912.

Bamford, H.A., D.L. Poster, and J. Baker, 2000. "Henry's Law constants of polychlorinated biphenyls and their variation with temperature." *Journal of Chemical and Engineering Data*, 49: 1069–1074.

Banic, C., S.T. Beauchamp, R.J. Tordon, W.H. Schroeder, A. Steffen, and K.A. Anlauf, 2001. "Vertical distribution of gaseous elemental mercury in Canada," submitted to *Journal of Geophysical Research*.

Barrie, L.A., 1986. "Arctic Air Pollution: an overview of current knowledge." *Atmospheric Environment*, 20(4): 643–63.

Barrie, L.A., and U. Platt., 1997. "Arctic tropospheric chemistry: an overview." *Tellus,* 49B: 450–454.

Barrie, LA, J.W. Bottenheim, R.C. Schnell, P.J. Crutzen, and R.A. Rasmussen, 1988. "Ozone destruction and photochemical reactions at polar sunrise in the lower Arctic atmosphere." *Nature,* 334: 138–141.

Barrie, L.A., D. Gregor, B. Hargrave, R. Lake, D. Muir, R. Shearer, B. Tracey, and T. Bidleman, 1992. "Arctic contaminants: sources, occurrence and pathways." *The Science of the Total Environment*, 122: 1–74.

Barrie, LA, R.W. Macdonald, T. Bidleman, M. Diamond, D. Gregor, R. Semkin, W. Strachan, M. Alaee, S. Backus, M. Bewers, C. Gobeil, C. Halsall, J. Hoff, Y.F. Li, L. Lockhart, D. Mackay, D. Muir, J. Pudykiewicz, K. Reimer, J. Smith, G. Stern, W. Schroeder, R. Wagemann, F. Wania, and M. Yunker, 1997. "Chapter 2. Sources, Occurrence and Pathways." In: Jensen, J., K. Adare, and R. Shearer, (eds), *Canadian Arctic Contaminants Assessment Report*, Department of Indian and Northern Affairs, Ottawa. pp. 25–182.

Barrie, L.A., E. Falck, D. Gregor, T. Iverson, H. Loeng, R. Macdonald, S. Pfirman, T. Skotvold, and E. Wartena, 1998. "The influence of physical and chemical processes on contaminant transport into and within the Arctic." In *The AMAP Assessment*, D. Gregor, L. Barrie, and H. Loeng, (eds), pp. 25–116.

Barron, M.G., and L. Ka'aihue, 2001. "Potential for photoenhanced toxicity of spilled oil in Prince William Sound and Gulf of Alaska waters." *Marine Pollution Bulletin* 43: 86–92.

Barry, R.G., and M.C. Serreze, 2000. "Atmospheric components of the Arctic Ocean freshwater balance and their interannual variability." pp. 45–56 in: E.L. Lewis, (ed.) *The Freshwater Budget of the Arctic Ocean*. Kluwer Academic Publishers, Netherlands.

Baumgartner, A., and E. Reichel, 1975. "The World Water Balance." Elsevier Scientific Publishing, New York, NY, 179 pp.

Beeman, R.W., and F. Matsumura, 1981. "Metabolism of *cis*- and *trans*-chlordane by a soil microorganism." *Journal of Agricultural Food and Chemistry*, 29: 84–89.

Beitz, H.A., 1993. "Use of pesticides in the former GDR," report prepared for Environment Canada, Downsview, Ontario, Canada. pp. 10.

Bennett, D.H., T.E. McKone, M. Matthies, and W.E. Kastenberg, 1998. "General formulation of characteristic travel distance for semivolatile organic chemicals in a multimedia environment." *Environmental Science and Technology* 32 : 4023–4030.

Bennett, D.H., T.E. McKone, and W.E. Kastenberg, 1999. "General formulation of characteristic time for persistent chemicals in a multimedia environment." *Environmental Science and Technology* 33: 503–509.

Bennett, D.H., T.E. McKone, and W.E. Kastenberg, 2000. "Evaluating multimedia chemical persistence: classification and regression tree analysis." *Environmental Toxicology and Chemistry,* 19: 810–819.

Bennett, D.H., T.E. McKone, and W.E. Kastenberg, 2001a. "ART screening level analysis of characteristic time — A case study." In: *American Chemical Society Symposium Series No. 773 Persistent, Bioaccumulative, and Toxic Chemicals,* Volume II, R. Lipnick, B Jansson, D. Mackay, M. Petreas (es.) pp. 29–41.

Bennett, D.H., M. Scheringer, T.E. McKone, and K. Hungerbühler, 2001b. "Predicting long-range transport: A systematic evaluation of two multimedia transport models." *Environmental Science and Technology,* 35: 1181–1189.

Berg, T., and A-G. Hjellbrekke, 1999. "AMAP Data Report, Atmospheric Subprogramme." Norwegian Institute for Air Research, Kjeller, Norway, 100 pp.

Berg, T, J. Bartnicki, J. Munthe, H. Lattila, J. Hrehornul, and A. Mazur, 2001. "Atmospheric mercury species in the European Arctic: measurements and modelling." *Atmospheric Environment,* 35: 2569–2582.

Bergmann, M.A., and H.E., 1985. "Welch Spring meltwater mixing in small Arctic lakes." *Canadian Journal of Fisheries and Aquatic Sciences,* 42: 1789–1798.

Bergander, L., A. Kierkegaard, U. Sellström, U. Wideqvist, and C. de Wit, 1995. "Are brominated flame–retardants present in ambient air?" Poster, 6th Nordic Symposium on Organic Pollutants, Smygehuk, September 17–20.

Bernhoft, A., J.U. Skaare, Ø. Wiig, A.E. Derocher, and H.J.S. Larsen, 2000. "Possible immunotoxic effects of organochlorines in polar bears (*Ursus maritimus*) at Svalbard." *Journal of Toxicology and Environmental Health,* 59: 561–574.

Berrow, S.D., S.C. Long, A.T. McGarry, D. Pollard, E. Rogan, and C. Lockyer, 1998. "Radionuclides (137Cs and 40K) in harbour porpoises (*Phocoena phocoena*) from British and Irish coastal waters." *Marine Pollution Bulletin,* 36: 569–576.

Beyer, A., and M. Matthies, 2001. "Long-range transport potential of semivolatile organic chemicals in coupled air-water systems." *Environmental Science and Pollution Research,* 8: 173–179 (2001).

Beyer, A., D. Mackay, M. Matthies, F. Wania, and E. Webster, 2000. "Assessing long-range transport potential of persistent organic pollutants." *Environmental Science and Technology,* 34: 699–703.

Beyer, A., M. Scheringer, C. Schulze,and M. Matthies, 2001. "Comparing assessment methods to determine the environmental spatial scale of organic chemicals." *Environmental Toxicology and Chemistry,* 20: 922–927.

Beyer, A., F. Wania, T. Gouin, D. Mackay, and M. Matthies, 2002. "Temperature dependence of the characteristic travel distance." Submitted to *Environmental Science and Technology.*

Beyer, A., F. Wania, T. Gouin, D. Mackay, and M. Matthies, 2003. "Selecting internally consistent physicochemical properties of organic compounds." *Environmental Toxicology and Chemistry,* in press.

Bidleman, T.F., 1988. "Atmospheric Processes." *Environmental Science and Technology,* 22: 361–367.

Bidleman, T.F., 1992. "Arctic contaminants: sources, occurrence and pathways." *The Science of the Total Environment,* 122: 1–74.

Bidleman, T.F., 1999. "Atmospheric transport and air-surface exchange of pesticides." *Water, Air, and Soil Pollution,* 115: 115–166.

Bidleman, T.F., *personal communication,* 2002. Meteorological Service of Canada.

Bidleman, T.F., U.B. Wideqvist, B. Jansson, and R. Soderlund, 1987. "Organochlorine pesticides and polychlorinated biphenyls in the atmosphere of southern Sweden." *Atmospheric Environment,* 21: 641–654.

Bidleman, T.F., R.L. Falconer,and M.D. Walla, 1995a. "Toxaphene and other organochlorine compounds in air and water at Resolute Bay, N.W.T., Canada." *The Science of the Total Environment,* 160/161: 55–63

Bidleman, T.F., L.M.M. Jantunen, R.L. Falconer, L.A.Barrie,and P. Fellin, 1995b. "Decline of hexachlorocyclohexanes in the Arctic atmosphere and reversal of air-sea gas exchange." *Geophysical Research Letters,* 22: 219–222.

Bidleman, T.F., H. Alegria, B. Ngabe, and C. Green, 1998a. "Trends of Chlordane and Toxaphene in Ambient Air of Columbia, South Carolina." *Atmospheric Environment,* 32: 1849–1856

Bidleman, T.F., L.M.M Jantunen, K. Wiberg, T. Harner, K. Brice, K. Su, R.L. Falconer, A.D. Leone, E.J. Aigner and W.J. Parkhurst, 1998b. "Soil as a source of atmospheric heptachlor epoxide." *Environmental Science and Technology* 32: 1546–1548.

Bidleman, T.F., L.M.M. Jantunen, T. Harner, K. Wiberg, J.L.Wideman, K. Brice, K. Su, R.L. Falconer, E.J. Aigner, A.D. Leone, J.J. Ridal, B. Kerman, A. Finizio, H. Alegria, W.J. Parkhurst, and S.Y. Szeto, 1998c. "Chiral pesticides as tracers of air-surface exchange." *Environmental Pollution* 102, 43–49.

Bidleman, T.F., D.C.G. Muir, and G. Stern, 1999. "New persistent chemicals in the Arctic Environment," pp. 17–25 in: *Synopsis of Research Conducted under the 1998–'99 Northern Contaminants Program*, S. Kalhok (ed.). Indian and Northern Affairs Canada, Ottawa, ON.

Bidleman, T.F., M. Alaee, and G. Stern, 2000. "New persistent chemicals in the Arctic Environment," in *Synopsis of Research Conducted under the 1999–2000 Northern Contaminants Program*, S. Kalhok (ed.); Indian and Northern Affairs Canada, Ottawa, ON. QS–8602–000–EF–A1, 87–93.

Bidleman, T.F., M. Alaee, and G. Stern, 2001. "New persistent chemicals in the Arctic Environment," in: *Synopsis of Research Conducted under the 2000–2001 Northern Contaminants Program*, S. Kalhok (ed.); Indian and Northern Affairs Canada, Ottawa.ON.

Bidleman, T.F, L.M.M Jantunen, P.A. Helm, E. Brorström-Lundén, and S. Juntto, 2002a. "Chlordane enantiomers and temporal trends of chlordane isomers in Arctic air." *Environmental Science and Technology*, 36: 539–544.

Bidleman, T.F., A.D. Leone, and R.L. Falconer, 2002b. "Vapor pressures and enthalpies of vaporization for toxaphene congeners." *Journal of Chemical and Engineering Data*, submitted.

Bindler, R., I. Renberg, P.G. Appleby, N.J. Anderson, and N.L. Rose, 2001. "Mercury accumulation rates and spatial patterns in lake sediments from West Greenland: A coast to ice margin transect." *Environmental Science and Technology,* 35: 1736–1741.

Bishop, K., Y.-H. Lee, c. Pettersson, and B. Allard. 1995. "Methylmercury output from the Svartberget catchment in norhtern Sweden during spring flood." *Water, Air, and Soil Pollution,* 80: 445–454.

Bisset & Associates, 1995. "Research of former military sites and activities in the Yukon. Action on Waste Program, Arctic Environmental Strategy (AES)." Indian and Northern Affairs, Canada, 312 pp.

Björk, G., S. Söderkvist, P. Winsor, A. Nikolopoulos, and M. Steele, 2002. "Partial return of the cold halocline layer to the Amundsen Basin of the Arctic Ocean: Implications for sea ice mass balance." *Geophysical Research Letters*, 10.1029/2001 GL014157 (07 June 2002), 8-1–8-4.

Bjornsson, H., L.A. Mysak, and R.D. Brown, 1995. "On the interannual variability of precipitation and runoff in the Mackenzie drainage basin." *Climate Dynamics,* 12: 67–76.

Blais, J. M., D.W. Schindler, D.C.G. Muir, L.E. Kimpe, D.B. Donald, and B. Rosenberg, 1998. "Accumulation of persistent organochlorine compounds in mountains of western Canada." *Nature,* 395: 585–588.

Blankenship, A.L., K. Kannan, S.A. Villalobos, D.L. Villeneuve, J. Falandysz, T. Imagawa, E. Jakobsson, and J.P. Giesy, 2000. "Relative potencies of individual polychlorinated naphthalenes and Halowax mixtures to induce Ah receptor-mediated responses." *Environmental Science and Technology,* 34: 3153–3158.

Bockstoce, J.R., 1986. *Whales, Ice and Men: The History of Whaling in the Western Arctic.* University of Washington Press, Seattle and London.

Bodaly, R.A., and T.A. Johnston, 1992. "The mercury problem in hydro-electric reservoirs with predictions of mercury burdens in fish in the proposed Grand Baleine complex, Québec." *James Bay Publication Series, Hydro-Electric Development: Environmental Impacts* 3.

Boethling, R.S., and D. Mackay, 2000. *Handbook of Property Estimation Methods for Chemicals*. Lewis, Boca Raton, FL.

Borgå, K., G.W. Gabrielsen, and J.U. Skaare, 2002. "Differences in contamination load between planktonic and ice-associated fauna in the Arctic Marginal Ice Zone: influence of habitat, diet and geography." *Marine Ecology Progress Series* submitted.

Botsford, L.W., J.C. Castilla, and C. H. Peterson, 1997. "The management of fisheries and marine ecosystems." *Science*, 277: 509–515.

Bottenheim, J.W., A.G. Gallant, and K.A. Brice, 1986. "Measurements of NO_y species and O_3 at 82°N latitude." *Geophysical Research Letters*, 22: 599–602.

Bourke, R.H., and R.P. Garrett, 1987. "Sea ice thickness distribution in the Arctic Ocean." *Cold Regions Science and Technology*, 13: 259–280.

Boutron, C.F., U. Gorlach, J.P. Candelone, M.A. Bolshov, and R.J. Delmas, 1991. "Decrease in anthropogenic lead, cadmium and zinc in Greenland snows since the late 1960s." *Nature*, 353: 153–156.

Boutron, C.F., J.P. Candelone, and S. Hong. 1995. "Greenland snow and ice cores: unique archives of large-scale pollution of the troposphere of the Northern Hemisphere by lead and other heavy metals." *The Science of the Total Environment*, 160/161: 233–241.

Boutron, C.F., G.M. Vandal, W.F. Fitzgerald, and C.P. Ferrari, 1998. "A forty year record of mercury in central Greenland snow." *Geophysical Research Letters*, 25: 3315–3318.

Boyd, T.J., M. Steele, R.D. Muench, and J.T. Gunn, 2002. "Partial recovery of the Arctic Ocean halocline." *Geophysical Research Letters*, 10.1029/2001 GL014047 (29(14)), 2-1–2-4.

Boyle, E.A., 1988. "Cadmium: chemical tracer of deep-water paleoceanography." *Paleoceanography*, 3: 471–489.

Bradley, R.S., and P.D. Jones, 1993. "'Little Ice Age' summer temperature variations: their nature and relevance to recent global warming trends." *The Holocene*, 3: 367–376.

Bradstreet, M.S.W., K.J. Finley, A.D. Sekerak, W.B. Griffiths, C.R. Evans, M.F. Fabijan, and H.E. Stallard, 1986. "Aspects of the biology of Arctic cod (*Boreogadus saida*) and its importance in Arctic marine food chains." *Canadian Technical Report of Fisheries and Aquatic Sciences*, 1–193.

Braekevelt, E., G.T. Tomy, and G.A. Stern, 2001. "Comparison of an individual congener standard and a technical mixture for the quantification of toxaphene in environmental matrices by HRGC/ECNI–HRMS." *Environmental Science and Technology* 35: 3513–3518

Braune, B., D. Muir, B. DeMarch, M. GAmberg, K. Poole, R. Currie, M. Dodd, W. Duschenko, J. Eamer, B. Elkin, M. Evans, S. Grudny, C. Hebert, R. Johnstone, K. Kidd, B. Koenig, L. Lockhart, H. Marshall, K. Reimer, J. Sanderson, and L. Shutt, 1999. "Spatial and temporal trends of contaminants in Canadian Arctic freshwater and terrestrial ecosystems: a review." *The Science of the Total Environment*, 230: 145–207.

Braune, B., T.F. Bidleman, A.T. Fisk, and D.C.G. Muir, 2001. "Assessment of spatial and temporal patterns of HCH isomers in the Arctic environment," In: *Synopsis of Research Conducted under the 2000–01 Northern Contaminants Program*, S. Kalhok, (ed.) Indian and Northern Affairs Canada, Ottawa, ON.

Breivik, K., J.M. Pacyna, and J. Munch, 1999. "Use of α-, β-, γ-Hexachlorocyclohexane in Europe, 1970–1996." *The Science of the Total Environment*, 239: 151–163.

Breivik, K., A. Sweetman, J.M. Pacyna, and K.C. Jones, 2002a. "Towards a global historical emission inventory for selected PCB-congeners — a mass balance approach. 1. Global production and consumption." *The Science of the Total Environment*, 290: 181–198.

Breivik, K., A. Sweetman, J.M. Pacyna, and K.C. Jones, 2002b. "Towards a global historical emission inventory for selected PCB-congeners — a mass balance approach. 2. Emissions." *The Science of the Total Environment*. 290: 199–224.

Brodeur, R.D., C.E. Mills, J.E. Overland, G.E. Walters, and J.D. Schumacher, 1999. "Evidence for a substantial increase in gelatinous zooplankton in the Bering Sea, with possible links to climate change." *Fisheries Oceanography*, 8: 296–306.

Brorström–Lundén, E, E. Junedal, H. Wingfors, and S. Juntto, 2000. "Measurements of the atmospheric concentrations and deposition fluxes of POPs at the Swedish west coast and in northern Fennoscandia." Swedish Environmental Research Institute, Report 2000-03-17.

Brown, J., A. Kolstad, A. Brungot, B. Lind, A. Rudjord, P. Strand, and L. Foeyn, 1999. "Levels of super(99)Tc in seawater and biota samples from Norwegian coastal waters and adjacent seas." *Marine Pollution Bulletin,* 38: 560–571.

Brown, R.D., and B.E. Goodison, 1996. "Interannual variability in reconstructed Canadian snow cover 1915–1992." *Journal of Climate,* 9: 1299–1318.

Brubaker, W.W., and R. A. Hites,1997. "Polychlorinated dibenzo–*p*–dioxins and dibenzofurans: gas-phase hydroxyl radical reactions and related atmospheric removal." *Environmental Science and Technology*, 32: 3913–3918.

Brubaker, W.W., and R. A. Hites. 1998a. "Gas-phase oxidation products of biphenyl and polychlorinated biphenyls." *Environmental Science and Technology*, 32: 3913–3918.

Brubaker, W.W., and R. A. Hites, 1998b. "OH reaction kinetics of polycyclic aromatic hydrocarbons and polychlorinated dibenzo-*p*-dioxins and dibenzofurans." *Atmospheric Environment,* 102: 915–921.

Brubaker, W.W., and R.A. Hites, 1998c. "OH reaction kinetics of gas-phase — and -hexachlorocyclohexane and hexachlorobenzene." *Environmental Science and Technology,* 32: 766–769.

Bruland, K.W., and R.P. Franksm, 1983. "Mn, Ni, Cu, Zn and Cd in the western North Atlantic," pp. 395–414 in: *Trace Metals in Sea Water.* C.S. Wong, E. Boyle, K. W. Bruland, J. D. Burton, and E.D. Goldberg, (eds). Plenum Press, New York, New York.

Brunner, S., E. Hornung, H. Santl, E. Wolff, O.G. Piringer, J. Altschuh, and R. Bruggemann, 1990. "Henry's law constants for polychlorinated biphenyls: experimental determination and structure-property relationships." *Environmental Science and Technology* 24: 1751–1754.

BSEF. 2000. Bromine Science and Environmental Forum, Brussels, Belgium. Data reported at http://205.232.112.21/bsef/docs/Major_Brominated.doc.

Burkhard, L.P., D.E. Armstrong, and A.W. Andren, 1985. "Henry's law constants for the polychlorinated biphenyls." *Environmental Science and Technology* 19: 590–596.

Burton, D.K., and J.R. Philogen, 1985. "An overview of pesticide usage in Latin America." A report to the Canadian Wildlife Service Latin American Program Contract # OST85–00181, 122 pp.

Buser, H.R., and M.D. Müller, 1994. "Identification of the (+) and (-) enantiomers of chiral chlordane compounds using chiral high performance liquid chromatography — chiroptic dtection and chiral high resolution gas chromatography — mass spectrometry." *Analytical Chemistry,* 66: 2155–2162.

Cabanes, A., L. Legagneux, and F. Dominé, 2002. "Evolution of the specific surface area and of crystal morphology of Arctic fresh snow during the ALERT 2000 campaign." *Atmospheric Environment,* in press.

Cai, D.J., 1997. "Based on the data from the Import and Export Commodity Inspection Bureau of the People's Republic of China (in Chinese).

Cai, D.J., P.Z. Yang, J.L. Wang, X.L. Jiang, J.D. Shen, R. Chen, and R.Z. Gong, 1985. "Contamination of ecosystem of the Taihu Lake basin and protection policy: reports for the agricultural modern scientific experimental fields in Taihu Lake basin, Jiangsu province." *Jiangsu Province Agricultural Academy*, Vol. 12.

Cai, D.J., L.J. Sun, J.L. Ke, and G.C. Tang, 1992. "Pesticide usage in China." Technical Report RP032, Report prepared for Environment Canada, Downsview, Ontario, Canada

Campbell, L.M., D.W. Schindler, D.C. G. Muir, D.B. Donald, and K.A. Kidd, 2000. "Organochlorine transfer in the food web of sub-alpine Bow Lake, Banff National Park." *Canadian Journal of Fisheries and Aquatic Science,* 57: 1258–1269.

CNSC (Canadian Nuclear Safety Commission), 2000. Radiation Protection Regulations. Canada Gazette, part II, vol. 134, no. 13.

Carmack, E.C., and R.W. Macdonald, 2002. "Oceanography of the Canadian shelf of the Beaufort Sea: A setting for marine life." *Arctic* in press.

Carmack, E.C., R.W. Macdonald, R.G. Perkin, F.A. McLaughlin, and R.J. Pearson, 1995. "Evidence for Warming of Atlantic Water in the Southern Canadian Basin of the Arctic Ocean: Results from the Larsen-93 Expedition." *Geophysical Research Letters*, 22: 1061–1064.

Carmack, E.C., Aagaard, K, J.H. Swift, R.W. Macdonald, F.A. McLaughlin, E.P. Jones, R. D. Perkin, J. Smith, K. Ellis, and K. L, 1997. "Rapid changes of water properties and contaminants within the Arctic Ocean." *Deep-Sea Research II,* 44: 1487–1502.

Carrera, G., P. Fernandez, R.M. Vilanova, and J.O. Grimalt, 2001. "Persistent organic pollutants in snow from European high mountain areas." *Atmospheric Environment,* 35: 245–254.

Carvalho, F.P., S. Montenegro-Guillen, J.P. Villeneuve, C. Cattini, J. Bartocci, M. Lacayo, and A. Cruz, 1999. "Chlorinated Hydrocarbons in Coastal Lagoons of the Pacific Coast of Nicaragua." *Archives of Environmental Contamination and Toxicology,* 36: 132–139.

Castillo, L.E., E. De La Cruz, and C. Ruepert, 1997. "Ecotoxicology and Pesticides in Tropical Aquatic Ecosystems of Central America", *Environmental Toxicology and Chemistry,* 16(1): 41–51.

Cavalieri, D.J., P. Gloersen, C.L. Parkinson, J.C. Comiso, and H.J. Zwally, 1997. "Observed hemispheric asymmetry in global sea ice changes." *Science*, 278: 1104–1106.

CCME (Canadian Council of Ministers of the Environment), 1999. Recommended Canadian Soil Quality Guidelines, Chap 6, Canadian Environmental Quality Guidelines.

Chaix, L., and F. Domine, 1997. "Effect of the thermal history of ice crushed at 77 K on its surface structure as determined by adsorption of CH_4 at low surface coverage." *Journal of Physical Chemistry,* B101: 6105–6108.

Chaix, L., J. Ocampo, and F. Domine, 1996. "Adsorption of CH_4 on laboratory-made crushed ice and on natural snow at 77 K. Atmospheric implications." *Comptes Rendus de la Academie des Sciences Serie II Fascicule A–Sciences de la Terre et des Planetes,* 322: 609–616.

Cheam, V., D. Muir, and D. Halliwell, 2001. "Spatial trends in loadings and historical inputs of metals inferred from Arctic lake sediment cores." in: *Synopsis of Research Conducted under the 2000–2001 Northern Contaminants Program.* S. Kalhok, (ed.). Indian and Northern Affairs Canada, Ottawa, ON.

Chemical and Engineering News, 1982. EPA bans most use of pesticide toxaphene, Oct. 25, pp. 6.

Chen, G., A.D. Konstantinov, B.G. Chittim, E.M. Joyce, N.C. Bols, and N.J. Bunce, 2001a. "Synthesis of polybrominated diphenyl ethers and their capacity to induce CYP1A by Ah receptor mediated pathway." *Environmental Science and Technology,* 35: 3749–3756.

Chen, J.W., X. Quan, Y.Z. Zhao, F.L. Yang, K.W. Schramm, and K. Kettrup, 2001b. "Quantitative structure-property relationships for octanol-air partition coefficients of PCDD/Fs." *Bulletin of Environmental Contamination and Toxicology*, 66: 755–761.

Cheng, M.D., and W.H. Schroeder, 2000. "Potential Atmospheric Transport Pathways for Mercury Measured in the Canadian High Arctic." *Journal of Atmospheric Chemistry,* 35: 101–107.

Chernyak, S.M., C.P. Rice, and L.L. McConnell, 1996. "Evidence of currently-used pesticides in air, ice, fog, seawater and surface microlayer in the Bering and Chukchi Seas." *Marine Pollution Bulletin,* 32: 410–419.

CIEL. 1998: Centre International d'Études du Lindane. Letter from J.W.G. Randall to Jorg Munch, dated April 14, 1998.

Coachman, L.K., and K. Aagaard, 1988. "Transports through Bering Strait: annual and interannual variability." *Journal of Geophysical Research,* 93: 15535–15539.

Cohen, S.J, 1997a. "Mackenzie Basin Impact Study." Environment Canada, Downsview, Ontario.

Cohen, S.J., 1997b. "What if and so what in Northwest Canada: Could climate change make a difference to the future of the Mackenzie Basin?" *Arctic,* 50: 293–307.

Coleman, P.J., R.G.M. Lee, R.E. Alcock, and K.C. Jones, 1997. "Observations on PAH, PCB, and PCDD/F trends in U. K. urban air, 1991–1995." *Environmental Science and Technology*, 31: 2120–2124.

Commoner, B., P.W. Bartlett, H. Eisl, and K. Couchot, 2000. "Long-range air transport of dioxin from North American sources to ecologically vulnerable receptors in Nunavut, Arctic Canada." Final report to the North American Commission for Environmental Cooperation. 101 pp.

Cooper, L.W., I.L. Larsen, T.M. Beasley, S.S. Dolvin, J.M. Grebmeier, J.M. Kelley, M. Scott, and A. Johnson-Pyrtle, 1998. "The distribution of radiocesium and plutonium in sea ice-entrained Arctic sediments in relation to potential sources and sinks." *Journal of Environmental Radioactivity,* 39: 279–303.

Cooper, L.W., T. Beasley, K. Aagaard, J.M. Kelley, I.L. Larsen, and J. M. Grebmeier, 1999. "Distributions of nuclear fuel-reprocessing tracers in the Arctic Ocean: Indications of Russian river influence." *Journal of Marine Research,* 57: 715–738.

Cortes, D.R., and R. A. Hites, 2000. "Detection of statistically significant trends in atmospheric concentrations of semivolatile compounds." *Environmental Science and Technology*, 34: 2826–2829.

Cortes, D.R., I. Basu, C.W. Sweet, K.A. Brice, R.M. Hoff, and R.A. Hites, 1998. "Temporal trends in gas-phase concentrations of chlorinated pesticides at the shores of the Great Lakes." *Environmental Science and Technology,* 32: 1920–1927.

Cotham, W.E. Jr., and T.F. Bidleman, 1991. "Estimating the atmospheric deposition of organochlorine contaminants to the Arctic." *Chemosphere,* 22: 165–188.

Couch, T.L., A.L. Sumner, T.M. Dassau, P.B. Shepson, and R.E. Honrath, 2000. "An investigation of the interaction of carbonyl compounds with the snowpack." *Geophysical Research Letters* 27: 2241–2244.

Council of Fire Marshals and Fire Commissioners, 1997. "Annual Report of Fire Losses in Canada." Edmonton, Alberta.

Cousins, I., N. Hartlieb, C. Teichmann and K.C. Jones. 1997. "Measured and predicted volatilisation fluxes of PCBs from contaminated sludge-amended soils." *Environmental Pollution*, 97: 229–238.

Cristol, S.J., 1947. "The kinetics of alkaline dehydrochlorination of the BHC isomers. The mechanism of second-order elimination reactions." *Journal of the American Chemical Society,* 69: 338–342.

Crookes, M.J., and P.D. Howe, 1993. "Environmental hazard assessment: halogenated naphthalenes," *Report TSD/13,* Department of the Environment, London, England.

Crosby, D.G., K.I. Beynon, P.A. Greve, F. Korte, G.G. Still, and J.W. Vonk, 1981. "Environmental chemistry of pentachlorophenol." *Pure and Appllied Chemistry,* 53: 1051–1080.

Croteau, M–N, L. Hare, and A. Tessier, 2002. "Increases in food web cadmium following reductions in atmospheric inputs to some lakes." *Environmental Science and Technology,* 36: 3079–3082.

Crowley, T.J., and T.S. Lowery, 2000. "How warm was the Medieval Warm Period?" *Ambi,* 29: 51–54.

Dahlgaard, H.,1995. "Transfer of European coastal pollution to the Arctic: radioactive tracers." *Marine Pollution Bulletin,* 31: 3–7.

Dai, M.H., and J.M. Martin, 1995. "First data on trace metal level and behaviour in two major Arctic river-estuarine systems (Ob and Yenisey) and in the adjacent Kara Sea, Russia." *Earth and Planetetary Science Letters,* 131: 127–141.

Dansgaard, W., S. J. Johnsen, H.B. Clausen, D. Dahl-Jensen, N.S. Gundesturp, C.U. Hammer, C.S. Hvidberg, J.P. Steffensen, A.E. Sveinibjornsdottir, J. Jouzel, and G. Bond, 1993. "Evidence for general instability of past climate from a 250-kyr ice-core record." *Nature,* 364: 218–220.

Darby, D.A., J.F. Bischof, and G.A. Jones, 1997. "Radiocarbon chronology of depositional regimes in the western Arctic Ocean." *Deep-Sea Research II,* 44: 1745–1757.

Darby, D., J. Bischof, G. Cutter, A. de Vernal, C. Hillaire–Marcel, G. Dwyer, J. McManus, L. Osterman, L. Polyak, and R. Poore, 2001. "New record shows pronounced changes in Arctic Ocean circulation and climate." *Eos, Transactions, American Geophysical Union,* 82: 601, 607.

Dasher, D., W. Hanson, S. Read, S. Faller, D. Farmer, W. Efurd, J. Kelley, and R. Patrick, 2002. "An assessment of the reported leakage of anthropogenic radionuclides from the underground nuclear test sites at Amchitka Island, Alaska to the surface environment." *Journal of Environmental Radioactivity,* 60(1–2): 165–188.

Datta, S., L.L. McConnell, J.E. Baker, J. Lenoir, and J.N. Seiber,1998. "Evidence for atmospheric transport and deposition of polychlorinated biphenyls to the Lake Tahoe Basin, California–Nevada." *Environmental Science and Technology,* 32: 1378–1385.

de Baar, H.J.W., P.M. Saager, R.F. Nolting, and J. van der Meer, 1994. "Cadmium versus phosphate in the world ocean." *Marine Chemistry,* 46: 261–281.

de Boer, J., 1989. "Organochlorine compounds and bromodiphenylethers in livers of Atlantic cod (*Gadus morhua*) from the North Sea." *Chemosphere,* 18: 2131–2140.

de Boer, J., 1990. "Brominated diphenyl ethers in Dutch freshwater and marine fish." *Organohalogen Compounds,* 2: 315–318.

de Carlo, V.J., 1979. "Studies on brominated chemicals in the environment." *Annals of the New York Academy of Sciences,* 320: 678–681.

de Graff, N., and L. Mychasiw, 1994. "Lake trout recovery and contaminant management program in Lake Laberge, Yukon Territory." pp. 301–303 in: *Synopsis of research conducted under the 1993/94 Northern Contaminants Program.* J. L. Murray and R.G. Shearer, eds. INAC, Ottawa, Ontario.

de March, B.G.E., C.A. de Wit, D.C.G. Muir, B.M. Braune, D.J. Gregor, R.J. Norstrom, M. Olsson, J.U. Skaare, and K. Stange, 1998. "Persistent Organic Pollutants." Chapter 6, pp. 183–371 in: *AMAP Assessment Report: Arctic Pollution Issues., AMAP,* Oslo, Norway. B.G.E. de March, C.A. de Wit, and D.C.G. Muir (eds.).

de Vernal, A., A. Rochon, C. Hillaire-Marcel, J.L. Turon, and J. Guiot, 1993. "Quantitative reconstruction of sea-surface conditions, seasonal extent of sea-ice cover and meltwater discharges in high latitude marine environments from dinoflagellate cycs assemblages." in: *Ice in the Climate System.* W.R. Peltier (ed.). Springer-Verlag, Heidelberg, Germany.

de Wit, C.A., K. Hobbs, A.T. Fisk, and D.C.G. Muir, 2003. "AMAP II Persistent organic pollutants report." In preparation.

Deser, C., 2000. "On the teleconnectivity of the "Arctic Oscillation." *Geophysical Research Letters,* 27: 779–782.

Dethleff, D., H. Nies, I.H. Harms, and M.J. Karcher, 2000a. "Transport of radionuclides by sea-ice and dense-water formed in western Kara Sea flaw leads." *Journal of Marine Systems,* 24: 233–248.

Dethleff, D., V. Rachold, M. Tintelnot, and M. Antonow. 2000b. "Sea-ice transport of riverine particles from the Laptev Sea to Fram Strait based on clay mineral studies." *International Journal of Earth Sciences,* 89: 496–502.

Diamond, M., 1999. "Development of a fugacity/aquivalence model of mercury dynamics in lakes." *Water, Air, and Soil Pollution,* 11: 337–357.

Dickson, D. L., and H. G. Gilchrist, 2002. "Status of marine birds of western Arctic Canada." *Arctic,* in press.

Dickson, R., 1999. "All change in the Arctic." *Nature,* 397: 389–391.

Dickson, R.R., T.J. Osborn, J.W. Hurrell, J. Meincke, J. Blindheim, B. Adlandsvik, T. Vinje, G. Alekseev, and W. Maslowski, 2000. "The Arctic Ocean response to the North Atlantic Oscillation." *Journal of Climate,* 13: 2671–2696.

Dietz, R., C.T. Ansen, and P. Have, 1989. "Clue to seal epizootic." *Nature,* 338: 627.

Dippner, J., and G. Ottersen. 2001. "Cod and climate variability in the Barents Sea." *Climate Research,* 17: 73–82.

Dominé, F., L. Chaix, and L. Hanot, 2000. "Reanalysis and new measurements of N_2 and CH_4 adsorption on ice and snow." *Journal of Colloid and Interface Science,* 227:104–110.

Dominé, F., A. Cabanes, A.S. Taillandier, and L. Legagneux, 2001. "Specific surface area of snow samples determined by CH_4 adsorption at 77 K and estimated by optical, microscopy and scanning electron microscopy." *Environmental Science and Technology,* 35, 771–780.

Dominé, F., A. Cabanes, and L. Legagneux, 2002. "Structure,microphysics,and surface area of the Arctic snowpack near Alert during the ALERT 2000 campaign." *Atmospheric Environment,* in press.

Donald, D.B., G.A. Stem, D.C.G. Muir, B.R Fowler, B.M. Miskirrunin, and R Bailey, 1998. "Chlorobornanes in water, sediment and fish from toxaphene-treated and untreated lakes in western Canada." *Environmental Science and Technology,* 32: 1391–1397.

Donald, D.B., J. Syrgiannis, R.W. Crosley, G. Holdsworth, D.C.G. Muir, B. Rosenberg, A. Sole, and D.W. Schindler, 2000. "Delayed deposition of organochlorine pesticides at a temperate glacier." *Environmental Science and Technology,* 33: 1794–1798 and *Environmental Science and Technology,* 34: 2873–3874.

Donaldson, D.J., and D. Anderson,1999. "Adsorption of atmospheric gases at the air-water interface. 2. C_1-C_4 alcohols, acids, and acetone." *Journal of Physical Chemistry A,* 103: 871–876.

Dörr, G., M. Hippelein, and O. Hutzinger,1996. "Baseline contamination assessment for a new resource recovery facility in Germany. Part V: analysis and seasonal/regional variability of ambient air concentrations of polychlorinated naphthalenes (PCN)." *Chemosphere,* 33: 1563–1568.

Douglas, M.S.V., J.P. Smol, and W.J. Blake, 1994. "Marked post-18th century environmental change in high Arctic ecosystems." *Science* 266: 416–419.

Dowdeswell, J.A., J.O. Hagen, H. Bjornsson, A.F. Glazovsky, W.D. Harrison, P. Holmlund, J. Jania, R.M. Koerner, B. Lefauconnier, C.S.L. Ommanney, and R.H. Thomas, 1997. "The mass balance of circum-Arctic glaciers and recent climate change." *Quarternary Research,* 48: 1–14.

Drouillard, K.G., G.T. Tomy, D.C.G. Muir, and K.J. Friesen,1998a. "Volatility of chlorinated n–alkanes (C_{10}-C_{12}): Vapor pressures and Henry's law constants." *Environmental Toxicology and Chemistry,* 17: 1252–1260.

Drouillard, K.G., T. Hiebert., T. Phuoung, Tomy, G.T., Muir, D.C.G., and K.J. Friesen, 1998b. "Estimating the aqueous solubilities of individual chlorinated *n*-alkanes (C_{10}-C_{12}) from measurements of chlorinated alkane mixtures." *Environmental Toxicology and Chemistry,* 17: 1261–1267.

Dunton, K., 1992. "Arctic biogeography: The paradox of the marine benthic fauna and flora." *Trends in Ecology and Evolution,* 7: 183–189.

Dunnivant, F.M., A. W. Eizerman, P.C. Jurs and M.N. Hasan, 1992. "Quantitative structure-property relationships for aqueous solubilities and henry's law constants of polychlorinated biphenyls." *Environmental Science and Technology* 26: 1567–1573.

Dyke, A.S., and J.M. Savelle, 2000. "Holocene driftwood incursion to southwestern Victoria Island, Canadian Arctic Archipelago, and its significance to paleoceanography and archaeology." *Quarternary Research,* 54: 113–120.

Dyke, A.S., and J.M. Savelle, 2001. "Holocene history of the Bering Sea bowhead whale (*Balaena mysticetus*) in its Beaufort Sea summer grounds off southwestern Victoria Island, western Canadian Arctic." *Quarternary Research,* 55: 371–379.

Dyke, A.S., J.E. Dale, and R.N. McNeeley, 1996a. "Marine molluscs as indicators of environmental change in glaciated North American and Greenland during the last 18,000 years." *Géographie physique et Quarternaire,* 50: 125–184.

Dyke, A.S., J. Hooper, and J.M. Savelle, 1996b. "A history of sea ice in the Canadian Arctic Archipelago based on postglacial remains of the Bowhead Whale (*Balaena mysticetus).*" *Arctic,* 49: 235–255.

Dyke, A.S.,W. England, J.E. Reimnitz, and H. Jette, 1997. "Changes in driftwood delivery to the Canadian Arctic Archipelago: The hypothesis of postglacial oscillations of the Transpolar Drift." *Arctic,* 50: 1–16.

Dyke, A.S., J.Hooper, C.R. Harington, and J.M. Savelle, 1999. "The late Wisconsinan and Holocene record of walrus (*Obobenus rosmarus*) from North America: A review with new data from Arctic and Atlantic Canada." *Arctic,* 52: 160–181.

Dynesius, M., and C. Nilsson, 1994. "Fragmentation and flow regulation of river systems in the northern third of the world." *Science,* 266: 753–761.

Dyurgerov, M.B., and M.F. Meier, 1997. "Year-to-year fluctuation of global mass balance of small glaciers and their contribution to sea level changes." *Arctic and Alpine Research,* 29: 392–402.

Easterling, D.R., G.A. Meehl, C. Parmesan, S.A. Changnon, T.R. Karl, and L.O. Mearns, 2000. "Climate extremes: observations, modelling and impacts." *Science,* 289: 2068–2074.

Ebinghaus, R.,S.G. Jennings, W.H. Schroeder, T. Berg, T. Donaghy, J. Guentzel, C. Kenny, H.H. Kock, K. Kvietkus, W. Landing, T. Mühleck, J. Munthe, E.M. Prestbo, D. Schneeberger, F. Slemr, J. Sommar, A. Urba, D. Wallschläger, and Z. Xiao, 1999. "International field intercomparison measurements of atmospheric mercury species at Mace Head, Ireland." *Atmospheric Enviornment,* 33: 3063–3073.

Ebinghaus, R, H.H. Kock, C. Temme, J.W. Einex, A.G. Lowe, A. Richter, J.P. Burrows, and W.H. Schroeder, 2002. "AntArctic springtime depletion of atmospheric mercury." *Environmental Science and Technology*, 36: 1238–1244.

Eckerman, D.E., 1984. Preliminary quantitative usage analysis of lindane, US Environmental Protection Agency; Economic Analysis Branch, Benefits and Use Division, Office of Pesticide Programs.

Egstrom, D.R., E.B. Swain, T.A.Henning, M.E. Brigham, and P.L. Brezonik, 1994. pp. 33–66, In: *Environmental Chemistry of Lakes and Reservoirs.* L.A. Backer (ed.). American Chemical Society, Washington, DC.

EHC (Environmental Health Criteria), 1995. Environmental Health Criteria 172, Brominated Dipheny Ethers. World Health Organization, Geneva.

EHC (Environmental Health Criteria), 1997. Environmental Health Criteria 192, Brominated Dipheny Ethers. World Health Organization, Geneva.

Eichers, T.R., P.A. Andrilinas, H. Black, R. Jenkins, and A. Fox, 1970. "Quantities of pesticides used by farmers in 1966." US Department of Agriculture, Economic Research Service, Economic Report No. 179, 61 pages.

Eichers, T.R., P.A. Andrilenas, and T.W. Anderson, 1978. "Farmers' Use of Pesticides in 1976." Economics, Statistics, and Cooperatives Service, US Department of Agricultural. Agricultural Economic Report No. 418. 58 pages.

Eicken, H., E. Reimnitz, V. Alexandrov, T. Martin, H. Kassens, and T. Viehoff, 1997. "Sea-ice processes in the Laptev Sea and their importance for sediment export." *Continental Shelf Research,* 17: 205–233.

Eicken, H., J. Kolatschek, F. Lindemann, I. Dmitrenko, J. Freitag, and H. Kassens, 2000. "A key source area and constraints on entrainment for basin scale sediment transport by Arctic sea ice." *Geophysical Research Letters,* 27: 1919–1922.

Ekwurzel, B., P. Schlosser, R.A. Mortlock, R.G. Fairbanks, and J.H. Swift, 2001. "River runoff, sea ice meltwater, and Pacific water distribution and mean residence times in the Arctic Ocean." *Journal of Geophysical Research*, 106: 9075–9092.

El–Sebae, A.H., M. Abou Zeid, and M.A. Saleh, 1993, "Status and Environmental Impact of Toxaphene in the Third World — A Case Study of African Agriculture." *Chemosphere* 27 (10): 2063–2072

Enge, E. K., E.S. Heimstad, and R. Kallenborn, 1998. "Distribution of polychlorinated bpphenyls (PCBs) in snow samples in northern Norway." *Organochlorine Compounds,* 39: 435–439.

Environment Canada, 1990. "The Climates of Canada." Ottawa, ON.

Environment Canada, 1993. Health and Welfare Canada, Priority substances list Assessment Report: Chlorinated Paraffins. Government of Canada, Catalogue No. En 40 215/17E, ISBN 0 662 20515 17E: 32 pp.

Erickson, M.D., 1997. "Anaytical Chemistry of PCBs," 2nd ed. Boca Raton: Lewis Publishers, Boca Raton, U. S. A., 667 pp.

Eriksson, P., E. Jakobsson, and A. Fredriksson, 1998. "Developmental neurotoxicity of brominated flame retardants, polybrominated diphenyl ethers and tetrabromo-bis-phenol A.." *Organohalogen Compounds,* 35: 375–377.

ESG (Environmental Sciences Group) Royal Military College, 1997. "Environmental Assessment of Saglek, Labrador (LAB-2)." Prepared for the Director, North Warning System Office, National Defence. Kingston ON.

ESG (Environmental Sciences Group) Royal Military College, 1999. "Saglek Food Chain Study Results Update." Prepared for the Director, North Warning System Office, National Defence. Kingston ON.

ESG (Environmental Sciences Group) Royal Military College, 2000a. "An Investigation of DDT Contamination, Kitigazuit, NWT." Prepared for the Contaminated Sites Office, Indian & Northern Affairs. Kingston ON.

ESG (Environmental Sciences Group) Royal Military College, 2000b. "Remediation of PCB Contaminated Soil at Saglek, Labrador (LAB-2)." Prepared for the Director, North Warning System Office, National Defence. Kingston ON.

ESG (Environental Sciences Group) Royal Military College, 2001a. "Ecological risk assessment of PCB contaminated sediments at Saglek, Labrador." Prepared for the Director, North Warning System Office, National Defence. Kingston ON.

ESG (Environmental Sciences Group) Royal Military College, 2001b. "Arsenic Levels in the Yellowknife Area: Distinguishing between Natural and Anthropogenic Inputs." Prepared for the Yellowknife Soil Remediation Committee (YSARC). Kingston ON.

ESG (Environmental Sciences Group) Royal Military College, 2001c. "Bioavailability of Arsenic in Yellowknife." Prepared for the Toxic Substances Research Initiative (Health Canada). Kingston ON.

ESG (Environmental Sciences Group) Royal Military College, 2001d. "Characterization of the Potential Human Health Risk from Consumption of Garden Produce in Yellowknife, NWT." Prepared for Yellowknife Soils Arsenic Soils Remediation Committee. Kingston ON.

Evans, M., and W.L. Lockhart, 1999. "An investigation of factors affecting high mercury concentrations in predatory fish in the Mackenzie River basin." Pages 103–110. In: *Synopsis of Research Conducted Under the 1998/99 Northern Contaminants Program*. S. Kalhok, (ed.). Indian and Northern Affairs Canada, Ottawa, ON.

Evans, M.S., R.A. Bourbonniere, D.C.G. Muir, W.L. Lockhart, P. Wilkinson, and B.N. Billeck, 1996. "Depositional history of sediment in Great Slave Lake: Spatial and temporal patterns in geochronology, bulk parameters, PAHs and chlorinated contaminants." Northern River Basins Study Project Report No. 99. Edmonton, Alberta, 173 pp.

Everett, J.T., and B.B. Fitzharris, 1998. The Arctic and the AntArctic. pp. 85–103. In: *The Regional Impacts of Climate Change. An Assessment of Vulnerability. A Special Report of IPCC Working Group II for the Intergovernmental Panel of Climate Change.* R.T. Watson, M. C. Zinyowera, R. H. Moss, and D.J. Dokken, eds. Cambridge University Press, Cambridge, United Kingdom and New York, NY, USA.

Ewald, G., P. Larsson, H. Linge, L. Okla, and N. Szarzi. 1998. "Biotransport of organic pollutants to an inland Alaska lake by migrating Sockeye salmon (*Oncorhyncus nerka*)." *Arctic*, 51: 40–47.

Fairbanks, R.G.,. 1989. "A 17,000-year glacio-eustatic sea level record: influence of glacial melting rates on the Younger Dryas event and deep-ocean circulation." *Nature,* 342: 637–642.

Falandysz, J., and C. Rappe, 1996. "Spatial distribution in plankton and bioaccumulation features of polychlorinated naphthalenes in a pelagic food chain in southern part of the Baltic Proper." *Environmental Science and Technology*, 30: 3362–3370.

Falandysz, J., L. Strandberg, P.A. Bergqvist, B. Strandberg, and C. Rappe, 1997. "Spatial distribution and bioaccumulation of polychlorinated naphthalenes (PCNs) in mussel and fish from the Gulf of Gdansk, Baltic SeA." *The Science of the Total Environment*. 203: 93–104.

Falandysz, J., 1998. "Polychlorinated naphthalenes: an environmental update." *Environmental Pollution,* 101: 77–90.

Falconer, R.L., and T.F. Bidleman, 1994. "Vapour pressures and predicted particle-gas distributions of polychlorinated biphenyl congerers as functions of temperature and *ortho*-chlorine substitution." *Atmospheric Environment,* 28: 547–554.

Falconer, R.L., and T. Harner, 2000. "Comparison of the octanol-air partition coefficient and liquid-phase vapour pressure as descriptors for particle/gas partitioning using laboratory and field data for PCBs and PCNs." *Atmospheric Environment.*, 34: 4043–4046.

Falconer, R.L., T.F. Bidleman, D.J. Gregor, R. Semkin, and C. Teixeira, 1995a. "Enantioselective breakdown of α-hexachlorocyclohexane in a small Arctic lake and its watershed." *The Science of the Total Environment*, 29: 297–1302.

Falconer, R.L., T.F. Bidleman, and D.J. Gregor, 1995b. "Air–water gas exchange and evidence for metabolism of hexachlorocyclohexanes in Resolute Bay, NT." *The Science of the Total Environment*, 160/161: 65–74.

Falkowski, P.G., R.T. Barber, and V. Smetacek, 1998. "Biogeochemical controls and feedbacks on ocean primary production." *Science* 281, 200–206.

FAO Production Yearbook,1993. *Food and Agriculture Organization Publication*. 47, Rome, Italy.

Fassnacht, S.R., J. Innes, N. Kouwen, and E. D, 1999. Soulis. "The specific surface area of fresh dendritic snow crystals." *Hydrological Processes*, 13: 2945–2962.

Fast, H., and F. Berkes. 1998. "Climate change, northern subsistence and land-based economies." Pages 206–226 in *Canada Country Study: Climate Impacts and Adaptation*. N. Mayer and W. Avis, eds. Environment Canada, Ottawa.

Fay, F. H., 1982. Ecology and biology of the Pacific walrus, *Odobenus rosmarus divergens* Illiger, North America Fauna, 74. US Department of the Interior, Fish and Wildlife Service, Washington DC.

Fedorov, L.A., 1999, "Persistent organic chemicals in the former Soviet Union." *Environmental Pollution*, 105: 283–287.

Fellin, P., L.A. Barrie, D. Dougherty, D. Toom, D. Muir, N. Grift, L. Lockhart, and B. Billeck, 1996. "Air Monitoring in the Arctic: Results foe Selected Persistent Organic Pollutants for 1992." *Environmental Toxicology and Chemistry*, 15: 253–261.

Fenner, K., M. Scheringer, and K. Hungerbühler, 2000. "Persistence of parent compounds and transformation products in a level IV multimedia model." *Environmental Science and Technology*, 34: 3809–3817.

Fingerling, G., N. Hertkorn, and H. Parlar, 1996. "Formation and spectroscopic investigation of two hexachlorobornanes from six environmentally relevant toxaphene components by reductive dechlorination in soil under anaerobic conditions." *Environmental Science and Technology*, 30: 2984.

Finizio, A., D. Mackay, T.F. Bidleman, and T. Harner, 1997a. "Octanol-air partition coefficient as a predictor of partitioning of semi-volatile organic chemicals to aerosols." *Atmospheric Environment,* 31: 2289–2296.

Finizio, A., M. Vighi, and D. Sandroni, 1997b. "Determination of n-octanol/water partition coefficient (K_{OW}) of pesticide: critical review and comparison of methods." *Chemosphere,* 34: 131–161.

Finley, K.J., G.W. Miller, R.A. Davis, and W.R. Koski, 1983. "A distinctive large breeding population of ringed seals (*Phoca hispida*) inhabiting the Baffin Bay pack ice." *Arctic,* 36: 162–163.

Finley, K.J., M.S.W. Bradstreet, and G.W. Miller, 1990. "Summer feeding ecology of harp seals (*Phoca groenlandica*) in relation to Arctic cod (*Boreogadus saida*) in the Canadian High Arctic." *Polar Biology*, 10: 609–618.

Finney, B.P., I. Gregory–Eaves, J. Sweetman, M.S.V. Douglas, and J.P. Smol, 2000. "Impacts of climatic change and fishing on Pacific salmon abundance over the past 300 years." *Science*, 290: 795–799.

Finney, B.P., I. Gregory–Eaves, M.S.V. Douglas, and J.P. Smol, 2002. "Fisheries productivity in the northeastern Pacific Ocean over the past 2,200 years." *Nature,* 416: 729–733.

Fischer, H., M. Werner, D. Wagenbach, M. Schwater, T. Thorsteinnson, F. Wilhelms, J. Kipfstuhl, and S. Sommer, 1998. "Little ice age clearly recorded in northern Greenland ice cores." *Geophysical Research Letters*, 25: 1749–1752.

Fisher, N.S., S.W. Fowler, F. Boisson, J. Carroll, K. Rissanen, B. Salbu, T.G. Sazykina, and K.L. Sjoeblom, 1999. "Radionuclide bioconcentration factors and sediment partition coefficients in Arctic seas subject to contamination from dumped nuclear wastes." *Environmental Science and Technology,* 33: 1979–1982.

Fisk, A.T, B. Rosenberg, C.D. Cymbalisty, G.A. Stern, and D.C.G. Muir, 1999. "Octanol/water partition coefficients of toxaphene congeners determined by the slow stirring method." *Chemosphere* 39, 2549–2562.

Fisk, A.T., K.A. Hobson, and R.J. Norstrom, 2001a. "Influence of chemical and biological factors on trophic transfer of persistent organic pollutants in the northwater polynya marine food web." *Environmental Science and Technology*, 35: 732–738.

Fisk, A.T., G.A. Stern, K.A. Hobson, W.J. Strachan, M.D. Loewen, and R.J. Norstrom, 2001b. "Persistent organic pollutants (POPs) in small, herbivorous, Arctic marine zooplankton (*Calanus hyperboreus*): trends from April to July and the influence of lipids and trophic transfer." *Marine Pollution Bulletin,* 43: 93–101.

Fitzgerald, W.F., D.R. Engstrom, R.P. Mason, and E.A. Nater, 1998. "The case for atmospheric mercury contamination in remote areas." *Environmental Science and Technology*, 32: 1–7.

Flato, G.M., and G.J. Boer, 2001. "Warming assymetry in climate change simulations." *Geophysical Research Letters,* 28: 195–198.

Forum 1997. "Mexico moves to phase out DDT and chlordane, Environ." *Health Prespectives*, 105: 788–789.

Frank, M., W.M. Smethie Jr, and R. Bayer, 1998. "Investigation of subsurface water flow along the continental margin of the Eurasian Basin using the transient tracers tritium, ^{3}He and CFCs." *Journal Geophysical Research,* 103 (C13): 30773–30792.

Franz, T.P., D.J. Gregor, and S.J. Eisenreich, 1997. "Snow deposition of atmospheric semivolatile organic chemicals." In: *Atmospheric deposition of contaminants to the Great Lakes and coastal waters.* Chap. 5. J.E. Baker (ed.) SETAC press, Pensacola, FL, pp. 73–107.

Franz, T.P., and S.J. Eisenreich, 1998. "Snow scavenging of polychlorinated biphenyls and polycyclic aromatic hydrocarbons in Minnesota." *Environmental Science and Technology,* 32: 1771–1778.

Freitas, H., M. Diamond, R. Semkin, and D. Gregor, 1997. "Contaminant fate in high Arctic lakes: development and application of a mass balance model." *The Science of the Total Environment* 1997, 201: 171–187.

Führer, U., A. Deibler, and K. Ballschmiter, 1996. "Determination of biogenic halogenated methyl-phenyl ethers (halogenated anisoles) in the picogram m3 range in air." *Fresenius' Journal of Analytical Chemistry,* 354: 333–343.

Fyfe, J., 2002. "Separating extratropical zonal wind variability and mean change." *Journal of Climate* submitted.

Fyfe, J.C., G.L. Boer, and G.M. Flato, 1999. "The Arctic and AntArctic oscillations and their projected changes under global warming." *Geophysical Research Letters,* 26: 1601–1604.

Gajewski, K., P.B. Hamilton, and R. McNeely, 1997. "A high-resolution proxy-climate record from an Arctic lake with annually-laminated sediments on Devon Island, Nunavut, Canada." *Journal of Paleolimnology.* 17: 215–225.

Gallaway, B.J., W.B. Griffiths, P.C. Craig, W.T. Gazey, and J.W. Helmericks, 1983. "An assessment of the Colville River delta stock of Arctic cisco — migrants from Canada." *Biol. Pap. Univ. Alaska,* 21: 4–23.

GLL (Gartner Lee Limited), 1999. "A Literature Review on Radionuclide, Metal and Organochlorine contamination in Great Bear Lake Water and Fish." Prepared for Indian and Northern Affairs Canada — NWT Region, Dec. 24, 1999.

Gevao, B., T. Harner, and K.C. Jones, 2000. "Sedimentary record of polychlorinated naphthalene concentrations and deposition fluxes in a dated lake core." *Environmental Science and Technology,* 34: 33–38.

Gianessi, L.P., 1986. A national pesticide usage data base, summary of report submitted to the office of standards and regulations, U.S. EPA under cooperative agreement CR 811858-01-0 by Resources for the Future, Washington, DC.

Giesy, J.P., D.J. Jude, D.E. Tillitt, R.W. Gale, J.C. Meadows, J.L. Zajieck, P.H. Peterman, D.A. Verbrugge, J.T. Sanderson, T.R. Schwartz, and M.L. Tuchman, 1997. "Polychlorinated dibenzo-*p*-dioxins, dibenzofurans, biphenyls and 2,3,7,8-tetrachlorodibenzo-*p*-dioxin equivalents in fishes from Saginaw Bay, Michigan." *Environmental Toxicology and Chemistry.,* 16: 713–724.

Gilchrist, H.G., and G.J. Roberston, 2000. "Observations of marine birds and mammals wintering at polynyas and ice edges in the Belcher Islands, Nunavut, Canada." *Arctic,* 53: 61–68.

Gilman, A., E. Dewailly, M. Feeley, V. Jerome, H. Kuhnlein, B. Kwavnick, S. Neve, B. Tracy, P. Usher, J. Van Oostdam, J. Walker, and B. Wheatley, 1997. "Chapter 4: Human Health." In: J. Jensen KA, R. Shearer, editor. *Canadian Arctic Contaminants Assessment Report,* Department of Indian and Northern Affairs, Ottawa, pp. 295–377.

Gips, T., 1991. "Breaking the Pesticide Habit Alternatives to 12 Hazardous Pesticides." IOCU. Penang, Malaysia. pp. 138–150.

Glassmeyer, S.T., D.S. DeVault, T.R. Myers, and R.A. Hites, 1997. "Toxaphene in Great Lakes fish: a temporal, spatial and trophicstudy."*Environmental Science and Technology,* 31: 84–88.

Glassmeyer, S.T., K.A. Brice, and R.A. Hites, 1999. "Atmospheric concentrations of toxaphene on the coast of Lake Superior."*Journal of Great Lakes Research,* 25: 492–499.

Gloersen, P., and W.J. Campbell, 1991. "Recent variations in Arctic and AntArctic sea-ice covers." *Nature,* 352: 33–35.

Gobeil, C, R.W. Macdonald, and J.N. Smith, 1999. "Mercury profiles in sediments of the Arctic Ocean basins." *Environmental Science and Technology*, 33: 4194–4198.

Gobeil, C., R.W. Macdonald, J.N. Smith, and L. Beaudin, 2001a. "Lead contamination in Arctic basin sediments tracks Atlantic water flow pathways." *Science*, 293: 1301–1304.

Gobeil, C., B. Sundby, R.W. Macdonald, and J.N. Smith, 2001b. "Recent change in organic carbon flux to Arctic Ocean deep basins: Evidence from acid volatile sulfide, manganese and rhenium discord in sediments." *Geophysical Research Letters*, 28: 1743–1746.

Golder Associates Ltd., 1995. "Port Radium Analytical Report." Second Draft, March.

Golder Associates Ltd., 1997. "Environmental Risk Assessment for Resolution Island, NWT." Submitted to the Indian and Northern Affairs Canada.

Gooch, J.W., F. Matsumura, and M.J. Zabik, 1990. "Chlordane residues in Great Lakes lake trout: acute toxicity and interaction at the Gaba receptor of rat and lake trout brain." *Chemosphere,* 21: 393–406.

Gordeev, V.V., and V.V. Tsirkunov, 1998. "River fluxes of dissolved and suspended substances." In: V. Kimstach, M. Meybeck, E. Baroudy (Eds). A Water Quality Assessment of the former Soviet Union. London. E&FN Spon. pp. 311–350.

Gorham, E., 1991. "Northern peatlands: role in the carbon cycle and probable responses to climatic warming." *Ecological Applications,* 1: 182–195.

Goss, K.U., 1997. "Conceptual model for the adsorption of organic compounds from the gas phase to liquid and solid surfaces." *Environmental Science and Technology* 31: 3600–3605.

Goss, K.U., and R.P. Schwarzenbach, 1999. "Empirical prediction of heats of vaporization and heats of adsorption of organic compounds." *Environmental Science and Technology* 33: 3390–3393".

Gough, W.A., and E. Wolfe, 2001. "Climate change scenarios for Hudson Bay, Canada, from general circulation models." *Arctic*, 54: 142–148.

Gouin, T., D. Mackay, E. Webster, and F. Wania.,2000. "Screening Chemicals for Persistence in the Environment." *Environmental Science and Technology*, 34: 881–884.

Graham, N.E., and H.F. Diaz, 2001. "Evidence for intensification of Pacific winter cyclones since 1948" *Bulletin of the American Meteorological Society,* 82: 1869–1893.

Grebmeier, J.M., and L. Cooper, 1995. "Infleunce of the St.Lawrence Island polynya upon the Bering Sea benthos." *Journal of Geophysical Research*, 100: 4439–4460.

Grebmeier, J.M., and K. Dunton, 2000. "Benthic processes in the northern Bering/Chukchi seas: Status and global change." Pages 80–93 in *Impacts of Changes in Sea Ice and Other Environmental Parameters in the Arctic*, H.P. Huntington (ed.), Girdwood, AK.

Grebmeier, J.M., C.P. McRoy, and H.M. Feder, 1988. "Pelagic benthic coupling on the shelf of the northern Bering and Chukchi Seas. I. Food supply, sources and benthic biomass." *Marine Ecological Progress Series,* 48: 57–67.

Gregor, D.J., 19991. "Organic micropollutants in seasonal snowcover and firn." In: Davies, T (ed.), *Seasonal Snowpacks.* Springer Verlag, Berlin/Heidelberg, pp. 323–358.

Gregor D.J., and W.D. Gummer, 1989. "Evidence of atmospheric transport and deposition of organochlorine pesticides and polychlorinated biphenyls in Arctic snow." *Environmental Science and Technology*, 23: 1528–1531.

Gregor, D.J., and A.J. Peters, 2000. Comment on "Delayed deposition of organochlorine pesticides at a temperate glacier" *Environmental Science and Technology*, 34, 2871–2872.

Gregor, D.J., A.J. Peters, C. Teixeira, N. Jones, and C. Spencer, 1995. "The historical residue trend of PCBs in the Agassiz Ice Cap, Ellesmere Island, Canada." *The Science of the Total Environment*, 160/161: 117–126.

Gregor, D.J., H. Loeng, and L. Barrie, 1998. "The influence of physical and chemical processes on contaminant transport into and within the Arctic." pp. 25–116. In: *Arctic Monitoring and Assessment Report. Arctic Pollution Issues. Arctic Monitoring and Assessment Program*, Oslo, Norway. xii + 859 pp.

Gribble, G.W., 1994. "The natural production of chlorinated compounds." *Environmental Science and Technology,* 28: 310A–319A.

Grimalt, J.O., P. Fernandez, L. Berdie, R.M. Vilanova, J. Catalan, R. Psenner, R. Hofer, P.G. Appleby, B.O. Rosseland, L. Lien, J.C. Massabuau, and R.W. Battarbee, 2001. "Selective trapping of organochlorine compounds in mountain lakes of temperate areas." *Environmental Science and Technology*, 35: 2690–2697 .

Gröllert, C., A. Kasper, and H. Puxbaum, 1997. "Organic compounds in high alpine snow." *International Journal of Environmental Analytical Chemistry,* 67: 213–222.

Guay, C.K., and K.K. Falkner, 1997. "Barium as a tracer of Arctic halocline and river waters." *Deep-Sea Research,* 44: 1543–1569.

Guay, C.K.H., K.K. Falkner, R.D. Muench, M. Mensch, M. Frank, and R. Bayer, 2001. "Wind-driven transport pathways for Eurasian Arctic river discharge." *Journal of Geophysical Research,* 106: 11469–11480.

Gubala, C.P., D.H. Landers, M. Monetti, M. Heit, T. Wade, B. Lasorsa, and S. Allen–Gil, 1995, "The rates of accumulation of atmospherically derived pollutants in Arctic Alaska, USA." *The Science of the Total Environment* 160/161: 347–361.

Gudkovich, Z.M., 1961. "Relation of the ice drift in the Arctic Basin to ice conditions in the Soviet Arctic seas (in Russian)." *Tr. Okeanogr. Kom. Akad. Nauk SSSR* ,11: 14–21.

Guieu, C., W.W. Huang, J.M. Martin, and Y.Y. Yong, 1996. "Outflow of trace metals into the Laptev Sea by the Lena River." *Marine Chemistry*, 53: 247–254.

Gupta, A., C.P. Kaushik, and A. Kaushik, 2001. "Degradation of hexachlorocyclohexane isomers by two stains of *Alcaligenes faecalis* isolated from a contaminated site." *Bulletin of Environmental Contamination and Toxicology*, 66: 794–800.

Gupta, P.K., 1986. Pesticides in the Indian Environment. *Interprint*, New Delhi, India, pp. 206.

Hainzl, D., J. Burhenne, and H. Parlar, 1994. "Theoretical consideration of the structural variety in the toxaphene mixture taking into account recent experimental results." *Chemosphere,* 28: 245–252.

Halsall, C.J., P.J. Coleman, B.J. Davis, V. Burnett, K.S. Waterhouse, P. Harding-Jones, and K.C. Jones, 1994. "PAHs in UK urban air." *Environmental Science and Technology*, 28: 2380–2386.

Halsall, C.J., L.A. Barrie, P. Fellin, D.C.G. Muir, B.N. Billeck, L. Lockhart, F.Y. Rovinsky, E.Y. Kononov, and B. Pastukhov, 1997. "Spatial and temporal variation of polycyclic aromatic hydrocarbons in the Arctic atmosphere." *Environmental Science and Technology*, 31: 3593–3599.

Halsall, C.J., R. Bailey, G.A. Stern, L.A. Barrie, P. Fellin, D.C.G. Muir, B. Rosenberg, F. Ya. Rovinsky, E.Ya. Kononov, and B. Pastukhov, 1998. "Multi-year observations of organohalogen pesticides in the Arctic atmosphere." *Environmental Pollution*. 102: 51–62.

Halsall, C.J., A.J. Sweetman, L.A. Barrie and K.C. Jones, 2001. "Modelling the behaviour of PAHs during atmospheric transport from the UK to the Arctic." *Atmospheric Environment*, 35: 255–267.

Hammar, J., 1989. "Freshwater ecosystems." *Ambio,* 18: 6–22.

Hanberg, A., F. Waern, L. Asplund, E. Haglund, and S. Safe, 1990. "Swedish dioxin survey: determination of 2,3,7,8-TCDD toxic equivalent factors for polychlorinated biphenyls and naphthalenes using biological tests." *Chemosphere,* 2: 1161–1164.

Hanberg, A., M. Ståhlberg, A. Georgellis, C. de Wit, and U.G. Ahlborg, 1991. "Swedish dioxin survey: evaluation of the H-4-II E bioassay for screening environmental samples for dioxin-like enzyme induction." *Pharmacology and toxicology*, 69: 442–449.

Hanot, L.,and F. Dominé, 1999. "Evolution of the surface area of a snow layer." *Environmental Science and Technology,* 33: 4250–4255.

Hart Hansen, J.P., J. Meldgaard, and J. Nordqvist, 1991. *The Greenland Mummies*. Smithsonian Institution Press, Washington, 192 pg.

Hare, S.R., and N.J. Mantua, 2000. "Empirical evidence for North Pacific regime shifts in 1977 and 1989." *Progress in Oceanography,* 47: 103–145.

Hargrave, B.T., L.A. Barrie, T.F. Bidleman, and H.E. Welch, 1997. "Seasonality in exchange of organochlorines between Arctic air and seawater." *Environmental Science and Technology*, 31: 3258–3266.

Hargrave, B.T., W.P. Vass, P.E. Erickson, and B.R. Fowler, 1988. "Atmospheric transport of organochlorines to the Arctic Ocean." *Tellus*, 40B: 480–493.

Harner, T., 1997. "Organochlorine contamination of the Canadian Arctic, and speculation on future trends." *International Journal of Environment and Pollution*, 8: 51–73.

Harner, T., 2001. "Measurements of octanol-air partition coefficient (K_{oa}) for polybrominated diphenyl ethers (PBDEs): Predicting partitioning in the enviroment."pp. 55–58 In: *The Second International Workshop on Brominated Flame Retardants,* BFR2001.

Harner, T., and T.F. Bidleman, 1996. "Measurements of octanol-air partition coefficients for polychlorinated biphenyls." *Journal of Chemical and Engineering Data*. 41: 895–899.

Harner, T., and T.F. Bidleman,1997. "Polychlorinated naphthalenes in urban air" *Atmospheric Environment,* 31: 4009–4016.

Harner, T., and T.F. Bidleman, 1998a. "Measurement of octanol-air partition coefficients for polycyclic aromatic hydrocarbons and polychlorinated naphthalenes." *Journal of Chemical and Engineering Data*, 43: 40–46.

Harner, T., and T.F. Bidleman, 1998b. "Octanol-air partition coefficient for describing particle/gas partitioning of aromatic compounds in air." *Environmental Science and Technology*." 32: 1494–1502.

Harner, T., and D. Mackay, 1995. "Measurements of octanol-air partition coefficients for chlorobenzenes, PCBs and DDT." *Environmental Science and Technology,* 29: 1599–1606.

Harner, T., and M. Shoeib, 2002. "Measurements of octanol-air partition coefficients (K_{OA}) for polybrominated diphenyl ethers (PBDEs): Predicting partitioning in the environment." *Journal of Chemical and Engineering Data,* 47: 228–232.

Harner, T., H. Kylin, T.F. Bidleman, C.J. Halsall, W.M.J. Strachan, L.A. Barrie, and P. Fellin, 1998. "Polychlorinated naphthalenes and coplanar polychlorinated biphenyls in Arctic air." *Environmental Science and Technology,* 32: 3257–3265.

Harner, T., H. Kylin, T.F. Bidleman, and W.M.J. Strachan, 1999. "Removal of α- and γ-hexchlorocycllohexane and enantiomers of α-hexachlorocyclohexane in the eastern Arctic Ocean." *Environmental Science and Technology,* 33: 1157–1164.

Harner, T., N.J.L. Green, and K.C. Jones, 2000a. "Measurements of octanol-air partition coefficients for PCCD/Fs: a tool in assessing air-soil equilibrium status." *Environmental Science and Technology*, 34: 3109–3114.

Harner, T., R.G.M. Lee, and, K.C. Jones, 2000b. "Polychlorinated naphtalenes in the atmosphere of the United Kingdom." *Environmental Science and Technology*, 34: 3137–3142.

Harner, T., L.M.M. Jantunen, T.F. Bidleman, L.A. Barrie, H. Kylin, W.M.J. Strachan, and, R.W. Macdonald, 2001a. "Microbial degradation is a key elimination pathway of hexachlorocyclohexanes from the Arctic Ocean." *Geophysical Research Letters*, 27: 1155–1158.

Harner, T., L.M.M. Jantunen, T.F. Bidleman, and D. Mackay, 2001b. "Soil-air exchange model of persistent pesticides in the United States cotton belt." *Environmental Toxicology and Chemistry*, 20: 1612–1621.

Harris, M.L., J.R. Van den Heuvel, P.A. Martin, J. Struger, C.A. Bishop, and P. Takacs, 2000. "Pesticides in Ontario: a critical assessment of potential toxicity of agricultural products to wildlife with consideration for endocrine disruption. Vol. 1. endosulfan, EBDC fungicides, dinitroaniline herbicides, 1,3-dichloropropane, azinphosmethyl and pesticide mixtures." Canadian Wildlife Service, Environmental conservation Branch, Ontario Region. Technical Report Series No. 340, Ministry of Government Services, Catalogue No. CW69-5/340E.

Harwood, L., T.G. Smith, and H. Melling, 2000. "Variation in reproduction and body condition of the ringed seal (*Phoca hispida*) in western Prince Albert Sound, NT, Canada, as assessed through a harvest-based sampling program." *Arctic,* 53: 422–431.

Hatfield Consultants Ltd., July 1985. "An valuation of Environmental Conditions Associated with the Abandoned Uranium Mines at Rayrock and Echo Bay." Science Institute of the Northwest Territories.

Haugen J.E, F. Wania, N. Ritter, M. Schlabach, 1998. "Hexachlorocyclohexanes in air in southern Norway: temporal variation, source allocation and temperature dependence." *Environmental Science and Technology*, 32: 217–224.

Health and Welfare Canada, 1977. "Toxaphene — Environmental health criteria document." Report 77-EHD-11). Health Protection Branch, Ottawa, ON.

Heide-Jørgensen, M.P., T. Härkönen, R. Dietz, and P.M. Thompson, 1992. "Retrospective of the 1988 European seal epizootic." *Diseases of Aquatic Organisms,* 13: 37–62.

Heinisch, E., A. Kettrup, S. Jumar, S. Wenzel-Klein, J. Stechert, P. Hartmann, and P. Schaffer, 1994. "Schadstoff Atlas Ost–Europa." Chapter 2.9, pp. 39–47. Ecomet/Landsberg–Lech, Germany.

Heintzenberg, J., 1989. "Arctic haze: air pollution in polar regions." *Ambio,*18: 50–55.

HELCOM, 2001. Helsinki Commission. Baltic Marine Environment Protection Commission. The pesticides selected for immediate priority action. Helsinki Commission, Helsinki, Finland, 77 pp.

Held, H., 2001. "Spherical spatial ranges of non-polar chemicals for reaction-diffusion type dynamics." *Applied Mathematics and Computation,* 124: 29–43 .

Held, H., 2002. "Robustness of spatial ranges against dimension." *Stoch. Env. Res. Risk Assess.,* 16: in press.

Helland–Hansen, B., and F. Nansen, 1909. "The Norwegian Sea. Its physical oceanography based upon the Norwegian researches 1900–1904." 390 pp. *Report on Norwegian Fishery and Marine Investigations,* Volume 2.

Helm, P.A., 2002. "The Influence of Sources, Source Regions and Fate and Transport Processes on the Occurrence of Polychlorinated Naphthalenes and Coplanar Polychlorinated Biphenyls in Urban and Arctic Environments." Ph.D. Thesis, University of Toronto, 214 pg.

Helm, P.A., and T.F. Bidleman, 2002. "Seasonal and spatial variation of polychlorinated naphthalenes and planar polychlorinated biphenyls in Arctic air." *Environmental Science and Technology*, submitted

Helm, P.A., T.F. Bidleman, L.M.M. Jantunen, and J. Ridal, 2000a. "Polychlorinated naphthalenes in Great Lakes air: source and ambient air profiles." *Organohalogen Compounds.*, 47: 17–20.

Helm, P.A., M.L. Diamond, R. Semkin, and T.F. Bidleman, 2000b. "Degradation as a loss mechanism in the fate of α-hexachlorocyclohexane in watersheds." *Environmental Science and Technology*, 34: 812–818.

Helm, P.A., T.F. Bidleman, G.A. Stern, and K. Koczanski, 2002a. "Polychlorinated naphthalenes and coplanar polychlorinated biphenyls in beluga whale (*Delphinapterus leucas*) and ringed seal (*Phoca hispida*) from the eastern Canadian Arctic." *Environmental Pollution*, in press.

Helm, P.A., M.L. Diamond, R. Semkin, W.M.J. Strachan, C. Teixeira, and D. Gregor, 2002b. "A mass balance model describing multi-year fate of organochlorine compounds in a high Arctic lake." *Environmental Science and Technology*, submitted.

Hequette, A., M.H. Ruz, and P.R. Hill, 1995. "The effects of the Holocene sea level rise on the evolution of the southeastern coast of the Canadian Beaufort Sea." *Journal of Coastal Research,* 11: 494–507.

Hermanson, M.H., 1993. "Historical accumulation of atmospherically derived pollutant trace metals in the Arctic as measured in dated sediment cores." *Water Science and Technology,* 8–9, 33–41.

Hileman, B., 1983. "Arctic haze." *Environmental Science and Technology* 17: 232A–236A.

Hillery, B.R., I. Basu, C.W. Sweet, and R.A. Hites, 1997. "Temporal and spatial trends in a long-term study of gas-phase PCB concentrations near the Great Lakes." *Environmental Science and Technology,* 31: 1811–1816.

Hinckley, D.A., T.F. Bidleman, W.T. Foreman, and J.R. Tuschall, 1990. "Determination of vapor pressures for nonpolar and semipolar organic compounds from gas chromatographic retention data." *Journal of Chemical and Engineering Data,* 35: 232–237.

Hinckley, D.A., T.F. Bidleman, and C.P. Rice, 1991. "Atmospheric organochlorine pollutants and air-sea exchange of hexachlorocyclohexanes in the Bering and Chukchi seas." *Journal of Geophysical Research*, 96: 7201–7213.

Hippelein, M., and M.S. McLachlan, 1998. "Soil/Air Partitioning of Semivolatile Organic Compounds. 1. Method Development and Influence of Physical — Chemical Properties." *Environmental Science and Technology*, 32: 310–316.

Hippelein, M., and M.S. McLachlan, 2000. "Soil/Air Partitioning of Semivolatile Organic Compounds. 2. Influence of temperature and relative humidity." *Environmental Science and Technology,* 34: 3521–3526.

Hobson, K., and H.E. Welch, 1992. "Determination of trophic relationships within a high Arctic marine food web using $\delta^{13}C$ and $\delta^{15}N$ analysis." *Marine Ecology Progress Series,* 84: 9–18.

Hodges, G., 2000. "The new cold war; stalking Arctic climate change by submarine." *National Geographic,* March: 30–41.

Hoekstra, P.F., T.M. O'Hara, C. Teixeira, S. Backus, A.T. Fisk, and D.C.G. Muir, 2002a. "Spatial trends and bioaccumulation of organochlorine pollutants in marine zooplankton from the Alaskan and western Canadian Arctic." *Environmental Science and Technology,* 21: 575–583

Hoekstra, P.F., T.M. O'Hara, S. Pallant, K.R. Solomon, and D.C.G. Muir, 2002b. "Bioaccumulation of organochlorine contaminants in bowhead whales (*Balaena mysticetus*) from Barrow, Alaska." *Archives for Environmental Contamination and Toxicology,* 42: 497–507.

Hoekstra, P.F., C.S. Wong, T.M. O'Hara, K.R. Solomon, S.A. Mabury, and D.C. G. Muir, 2002c. "Enantiomer-Specific Accumulation of PCB Atropisomers in the Bowhead Whale (*Balaena mysticetus*)", *Environmental Science and Technology,* 36: 1419–1425.

Hofer, R., R. Lackner, J. Kargl, B. Thaler, D. Tait, L. Bonetti, R. Vistocco, and G. Flaim, 2001. "Organochlorine and metal accumulation in fish (*Phoxinus phoxinus*) along a North-South transect in the Alps." *Water, Air, and Soil Pollution,* 125: 189–2000.

Hoff, J.T., F. Wania, D. Mackay, and R. Gilham, 1995. "Sorption of non-polar organic vapors by ice and snow." *Environmental Science and Technology*, 29: 1982–1989.

Hoff, J.T., D.J. Gregor, D. Mackay, F. Wania, and C.Q. Jia, 1998. "Measurement of the specific surface area of snow with the nitrogen adsorption technique." *Environmental Science and Technology,* 32: 58–62.

Hoff, J.T., F. Wania, D. Mackay, R. Semkin, and C.Q. Jia, 1999. "Quantifying the atmospheric delivery of organic contaminants into the high Artcic marine ecosystem as influenced by snow and ice," pp. 39–46 In: *Synopsis of Research Conducted under the 1997/98 Northern Contaminants Program.* Indian and Northern Affairs Canada, J. Jensen (ed.) Environmental Studies No. 75, Ottawa, Ontario.

Hoff, R.M., T.F. Bidleman, and S.J. Eisenreich, 1993. "Estimate of PCC loadings from the atmosphere to the Great Lakes." *Chemosphere, 27:* 2047–2055.

Holden, A.V., 1970. "Monitoring organochloring contamination of the marine environment by the analysis of residues in seals." In: *Marine Pollution and Sea Life*, pp. 266–272. M.Ruivo (ed.). Fishing News Books Ltd., England.

Holloway, G., and T. Sou, 2001. "Has Arctic sea ice rapidly thinned?" *Journal of Climate*, submitted.

Holtan, G,D. Berge, H. Holtan, and T. Hopen, 1994. "Direct and riverine inputs for Norwegian coastal waters during the year 1993." Norwegian State Pollution Control Authority Report 580/94, Norwegian Institute for Water Research (NIVA), Oslo.

Holtz, A., M.A. Sharpe, M. Constable, and W. Wilson, 1987. "Removal of contaminants from Distant Early Warning sites in Canada's Arctic." Edmonton, Alberta: Environmental Protection Service. 111 pp.

Holzer, M., and G.J. Boer, 2001. "Simulated changes in atmospheric transport climate." *Journal of Climate,* 14: 4398–4420.

Hong, S., J.P. Candelone, C.C. Patterson, and C.F. Boutron. 1994. "Greenland ice evidence of hemispheric lead pollution two millennia ago by Greek and Roman civilizations." *Science*, 265: 1841–1843.

Hönninger, G, 2002. "Halogen oxide studies in the boundary layer by multi axis differential optical absorption spectroscopy and active longpath-DOAS." Dissertation submitted to the combined faculties for the natural sciences and for mathematics of the Rupert Carola University of Heidelberg, Germany, for the degree of Doctor of Natural Sciences, February 14, 262 pp.

Hönninger, G, and U. Platt, 2002. "Observations of BrO and its vertical distribution during surface ozone depletion at Alert." accepted to *Atmospheric Environment.*

Honrath, R.E., M.C. Peterson, S. Guo, J.E. Dibb, P.B. Shepson, and B. Campbell, 1999. "Evidence of NO_x production within or upon ice particles in the Greenland snowpack." Geophysical *Research Letters,* 26: 695–698.

Hopkins, D.M., 1979. "Landscape and climate of Beringia during the late Pleistocene and Holocene time." Pages 15–41 in *The First Americans: Origins, Affinities and Adaptations.* W. S. Laughlin and A.B. Harper, eds. Gustav Fisher, New York.

Hough, C., 1999. "Characterization of Arsenic in a Short Terrestrial Food Chain, Yellowknife, NWT." MSc Thesis, Royal Military College of Canada.

Houghton, J.T., L.G. Meira Filho, B.A. Callander, N. Harris, A. Kattenberg, and K. Maskell, 1995. "Climate Change 1995: The Science of Climate Change." Intergovernmental Panel on Climate Change, Cambridge, 572 pp.

Howard, P.H., J. Santadonato, and J. Saxena, 1975. "Investications of selected potential environmental contaminants: Chlorinated paraffins. Rep. 68-01-3101." Syracuse University Research Corporation, Syracuse, New York, US

Howdeshell, M.J., and R.A. Hites, 1996, "Historical Input and Degradation of Toxaphene in Lake Ontario Sediment." *Environmental Science and Technology*, 30: 220–224

Huang, H.X.,1989. "Discussion on environmental contamination and related problems caused by organochlorine pesticides." *Science and Management of Pesticides*, 2: 26–29 (in Chinese)

Huh, C.A., N.G. Pisias, J.M. Kelley, T.C. Maiti, and A. Grantz, 1997. "Natural radionuclides and plutonium in sediments from the Western Arctic Ocean: sedimentation rates and pathways of radionuclides." *Deep-Sea Research II*, 44(8): 1725–1745.

Hung, H, C.J. Halsall, P. Blanchard, H.H. Li, P. Fellin, G. Stern, and B. Rosenberg, 2001. "Are PCBs in the Canadian Arctic atmosphere declining? Evidence from 5 years of monitoring." *Environmental Science and Technology*, 35: 1303–1311.

Hung, H, C.J. Halsall, P. Blanchard, H.H. Li, P. Fellin, G. Stern, and B. Rosenberg, 2002a. "Temporal trends of organochlorine pesticides in the Canadian Arctic atmosphere." *Environmental Science and Technology*, In press

Hung, H, P. Blanchard, G. Poole, B. Thibert, and C.H. Chiu, 2002b. "Measurement of Particle-Bound PCDD/PCDFs in Arctic Air at Alert, Canada." Accepted for publication in *Atmospheric Environment.*

Hunt, G.L., C.L. Baduini, R.D. Brodeur, K.O. Coyle, N.B. Kachel, J.M. Napp, S.A. Salo, J.D. Schumacher, P.J. Stabeno, D.A. Stockwell, T.E. Whitledge, and S.I. Zeeman, 1999. "The Bering Sea in 1998: The second consecutive year of extreme weather-forced anomalies." *Eos, Transactions, American Geophysical Union*, 80: 561: 565–566.

Hurrell, J.W., 1995. "Decadal trends in the North Atlantic Oscillation: regional temperatures and precipitation." *Science,* 269: 676–679.

Hurrell, J., 2002. North Atlantic Oscillation (NAO) Indices Information. http://www.cgd.ucar.edu/~jhurrell/nao.html.

Hutterli, M.A., R. Rothlisberger, and R.C. Bales, 2001. "Atmosphere-to-snow-to-firn transfer studies of HCHO at Summit, Greenland." Geophysical Research Letters, 26: 1691–1694.

Hutzinger, O., and H. Thoma, 1987. "Polybrominated dibenzo-*p*-dioxins and dibenzofurans: The flame retardant issue." *Chemosphere*, 16: 1877–1880.

Hutzinger, O., G. Sundstrom, and S. Safe, 1976. "Environmental chemistry of flame retardants. Part I. Introduction and principles." *Chemosphere*, 1: 3–10.

Ikonomou, M.G., S. Rayne, and R.F. Addison, 2002. "Exponential increases of the brominated flame retardants, polybrominated diphenyl ethers, in the Canadian Arctic from 1981–2000." *Environmental Science and Technology*, 36: 1886–1892.

IANC (Indian and Northern Affairs Canada), 2000. "Resolution Island Removal and Disposal of PCB — Contaminated Soils Project — Environmental Impact Statement." Indian and Northern Affairs Canada, pp. 80.

INAC (Indian and Northern Affairs Canada), 2000. "Summary of Northern Contaminants Program, Projects for 2000–2001." Ottawa, ON.

IANC (Indian and Northern Affairs Canada), 2002. Waste Sites Database: Yukon, NWT and Nunavut — an electronic inventory of INAC waste sites north of 60.

IPCC. 1990. "Climate Change, The IPCC Scientific Assessment." Press Syndicate, University of Cambridge, Bracknell, UK.

IPCC. 1995. "Climate Change 1995: the Science of climate change (summary for policymakers)." Pages 55. WMO/UNEP.

IPCC. 2002. The Third Assessment, Http://www.ipcc.ch/pub/tar/wg1 (and wg2, wg3). Http://www.ipcc.ch/pub/tar/wg1 (and wg2, wg3).

IRPTC. International Registry of Potentially Toxic Chemicals. http://irptc.unep.ch/irptc

Ishaq, R., K. Karlson, and C. Näf, 2000. "Tissue distribution of polychlorinated naphthalenes (PCNs) and non-*ortho* chlorinated biphenyls (non-*ortho* CBs) in harbour porpoises (*Phocoena phocoena*) from Swedish waters." *Chemosphere*, 41: 1913–1925.

Ivanov, V., and E. Sandell, 1992. "Characterization of polychlorinated biphenyl isomers in Sovol and trichlorodiphenyl formulations by high-resolution gas chromatography with electron capture detection and high resolution gas chromatography-mass spectrometery techniques." *Environmental Science and Technology, 26*: 2012–2017.

Iversen, T., 1996. "Atmospheric transport pathways for the Arctic." pp. 71–92 in *Chemical exchange between the atmsophere and polar snow. Global Environmental Change. NATO ASI Series I.* E. Wolff and R.C. Bales, eds. Springer-Verlag, Berlin and Heidelberg.

Iwata, H., S. Tanabe, and N. Sakai , and R. Tatsukawa, 1993a. "Distribution of persistent organochlorines in the oceanic air and seawater and the role of ocean on their global transport and fate." *Environmental Science and Technology, 27*: 1080–1098.

Iwata, H., S. Tanabe, and R. Tatsukawa, 1993b. "A new view on the divergence of HCH isomer composition in oceanic air." *Marine Pollution Bulletin*, 26: 302–305.

Iwata, H., S. Tanabe, N. Sakai, A. Nishimura, and R. Tatsukawa, 1994. "Geographical distribution of persistent organochlorines in air, water and sediments from Asia and Oceania and their implications for global redistribution from lower latitudes." *Environmental Pollution*, 85: 13–33.

Jakobsson, E., and L. Asplund, 1999. "Chapter 5. Polychlorinated naphthalenes (PCNs)." In: *The Handbook of Environmental Chemistry Vol. 3* Part K: new types of persistent halogenated compounds. Paasivirta, J. (Ed.), Springer-Verlag, New York, pp. 97–126.

James, R.R., and R.A. Hites, 2002. "Atmospheric transport of toxaphene from the southern United States to the Great Lakes region." *Environmental Science and Technology,* 36 (16): 3474–3481.

Jansson, B., L. Asplund, and M. Olsson, 1987. "Brominated flame retardants — ubiquitous environmental pollutants? " *Chemosphere*, 16: 2343–2349.

Jansson, B., R. Andersson, L. Asplund, K. Litzén, K. Nylund, U. Sellström, U.B. Uvemo, C. Wahlberg, U. Wideqvist, T. Odsjö, and M. Olsson, 1993. "Chlorinated and brominated perstistent organic compounds in biological samples from the environment." *Environmental Toxicology and Chemistry*, 12: 1163–1174.

Jantunen, L.M., and T.F. Bidleman, 1995. "Reversal of the air–sea gas exchange direction of hexachlorocyclohexanes in the Bering and Chukchi seas: 1993 vs. 1988." *Environmental Science and Technology*, 29: 1081–1089.

Jantunen, L.M., and T.F. Bidleman, 1997. "Air–water gas exchange of hexachlorocyclohexanes (HCHs) and the enantiomers of α-HCH in Arctic regions." *Journal of Geophysical Res*earch, 1996; 101: 28837–28842; corrections 1997 Ibid. 102: 19279–19282.

Jantunen, L.M., and T.F. Bidleman, 1998. "Organochlorine pesticides and enantiomers of chiral pesticides in Arctic Ocean water." *Archives of Environmental Contamination and Toxicology,* 35: 218–228.

Jantunen, L.M., and T.F. Bidleman, 2000. "Temperature-dependent Henry's law constant for technical toxaphene." *Chemosphere*, 2: 225–231.

Jantunen, L.M., and T.F. Bidleman, 2002. "Air-water gas exchange of toxaphene in Lake Superior." *Environmental Toxicology and Chemistry*, in press.

Jantunen, L.M., T.F. Bidleman, T. Harner, and W.J. Parkhurst, 2000. "Toxaphene and other organochlorine pesticides in Alabama air." *Environmental Science and Technology,* 34: 5097–5105.

Järnberg, U., L. Asplund, C. de Wit, A. Grafström, P. Haglund, B. Janson, K. Lexén, M. Strandell, M. Osson and B. Jonsson, 1993. Poychlorinated biphenyls and polychorinated napthalenes in Swedish sediment and biota: levels, patterns and time trends. *Environmental Science and Technology*, 27: 1364–1374.

Järnberg, U., L. Asplund,C. de Wit, A.L. Egebäck, U. Wideqvist, E. Jakobsson, 1997. "Distribution of polychlorinated naphthalene congeners in environmental and source-related samples."*Archives of Environmental Contamination and Toxicology,* 32: 232–245.

Jeffries, D.S., *personal communication*, 1999. Backus Consulting for Environment Canada.

Jeffries, D., J. Carey, and M. Swyripa, 1996. "Riverine inputs of contaminants to the Arctic marine environment." In: J.L. Murray and R.G. Shearer (eds.). *Synopsis of Research Conducted Under the 1994/95 Northern Contaminants Program*, Environmental Studies No. 73. Ottawa: Indian and Northern Affairs Canada. pp. 85–96.

Jensen, J., K. Adare, and R. Shearer (eds.). 1997. *Canadian Arctic Contaminants Assessment Report.* Department of Indian and Northern Affairs, Ottawa, Canada.

Jeremiason, J.D., K.C. Hornbuckle, and S.J. Eisenreich, 1994. "PCBs in Lake Superior, 1978–1972: Decreases in water concentrations reflect loss by volatilization." *Environmental Science and Technology,* 28: 903–914.

Johannessen, O.M., and M.W. Miles, 2001. "Arctic sea ice and climate change — will the ice disappear in this century?" *Science and Technology Letters Reviews,* in press.

Johannessen, O.M., M. Miles, and E. Bjorgo, 1995. "The Arctic's shrinking sea ice." *Nature,* 376: 126–127.

Johannessen, O.M., E.V. Shalina, and M.W. Miles,1999. "Satellite evidence for an Arctic sea ice cover in transformation." *Science,* 286: 1937–1939.

Johns, D.G., M. Edwards, and S.D. Batten, 2001. "Arctic boreal plankton species in the Northwest Atlantic." *Canadian Journal of Fisheries and Aquatic Sciences*, 58: 2121–2124.

Johnson, M.A., A.Y. Proshutinsky, and I.V. Polyakov, 1999. "Atmospheric patterns forcing two regimes of Arctic circulation: A return to anticyclonic conditions?" *Geophysical Research Letters*, 26: 1621–1624.

Johnson, M.A., and I.V. Polyakov, 2001. "The Laptev Sea as a source for recent Arctic Ocean salinity changes." *Geophysical Research Letters*, 28: 2017–2020.

Johnson, H.L., and D.P. Marshall, 2002. "Localization of abrupt change in the North Atlantic thermohaline circulation." *Geophysical Research Letters, 29*: 7–1 — 7–4 DOI 10.1029/2001GL014140.

Josefsson, D., 1998. "Anthropogenic radionuclides in the Arctic Ocean: Distribution and pathways." *Department of Radiation Physics.* Lund University, Lund 159 pp.

Kabat, P., H. Hoff, R. Hutjes, M. Claussen, P. Viterbo, M. Meybeck, and C. Vorosmarty, 2001. *Global Change NewsLetter,* 46: 31–34.

Kahl, J.D.W., J.A. Galbraith, and D.A. Martinez, 1999. "Decadal-scale variability in long-range atmospheric transport to the Summit of the Greenland Ice Sheet." *Geophysical Research Letters,* 26: 481–484.

Kalff, J., and B.E. Welch, 1974. "Phytoplankton production in Char Lake, a natural polar lake, and in Meretta Lake, a polluted polar lake." Cornwallis Island, Northwest Territories. *J. Fish. Res. Board Can.*, 31: 621–636.

Kalin, M., 1983. "Port Radium: An Environmental Survey of Mining Waste." Institute of Environmental Studies, University of Toronto, Toronto, Ontario.

Kannan, K., S. Tanabe, and R. Tatsukawa 1995. "Geographical distribution and accumulation features of organochlorine residues in fish in tropical Asia and Oceania." *Environmental Science and Technology,* 29: 2673–2683

Kannan, K., N.Yamashita, T. Imagawa, W. Decoen, J.S. Khim, R.M. Day, C.L.Summer,J.P. Giesy, 2000. "Polychlorinated naphthalenes and polychlorinated biphenyls in fishes from Michigan waters including the Great Lakes." *Environmental Science and Technology*, 34: 566–572.

Kannan, K., K. Hilscherova, T. Imagawa, N. Yamashita, L.L. Williams, and J.P. Giesy, 2001. "Polychlorinated naphthalenes, -biphenyls, -dibenzo-*p*-dioxins, and dibenzofurans in double-crested cormorants and herring gulls from Michigan waters of the Great Lakes." *Environmental Science and Technology*, 35: 441–447.

Karlsson, H., D.C.G. Muir, C.F. Teixiera, D.A. Burniston, W.M.J. Strachan, R.E. Hecky, J. Mwita, H.A. Bootsma, N.P. Grift, K.A. Kidd, and B. Rosenberg, 2000. "Persistent chlorinated pesticides in air, water, and precipitation from the Lake Malawi area, southern Africa." *Environmental Science and Technology,* 34: 4490–4495.

Kato Y., and K. Kenne, 1996. "Inhibition of cell-cell communication by commercial chlorinated paraffins in rat liver epithelial IAR 20 cells." *Pharmacology and Toxicology, 79*: 23–28.

Kawai, Y., Y. Hori, and Y. Nigawa, 1973. "On the pollution of mother's milk by pesticides and PCB in Hokkaido." *J. Food Hyg. Soc. Jpn.* 14, 302 (in Japanese)

Kawai, Y., 1995. "Preliminary model to estimate chemical behaviour in lakes of the Upper Yukon River Basin." *M.Sc. Eng. Thesis.* Department of Chemical Engineering and Applied Chemistry, University of Toronto.

Kawano, M., T. Inour, T. Wada, H. Hidaka, and R. Tatsukawa, 1988. "Bioconcentration and residue patterns of chlordane compounds in marine mammals: invertebrates, fish, mammals and seabirds." *Environmental Science and Technology*, 22: 792–797.

Kawano, M., A. Ramesh, V.D. Thao, R. Tatsukawa, and A.N. Subramanian, 1992, "Persistent organochlorine insecticide residues in some paddy, upland and urban soils of India." *International Journal of Environmental Analytical Chemistry*, 48: 163–174

Keeling, C.D., and M. Heimann, 1986. "Meridional eddy diffusion model of the transport of atmospheric carbon dioxide 2. Mean annual carbon cycle." *Journal of Geophysical Research,* 91: 7782–7796.

Kendall, R., R. Dickerson, J. Giesy, and W. Suk, 1998. "Principles and Processes for Evaluating Endocrine Disruption in Wildlife." *SETAC Press*, Pensacola, Florida, 491 pp.

Kenna, T.C., and F.L. Sayles, 2002. "The distribution and history of nuclear weapons related contamination in sediments from the Ob River, Siberia as determined by isotopic ratios of plutonium and neptunium" *Journal of Environmental Radioactivity*, 60(1–2): 105–138.

Kerr, R.A., 1999. "A new force in high-latitude climate." *Science,* 284: 241–242.

Kershaw, P.J., and A.J. Baxter, 1995. "The transfer of reprocessing wastes from NW Europe to the Arctic." *Deep-Sea Research II,* 42(6): 1413–1448.

Kidd, K.A., J.E. Eamer, and D.C.G. Muir, 1993. "Spatial variability of chlorinated bornanes (toxaphene) in fishes from Yukon lakes." *Chemosphere,* 27: 1975–1993.

Kidd, K.A., R.H. Hesslein, R.J.P. Fudge, and K.A. Hallard, 1995a. "The influence of trophic level as measured by δ15N on mercury concentrations in freshwater organisms." *Water, Air, and Soil Pollutiotion,* 80: 1011–1015.

Kidd, K.A., D.W. Schindler, D.C.G. Muir, W.L. Lockhart, and R.H. Hesslein, 1995b. "High concentrations of toxaphene in fishes from a subArctic lake." *Science,* 269: 240–242.

Kinloch, D., H. Kuhnlein, and D.C.G. Muir, 1992. "Inuit foods and diet: a preliminary assessment of benefits and risks." *The Science of the Total Environment,* 122: 247–278.

Kinney, S.P., 1997. "A physical-based hydrologic model of a high Arctic watershed and evaluation using isotope tracers." *M.Sc. Thesis*, University of Waterloo, Waterloo, Ontario, Canada.

Kjeller, L.O., K.C. Jones, A.E. Johnston, and C. Rappe, 1996. "Evidence for a decline in atmospheric emissions of PCDD/Fs in the U.K." *Environmental Science and Technology*, 30:1398–1403.

Klein, M., 1999. *Bestimmung von Gesamthalbwertszeiten von POPs (Persistente Organische Stoffe) in der Umwelt.* Vorläufiger Endbericht, Fraunhofer Institut für Umweltchemie and Ökotoxikologie, Schmallenberg, Germany, pp. 67+58.

Klyashtorin, L.B., 1998. "Long-term climate change and main commercial fish production in the Atlantic and Pacfic." *Fisheries Research*, 37: 115–125.

Koch, I., C.A. Ollson, J. Potten, and K.J Reimer, 2002. "Arsenic in Vegetables: An Evaluation of Risk from the Consumption of Produce from Residential and Mine Gardens in Yellowknife, Northwest Territories, Canada." In *Annual Reviews in Food and Nutrition: Toxic and Pathological Aspects*, London: Taylor & Francis.

Kömp, P., and M.S. McLachlan, 1997. "Octanol/air partitioning of polychlorinated biphenyls." *Environmental Toxicology and Chemistry*, 16: 2433–2437.

Kovda, E.N., I.V. Yakushevskaya, and A.N. Tuoruokanov, 1959. "Microelements in some soils of the USSR." *Moscow State University*, Moscow.

Koziol, A.S., and J. Pudykiewicz, 2001. "Global-scale environmental transport of persistent organic pollutants." *Chemosphere,* 45: 1181–1200.

Krabill, W., W. Abdalati, E. Frederick, S. Manizade, C. Martin, J. Sonntag, R. SWift, R. Thomas, W. Wright, and J. Yungel, 2000. "Greenland Ice Sheet: High-elevation balance and peripheral thinning." *Science*, 289: 428–430.

Krebs, C.J., S. Boutin, and R. Boonstra (eds.), 2001. "Ecosystem Dynamics of the Boreal Forest." *The Kluane Project*, New York, New York.

Krupnik, I.I., 2000. "Humans in the Bering Strait region: responses to environmental change and implications for the future," pp. 48–60 In: *Impacts of Changes in Sea Ice and Other Environmental Parameters in the Arctic*, H. P. Huntington (ed.), Girdwood, AK.

Kucklick J.R., D.A. Hinckley, and T.F. Bidleman, 1991. "Determination of Henry's law constants for hexachlorocyclohexanes in distilled water and artificial seawater as a function of temperature." *Marine Chemistry*, 34: 197–209.

Kundiev, Y.I., and Y.S. Kagan, 1993. Pesticide usage in the former Soviet Union, Report prepared for Environment Canada, Downsview, Ontario, Canada.

Kutz, F.W., P.H. Wood, and D.P. Bottimore, 1991. "Organochlorine pesticides and polychlorinated biphenyls in human adipose tissue." *Reviews of Environmental Contaminants and Toxicology*, 120: 1–82.

Kuzyk, Z.A., N.M. Burgess, J.P. Stow, and G.A. Fox, 2003. "Bilogical effects of marine PCB contamination on Black Guillemot Nestlings at Saglek, Labrador: liver biomarkers." *Ecotoxicology*, in Press.

Kuzyk, Z.A., *personal communication*, 2002. Environmental Sciences Group, Kingston Ontario.

Kwok, E.S., and R. Atkinson, 1995. "Estimation of hydroxyl radical reaction rate constants for gas-phase organic compounds using a structure-reactivity relationship: an update." *Atmospheric Environment*, 29: 1685–1695.

Lakaschus, S., K. Weber, F. Wania, R. Bruhn, O. Schrems, 2002. "The air-sea equilibrium and time trend of hexachlorocyclohexanes in the Atlantic Ocean between the Arctic and AntArctica." *Environmental Science and Technology*, 36: 138–145.

Lalonde, J.D., A.J. Poulain, and M. Amyot, 2002. "The role of mercury redox reactions in snow on snow-to-air mercury transfer." *Environmental Science and Technology*, 36: 174–178.

Lambert, S., 1995. "The effect of enhanced greenhouse warming on winter cyclone frequencies and strengths." *Journal of Climate*, 8: 1447–1452.

Lamborg, C.H., W.F. Fitzgerald, J. O'Donnell, and T. Torgerson, 2002. "A non-steady-state compartmental model of global-scale mercury biogeochemistry with interhemispheric gradients." *Geochimica et Cosmochimica Acta*, 66: 1105–1118.

Lammers, R.B., A.I. Shiklomanov, C.J. Vörösmarty, B.M. Fekete, and B.J. Peterson, 2001. "Assessment of contemporary Arctic river runoff based on observational discharge records." *Journal of Geophysical Research*, 106: 3321–3334.

Landa, E.R., E. Reimnitz, D.M. Beals, J.M. Pochkowski, W.G. Winn, and I. Rigor. 1998. "Transport of ^{137}Cs and $^{239,240}Pu$ with ice-rafted debris in the Arctic Ocean." *Arctic*, 51: 27–39.

Landers, D.H., J. Ford, C. Gubala, M. Monetti, B.K. Lasorsa, and J. Martinson. 1995, "Mercury in vegetation and lake sediments from the US Arctic." *Water, Air, and Soil Pollution*, 80: 591–601.

Lange, M.A., and S.L. Pfirman, 1998. "Arctic sea ice contamination: major characteristics and consequences." In: *Physics of Ice-covered Seas.* M. Lepparanta (ed.). Helsinki University Printing House, Helsinki., Finland.

Laniewski, K., H. Boren, A. Grimvall, 1998. "Identification of volatile and extractable chloro-organics in rain and snow." Environmental Science and *Technology*, 32: 3935–3940.

Larson, R.A., and K.A. Marley, 1988. "Sunlight photochlorination and dark chlorination of monoterpenes." *The Science of the Total Environment*, 77: 245–252

Laurier, P., and C.T. Driscoll, 1999. "Historical trends of mercury deposition in Adirondack lakes." *Environmental Science and Technology*,33: 718–722.

Lavigne, D.M., and O.J. Schmitz, 1990. "Global warming and increasing population densities: a prescription for seal plagues." *Marine Pollution Bulletin*, 21: 280–284.

Law, S.A., M.L. Diamond, P.A. Helm, L.M. Jantunen, and M. Alaee, 2001. "Factors affecting the occurence and enantiomeric degradation of hexachlorocyclohexane isomers in northern and temperate aquatic systems." *Environmental Toxicology and Chemistry*, 20: 2690–2698.

Layton, D., R. Edson, M. Varela, and B.E. Napier, 1997. "Radionuclides in the Arctic Seas from the Former Soviet Union: Potential Health and Ecological Risks." *Arctic Nuclear Waste Assessment Program (ANWAP)*, Office of Naval Research (ONR).

Lee, S.C., H. Hung, W.Y. Shiu, and D. Mackay, 2000. "Estimations of vapor pressure and activity coefficients in water and octanol for selected aromatic chemicals at 25°C." *Environmental Toxicology and Chemistry,* 19: 2623–2630.

Legagneux, L., A. Cabanes, and F. Dominé, 2002. "Measurement of the Specific Surface Area of 176 Snow Samples Using Methane Adsorption at 77 K." *Journal of Geophysical Research,* in press.

Legates, D.R., and C. Wilmott, 1990. "Mean seasonal and spatial variability in global surface air temperature." *Theoretical and Applied Climatology,* 41: 11–21.

Lei, Y.D., F. Wania, and W.Y. Shiu, 1999. "Vapor pressures of the polychlorinated naphthalenes." *Journal of Chemical and Engineering Data*, 44: 577–582.

Lei, Y.D., F. Wania, W.Y. Shiu, and D.C.B. Boocock, 2000. "HPLC-based method for estimating the temperature dependence of n-octanol-water partition coefficients." *Journal of Chemical and Engineering Data*, 45: 738–742.

Lei, Y.D., Chankalal, R.A. Chan, and F. Wania, "Supercooled liquid vapor pressures of the polycyclic aromatic hydrocarbons." Submitted to *Journal of Chemical and Engineering Data*

Levi, B.G., 2000. "The decreasing Arctic ice cover." *Physics Today,* January 2000: 19–20.

Lewis, D.L., A.W. Garrison, K.E. Wommack, A. Whittemore, P.S. Steudler, and J. Melillo, 1999. "Influence of environmental changes on degradation of chiral pollutants in soil." *Nature*, 401: 898–901.

Li Y-F., 1999a. "Global gridded technical hexachlorocylohexane usage inventory using a global cropland as a surrogate." *Journal of Geophysical Research, 104*(D19): 23785–23797.

Li, Y-F., 1999b. "Global technical hexachlorocyclohexane usage and its contamination consequences in the environment, from 1948 to 1997." *The Science of the Total Environment,* 232: 121–158.

Li, Y-F., 2001. "Toxaphene in the United States: (1) Usage gridding."*Journal of Geophysical Research*, 106: D16,17919–17927.

Li, Y-F., 2002, "Global gridded emission inventory of β-hexachlorocyclohexane," to be submitted.

Li, Y-F., A. McMillan, and M.T. Scholtz,1996. "Global HCH usage with 1°X1°longitude/latitude resolution." *Environmental Science and Technology,* 30: 3525–3533.

Li, Y-F., T.F. Bidleman, L.A. Barrie, and L.L. McConnell, 1998a. "Global hexachlorocyclohexane use trends and their impact on the Arctic atmospheric environment." *Geophysical Research Letters*, 25: 39–41.

Li, Y-F., D.J. Cai, and A. Singh, 1998b "Technical hexachlorocyclohexane use trends in China and their impact on the environment." *Archives of Environmental Contamination and Toxicology*, 35: 688–697.

Li, Y-F., D.J. Cai, and A. Singh,1999. "Historical DDT use trend in China and usage data gridding with 1/4°by 1/6°longitude/latitude resolution." *Advances in Environmental Research*, 2: 497–506.

Li, Y-F., M.T. Scholdz, and B.J. van Heyst, 2000. "Global gridded emission inventory of α–hexachlorocyclohexane." *Journal of Geophysical Research*, 105: D5, 6621–6632.

Li, Y-F., D.J. Cai, Z.J. Shan, and Z.L. Zhu, 2001a. "Gridded usage inventories of technical hexachlorocyclohexane and lindane for China with 1/6°latitude by 1/4°longitude resolution." *Archives of Environmental Contamination and Toxicology*, 41: 261–266.

Li, Y-F., T.F. Bidleman, and L.A. Barrie, 2001b. "Toxaphene in the United States: (2) Emissions and Residues." *Journal of Geophysical Research*, 106: D16, 17929–17938.

Li, Y-F., R.W. Macdonald, L.M.M. Jantunen, T. Harner, T.F. Bidleman, and W.M.J. Strachan, 2002. "The transport of ß-hexachlorocyclohexane to the western Arctic Ocean: a contrast to α-HCH." *The Science of the Total Environment,* 291, 229–246.

Lin, C.J., M.D. Cheng, and W.H. Schroeder, 2001. "Transport patterns and potential sources of total gaseous mercury measured in Canadian high Arctic in 1995." *Atmospheric Environment*, 35: 1141–1154.

Lindberg, S.E., S. Brooks, C.J. Lin, K. Scott, T. Meyers, L. Chambers, M. Landis, and R. Stevens, 2001. "Formation of reactive gaseous mercury in the Arctic: evidence of oxidation of Hg^0 to gas-phase Hg-II compounds after Arctic sunrise." *Water, Air, and Soil Pollution*, 1: 295–302.

Lindberg, S.E., S. Brooks, C.J. Lin, K.J. Scott, M.S. Landis, R.K. Stevens, M. Goodsite, and A. Richter. 2002. "Dynamic oxidation of gaseous mercury in the Arctic troposphere at polar sunrise." *Environmental Science and Technology,* 36: 1245–1256.

Lisitzin, A.P. 1995. "The marginal filter of the ocean." *Oceanology,* 34: 671–682.

Livingston, H.D.,1988. "The use of Cs and Sr isotopes as tracers in the Arctic Mediterranean Seas." *Philosophical Transactions of the Royal Society*, *325*: 161–176.

Lockhart, W.L., and M. Evans, 2000. "Mercury in fish from stock surveys of lakes in the western Northwest Territories: Investigations into the factors affecting mercury levels." In: *Synopsis of Research Conducted Under the 1999/2000 Northern Contaminants Program.* S. Kalhok, (ed,). R71-19/76-2000E. Indian and Northern Affairs Canada, Ottawa, ON.

Lockhart, W.L., P. Wilkinson, B.N. Billeck, R.V. Hunt, R. Wagemann, and G.J. Brunskill, 1995. "Current and historical inputs of mercury to high-latitude lakes in Canada and to Hudson Bay." *Water, Air, and Soil Pollution,* 80: 603–610.

Lockhart, W.L., P. Wilkinson, B.N. Billeck, R.A. Danell, R.V. Hunt, G.J. Brunskill, J. Delaronde, and V. St. Louis, 1998. "Fluxes of mercury to lake sediments in central and northern Canada inferred from dated sediment cores." *Biogeochemistry,* 40: 163–173.

Lockhart, W.L., G.A. Stern, and P. Outridge, 2000. "Mercury and other contaminants in a laminated sediment core from Devon Island." In: *Synopsis of Research Conducted under the 1999–2000 Northern Contaminants Program,* S. Kalhok. (ed.). Minister of Public Works and Government Services Canada, QS-8602-000-EF-A1, Catalogue No. R71-19/76-2000E, 104–112.

Loeng, H., 2001. "Klima og fisk — hava vet vi og hva tror vi." *Naturen,* 2001: 1332–140.

Loganathan, B.G., S. Tanabe, M. Goto, and R. Tatsukawa, 1989, "Temporal trends of organochlorine residues in lizard goby, Rhinogobdus flumineus, from the RiverNagaragawa, Japan." *Environmental Pollution,* 62: 237–251.

Lorber, M.N., R.G. Barton, D.L. Winters, K.M. Bauer, M. Davis, and J. Palausky. 2002. "Investigation of the potential release of polychlorinated dioxins and furans from PCP-treated utility poles." *The Science of the Total Environment*, 290: 15–39.

Loring, D.H., K. Naes, S. Dahle, G.G. Matishov, and G. Illin, 1995. "Arsenic, trace metals, and organic micro contaminants in sediments from the Pechora Sea, Russia." *Marine Geology,* 128: 153–167.

Loring, D., S. Dahle, K. Naes, J. Dos Santos, J. Skei, and G. Matishov, 1998. "Arsenic and other trace metals in sediments from the Kara Sea and the Ob and Yenisey Estuaries, Russia." *Aquatic Geochemistry*, 4: 233–252.

Lowry, L. F. 2000. "Marine mammal-sea ice relationships." pp. 91–96 In: *Impacts of Changes in Sea Ice and Other Environmental Parameters in the Arctic*, H. P. Huntington, (ed.), Girdwood, AK.

Lu, J.Y., W.H. Schroeder, T. Berg, J. Munthe, D. Schneeberger, and F.A. Schaedlich, 1998. "A device for sampling and determination of total particulate mercury in ambient air." *Analytical Chemistry*, 70:2403–2408.

Lu, J.Y., W.H. Schroeder, L.A. Barrie, A. Steffen, H.E. Welch, K. Martin, W.L. Lockhart, R.V. Hunt, G. Boila, and A. Richter, 2001. "Magnification of atmospheric mercury deposition to polar regions in springtime: the link to tropospheric ozone depletion chemistry." *Geophysical Research Letters*, 28:3219–3222.

Lugar, R.M., Harless, R.L., Dupuy, A.E. Jr., and D.D. McDaniel, 1996. "Results of monitoring for polychlorinated dibenzo-*p*-dioxins and dibenzofurans in ambient air at McMurdo Station, AntArctica." *Environmental Science and Technology,* 30: 555–561.

Lundén, Å., and K. Norén, 1998. "Polychlorinated naphthalenes and other organochlorine contaminants in Swedish human milk, 1972–1992." *Archives of Environmental Contamination and Toxicology*, 34: 414–423.

Luross, J.M., M. Alaee, D.B. Sergeant, C.M. Cannon, D.M. Whittle, K.R. Solomon, and D.C.G. Muir, 2002. "Spatial distribution of polybrominated diphenyl ethers and polybrominated biphenyls in lake trout from the Laurentian Great Lakes." *Chemosphere*, 46(5): 665–672.

MacDonald, G.M., T.W.D. Edwards, K.A. Moser, R. Pienitz, and J.P. Smol, 1993. "Rapid response of treeline vegetation and lakes to past climate warming." *Nature,* 361: 243–246.

Macdonald, R.W., 1996. "Awakenings in the Arctic." *Nature*, 380: 286–287.

Macdonald, R.W., 2000. "Arctic estuaries and ice: a positive-negative estuarine couple," pp. 383–407 In: *The Freshwater Budget of the Arctic Ocean*, E.L. Lewis *et al.,* (eds.), Kluwer Academic Publishers, Netherlands,.

Macdonald, R.W., 2001. "The Arctic Ocean and Contaminants: Pathways that lead to us." pp. 135–149 In: *Waters In Peril.* L. Bendell-Young and P. Gallaugher, (eds.) Kluwer Publishers, Norwell, Massachusetts.

Macdonald, R.W., and J.M. Bewers,1996. "Contaminants in the Arctic marine environment: priorities for protection." *ICES Journal of Marine Science,* 53: 537–563.

Macdonald, R.W., F.A. McLaughlin, and E.C. Carmack, 1995. "Long-Range transport of contaminants to the Canadian Basin." pp. 45–50 In: *Synopsis of Research Conducted under the 1994/95 Northern Contaminants Program*, Murray JL *et al.* editors. Environmental Studies no. 73. Ottawa, Indian and Northern Affairs Canada, Ottawa, Ontario.

Macdonald, R.W., F.A.McLaughlin, and L. Adamson, 1997a. "The Arctic Ocean — The last refuge of volatile organochlorines." *Canadian Chemical News,* 28–29.

Macdonald, R.W., F.A. McLaughlin, and E.C. Carmack, 1997b. "Long-range transport of contaminants to the Canadian Basin." pp. 89–95 In: *Synopsis of Research Conducted under the 1998/99 Northern Contaminants Program*, J. Jensen(ed.) Environmental Studies no. 74. Ottawa, Indian and Northern Affairs Canada, Ottawa, Ontario.

Macdonald, R.W., E.C. Carmack, F.A. McLaughlin, K.K. Falkner, and J.H. Swift, 1999a. "Connections among ice, runoff and atmospheric forcing in the Beaufort Gyre." *Geophysical Research Letters,* 26: 2223–2226.

Macdonald, R.W., F.A. McLaughlin, E.C. Carmack, and G. Stern, 1999b. "Long-range transport of contaminants to the Canadian Basin and selective withdrawal through the Canadian Archipelago." pp. 39–42 In: *Synopsis of Research Conducted under the 1998/99 Northern Contaminants Program*, S. Kalhok, (ed.). Environmental Studies no. 76. Indian and Northern Affairs Canada, Ottawa, ON.

Macdonald, R.W., F.A. McLaughlin, and G. Stern, 1999c. "The seasonal cycle of organochlorine concentrations in the Canada Basin." pp. 43–47 In: *Synopsis of Research Conducted under the 1998/99 Northern Contaminants Program*, S. Kalhok, (ed.).Environmental Studies no. 76. Ottawa, Indian and Northern Affairs Canada, Ottawa, ON.

Macdonald, R.W., L.A. Barrie, T.F. Bidleman, M.L. Diamond, D.J. Gregor, R.G Semkin, W.M.J. Strachan,Yi-F. Li , F. Wania, M. Alaee, L.B. Alexeeva, S.M. Backus, R. Bailey, J.M. Bewers, C. Gobeil, C.J. Halsall, T. Harner, J.T. Hoff, L.M.M. Jantunen, Lockhart W.L., D. Mackay, D.C.G. Muir, J. Pudykiewicz, K.J. Reimer, J.N. Smith, G.A. Stern, W.H. Schroeder, R.Wagemann, and M.B.Yunker, 2000. "Sources, occurrence and pathways of contaminants in the Canadian Arctic: A review." *The Science of the Total Environment*, 254: 93–236.

Macdonald, R.W., S.J. Eisenreich, T.F. Bidleman, J. Dachs, J. Pacyna, K. Jones, B. Bailey, D. Swackhamer, and D.C.G. Muir, 2001a. "Case studies on persistence and long-range transport of persistent organic pollutants," pp. 245–314 In: *Evaluation of persistence and long-range transport of organic chemicals in the environment.* Klecka G, *et al.,*(eds.), SETAC Press, Pensacola.

Macdonald, R.W., G. Stern, and F.A. McLaughlin. 2001b. "The seasonal cycle of organochlorine concentrations in the Canada Basin." pp. 118–124 In: *Synopsis of Research Conducted under the 1998/99 Northern Contaminants Program.* S. Kalhok, (ed.). Indian and Northern Affairs Canada, Ottawa, Ontario.

Macdonald, R.W., F.A. McLaughlin, and E.C. Carmack. 2002a. "Freshwater and its sources during the SHEBA drift." *Deep-Sea Research II* submitted.

Macdonald, R.W., B. Morton, R.F. Addison, and S.C. Johannessen. 2002b. "Marine environmental contaminant issues in the North Pacific: what are the dangers and how do we identify them?" In: A. *PICES Scientific Report No. 18*. Bychkov and S. M. McKinnell, (eds.). North Pacific Marine Science Organization, PICES, Sidney, BC.

Macdonald, R.W., D.Mackay, and B. Hickie, 2002c. "Contaminant amplification in the environment: Revealing the fundamental mechanisms." *Environmental Science and Technology*, 36(23), A457–A462.

Mackay, D., 1991. "Multimedia Environmental Models." Lewis Publishers., Boca Raton, FL.

Mackay, D., and M.L. Diamond. 1989. "Application of the QWASI (Quantitative Water Air Sediment Interaction) fugacity model to the dynamics of organic and inorganic chemicals in lakes." *Chemosphere, 18*: 1343–1365.

Mackay, D., and F. Wania,1995. "Transport of contaminants to the Arctic: partitioning, processes and models." *The Science of the Total Environment,* 160/161: 25–38.

Mackay D., W-Y. Shiu , and K-C. Ma, 1997. "Illustrated Handbook of Physicochemical Properties and Environmental Fate for Organic Chemicals, Volume 5, Pesticide Chemicals." Lewis Publishers, Boca Raton, FL.

Mackay, D., A. Di Guardo, S. Paterson, and C.E. Cowan, 1996. "Evaluating the environmental fate of a variety of types of chemicals using the EQC model." *Environmental Toxicology and Chemistry,* 15: 1627–1637.

Mackay, D., E. Webster, A. Beyer, M. Matthies, and F. Wania, 2001. "Defining the bioaccumulation, persistence, and transport attributes of priority chemicals." In: *American Chemical Society Symposium Series No. 773 Persistent, Bioaccumulative, and Toxic Chemicals*, Volume II, R. Lipnick, B. Jansson, D. Mackay, M. Petreas (Eds.) 14–28.

Magnuson, J.J., K.D.M. Robertson, B.J. Benson, R.H. Wynne, D.M. Livingstone, T. Arai, R.A. Assel, R.G. Barry, V. Card, E. Kuusisto, N.G. Granin, T.D. Prouse, K.M. Stewart, and V.S. Vuglinski, 2000. "Historical trends in lake and river ice cover in the northern hemisphere." *Science,* 289: 1743–1746.

Majewski, M.S., W.T. Foreman, D.A. Goolsby, and N. Nakagaki, 1998. "Airborne pesticide residues along the Mississippi River." *Environmental Science and Technology*, 32: 3689–3698

Malcolm, E.G., and G.J. Keeler, 2002. "Measurements of mercury in dew: atmospheric removal of mercury species to wetted surfaces." *Environmental Science and Technology,* 36: 2815–2821.

Maltby, C., 1980. "Report on the use of pesticides in Latin America", UNID/IOD.353, 139 pp. [Rp013].

Mamontova, E.A., A.A. Mamontov, N.I. Matorova, E.N. Tarsova, and U.A. Chuvashev, 1997. "PCBs in snow of the Baikal region." Organohalogen Compounds, 32: 72–75.

Manabe, S., M.M. Spelman, and R.J. Stouffer, 1992. "Transient responses of a coupled ocean-atmosphere model to gradual changes in atmospheric CO2." *Journal of Climate*, 5: 105–126.

Marteinsdottir, G., and K. Thorarinsson, 1998. "Improving the stock-recruitment relationship in Icelandic cod (*Gadus morhua*) in including age diversity of spawners." *Canadian Journal of Fishery and Aquatic Sciences*, 55: 1372–1377.

Martin, A.J., and T.F. Pedersen, 2002. "Seasonal and interannual mobility of arsenic in a lake impacted by metal mining." *Environmental Science and Technology*, 36: 1516–1523.

Masclet, P., and V. Hoyau, 1994. "Evidence for the presence of polycyclic aromatic hydrocarbons in the polar atmosphere and in polar ice." *Analusis Magazine,* 22: M31–M33.

Masclet, P.P, V. Hoyau, J.L. Jaffrezo, and H. Cachier, 2000. "Polycyclic aromatic hydrocarbon deposition on the ice sheet of Greenland. Part I: Superficial snow." *Atmospheric Environment*, 34: 3195–3207.

Maslanik, J.A., M.C. Serreze, and R.G. Barry, 1996. "Recent decreases in Arctic summer ice cover and linkages to atmospheric circulation anomalies." *Geophysical Research Letters,* 23: 1677–1680.

Maslanik, J.A., M.C. Serreze, and T. Agnew, 1999. "On the record reduction in 1998 western Arctic sea-ice cover." *Geophysical Research Letters*, 26: 1905–1908.

Maslowski, W., J. McLean, R. Newton, and P. Schlosser, 1998. "Modelling the interannual variability of the Arctic Ocean and sea ice circulation." *Eos (supplement),* 79: F414.

Maslowski, W., B. Newton, P. Schlosser, A. Semtner, and D. Martinson, 2000. "Modelling recent climate variability in the Arctic Ocean." *Geophysical Research Letters,* 27: 3743–3746.

Mason, R.P., and W.F. Fitzgerald, 1996. "Sources, sinks and biogeochemical cycling of mercury in the ocean." pp. 249–272 In: *Global and Regional Mercury Cycles: Sources, Fluxes and Mass Ballances.* W. B. e. al., ed. Kluwer Academic Publishers, Netherlands.

Mason, R.P., W.F. Fitzgerald, and F.M.M. Morel, 1994. "The biogeochemical cycling of elemental mercury: Anthropogenic influences." *Geochimica et Cosmochimica Acta,* 58:3191–3198.

Matishov, G.G., D.G. Matishov, A.A. Namjatov, J. Carroll, and S. Dahle, 1999. "Anthropogenic radionuclides in Kola and Motovsky Bays of the Barents Sea, Russia." *Journal of Environmental Radioactivity,* 43: 77–88.

Matishov, G.G., D.G. Matishov, A.E. Namjatov, J.N. Smith, J. Carroll and S. Daahle, 2002. "Radioactivity near the sunken submarine "Kursk" in the Southern Barents Sea." *Environmental Science and Technology*, 36: 1919–1922.

Mattina, M.J.I., W. Iannucci–Berger, L. Dykas, and J. Pardus, 1999. "Impact of long-term weathering, mobility and land use on chlordane residues in soils." *Environmental Science and Technology*, 33: 2425–2431.

McConnell, L.L., Kucklick, J.R., Bidleman, T.F., Ivanov, G.P. and S.M.Chernyak, 1996. "Air-Water Gas Exchange of Organochlorine Compounds in Lake Baikal, Russia." *Environmental Science and Technology,* 30: 2975–2983

McConnell, L.L., T.F. Bidleman, W.E.Cotham, and M.D.Walla, 1998, "Air concentrations of organochlorine insecticides and polychlorinated biphenyls over Green Bay, WI, and the four lower Great Lakes." *Environmental Pollution*, 101: 391–399.

McDonald, M.E., A.E. Hershey, and M.C. Miller,1996. "Global warming impacts on trophic structure in Arctic lakes." *Limnology and Oceanography,* 41: 1102–1108.

McElroy, C.T.,C.A McLinden, and J.C. McConnell, 1999. "Evidence for bromine monoxide in the free troposphere during the Arctic polar sunrise." *Nature*, 397: 338–341.

McGhee, R., 1996. "Ancient People of the Arctic." *UBC Press* in association with Canadian Museum of Civilization, Vancouver, British Columbia.

McGowan, J. A., 1990. "Climate and change in oceanic ecosystems: the value of time-series data." *Trends in Ecology and Evolution, 5*: 293–299.

McKone, T.E., J.I. Daniels, and M. Goldman.1996, "Uncertainties in the link between global climate change and perdicted health risks from pollution: Hexachlorobenzene (HCB) case study using a fugacity model." *Risk Analysis*, 16: 377–393.

McLaughlin, F.A., E.C. Carmack, R.W. Macdonald, and J.K. B. Bishop. 1996. "Physical and geochemical properties across the Atlantic/Pacific water mass boundary in the southern Canadian Basin." *Journal of Geophysical Research,* 101(C1): 1183–1197.

McLaughlin, F.A., E.C. Carmack, R.W. Macdonald, A.Weaver, J.N. Smith, and K. Aagaard. 2002. "The Canada Basin 1989–1995: Upstream change and far-field effects of the Barents Sea Branch." *Journal of Geophysical Research,* 10.1029/2001 JC000904 (26 July 2002), 19-1–19-20.

McPhee, M.G., T.P. Stanton, J.H. Morison, and D.G. Martinson, 1998. "Freshening of the upper ocean in the Arctic: Is perennial sea ice disappearing?" *Geophysical Research Letters,* 25: 1729–1732.

Meerts, I.A.T.M., R.J. Letcher, S. Hoving, G. Marsh, Å. Bergman, J.G. Lemmen, B. van der Burg, and A. Brouwer, 2001, "In Vitro Estrogenicity of Polybrominated Diphenyl Ethers, Hydroxylated PBDEs, and Polybrominated Bisphenol A Compounds." *Environmental Health Perspectives,* 109: 399–407.

Meese, D.A., E. Reimnitz, W.B. I. Tucker, A.J. Gow, J. Bischof, and D. Darby, 1997. "Evidence for radionuclide transport by sea ice." *The Science of the Total Environment*, 202: 267–278.

Meijer, S.N., C.J. Halsall, T. Harner, A.J. Peters, W.A. Ockenden and K.C. Jones,2001. "Organochlorine pesticide residues in archived UK soil." *Environmental Science and Technology, 35*(10); 1989–1995.

Meijer, S.M., M. Shoeib, K.C. Jones, and T. Harner, 2002a. "Air-soil exchange of organochlorine pesticides in agricultural soil. 1. Field measurements of the soil/air partition coefficient (K_{SA})." *Environmental Science and Technology*, in press.

Meijer, S.M., M. Shoeib, K.C. Jones, and T. Harner, 2002b. "Air-soil exchange of organochlorine pesticides in agricultural soil. 2. Laboratory measurements of the soil/air partition coefficient (K_{SA})." *Environmental Science and Technology,* in press.

Melnikov, I.A., E.G. Kolosova, H.E. Welch, and L.S. Zhitina, 2002. "Sea ice biological communities and nutrient dynamics in the Canadian Basin of the Arctic Ocean." *Deep-Sea Research II,* submitted.

Mercier, G., C. Gariépy, L.A. Barrie, and A. Simonetti, 1999, "Source discrimination of atmospheric aerosols at Alert, Arctic Canada during 1994–95 using a year long record of Pb isotope and trace element data." Ninth Annual V.M. Goldschmidt Conference, Cambridge, MA, August.

Miller, G. H., A. Geirsdottir, and R.M. Koerner, 2001. "Climate implications of changing Arctic sea ice." *Eos, Transactions, American Geophyiscal Union,* 82: 97,103.

Miller, J.R., and G.L. Russell, 1992. "The impact of global warming on river runoff." *Journal of Geophysical Research,* 97: 2757–2764.

Miskimmin, B.M., D.C.G. Muir, D.W. Schindler, and G.A. Stern, N.P. Grift, 1995. "Chlorobornanes in sediment and fish 30 years after toxaphene treatment of lakes." *Environmental Science and Technology,* 29: 2490–2495

Mitchell, J.F.B., T.C. Johns, J.M. Gergory, and S.F.B. Tett, 1995. "Climate response to increasing levels of greenhouse gases and sulphate aerosols." *Nature,* 376: 5010504.

Mmereki, B.T., J.M. Hicks, and J.D. Donaldson, 2000. "Adsorption of atmospheric gases at the air-water interface. 3. Methylamines." *Journal of Physical Chemistry* A, 104: 10789–10793.

Moisey, J, A.T. Fisk, K.A. Hobsen, and R.J. Norstrom, 2001. "Hexachlorocyclohexane (HCH) isomers and chiral signatures in the Arctic marine food web of the Northwater Polynya." *Environmental Science and Technology,* 35: 1920–1927.

Moore, S.E., J.C. George, K.O. Coyle, and T.J. Weingartner,1995. "Bowhead whales along the Chukotka coast in autumn." *Arctic,* 48: 155–160.

Moore, S.E., and J.T. Clarke, 1986. "A comparison of gray whale (*Eschrichtius robustus*) and bowhead whale (*Balaena mysticetus*) distribution, abundance, habitat preference and behavior in the northeastern Chukchi Sea, 1982–84." *Reports of the International Whaling Commission,* 36: 273–279.

Moran, S.A. and W.L.Woods, 1997. "Cd, Cr, Cu, Ni and Pb in the water column and sediments of the Ob-Irtysh Rivers, Russia." *Marine Pollution Bulletin,* 35: 270–279.

Morison, J., M. Steele, and R. Andersen, 1998. "Hydrography of the upper Arctic Ocean measured from the nuclear submarine USS, Pargo." *Deep-Sea Research, I* 45: 15–38.

Morison, J., K.Aagaard, and M. Steele. 2000. "Recent environmental changes in the Arctic: A review." *Arctic,* 53: 359–371.

Morrison, H.A., F.A. P.C. Gobas, R. Lazar, D.M. Whittle, and G.D. Haffner, 1998. "Projected changes to the trophodynamics of PCBs in the western Lake Erie ecosystem attributed to the presence of zebra mussels (*Dreissena polymorpha*)." *Environmental Science and Technology,* 32: 3862–3867.

Morrison, H.A., M.D. Whittle, and D.G. Haffner, 2000. "The relative importance of species invasions and sediment disturbance in regulating chemical dynamics in western Lake Erie." *Ecological Modelling,* 125: 279–294.

Moritz, R.E., C.M.Bitz, and E.J. Steig, 2002. "Dynamics of recent climate change in the Arctic." *Science, 297*: 1497–1502.

Mossner, S., and K. Ballschmiter, 1997, "Marine Mammals as Global Pollution Indicators for Organochlorines", *Chemosphere, 34*: 1285–1296.

Mossner, S., T.R. Spraker, P.R. Becker, and K. Ballschmiter, 1992, "Ratios of enantiomers of –HCH and determination of α-, β-, and γ-HCH isomers in brain and other tissues of neonatal northern fur seals (*callorhinus ursinus*)." *Chemosphere, 24*: 1171–1180.

Muir, D.C.G., and deBoer, J, 1993. "Toxaphene: analytical chemistry." *Chemosphere,* 27, 1827–1834.

Muir, D.C.G., and R.J. Norstrom,1994. "Persistent organic contaminants in Arctic marine and freshwater ecosystems." *Arctic Research,* 8: 136–146.

Muir, D.C.G., N.P. Grift, W.L. Lockhart, P. Wilkinson, B.N. Billeck, and G.J. Brunskill, 1995. "Spatial trends and historical profiles of organochlorine pesticides in Arctic lake sediments." *The Science of the Total Environment,* 160/161: 447–457.

Muir, D.C.G., A. Omelchenko, N.P. Grift, D.A. Savoie, W.L. Lockhart, P. Wilkinson, and G.J. Brunskill, 1996a. "Spatial trends and historical deposition of polychlorinated biphenyls in Canadian midlatitude and Arctic lake sediments." *Environmental Science and Technology*, 30: 3609–3617.

Muir D, B. Braune, B. DeMarch, R. Norstrom, R. Wagemann, M. Gamberg, K. Poole, D. Addison, D. Bright, M. Dodd, W. Dushenko, J.Eamer, M. Evans, B. Elkin, S. Grundy, B. Hargrave, C. Hebert, R. Johnstone, K. Kidd, B. Koenig, L. Lockhart, J. Payne, J. Peddle, and K. Reimer, 1996b. "Chapter 3. Ecosytem uptake and effects." In: J. Jensen, K. Adare, and R. Shearer, editors. *Canadian Arctic Conaminants Assessment Report.* Indian and Northern Affairs Canada, Ottawa, 1997, pp. 191–294.

Muir, D., B. Braune, B. DeMarch, R. Norstrom, R. Wagemann, L. Lockhart, B. Hargrave, D. Bright, R. Addison, J. Payne, and K. Reimier. 1999a. "Spatial and temporal trends and effects of contaminants in the Canadian Arctic marine ecosystem: a review." *The Science of the Total Environment*, 230: 83–144.

Muir, D.C.G., G. Stern, and H. Karlsson, 1999b. "Congener and enantiomer specific studies on the environmental fate of toxaphene." *Organohalogen Compounds*, 41: 565–568.

Muir, D.C.G., D. Bennie, C. Teixeira, A. Fisk, G. Tomy, G. Stern, and M. Whittle, 2000a. "Short-chain chlorinated paraffins: Are they persistent and bioaccumulative?" In: *Persistent, bioaccumulative , and Toxic Chemicals, Volume II: Assessment and New Chemicals*, 0841236755.

Muir D.C.G., F. Riget, M. Cleeman, J. Skaare, L. Kleivane, H. Nakata, R. Dietz, T. Severinsen, and S. Tanabe, 2000b. "Circumpolar trends of PCBs and organochlorine pesticides in the Arctic marine environment inferred from levels in ringed seals." *Environmental Science and Technology*, 34: 2431–2438.

Muir, D.C.G., D.M.D. Whittle, D. deVault, S. Backus, M. Comba, and H. Karlsson, 2001. "Toxaphene in the Food Webs of Lake Superior and Nearby Lakes." Report prepared for the US EPA Great Lakes National Program Office and US EPA Region V, Chicago, IL , National Water Research Institute, Burlington, ON. 50 pp.

Muir, D.C.G., T. Savinova, V. Savinov, and L. Alexeeva, 2002. Unpublished data, National Water Research Institute, Burlington, ON.

Müller-Herold, U., 1996. "A simple general limiting law for the overall decay of organic compounds with global pollution potential." *Environmental Science and Technology*, 30: 586–591.

Müller-Herold , U., D. Caderas, and P. Funck, 1997. "Validity of global life-time estimates by a simple general limiting law for the decay of organic compounds with long-range pollution potential." *Environmental Science and Technology*, 31: 3511–3515.

Müller-Herold , U., and G.A. Nickel, 2000. "Closed analytical formula for the characteristic spatial range of persistent organic pollutants." *Ecological Modelling*, 126: 191–200.

Murphy, T. J., M.D. Mullin, and J.A. Meyer, 1987. "Equilibration of polychlorinated biphenyls and toxaphene with air and water." *Environmental Science and Technology*, 21: 155–162.

Mysak, L.A., 2001. "Patterns of Arctic circulation." *Science*, 239: 1269–1270.

NAFAP, 2000. "Pesticide use survey record detail." *National Center for Food and Agricultural Policy*, Washington, DC.

Nair, A., and M.K.K. Pillai,1992. "Trends in ambient levels of DDT and HCH residues in humans and the environment of Delhi, India." *The Science of the Total Environment*, 121: 145–157

Nakazawa, T; M. Ishizawa, K. Higuchi, and N.B.A. Trivett. 1997. "Two curve fitting methods applied to CO_2 flask data." *Environmetrics*, 8: 197–218.

Nash, R.G., W.G. Harris, P.D. Ensor, E.A. Woolson, 1973. "Comparative extraction of chlorinated hydrocarbon insecticides from soils 20 years after treatment." *Journal of the Association of Official Analytic Chemists*, 56: 728–732.

Newell, J.P., 1996. "Sea-ice and atmospheric circulation anomalies in the Labrador Sea region associated with extremes of the southern oscillation." *International Journal of Climatology*, 16: 63–71.

Newton, J.L., and B.J. Sotirin, 1997. "Boundary undercurrent and water mass changes in the Lincoln Sea." *Journal of Geophysical Research*, 102: 3393–3403.

Ngabe, B., T.F. Bidleman, and R.L. Falconer, 1993. "Base hydrolysis of α- and β- hexachlorocylcohexanes." *Environmental Science and Technology,* 27: 1930–1933.

Niebauer, N.J., and V. Alexander, 1985. "Oceanographic frontal structure and biological production at an ice edge." *Continental Shelf Research,* 4: 367–388.

Niebauer, H.J., and R.H. Day, 1989. "Causes of interannual variability in the sea ice cover of the eastern Bering Sea." *GeoJournal,*18: 45–59.

Nielsdottir, M.C., H. Skov, M.E. Goodsite, J. Christensen, C.A. Skjøth, G. Geernaert, O. Hertel, and J. Olsen, 2002. "Measurements of gaseous elemental mercury on Faroe Islands." Submitted to *Atmpheric Envvironment,* 2002.

Nirwal, A., B. Zeeb, and K. Reimer, 2000. "Ecological risk assessment of DDT contamination at Kittigazuit, NWT." 83rd Canadian Society for Chemistry and Chemical Engineering Conference, Calgary, Alberta, May.

Nordin, K., B. Burns, and B. Lendrum, 1993. "Use, disposal and transportation of selected contaminants in Yukon." Indian and Northern Affairs Canada. 114 pp. Ottawa, Ontario.

Nørgaard-Petersen, N., R.F.Hagen, J. Thiede, and H. Kassens, 1998. "Central Arctic surface ocean environment during the past 80,000 years." *Paleoceanography,* 13: 193–204.

Norstrom, R.J., S.E. Belikov, E.W. Born, G. W. Garner, B. Malone, S. Olpinski, M.A. Ramsay, S. Schlieve, I. Stirling, M.S. Stishov, M.K. Taylor, and Ø. Wiig, 1998. "Chlorinated hydrocarbon contaminants in polar bears from eastern Russia, North America, Greenland, and Svalbard: Biomonitoring of Arctic pollution." *Archives of Environmental Contamination and Toxicology*, 35: 354–367.

Nriagu, J.O. 1989. "A global assessment of natural sources of atmospheric trace metals." *Nature,* 338: 47–49.

Nriagu, J.O., 1996. "History of Global Metal Pollution." *Science*, 272: 223–224.

Nriagu, J.O., and J.M. Pacyna, 1988. "Quantitative assessment of worldwide contamination of air, water and soils by trace metals." *Nature,* 333: 134–139.

Nürnberg, D., I. Wollenburg, D. Dethleff, H. Eicken, H. Kassens, T. Letzig, E. Reimnitz, and J. Thiede, 1994. "Sediments in Arctic sea ice: implications for entrainment, transport and release." *Marine Geology,* 119: 185–214.

Ockenden, W.A., A.J. Sweetman, H.F. Prest, E. Steinnes, and K.C. Jones, 1998. "Toward an understanding of the global atmospheric distribution of persistent organic pollutants: The use of semipermeable membrane devices as time-integrated passive samplers." *Environmental Science and Technology,* 32: 2795–2803.

Oehme, M., 1991. "Further evidence for long-range air transport of polychlorinated aromates and pesticides, North America and Eurasia to the Arctic." *Ambio,* 20: 293–297.

Oehme, M., and S. Mano, 1984. "The long-range transport of organic pollutants to the Arctic." *Fresenius' Journal of Analytical Chemistry,* 319: 141–146.

Oehme, M., and B. Ottar, 1984. "The long-range transport of polychlorinated hydrocarbons to the Arctic." *Geophysical Research Letters,* 11: 1133–1136.

Oehme, M., S. Manø, and A. Mikalsen, 1987. "Formation and presence of polyhalogenated and polycyclic compounds in the emissions of small and large scale municipal waste incinerators." *Chemosphere*, 16: 143–153.

Oehme, M., J-E Haugen, and M. Schlabach, 1995. "Ambient air levels of persistent organochlorines in spring1992 at Spitsbergen and the Norwegian mainland: comparison with 1984 results and quality control measures." *The Science of the Total Environment*, 160/161: 139–152

Oehme, M, J-E. Haugen, and M. Schlabach, 1996. "Seasonal changes and relations between organochlorines in Arctic ambient air: first results of an all-year monitoring program at Ny Ålesund, Svalbard, Norway." *Environmental Science and Technology*, 30: 2294–2304.

Ogilvie, A.E.J., and T. Junsson, 2000. "The Iceberg in the Mist: Northern Research in Pursuit of a "Little Ice Age"." Kluwer Academic Publishers, Dordrecht.

Okubo, A. "Oceanic diffusion diagrams." *Deep Sea Research,* 18:789–802.

Olafsdottir, K., K. Skirnisson, G. Gylfadottir, and T. Johannesson, 1998. "Seasonal fluctuations of organochlorine levels in the common eider (*Somateria mollissima*) in Iceland." *Environmental Pollution*, 103: 153–158.

Ollson, C.A., 1999. "Arsenic contamination of the terrestrial and freshwater environment impacted by gold mining operations, Yellowknife, NWT." MSc. Thesis, Royal Military College of Canada.

Ollson, C.A., 2002. "A novel approach to characterizing potential ecological and human health risks associated with arsenic contaminated sites: Yellowknife, NWT, Canada." Ph.D. thesis. Royal Military College of Canada.

Olson, M.P., K.K. Oikawa, and A.W. MacAfee. 1978. "A trajectory model applied to the long-range transport of air pollutants: a technical description and some model intercomparisons." Environment Canada/AES, Toronto, Canada, Internal report LRTAP 78–4.

Oreskes, N., K. Shrader Frechette, and K. Belitz, 1994. "Verification, validation, and confirmation of numerical models in the earth sciences." *Science,* 263: 641–646.

Osterkamp, T.E.,L. Viereck, Y. Shur, M.T. Jorgenson, C.H. Racine, A.P. Doyle, and R. D. Boone, 2000. "Observations of thermokarst in boreal forests in Alaska." *Arctic, AntArctic, and Alpine Research,* 32: 303–315.

Overpeck, J., K. Hughen, D. Hardy, R. Bradley, R. Case, M. Douglas, B. Finney, K. Gajewski, G. Jacoby, A. Jennings, S. Lamoureux, A. Lasca, G. MacDonald, J. Moore, M. Retelle, S. Smith, A. Wolfe, and G. Zielinski, 1997. "Arctic environmental change of the last four centuries." *Science,* 278: 1251–1256.

Paasivirta, J., S. Sinkkonen, P. Mikkelson, T. Rantio and F. Wania. 1999. "Estimation of vapor pressures and Henry's Law constants of selected persistent organic pollutants as functions of temperature." *Chemosphere*, 39: 811–832.

Pacyna, E.G. and J.M Pacyna, 2002. "Global emission of mercury from anthropogenic sources in 1995." *Water. Air, and. Soil Pollution,* 137: 149–165.

Pacyna, J.M, 1995. "The origin of Arctic air pollutants: lessons learned and future research." *The Science of the Total Environment*, 160/161: 39–53.

Pacyna, J.M., and M. Oehme, 1988. "Long-range transport of some organic compounds to the Norwegian Arctic." *Atmospheric Environment, 22* : 243–257.

Pacyna, J.M., and G.J. Keeler, 1995. "Sources of mercury to the Arctic." *Water, Air, Soil Pollut.*, 80:621–63.

Pcyna, J.M., and E.G. Pacyna, 2001. "An assessment of global and regional emissions of trace metals to the atmosphere from anthropogenic sources worldwide." *Canadian Journal of Environmental Reviews,* 9: 269–298.

Pacyna, J.M., M.T. Scholtz, and Y-F. Li, 1995. "Global budget of trace metal sources." *Environmental Reviews,* 3: 145–159.

Painter, S., E.M. Cameron, R. Allan and J. Rouse, 1994. "Reconnaissance geochemistry and its environmental relevance." *Journal of Geochemical Exploration,* 51: 213–246.

Paluszkiewicz, T., L.F. Hibler, L.F. Richmond, M.C. Richmond, D.J. Bradley, and S.A. Thomas, 2001. "Modelling the potential radionuclide transport by the Ob and Yenisey Rivers to the Kara Sea." *Marine Pollution Bulletin,* 43: 111–121.

PAN, 1995. "Demise of the dirty dozen," compiled by the Pesticide Action Network North American Regional Center for PAN International, San Francisco, CA, USA, 1989, 1992, 1993, 1995.

Pankow, J.F., 1998. "Further discussion of the octanol/air partition coefficient KOA as a correlating parametrer for gas/particle partition coefficients." *Atmospheric Environment*, 32: 1493–1497.

PANNA Outlook, 1990. "India to Ban Agricultural Use of DDT." Pesticide Action Network North America Regional Chapter, San Francisco, California.

Pannatier, E. G., 1997, "Sediment accumulatikon and historical deposition of trace metals and trace organic compounds in the Mackenzie Delta (Northwest Territories, Canada)." Thesis, No 2952, Universite de Geneve, Institut F. -A. Forel, Switzerland, 222 pg.

Parkinson, C.L., 1992. "Spatial patterns of increases and decreases in the length of sea ice season in the north polar region, 1979–1986." *Journal of Geophysical Research*, 97: 14377–14388.

Parkinson, C.L., D.J. Cavalieri, P. Gloersen, J.H. Zwally, and J.C. Comiso, 1999. "Arctic sea ice extents, areas, and trends, 1978–1996." *Journal of Geophysical Research*, 104: 20837–20856.

Parsons, T.R., 1979. "Some ecological, experimental and evolutionary aspects of the upwelling ecosystem." *South African Journal of Science,* 75: 536–540.

Parsons, T.R., 1992. "The removal of marine predators by fisheries and the impact of trophic structure." *Marine Pollution Bulletin,* 25: 51–53.

Parsons, T. R., 1996. "The impact of industrial fisheries on the trophic structure of marine ecosystems." pp. 352–357 In: *Food Webs — Integration of Patterns and Dynamics.* K. O. W. G.A. Polis, ed. Chapman Hall, New York, NY.

Patin, S., 1999. "*Environmental impact of the offshore oil and gas industry.*" Ecomonitor Publishing.

Patton, G.W., D.A. Hinckley, M.D. Walla, and T.F. Bidleman, 1989. "Airborne organochlorines in the Canadian High Arctic." *Tellus,* 41B: 243–255.

Patton, G.W., M.D. Walla, T.F. Bidleman, and L.A. Barrie, 1991. "Polycyclic aromatic and organochlorine compounds in the atmosphere of Northern Ellesmere Island, Canada." *Journal of Geophysical Research,* 96: 10867–10877.

Pauly, D., V. Christensen, J. Dalsgaard, R. Froese, and F. Torres, 1998. "Fishing down marine food webs." *Science,* 279: 860–863.

Pauly, D., M.L. Palomares, R. Fooese, P. Sa-a, M. Vakily, D. Preikshot, and S. Wallace, 2001. "Fishing down Canadian aquatic food webs." *Canadian Journal of Fisheries and Aquatic Sciences,* 58: 51–62.

Peinert, R., A. Antia, E. Baurefeind, B. v. Bodungen, O. Haupt, M. Krumbholz, I. Peeken, R. Ramseier, M. Voss, and B. Zeitschel, 2001. "Particle flux variability in the polar and Atlantic biogeochemical provinces of the North Atlantic."pp. 53–68 In: *The Northern North Atlantic: A Changeing Environment.* P. Schafer, W. Ritzrau, M. Schluter, and J. Thiede, (eds.) Springer Verlag, Berlin.

Pelletier, M., R. Burgess, K.Ho, A. Kuhn, R. McKinney, and S. Ryba, 1997. "Phototoxicity of individual polycyclic aromatic hydrocarbons and petroleum to marine invertebrate larvae and juveniles." *Environmental Toxicology and Chemistry,* 16: 2190–2199.

Pennington, D.W., 2001a. "An evaluation of chemical persistence screening approaches." *Chemosphere,* 44: 1589–1601.

Pennington, D.W., 2001b. "Relationship of approaches and a tiered methodology for screening chemicals in the context of long-range transport." *Chemosphere,* 44: 1617–1631.

Perovich, D.K., E.L. Andreas, J.A. Curry, H. Eicken, C.W. Fairall, T.C. Grenfell, P.S. Guest, J. Intrieri, D. Kadko, R.W. Undsay, M.G. McPhee, J. Morison, R.E. Moritz, C.A. Paulson, W.S. Pegau, P.O.G. Persson, R. Pinkel, J.A. Richter-Menge, T. Stanton, H. Stern, M. Sturm, W.B. Tucker, and T. Ottal. 1999. "Year on ice gives climate insights." *Eos, Transactions, American Geophysical Union* , 80: 481, 485–6.

Pesticide Action Network North America (PANA). 1997. http://www.igc.org/panna/resources/_pestis/PESTIS.1997.43.html (May 23, 1997).

Pesticide Action Network (PAN), 1993. "Demise of the dirty dozen." North American Regional Center for the Pesticide Action Network International, San Francisco, California, USA.

Peterjohn, W.T., 1994. "Responses of trace gas fluxes and N availability to experimentally elevated soil temperatures." *Ecology Applied,* 4: 617–625.

Peters, A.J., D.J. Gregor, C.F. Teixeira, N.P. Jones, C. Spencer, 1995. "The recent depositional trend of polycyclic aromatic hydrocarbons and elemental carbon to the Agassiz Ice Cap, Ellesmere Island, Canada." *The Science of the Total Environment,* 160/161: 167–179.

Petersen S.M., G.E. Batley, 1993. "The fate of endosulfan in aquatic ecosystems." *Environmental Pollution,* 82: 143–152.

Petit, J.R., J. Jouzel, D. Raynout, N.I. Barkov, J.M. Barnola, I. Basile, M. Bender, J. Chappellaz, M. Davis, G. Delaygue, M. Delmotte, V.M. Kotlyakov, M. Legrand, V.Y. Lipenkov, C. Lorius, L. Pépin, C. Ritz, E. Saltzman, and M. Stievenard, 1999. "Climate and atmospheric history of the past 420,000 years from the Vostok ice core, AntArctica." *Nature,* 399: 429–436.

Petrick, G., D.E.Schultz–Bull, V. Martens, K. Scholz and J.C. Duinker. 1996. "An *in-situ* filtration/extraction system for the recovery of trace organics in solution and on particles tested in deep ocean water." *Marine Chemistry* 54: 97–105.

Petrosyan, V.S., Y.Y. Vinnikov, V.A. Kimstach, L.V. Boeva and T.A. Sergeeva, 1998. "Organic pollutants." In: V. Kimstach, M. Meybeck, E. Baroudy,(Eds). *A Water Quality Assessment of the former Soviet Union*. London. E&FN Spon. pp. 211–266.

Pfirman, S.L., H. Eicken, D. Bauch, and W. Weeks. 1995. "The Potential transport of pollutants by Arctic sea ice." *The Science of the Total Environment*, 159: 129–146.

Pfirman, S.L., Kogeler, J.W., and I. Rigor, 1997. "Potential for rapid transport of contaminants from the Kara Sea." *The Science of the Total Environment*, 202: 111–122.

Phillips, R.L., and A. Grantz, 1997. "Quarternary history of sea ice and paleoclimate in the Amerasia Basin, Arctic Ocean, as recorded in the cyclical strata of Northwind Ridge." *Geological Society of America Bulletin*, 109: 1101–1115.

Pickhardt, P.C., C.L. folt, C.Y. Chen, B. Klaue, and J.D. Blum, 2002. "Algal blooms reduce uptake of toxic methylmercury in freshwater food webs." *PNAS*, 99: 4419–4423.

Pier, M., Dawn, A.A. Betts-Piper, C.C. Knowlton, B.A. Zeeb and K.J. Reimer, 2002a. "Redistribution of Polychlorinated Biphenyls from a Local Point Source: Terrestrial soil, Vascular Plants and Freshwater Sediment as Indicators." *The Science of the Total Environment,* In press.

Pier, M.D., B.A. Zeeb, and K.J. Reimer, 2002b. "Patterns of contamination among vascular plants exposed to local sources of polychlorinated biphenyls in the Canadian Arctic and Subarctic." *The Science of the Total Environment,* 297: 215–227.

PMRA (Pesticide Management Regulatory Agency), 2000. *POINTS*, Issue 8, January 19, 2000. Health Canada, Ottawa, ON.

Poissant, L., 1998. "Preferred source of atmospheric total gaseous mercury in the St-Lawrence river valley." *Atmospheric Environment*, 33: 2537–2547.

Poissant, L., 2000. "Atmospheric mercury transport, oxidation and fallout in northern Quebec (Nunavik): an important potential route for contamination." In: *Synopsis of Research Conducted under the 1999–2000. Northern Contaminants Program*. S. Kalhok, (ed.). Indian and Northern Affairs Canada: Ottawa, ON.

Poissant, L., 2001. "Atmospheric mercury transport, oxidation and fallout in northern Quebec (Nunavik): an important potential route for contamination." In: *Synopsis of Research Conducted under the 2000–2001 Northern Contaminants Program*. S. Kalhok, (ed.). Indian and Northern Affairs Canada: Ottawa, ON.

Poissant, L., and J.F. Kogavnjak, 1996, "Fate and Atmospheric Concentrations of α- and γ-Hexachlorocyclohexane in Quebec, Canada." *Environmental Science and Technology*, 30: 845–851.

Poissant, L., J-F. Kogavnjak, and B. Harvey, 1994, "Atmospheric Contribution to the St. Lawrence River Pollution (Quebec,Canada): On-Going Research. Internal Report." Environment Canada, (Quebec region), 13 pp.

Poissant, L., M. Pilote, M. Amyot, and J.D. Lalonde, 2001. "Atmospheric mercury depletion events, fall-out and snow-air exchanges in northern environment." Submitted to *Geophysical Research Letters.*

Poliakova, O.V., A.T. Lebedev, and O. Hanninen, 2000. "Organic pollutants in snow of urban and rural Russia and Finland." *Environmental Toxicology and Chem*istry, 75: 181–194.

Polyakov, I.V., and M.A. Johnson, 2000. "Arctic decadal and interdecadal variability." *Geophysical Research Letters,* 27: 4097–4100.

Predavec, M., C.J. Krebs, K. Danell, and R. Hyndman, 2001. "Cycles and synchrony in the collared lemming (*Dicrostonyx groenlandicus*) in Arctic north America." *Oecologia,* 126: 216–224.

Prinsenberg, S.J., 1991. "Effects of hydro-electric projects on Hudson Bay's marine and ice environments." *James Bay Publication Series* Paper No. 2: 1–8.

Proshutinsky, A.Y., and M.A. Johnson, 1997. "Two circulation regimes of the wind-driven Arctic Ocean." *Journal of Geophysical Research*, 102: 12493–12514.

Proshutinsky, A., V. Pavlov, and R.H. Bourke, 2001. "Sea level rise in the Arctic Ocean." *Geophysical Research Letters,* 28: 2237–2240.

Quadfasel, D., A. Sy, D. Wells, and A. Tunik, 1991. "Warming in the Arctic." *Nature* 350: 385.

Quartier, R., and U. Müller-Herold, 2000. "On secondary spatial ranges of transformation products in the environment." *Ecological Modeling,* 135: 187–198.

Rahmstorf, S., and R. Alley, 2002. "Stochastic resonance in glacial climate." *Eos, Transactions of the American Geophysical Union,* 83: 129, 135.

RAIPON/AMAP/GEF Project, 2001. "Persistent Toxic Substances (PTS), Food Security and Indigenous Peoples of the Russian North." Contact Vitaly Kimstach, AMAP Secretariat, Olso, Norway.

Rajendran, R.B., V.K. Venugopalan, and R. Ramesh, 1999. "Pesticide residues in air from coastal environment, south India." *Chemosphere*, 39: 1699–1706.

Ramdahl, T., 1983. "Retene-a molecular marker of wood combustion in ambient air." *Nature*, 306: 580–582.

Ramesh, A., S. Tanabe, R. Tatsukawa, A.N. Subramanian, S. Palanichamy, D. Mohan, and V.K. Venugopalan, 1989. "Seasonal variations of organochlorine insecticide residues in air from Porto Novo, South India." *Environmental Pollution,* 62: 213–222.

Ramseier, R.O., C. Garrity, E. Bauerfeind, and R. Peinert, 1999. "Sea-ice impact on long-term particle flux in the Greenland Sea's Is Odden-Nordbukta region, 1985–1996." *Journal of Geophysical Research*, 104: 5329–5343.

Rapaport, R.A., and S.J. Eisenreich, 1988. "Historical atmospheric inputs of high molecular weight chlorinated hydrocarbons to eastern North America." *Environmental Science and Technology*, 22: 931–947.

Rapaport, R.A., and S.J. Eisenreich, 1986. "Atmospheric Deposition of Toxaphene to Eastern North America Derived from Peat Accumulation." *Atmospheric Environment,* 20: 2367–2379.

Ravenschlag, K., K. Sahm, and R. Amann, 2001. "Quantitative molecular analysis of the microbial community in marine Arctic sediments (Svalbard)." *Applied Environmental Microbiology*, 67: 387–395.

Rawn, D.F.K., W.L. Lockhart, P. Wilkinson, D.A. Savoie, B. Rosenberg, and D.C.G. Muir, 2001. "Historical contamination of Yukon Sediments by PCBs and organochlorine pesticides: Influence of local sources and watershed characteristics." *The Science of the Total Environment,* 280: 17–37.

Reimer, K.J., Ollson, C.A., and I. Koch, 2003. "An Approach for Characterizing Arsenic Sources and Risk at Contaminated Sites: Application to Gold Mining Sites in Yellowknife, NWT Canada." In *ACS Symposium Series No. 835 Biogeochemistry of Environmentally Important Trace Elements,* Y. Cai and O.C. Braids (Eds.), Chapter 13.

Reimnitz, E., L. Marincovich, M. McCormick, and W.M. Briggs. 1992. "Suspension freezing of bottom sediment and biota in the Northwest Passage and implications for Arctic Ocean sedimentation." *Canadian Journal of Earth Sciences,* 29: 693–703.

Reimnitz, E., M. McCormick, K. McDougall, and E. Brouwers, 1993. "Sediment export by ice rafting from a coastal polynya, Arctic Alaska, USA." *Arctic and Alpine Research,* 25: 83–98.

Reimnitz, E., D. Dethleff, and D. Nürnberg, 1994. "Contrasts in Arctic shelf sea-ice regimes and some implications: Beaufort Sea versus Laptev Sea." *Marine Geology*, 119: 215–225.

Rice, C.P., and M.S. Evans, 1984. In: "Toxic Contaminants in the Great Lakes," J.O. Nriagu and M.S. Simmons (eds.) Adv. *Environmental Science and Technology*, 14: 163–194.

Rice, C.P., and V.V. Shigaev, 1997. "Spatial distribution of hexachlorocyclohexane isomers in the Bering and Chukchi sea shelf ecosystem." *Environmental Science and Technology*, 31: 2092–2097.

Rice, C.P., S.M. Chernyak, C.J. Hapeman, and S. Bilboulian, 1997. "Air-water distribution of the endosulfan isomers." *Journal of Environmental Quality.*, 26: 1101–1106.

Richter, A.F., A. Wittrock, M. Eisenger, and J.P. Burrow, 1998. "GOME Observation of tropospheric BrO in Northern hemispheric Spring and Summer of 1997." *Geophysical Research Letters*, 25: 2683–2686.

Riedlinger, D, 2001. "Responding to climate change in northern communities: Impacts and adaptations." *Arctic*, 54: 96–98.

Rigor, I.G., R.L. Colony, and S. Martin, 2000. "Variations in surface air temperature observations in the Arctic, 1979–97." *Journal of Climate,* 13: 896–914.

Rigor, I.G., J.M. Wallace, and R. Colony, 2002. "On the response of sea ice to the Arctic Oscillation." *Journal of Climate,* 15: 2648–2663.

Risklogic. 2002. "Assessment of Human Health Risks posed by Arsenic Contamination in Yellowknife, NWT." Prepared for the Yellowknife Soil Arsenic Remediation Committee (YSARC).

Roach, A.T., K. Aagaard, C.H. Pease, S.A. Salo, T. Weingartner, V. Pavlov, and M. Kulakov, 1995. "Direct measurements of transport and water properties through the Bering Strait." *Journal of Geophysical Research,* 100: 18,443–18,457.

Roach, P.S. Mitchell, and G. Stern, 1999. "Carcross organochlorine contamination report." Whitehorse: Indian and Northern Affairs Canada. June, 21 pp.

Rodan, D., D.W. Pennington, N. Eckley, and R.S. Boethling, 1999. " Screening for persistent organic pollutants techniques to provide a scientific basis for POPs criteria in international negotiations." *Environmental Science and Technology,* 33: 3482–3488.

Rodriguez, R.R., N.T.Basta, S.W. Casteel, and L.W. Pace, 1999. "An in vitro gastrointestinal method to estimate bioavailable arsenic in contaminated soils and solid media." *Environmental Science and Technology,* 33(4): 742–749.

Ruddiman, W.F., 2000. "Earth's Climate: Past and Future." W.H. Freeman and Company, New York.

Rosman, K.J.R., W. Chisolm, C.F. Boutron, J.P. Candelone, and U. Gorlach, 1993. "Isotopic evidence for the source of lead in Greenland snows since the late 1960s." *Nature,* 362: 333–334.

Rosman, K.J.R., W. Chisholm, S. Hong, J.P. Candelone, and C.F. Boutron,1997. "Lead from Carthaginian and Roman spanish mines isotopically identified in Greenland ice dated from 600 B.C. to 300 A.D." *Environmental Science and Technology*, 31: 3413–3416.

Ross, P.S., 2002. "The role of immunotoxic environmental contaminants in facilitating the emergence of infectious diseases in marine mammals." *Human and Ecological Risk Assessment,* 8: 277–292.

Ross, P.S., J.G. Vos, L.S. Birnbaum, and A.D.M.E. Osterhaus, 2000. "PCBs are a health risk for humans and wildlife." *Science*, 289: 1878–1879.

Rothrock, D.A., Y. Yu, and G.A. Maykut, 1999. "Thinning of the Arctic sea-ice cover." *Geophysical Research Letters,* 26: 3469–3472.

Rouse, W.R., Douglas, R.E. Hecky, A.E. Hershey, G.W. Kling, L. Lesack, P. Marsh, M. McDonald, B.J. Nicholson, N.T. Roulet, and J.P. Smol, 1997. "Effects of climate change on the fresh waters of Arctic and sub-Arctic North America." *Hydrological Processes,* 11: 873–902.

Royer T.C., 1982. "Coastal freshwater discharge in the N.E. Pacific." *Journal of Geophysical Research,* 87(C3): 2017–2021.

Ruby, M.V., Davis, A., Schoof, R., Eberla, S., Sellstone, C.M, 1996. "Estimation of lead and arsenic bioavailabilty using a physiologically based extraction test." *Environmental Science and Technology*, 30(2): 422–430.

Rudels, B., E.P. Jones, L.G. Anderson, and G. Katner, 1994. "On the intermediate depth waters of the Arctic Ocean." pp. 33–46 In: *The Polar Oceans and Their Role in Shaping the Global Environment.* R.D.M. O.M. Johannessen, and J.E. Overland, (ed.) American Geophysical Union, Washington, D.C.

Rugh, D.J., M.M. Muto, S.E. Moore, and D.P. DeMaster, 1999. "Status review of the eastern North Pacific stock of gray whales." NOAA Technical Memorandum, NMFS-AFSC-103, 96 pp.

Ryan, J., Patry, B., 2000. "Determination of Brominated Diphenyl Ethers (BDEs) and Levels in Canadian Human Milk." *Organohalogen Compounds*, 47: 57–60.

Saager, P.M., and H.J.W. deBaar, 1993. "Limitations to the quantitative application of Cd as a paleoceanographic tracer, based on results of a multi-box model (MENU) and statistical considerations." *Global and Planetary Change,* 8: 69–92.

Saar, R.A., 2000. "The unbearable capriciousness of Bering." *Science*, 287: 1388–1389.

Sagemann, J., B. Jørgensen, and O. Greeff, 1998. "Temperature dependence and rates of sulfate reduction in cold sediments of Svalbard, Arctic Ocean." *Geomicrobiology Journal,*15: 85–100.

Sahm, K., and U.G. Berninger, 1998. "Abundance, vertical distribution, and community structure of benthic prokayotes from permanently cold marine sediments (Svalbard, Arctic Ocean)." *Marine Ecology Progress Series,* 165: 71–80.

Sahsuvar, L., P.A. Helm, L.M. Jantunen, and T.F. Bidleman, 2002. "Henry's law constants for α, β, and γ-hexachlorocyclohexanes (HCHs) as a function of temperature and gas exchange in Arctic regions." *Atmospheric Environment*, submitted.

Sakshaug, E., and D. Slagstad, 1992. "Sea ice and wind: Effects on primary productivity in the Barents Sea." *Atmosphere-Ocean,* 30: 579–591.

Sakshaug, E., C.C.E. Hopkins, and N.A. Øritsland, 1991. "Proceedings of the Pro Mare Symposium on Polar Marine Ecology, Trondheim, 12–16 May, 1990." *Polar Research* : 1–610.

Sakshaug, E., A. Bjørge, B. Gulliksen, H. Loeng, and F. Mehlum, (eds.), 1994. *Økosystem Barentshavet.* Univesitetsforlaget, Oslo.

Salbu, B., A.I. Nikitin,P. Strand, G.C. Christensen, V.B. Chumichev, B.Lind, H. Fjelldal, T.D. Bergan, A.L.Rudjord, M. Sickel, N.K. Valetova, and L. Foyn, 1997. "Radioactive contamination from dumped nuclear waste in the Kara Sea — results from the joint Russian–Norwegian expeditions in 1992–1994." *The Science of the Total Environment*, 202: 185–198.

Sangster, D.F., P.M. Outridge, and W.J. Davis, 2000. "Stable lead isotope characteristics of lead ore deposits of environmental significance." *Environmental Reviews,* 8: 115–147.

Savinova, T.N., A. Polder, G.W. Gabrielson, and J.U. Skaare, 1995. "Chlorinated hydrocarbons in seabirds from the Barents Sea area." *The Science of the Total Environment,* 160/161: 497–504.

Sayles, F.L., H.D. Livingston, and G.P. Panteleyev, 1997. "The history and source of particulate ^{137}Cs and $^{239,240}Pu$ deposition in sediments of the Ob River Delta, Siberia." *The Science of the Total Environment,* 202: 25–41.

Schell, D.M., 2000. "Declining carrying capacity in the Bering Sea: Isotopic evidence from whale baleen." *Limnology and Oceanography,* 45: 459–462.

Scheringer, M., 1996. "Persistence and spatial range as endpoints of an exposure-based assessment of organic chemicals." *Environmental Science and Technology,* 30: 1652–1659.

Scheringer, M., 1997. "Characterization of the environmental distribution behavior of organic chemicals by means of persistence and spatial range." *Environmental Science and Technology,* 31: 2891–2897.

Scheringer, M., and F. Wania, 2003. "Multimedia models of global transport and fate of persistent organic pollutants." In: Fiedler, H. (Ed.) *The Handbook of Environmental Chemistry*, Vol. 3, Part O, Springer–Verlag, Berlin, Heidelberg, pp. 238–269.

Scheringer, M., F. Wegmann, K. Fenner, and K. Hungerbühler, 2000. "Investigation of the cold condensation of persistent organic pollutants with a global multimedia fate model" *Environmental Science and Technology,* 34: 1842–1850.

Scheringer, M., K. Hungerbühler, and M. Matthies, 2001a. "The spatial scale of organic chemicals in multimedia fate modelling — Recent developments and significance for chemical assessment." *Environmental Science and Pollution Research*, 8: 150–155 .

Scheringer, M., D.H. Bennett, T.E. McKone, and K. Hungerbühler, 2001b. "Relationship between persistence and spatial range of environmental chemicals." pp. 52–63. In: *American Chemical Society Symposium Series No. 773 Persistent, Bioaccumulative, and Toxic Chemicals*, Volume II, R. Lipnick, B Jansson, D. Mackay, and M. Petreas (Eds.).

Schindler, D.W., 1997. "Widespread effects of climate warming on freshwater ecosystems in North America." *Hydrological Processes*, 11: 1043–1067.

Schlabach, M., A. Biseth, and H. Gundersen, 1996. "Sampling and Measurement of PCDD/PCDF and non-*ortho* PCB in Arctic Air at Ny-Ålesund, Spitsbergen." *Organohalogen Compounds*, 28: 325–329.

Schlosser, P., R. Newton, B. Ekwurzel, S. Khatiwala, R. Mortlock, and R. Fairbanks, 2002. "Decrease of river runoff in the upper waters of the Eurasian Basin, Arctic Ocean, between 1991 and 1996: evidence from d18O data." *Geophysical Research Letters,* submitted.

Schreitmüller, J., K. Ballschmiter, 1995. "Air-water equilibrium of hexachlorocyclohexanes and chloromethoxybenzenes in the North and South Atlantic." *Environmental Science and Technology*, 29: 207–215.

Schroeder, W.H., 1996. "Atmospheric mercury measurements in the Canadian Arctic." In: J.L. Murray, R.G. Shearer, and S.L. Han (eds.), *Synopsis of Research Conducted Under the 1994/95 Northern Contaminants Program.* Environmental Studies No. 73. Ottawa: Indian and Northern Affairs Canada. pp. 21–30.

Schroeder, W.H., and R. Jackson, 1984. "An instrumental analytical technique for vapour-phase mercury species in the air." *Chemosphere,* 13: 1044–1051.

Schroeder, W.H., and J. Munthe, 1998. "Atmospheric mercury — an overview." *Atmospheric Environment,* 32(5): 809–822.

Schroeder, W.H., G.Yarwood, and H. Niki, 1991. "Transformation processes involving mercury species in the atmosphere-results from a literature survey." *Water, Air , and Soil Pollution*, 56: 653–666.

Schroeder, W.H., R. Ebinghaus, M. Shoeib, K. Timoschenko, and L.A. Barrie, 1995a. "Atmospheric mercury measurements in the Northern Hemisphere from 56°E to 82.5°N latitude." *Water, Air , and Soil Pollution,* 80: 1227–1236.

Schroeder, W.H., C. Lamborg, D. Schneeberger, W.F. Fitzgerald, and B. Srivastava, 1995b. "Comparison of a manual method and an automated analyser for determining total gaseous mercury in ambient air." In: *Proceedings of the International Conference for Heavy Metals in the Environment,* R.D. Wilkin, U. Förstner,and A. Knöchel, (eds), 2: 57–60.

Schroeder, W.H., K.G. Anlauf, L.A. Barrie, J.Y. Lu, A. Steffen, and D.R. Schneeberger, 1998. "Arctic springtime depletion of mercury." *Nature*, 394: 331–332.

Schroeder, W.H., K.G. Anlauf, L.A.Barrie, A. Steffen, and J.Y. Lu, 1999a. "Depletion of Mercury Vapour in the Arctic Troposphere after Polar Sunrise." In: *Proceedings of the 1998 EUROTRAC Symposium on Transport & Chemical Transformation in the Troposphere*; Garmisch–Partenkirchen, pp. 358–362.

Schroeder, W.H., A. Steffen, J.Y. Lu, D.R. Schneeberger, C. Scherz, K. Uyede, C. Lamborg, and L.A. Barrie, 1999b. "Mercury measurements at Alert." pp. 69–76 In: *Synopsis of Research Conducted under the 1997–1998 Northern Contaminants Program.* J. Jensen, (ed.) Indian and Northern Affairs Canada: Ottawa, ON.

Schroeder, W.H., A. Steffen, J.Y. Lu, P. Blanchard, and L.A. Barrie, 1999c. "Mercury measurements at Alert."pp. 49–53 In: *Synopsis of Research Conducted under the 1998–1999 Northern Contaminants Program.* S. Kalhok, (ed.). Indian and Northern Affairs Canada: Ottawa, ON.

Schroeder, W.H., A. Steffen, J.Y. Lu, and L.A. Barrie, 2000. "Mercury measurements at Alert." pp. 137–141. In: *Synopsis of Research Conducted under the 1999–2000 Northern Contaminants Program.* S. Kalhok, (ed.). Indian and Northern Affairs Canada: Ottawa, ON.

Schroeder, W.H., A. Steffen, G. Lawson, and W. Strachan, 2001a. "Mercury measurements at Alert." pp. 130–135 In: *Synopsis of Research Conducted under the 2000–2001 Northern Contaminants Program.* S. Kalhok, (ed.). Indian and Northern Affairs Canada: Ottawa, ON.

Schroeder, W.H., A. Steffen, G. Lawson, and W. Strachan, 2001b. "Mercury measurements at Amderma, Russia." pp. 136–137. In: *Synopsis of Research Conducted under the 2000–2001 Northern Contaminants Program.* S. Kalhok (ed.). Indian and Northern Affairs Canada: Ottawa, ON.

Scoresby, W., 1969. *An Account of the Arctic Regions.* David and Charles Reprints, Trowbridge, Great Britain.

Scott, K.J., 2001. "Bioavailable mercury in arctic snow determined by light-emitting mer-lux bioreporter." *Arctic,* 54: 92–101.

Seiber, J.N., T. Cahill, and J.E. Woodrow, 2001. "Distribution of some organic contaminants in selected western USA. atmsopheric and surface waters." *Abstr. Pap. — Am. Chem.* Soc., 221st.

Sellström, U., B. Jansson, A. Kierkegaard, C. De Wit, T. Odsjö, and M. Olsson. 1993. "Polybrominated diphenyl ethers (PBDE) in biological samples from the Swedish environment." *Chemosphere*, 26: 1703–1718.

Semiletov, I.P., N.I. Savelieva, G.E. Weller, I.I. Pipko, S.P. Pugach, A.Y. Gukov, and L. N. Vasilevskaya, 2000. "The dispersion of Siberian river flows into coastal waters: meteorological, hydrological and hydrochemical aspects." pp. 323–366. In: *The Freshwater Budget of the Arctic Ocean.* E.L. Lewis, E.P. Jones, P. Lemke, T.D. Prowse, and P. Wadhams, (eds.). Kluwer Academic Publishers, London.

Semkin, R., 1996. "Processes and fluxes of contaminants in aquatic systems — 1994/95." In: *Synposis fo Research Conducted under the 1994/95 Northern Contaminants Program.* QS-8519-000-EF-A1, Catal. NO. 71-19/83-1995E. J.L. Murray, R.G. Shearer, and S.L. Han (eds.). pp. 105–118.

Semkin, R., 1997. "Processes and fluxes of contaminants in aquatic systems." In: *Synopsis of Research Conducted under the 1995–1997 Northern Contaminants Program.* Environmental Studies No. 74. pp. 127–135.

Semkin, R., G. Mierle, and R.J. Neureuther, 2003. "Hydrochemistry of the Amituk Lake watershed: mercury dynamics in the high Arctic." In preparation.

SENES Consultants Ltd, 2001. "Preliminary Assessment of Environmental Conditions at the former Port Radium Mine Site." Richmond Hill, Ontario.

SENES Consultants Ltd, 2002a. "Port Radium Site Assessment — Human Health and Ecological Risk Assessment." Draft Report. Richmond Hill, Ontario.

SENES Consultants Ltd, 2002b. "Port Radium Site Assessment — Summary Report on Site Conditions and Decommissioning Considerations." Draft Report. Richmond Hill, Ontario.

Sericano, J.L., J.M. Brooks, M.A. Champ, M.C. Kennicutt II, and V.V. Makeyev, 2001. "Trace contaminant concentrations in the Kara Sea and its adjacent rivers, Russia." *Marine Pollution Bulletin*, 42: 1017–1030.

Serreze, M.C., J.A. Maslanik, J.R. Key, and R.F. Kokaly, 1995. "Diagnosis of the record minimum in Arctic sea ice area during 1990 and associated snow cover extremes." *Geophysical Research Letters*, 22: 2183–2186.

Serreze, M.C., J.E. Walsh, F.S. I. Chapin, T. Osterkamp, M. Dyurgerov, V. Romanovsky, W.C. Oechel, J. Morison, T. Zhang, and R.G. Barry, 2000. "Observational evidence of recent change in the northern high-latitude environment." *Climatic Change,* 46: 159–207.

Seth, R., D. Mackay, and J. Muncke. 1999. "Estimating the organic carbon partition coefficient and its variability for hydrophobic chemicals." *Environmental Science and Technology,* 33: 2390–2394.

Sharma, S., J.R. Brook, H. Cachier, J. Chow, A. Gaudenzi, and G. Lu, 2002. "Light absorption and thermal measurements of black carbon in different regions of Canada." *Journal of Geophysical Research, (submitted).*

Shen, G.T., E.A. Boyle, and D.W. Lea, 1987. "Cadmium in corals as a tracer of historical upwelling and industrial fallout." *Nature,* 328: 794–796.

Sherrell, R.M., E.A. Boyle, K.K. Falkner, and N.R. Harris, 2000. "Temporal variability of Cd, Pb and Pb isotope deposition in central Greenland snow." *Geochemisty, Geophysics, Geosystems*, 1: 29 pp.

Sherwood, G.D., J.B. Rasmussen, D.J. Rowan, J. Brodeur, and A. Hontela, 2000. "Bioenergetic costs of heavy metal exposure in yellow perch (*Perca flavescens*): *in-situ* estimates with a radiotracer (137Cs) technique." *Canadian Journal of Fisheries and Aquatic Science*, 57: 441–450.

Sherwood, G.D., J. Kovecses, A. Iles, J. Rasmussen, A. Gravel, H. Levesque, A. Honteal, A. Giguere, L. Kraemer, and P. Campbell, 2001."The bigger the bait." *MITE–RN News*, Winter: 4.

Shiklomanov, I.A., and B.G. Skakalsky, 1994. "Studying water, sediment and contaminant runoff of Siberian rivers. Modern status and prospects." *Arctic Research of the U S A,* Arlington, 8: 295–306.

Shiklomanov, I. A., R. B. Lammers, B. J. Peterson, and C. J. Vörösmarty, 2000. "The dynamics of river water inflow to the Arctic Ocean." pp. 281–296. In: *The Freshwater Budget of the Arctic Ocean.* E.L. Lewis, E.P. Jones, P. Lemke, T.D. Prowse, and P. Wadhams, (eds.). Kluwer Academic Publishers, Netherlands.

Shindell, D.T., R.L. Miller, G.A. Schmidt, and L. Pandolfo, 1999. "Simulation of recent northern winter climate trends by greenhouse-gas forcing." *Nature,* 399: 452–455.

Schindler, D.W., G.E. Welch, J. Kalff, G.J. Brunskill, and N.J. Kritsch, 1974. "Physical and chemical limnology of Char Lake, Cornwallis Island (75°N Lat.)." *J. Fish. Re. Board Canada,* 31: 585–607.

Shiu, W-Y., F. Wania, H. Hung, and D. Mackay, 1997. "Temperature dependence of aqueous solubility of selected chlorobenzenes, polychlorinated biphenyls, and dibenzofuran." *Journal of Chemical and Engineering Data*, 42: 293–297.

Shiu, Y-S., and K-C. Ma, 2000a. "Temperature dependence of physicochemical properties of selected chemicals of environmental interest. 1. Mononuclear and polynuclear aromatic hydrocarbons." *Journal of Physical Chemical Reference Data*, 29: 41–130.

Shiu, Y-S., and K-C. Ma, 2000b. "Temperature dependence of physicochemical properties of selected chemicals of environmental interest. 2. Chlorobenzenes, polychlorinated biphenyls, polychlorinated dibenzo-*p*-dioxins, and dibenzofurans." *Journal of Physical Chemical Reference Data*, 29: 387–462.

Shoeib, M.T., and T.Harner, 2002. "Using measured octanol–air partition coefficients (KOA) to explain environmental partitioning of organochlorine pesticides." *Environmental Toxicology and Chemistry*, 21: 984–990.

Shoeib, M, K.A. Brice, R.M., and Hoff, 1999. Airborne concentrations of toxaphene congeners at Point Petre (Ontario) using gas chromatography — electron capture negative ion mass spectrometry (GC-ECNI-MS)." *Chemosphere,* 45, 849–871.

Showstack, R, 2001. "IPCC report cautiously warns of potentially dramatic climate change impacts." *Eos, Transactions of the American Geophysical Union*, 82: 113, 114, 120.

Schulz, D.E., G. Petrick, and J.C. Duinker, 1989. "Complete characterization of polychlorinated biphenyl congeners in commercial aroclor and clophen mixtures by multidimensional gas chromatography-electron capture detection." *Environmental Science and Technology,* 23: 852–859.

Schultz–Bull, D.E., G. Petrick, R.Bruhn, and J.C. Duinker, 1998. Chlorobiphenyls (PCB) and PAHs in water masses of the northern North Atlantic. Marine Chemistry, 61: 101–114.

Siegel, F.R., J.H. Kravitz, and J.J. Galasso, 2001. "Arsenic and mercury contamination in 31 cores taken in 1965, St. Anna Trough, Kara Sea, Arctic Ocean." *Environmental Geology,* 40: 528–542.

Siegenthaler, U., and F. Joos, 1992. "Use of a simple model for studying oceanic tracer distribution and the global carbon cycle." *Tellus,* 44B: 186–207.

Simcik, M.F., I. Basu, C.W. Sweet, and R.A. Hites, 1999. "Temperature dependence and temporal trends of polychlorinated biphenyl congeners in the Great Lakes atmosphere." *Environmental Science and Technology*, 33: 1991–1995.

Simonich, S.L., and R. Hites, 1994. "Importance of vegetation in removing polycyclic aromatic hydrocarbons from the atmosphere." *Nature,* 370: 49–51.

Simpson, C.D., R.J. Wilcock, T.J. Smith, A.L. Wilkins, and A.G. Langdon, 1995. "Determination of octanol-water partiton coefficients for the major components of technical chlordane." *Bulletin of Environmental Contamination and Toxicology*, 55: 149–153.

Singh, P.P., R.S. Battu, and R.L. Kalra, 1988. "Insecticide residues in wheat grains and straw arising from their storage in premises treated with BHC and DDT under malaria control program." *Bulletin of Environmental Contamination and Toxicology*, 40: 696–702

Sirois, A., and L.A. Barrie, 1999. "Arctic lower tropospheric aerosol trends and composition at Alert, Canada: 1980–1995." *Journal of Geophysical Research,* 104(D9): 11599–11618.

Sjödin, A., 2000. Doctoral Dissertation, Stockholm University, Stockholm, Sweden.

Skaare, J.U., A. Bernhoft, Ø. Wiig, K.R. Norum, D.M. Eide, and A.E. Derocher, 2001. "Relationships between plasma levels of organochlorines, retinol and thyroid hormones from polar bears (*Ursus maritimus*) at Svalbard." *Journal of Toxicology and Environmental Health*, 62: 227–241.

Skinner, W.R., R.L. Jefferies, T.J. Carleton, R.F. Rockwell, and K.F. Abraham,1998. "Prediction of reproductive success and failure in lesser snow geese based on early season climate variables." *Global Change Biology,* 4: 3–16.

Skov, H., J.H. Christensen, M.E. Goodsite, M.C. Petersen, N. Zeuthen-Heidam, G. Geernaert, and J. Olsen, 2001. "Near surface conversion and fluxes of gaseous elemental mercury to reactive gaseous mercry in the Arctic." In: *Dynamics and chemistry of atmospheric mercury;* Danish contribution to EUROTRAC MEPOP report.

Slemr, F, W. Junkermann, R.W.H. Schmidt, and R. Sladkovic, 1995. "Indication of change in global and regional trends of atmospheric mercury concentrations." *Geophysical Research Letters*, 22: 2143–2146.

Slorach, A.S., and R. Vaz ,1983. "Assessment of human exposure to selected organochlorine compounds through biological monitoring." *Global Environment Monitoring System (GEMS) UNEP/ WHO.* Prepared by the Swedish National Food Administration, Uppsala, Sweden.

Smith, D., 1998. "Recent increase in the length of the melt season of perennial Arctic sea ice." *Geophysical Research Letters*, 25: 655–658.

Smith, J.N., and K.M. Ellis, 1995. "Radionuclide tracer profiles at the CESAR Ice Station and Canadian Ice Island in the Western Arctic Ocean." *Deep-Sea Research II*, 42(6): 1447–1470.

Smith, J.N., K.M. Ellis, and E.P. Jones, 1990. "Cesium-137 transport into the Arctic Ocean through Fram Strait." *Journal of Geophysical Research*, 95(C2): 1693–1701.

Smith, J.N., K.M. Ellis, K. Naes, S. Dahle, and D. Matishov, 1995. "Sedimentation and mixing rates of fallout radionuclides in Barents Sea sediments off Novaya Zemlya." *Deep-Sea Research II*, 42 (6): 1471–1493.

Smith, J.N., K.M. Ellis, and L.R. Kilius,1998. "^{129}I and ^{137}Cs tracer measurements in the Arctic Ocean." *Deep-Sea Research I,* 45: 959–984.

Smith, J.N., K.M. Ellis, and T.M. Boyd, 1999. "Circulation features in the Central Arctic Ocean revealed by nuclear fuel reprocessing tracers from SCICEX 95 and 96, 1999." *Journal of Geophysical Research*, 104, No. C12,29, 663–29,677.

Smith, J.N., K.M. Ellis, L. Polyak, G. Ivanov, S.L. Forman, and S.B. Moran, 2000. "$^{239,240}Pu$ transport into the Arctic Ocean from underwater nuclear tests in Chernay Bay, Novaya Zemlya." *Continental Shelf Research,* 20: 255–279.

Smith, T.G., and L. A. Harwood, 2001."Observations of neonate ringed seals, *Phoca hispida*, after early break-up of the sea ice in Prince Albert Sound, Northwest Territories, Canada, spring 1998." *Polar Biology,* 24: 215–219.

Sobek, A., and Ö. Gustafsson, Unpublished data. Institute of Applied Env. Res. (ITM), Stockholm University, Stockholm

Sokolov A.A., 1952. "Hvdroqraphv of the USSR. HYDROMETEOIZDAT, Leningrad."

Solomon, S.M., 2001. "The distribution and mobility of sediments in Saglek Bay at the North Warning System Site (LAB–2), Labrador." Geological Survey of Canada (Atlantic), draft report. 33 pp.

Sowers, T., and M. Bender, 1995. "Climate records covering the last deglaciation." *Science,* 269: 210–214.

SRI International, 1994. "Pesticides, the Chemical Economics Handbook." February.

Spiegelstein, M., 2000. "Brominated Compounds, Addressing the gap in scientific knowledge." In: *Proceedings of BSEF Workshop on Polybrominated Diphenyl Ethers (PBDEs): Global Distribution of PBDEs*. J. de Boer, P.E.G. Leonards, J.P. Boon, and R.J. Law, (eds.). IJmuiden, Netherlands.

Springer, A.M., 1998. "Is it all climate change? Why marine bird and mammal populations fluctuate in the North Pacific." pp. 109–119. In: *Biotic impacts of extratropical climate variability in the Pacific, Proceedings 'Aha Huliko'a Hawaiin Winter Workshop*. G. Holloway, P. Muller, and D. Henderson, (eds.) University of Hawaii, Honalulu.

Stabeno, P., and J.E. Overland, 2001. "Bering Sea shifts toward an earlier spring transition." *Eos, Transactions of the American Geophysical Union,* 82: 317, 321.

Stafford, C.J., 1983. "Halogenated diphenyl ethers identified in avian tissues and eggs by GC/MS." *Chemosphere*, 12: 1487–1495.

Stanley, J.B., P.F. Schuster, M.M. Reddy, D.A. Roth, H.E. Taylor, and G.R. Aiken, 2002. "Mercury on the move during snowmelt in Vermont."*Eos, Transactions of the American Geophysical Union*, 83: 45, 47–48.

Staudinger, J., and P.V. Roberts, 2001. "A critical compilation of Henry's law constant temperature dependence relations for organic compounds in dilute aqueous solution." *Chemosphere*, 44, 561–576.

Steele, M., and T. Boyd, 1998."Retreat of the cold halocline layer in the Arctic Ocean." *Journal of Geophysical Research,* 103: 10419–10435.

Steffen, A., W.H. Schroeder, J. Bottenheim, J. Narayan, and J.D. Fuentes, 2002. "Atmospheric Mercury Concentrations: Measurements and Profiles near Snow and Ice Surfaces in the Canadian Arctic during Alert 2000." *Atmospheric Environment,* in press.

Stein, R., 2000. "Circum-Arctic river discharge and its geological record: an introduction." *International Journal of Earth Sciences,* 89: 447–449.

Stein, R., B. Boucsein, K. Fahl, T. Garcia de Oteyza, J. Knies, and F. Niessen, 2001. "Accumulation of particulate organic carbon at the Eurasian continental margin during late Quarternary times: controlling mechanisms and paleoenvironmental signficance." *Global Planetary Change,* 31: 87–102.

Stein, R., C. Schubert, C. Vogt, and D. Fütterer, 1994. "Stable isotope stratigraphy, sedimentation rates, and salinity changes in the Latest Pleistocene to Holocene eastern central Arctic Ocean." *Marine Geology*, 119: 333–355.

Stern, G.A., M.D. Loewen, B.M. Miskimmin, D.C.G. Muir, and J.B. Westmore, 1995. "Characterization of two major toxaphene components in treated lake sediment." *Environmental Science and Technology*, 30: 2251–2258

Stern, G.A., C.J. Halsall, L.A. Barrie, D.C.G. Muir, P. Fellin, B. Rosenberg, F. Ya. Rovinsky, E.Ya. Kononov, and B. Pastuhov, 1997. "Polychlorinated biphenyls in Arctic air. 1. Temporal and spatial trends: 1992–1994." *Environmental Science and Technology*, 31: 3619–3628.

Stern, G.A., P. Roach, and M Diamond, 2000. "Temporal and spatial trends of mercury in fish from Yukon lakes and sediments cores from Lake Laberge." pp. 243–250. In: *Synopsis of Research Conducted under the 1998/99 Northern Contaminants Program*. S. Kalhok, (ed.), Indian and Northern Affairs Canada, Ottawa, ON.

Stern, G.A., P. Roach, M. Diamond, L. Lockhart, M. Ryan, P. Wilkinson, J. DeLaronde, G. Boila, S. Gewurtz, and R. Ricks, 2001b. "Temporal and spatial trends of mercury in fish from Yukon lakes and in sediment cores from Lake Laberge." pp. 243–250. In *Synopsis of Research Conducted under the 2000–2001 Northern Contaminants Program*. S. Kalhok (ed.), Indian Affairs and Northern Development, Ottawa, ON.

Stern, G.A., A. Fisk, L. Lockhart, 2002, (unpublished data). Department of Fisheries and Oceans, Freshwater Institute, 501 University Crescent, Winnipeg, Manitoba.

Stern, G.A, E. Braekevelt, P.A. Helm, W.L. Lockhart, B. Rosenberg, G.T. Tomy, P. Wilkinson, P.M. Outridge, R. MacNeeley, M.G. Ikonomou, and T.F. Bidleman, 2003. "Organohalogen contaminants in a laminated sediment core from Devon Island (Lake DV09)." Submitted to *Environmental Science and Technology*

Stewart, R.E., 2000. "The variable climate of the Mackenzie River Basin: Its water cycle and fresh water discharge." pp. 367–381 In: *The Freshwater Budget of the Arctic Ocean*. E.L. Lewis, E.P. Jones, P. Lemke, T.D. Prowse, and P. Wadhams, (eds.). Kluwer Academic Publishers, Netherlands.

Stewart, R.E., H.G. Leighton, P. Marsh, G.W.K. Moore, H. Ritchie, W.R. Rouse, E.D. Soulis, G.S. Strong, R.W. Crawford, and B. Kochtubajda, 1998. "The Mackenzie GEWEX study: The water and energy cycles of a major North American River Basin." *Bulletin of the American Meteorological Society*, 79: 2665–2683.

Stirling, I., 1997. "The importance of polynyas, ice edges, and leads to marine mammals and birds." *Journal of Marine Systems*, 10: 9–21.

Stirling, I., 2002. "Polar bears and seals in the eastern Beaufort Sea and Amundsen Gulf: A synthesis of population trends and ecological relationships over three decades." *Arctic,* 55 (Supp. 1), 59–76.

Stirling, I., and A.E. Derocher, 1993. "Possible impacts of climatic warming on polar bears." *Arctic,* 46: 240–245.

Stirling I., and N.J. Lunn, 1997. "Environmental fluctuations in arctic marine ecosystems as reflected by variability in reproduction of polar bears and ringed seals." In: *Ecology of Arctic Environments,* 13, S.J. Woodin, M. Marquiss, (eds.). The British Ecological Society, pp. 167–181.

Stirling, I., N.J. Lunn, and J. Iacozza, 1999. "Long-term trends in the population ecology of polar bears in western Hudson Bay in relation to climatic change." *Arctic,* 52: 294–306.

Stockwell, D.A., T.E. Whitledge, T. Rho, P.J. Stabeno, K.O. Coyle, and J.M. Napp., 2000. "Field observations in the southeastern Bering Sea during three years with extensive coccolithophorid blooms." *Eos, Transactions of the American Geophysical Union,* 80.

Strachan, W.M.J., A. Fisk, C.F. Teixeira, D.A. Burniston, and R. Norstrom, 2000. "PCBs and organochlorine pesticide concentrations in the water of the Canadian archipelago and other Arctic regions." In: *Workshop on Persistent Organic Pollutants in the Arctic: Human health and Environmental concerns.* Rovaneimi Finland, Jan 2000. Arctic Monitoring and Assessment Program. Abstract 13. Available from http://www.amap.no.

Strachan, W.M.J., D.A. Burniston, M. Williamson, and H. Bohdanowicz, 2001. "Spatial differences in persistent organochlorine pollutant concentrations between the Bering and Chukchi Seas (1993)." *Marine Pollution Bulletin,* 43: 133–142.

Strand, A., and Ø . Hov, 1996. "A model strategy for the simulation of chlorinated hydrocarbon distribution in the global environment." *Water, Air, and Soil Pollution,* 86, 283–316.

Strand, A., and Ø. Hov,1993. "A two-dimensional zonally averaged transport model including convective motions and a new strategy for the numerical solution." *Journal of Geophysical Research,* 98: 9023–9037.

Strand, P., B.J. Howard, A. Aarkrog, M. Balonov, Y. Tsaturov, J.M. Bewers, A. Salo, M. Sickel, R. Bergman, and K. Rissanen, 2002. "Radioactive contamination in the Arctic-sources, dose assessment and potential risks." *Journal of Environmental Radioactivity,* 60(1–2): 5–22.

Strandberg, B., N. Dodder, I. Basu, and R. Hites, 2001. "Concentration and spatial variation of polybrominated diphenyl ethers and other organohalogen compounds in Great Lakes air." *Environmental Science and Technology,* 13: 1494–1502.

Stringer, O.E., and R.B McMynn, 1960. "Three year's use of toxaphene as a fish toxicant in British Columbia." *The Canadian Fish Culturist,* 28: 37–44.

Stull, J.K., C.I. Haydock, and D.E. Montagne, 1986. "Effects of *Listriolobus pelodes* (Echiura) on coastal shelf benthic communities and sediments modified by a major California wastewater discharge." *Estuarine, Coastal and Shelf Science,* 22: 1–17.

Sturges, W.T., and L.A. Barrie, 1989. "Stable lead isotope ratios in Arctic aerosols: evidence for the origin of Arctic air pollution." *Atmospheric Environment,* 23: 2513–2519.

Sturges, W.T., J.F. Hopper, L.A. Barrie, and R.C. Schnell, 1993. "Stable lead isotope ratios in Alaskan Arctic aerosols." *Atmospheric Environment,* 27A: 2865–2871.

Sturm, M., J.P. McFaden, G.E. Liston, I. Chapin, F.S., J. Holmgren, and M. Walker, 2001. "Snow-shrub interactions in Arctic tundra: A hypothesis with climatic implications." *Journal of Climate,* 14: 336–344.

Su, Y., Y.D. Lei, G. Daly, F. Wania. "Determination of octanol-air partition coefficients (K_{OA}s) for chlorobenzenes and polychlorinated naphthalenes from gas chromatographic retention times." Submitted to *Journal of Chemical and Engineering Data.*

Sullivan, K.A., and R.C. Aller,1996. "Diagenetic cycling of arsenic in Amazon shelf sediments." *Geochimica et Cosmochimica Acta,* 60: 1465–1477.

Sumner, A.L., P.B. Shepson, 1999. "Snowpack production of formaldehyde and its effect on the Arctic troposphere." *Nature,* 398: 230–233.

Suydam, R.S., D.L. Dickson, J.B. Fadely, and L.T. Quakenbush, 2000. "Population declines of King and Common Eiders of the Beaufort Sea." *Condo,* 102: 219–222.

Suzuki, M, Y.Yamato, and T. Watanabe, 1977. "Organochlorine insecticide residues in field soils of the Kitakyushu district — Japan 1970–74." Pest Monitor J., 11: 88–93.

Suzuki, Y., H. Sugaya, and S. Sasaki, 1973. "Results of 3 years tracing investigation of organochlorine pesticide residues in humans." *J. Jpn. Assoc. Rural Med.* 22: 276–280 (in Japanese).

Swackhamer, D.L., R.F. Pearson, and S.P. Schlotter, 1998. "Toxaphene in the Great Lakes." *Chemosphere, 37:* 2547–2561.

Swain, E.B., D.R.Engstrom, M.E. Brigham, T.A. Henning, P.L. Brezonik, 1992. "Increasing rates of atmospheric mercury deposition in midcontinental North America." *Science,* 257: 784–787.

Swift, J.H., E.P. Jones, K. Aagaard, E.C. Carmack, M. Hingston, R.W. Macdonald, F.A. McLaughlin, and R.G. Perkin, 1997. "Waters of the Makarov and Canada basins." *Deep-Sea Research II,* 44: 1503–1529.

Tanabe, S., and R. Tatsukawa, 1980. "Chlorinated hydrocarbons in the North Pacific and Indian Oceans." *Journal of the Oceanographic Society of Japan,* 36: 217–226.

Tanabe, S., and R. Tatsukawa, 1983. "Vertical transport and residence time of chlorinated hydrocarbons in the open ocean water column." *Journal of the Oceanographic Society of Japan,* 39: 533–562.

Tanabe, S., H. Iwata, and, R. Tatsukawa, 1994. "Global contamination by persistent organochlorines and their ecotoxicological impact on marine mammals." *The Science of the Total Environment,* 154: 163–177

Telang, S. A., R. Pocklington, A.S. Naidu, E.A. Romankevich, I.I. Gitelson, and M.I. Gladyshev, 1991. "Carbon and mineral transport in major North American, Russian Arctic, and Siberian rivers: the St. lawrence, the Mackenzie, the Yukon, the Arctic Alaskan rivers, the Arctic basin rivers in the Soviet Union and the Yenisei." In: *Biogeochemistry of Major World Rivers,* E.T. Degens, S. Kempe, J.E. Richey, eds. pp. 75–104, J. Wiley & Sons, New York.

Temman, M., J. Paquette, and H. Vali, 2000. "Mn and Zn incorporation into calcite as a function of chloride aqueous concentration." *Geochimica et Cosmochemica Acta,* 64: 2417–2430.

ten Hulscher Th.E.M., L.E. van der Velde, and W.A. Bruggeman, 1992. "Temperature dependence of henry's law constants for selected chlorobenzenes, polychlorinated biphenyls and polycyclic aromatic hydrocarbons." *Environmental Toxicology and Chemistry,* 11: 1595–1603.

Thompson, D.W.J., and J.M. Wallace,1998. "The Arctic Oscillation signature in the wintertime geopotential height and temperature fields." *Geophysical Research Letters,* 25: 1297–1300.

Tittlemier, S.A., and G.T. Tomy, 2001. "Vapor pressures of six brominated diphenyl ethers." *Environmental Toxicology and Chemistry*, 20: 146–148.

Tomy, G.T., G.A. Stern, D.C.G. Muir, A.T. Fisk, C.D. Cymbalisty, and J.B. Westmore, 1997. "Quantifying C_{10}-C_{13} polychloroalkanes in environmental samples by high resolution gas chromatography/electron capture negative ion high resolution mass spectrometry." *Analytical Chemistry*, 69: 2762–2771.

Tomy, G.T., G.A. Stern, W.L. Lockhart, and D.C.G. Muir, 1999. "Occurrence of C_{10}-C_{13} polychlorinated n–alkanes in Canadian midlatitude and Arctic lake sediments." *Environmental Science and Technology* *33*(17): 2858–2863.

Tratnyek, P.G., and D.L. Macalady, 2000. "Oxidation-reduction reactions in the aquatic environment." pp. 383–419. In: *Property Estimation Methods for Chemicals*. R.S. Boethling and D. Mackay, eds. Lewis, New York.

Tremblay, L.B., L.A. Mysak, and A.S. Dyke, 1997. "Evidence from driftwood records for century-to-millennial scale variations of the high latitude atmospheric circulation during the Holocene." *Geophysical Rresearch Letters,* 24: 2027–2030.

Troll, C., 1966. "Seasonal climates of the earth: The seasonal course of natural phenomena in the different climatic zones of the earth." In: *World Maps of Climatology*. E. Rodenwald and H. Jusatz, eds. Springer–Verlag, Berlin, Germany.

Tullo, A.H., 2000. "Plastic additives steady evolution." *Chemical and Engineering News*, 78(49). December 4.

Tynan, C., and D.P. DeMaster,1997. "Observations and predictions of Arctic climate change: Potential effects on marine mammals." *Arctic,* 50: 308–322.

Tysklind, M., I. Fängmark, S. Marklund, A. Lindskog, L. Thaning, and C. Rappe, 1993. "Atmospheric transport and transformation of polychlorinated dibenzo-*p*-dioxins and dibenzofurans." *Environmental Science and Technology,* 27: 2190–2197.

USEPA, (United States Environmental Protection Agency), 1986. "Information on Lead, Mercury, PCBs, BaP, DDT, Dieldrin/Aldrin, Toxaphene, Mirex, TCDD, TCDF," *Federal Register.*

USEPA, (United States Environmental Protection Agency), 1991. *Federal Registry,* 56 (20), 3526–3597.

Ueda, K., S. Nanbu, T. Yoshida, R. Tatsukawa, M. Fujii, and K. Oikawa, 1976. "The utilization and environmental levels of certain chemical substances. II. Benzene hexachloride (BHC) and dichlorodiphenyltrichloroethane (DDT)." *J Environ Pollut Control,* 12: 1419–1429.

Ulrich, E.M., and R.A. Hites. 1998. "Enantiomer ratios of chlordane-related compounds in air near the Great Lakes." *Environmental Science and Technology,* 32: 1870–1874.

UN, (United Nations), 1994. "Consolidated List of Products Whose Consumption and/or Sale have been Banned, Withdrawn, Severely Restricted or not Approved by Governments," 5th Issue, Department for Policy Coordination and Sustainable Development, United Nations, New York, 935 pp.

UNEP, (United Nations Environment Program), 1996. "UNEP Survey on Sources of POPs." A report prepared for an IFCS Expert Meeting on Persistent Organic Pollutants, Manila, the Philippines, 17–19 June 1996. http://www.chem.unep.ch/pops/indhtms/manexp3/html.

UNEP, (United Nations Environment Program), 2001. "Final Act of the conference of plenipotentiaries on the stockholm convention on persistent organic pollutants." UNEP, Geneva Switzerland, 44 pp.

United States Army Corp of Engineers (U.S. ACE), 2001. "Remedial Investigation Report for the Former Schenectady Army Depot — Voorheesville Area."

UNSCEAR, (United Nations Scientific Committee on the Effects of Atomic Radiation), 2000. "Sources and Effects of Ionizing Radiation. Volume 1 Sources, Volume II Effects." Report E.00.IX.4, New York, United Nations.

Van de Meent, D., T.E. McKone, T. Parkerton, M. Matthies, M. Scheringer, F. Wania, R. Purdy, D.H. Bennett, 2001. "Persistence and transport potential of chemicals in a multimedia environment." pp. 169–204. In: *Evaluation of Persistence and Long-Range Transport of Organic Chemicals in the Environment,* G.Kle_ka, and D. Mackay eds. Proceedings of a SETAC Pellston Workshop, Fairmont Hot Springs, BC, Canada, 13–19 July, 1998, SETAC Press, Pensacola, FL.

Van Oostdam, J., A. Gilman, E. Dewailly, P. Usher, B. Wheatley, H. Kuhnlein, S. Neve, J. Walker, B. Tracy, M. Feeley, V. Jerome, and B. Kwavnick, 1999. "Human health implications of environmental contaminants in Arctic Canada: a review." *The Science of the Total Environment,* 230: 1–82.

van Pul, W.A.J., F.A.A.M. de Leeuw, J.A. van Jaarsveld, M.A. van der Gaag, C.J. Sliggers, 1998. "The potential for long-range transboundary atmospheric transport." *Chemosphere,* 37: 113–141.

Vandal, G.M., W.F. Fitzgerald, C.F. Boutron, and J.P. Candelone, 1993. "Variations in mercury deposition to AntArctica over the past 34,000 years." *Nature,* 362: 621–623.

Vanegas, S.A., and L.A. Mysak, 2000. "Is there a dominant timescale of natural climate variability in the Arctic?" *Journal of Climate,* 13: 3412–3434.

Verigina, K.V., 1964. "The role of microelements (Zn, Cu, Co and Mo) in the life of plants and their content in soil and rocks." pp. 3–26, Chapter 1. In: *Microelements in Some Soils of the USSR,* Nauka, Moscow

Vetter, W., 1993. "Toxaphene: theoretical aspects of the distribution of chlorinated bornanes including symmetry aspects." *Chemosphere,* 26, 1079–1084.

Vetter, W., C. Natzeck, B. Luckas, G. Heidemann, B. Kiabi, and M. Karami, 1995. "Chorinated Hydrocarbons in the Blubber of A Seal (*Phoca caspica*) from the Caspian Sea." *Chemosphere,* 30:1685–1696.

Vetter, W., R. Bartha, G.A. Stern, G.T. Tomy, 1999. "Enantioselective determination of two persistent chlorobornane congeners in sediment from a toxaphene-treated Yukon lake." *Environmental Toxicology and Chem*istry, 18: 2775–2781.

Vibe, C., 1967. *Arctic Animals in Relation to Climatic Fluctuations.* C.A. Reitzels Forlag, Copenhagen.

Viessman, W., and M.J. Hammer, 1985. "Biological treatment processes." pp. 428–568. In: *Water Supply and Pollution Control,* H. a. Row, (ed.) New York, New York.

Villeneuve, D.L., K. Kannan, J.S. Khim, J. Falandysz, V.A. Nikiforov, A.L. Blankenship, and J.P. Giesy, 2000. "Relative potencies of individual polychlorinated naphthalenes to induce dioxin-like responses in fish and mammalian *in vitro* bioassays." *Archives of Environmental Contamination and Toxicology,* 39: 273–281.

Vinnikov, K.Y., A. Robock, R.J. Stouffer, J.E. Walsh, C.L. Parkinson, D.J. Cavalieri, J. F.B. Mitchell, D. Garrett, and V.F. Zakharov, 1999. "Global warming and northern hemisphere sea ice extent." *Science,* 286: 1934–1937.

Vinogradov, A.P., 1957. "Geochemistry of Sparse and Dissipated Elements in Soil." *Academia Nauk,* Moscow.

Vinogradova, A.A., 1996. "Elemental composition of atmospheric aerosol in eastern Arctic regions, Izvestiya." *Atmospheric and Oceanic Physics,* 32(4): 440–447

Vintro, L.L., C.A. McMahon, P.I. Mitchell, D. Josefsson, E. Holm, and P. Roos, 2002. "Transport of plutonium in surface and subsurface waters from the Arctic shelf to the North Pole via the Lomonosov Ridge." *Journal of Environmental Radioactivity,* 60(1–2): 73–90.

Voldner, E.C., and L. Smith, 1986. "Production, usage and atmospheric emissions of 14 priority toxic chemicals." In: *Proceedings of a Workshop on Great Lakes Atmospheric Deposition,* International Joint Commission, Scarborough, Ontario, Oct. 29–31, Appendix 2.

Voldner, E.C., and G. Ellenton, 1987. "Production, usage and atmospheric emissions of priority toxic chemicals with emphasis on North America," prepared for the International Joint Commissions, Environment Canada, Atmospheric Environment Service, Report: ARD-88-4, 215 p.

Voldner, E.C., and Y.F. Li, 1993. "*Global usage of toxaphene.*" Chemosphere, 10: 2073–2078.

Voldner, E.C., and Y.F. Li, 1995. "*Global usage of selected persistent organochlorines.*" *The Science of the Total Environment*, 160/161: 201–210.

Voldner, E.C., and W.H. Schroeder, 1989. "Modelling of Atmospheric transport and deposition of toxaphene into the Great Lakes ecosystem." *Atmospheric Environment*, 23: 1949–1961.

von Sydow, L.M., A.T. Nielsen, A.B. Grimvall, H.B Borén, 2000. "Chloro- and bromoacetates in natural archives of firm from AntArctica." *Environmental Science and Technology*, 34: 239–245.

Vörösmarty, C.J., L.D. Hinzman, B.J. Peterson, D.H. Bromwich, L.C. Hamilton, J. Morison, V.E. Romanovsky, M. Sturm, and R.S. Webb, 2001. "The Hydrologic Cycle and its Role in the Arctic and Global Environmental Change: A Rationale and Strategy for Synthesis Study." *Arctic Research Consortium of the US*, Fairbanks, Alaska, 84 pp.

Vos, J.G., and M.I. Luster, 1989. "Immune alterations." pp. 295–322. In: *Halogenated biphenyls, terphenyls, naphthalenes, dibenzodioxins and related products.* R. D. Kimbrough and S. Jensen, (eds.). Elsevier Science Publishers, Amsterdam.

Wadhams, P., 1997. "Variability of Arctic sea ice thickness — statistical significance and its relationship to heat flux." pp. 368–384. In: *Operational Oceanography. The Challenge for European Co-operation.* J.H. Stel, H.W.A. Behrens, J.C. Borst, L.J. Droppert, and J.v.d. Meulen, (eds.). Elsevier Science.

Wadhams, P., 2000. "Ice in the Ocean." Gordon and Breach Science Publishers.

Wagemann, R, W.L. Lockhart, H. Welch, S. Innes, 1995. "Arctic marine mammals as integrators and indicators of mercury in the Arctic." *Water, Air, and Soil Pollution*, 80: 683–693.

Wagemann, R., S. Innes, and P.R. Richard, 1996. "Overview of regional and temporal differences of heavy metals in Arctic whales and ringed seals in the Canadian Arctic." *The Science of the Total Environment,* 186: 41–66.

Wagemann, R., E. Trebacz, R. Hunt, and G. Boila,1997. "Percent methylmercury and organic mercury in tissues of marine mammals and fish using different experimental and calculation methods." *Environmental Toxicology and Chemistry,* 16:1859–1866.

Waite, D., 2002. *personal communication.* Environment Canada.

Waite, D.T., N.P. Gurprasad, J.F. Sproull, D.V. Quiring, and M.W. Kotylak, 2000. "Atmospheric Movements of Lindane (γ-hexachlorocyclohexane) from Canola Fields Planted With Treated Seed." *Journal of Environmental Quality* 30:768–775.

Walker, C.E., J.E. Bryan, and R.F. Brown, 1973. "Report on Rainbow Trout Planting and Lake Survey Program in Yukon Territories 1956–1971." Environment Canada, Fisheries and Marine Services, Northern Operations Branch, Pacific Region, PCA/T-73-12, 25 pp.

Walker, K, D.A. Vallero, and R.G. Lewis, 1999. "Factors influencing the distribution of lindane and other hexachlorocyclohexanes in the environment." *Environmental Science and Technology*, 33: 4373–4378.

Wallace, J.M., and D.W.J. Thompson, 2002. "Annual modes and climate prediction." *Physics Today*, February: 28–33.

Walsh, J.E., 1991. "The Arctic as a bellwether." *Nature,* 352: 19–20.

Walsh, J.E., 1995. "Long-term observations for monitoring of the cryosphere." *Climatic Change*, 31: 369–394.

Walsh, J.E., 2000. "Global atmospheric circulation patterns and relationships to Arctic freshwater fluxes." pp. 21–43. In: *The Freshwater Budget of the Arctic Ocean.* E.L. Lewis, E.P. Jones, P. Lemke, T.D. Prowse, and P. Wadhams (eds.). Kluwer Academic Publishers, Netherlands.

Walsh, J.E., and R.G. Crane, 1992. "A comparison of GCM simulations of Arctic climate." *Geophysical Research Letters*, 19: 29–32.

Walsh, J.E., and W.L. Chapman, 2000. "20th-century sea ice variations from observational data." *Annals of Glaciology*, 33: 444–448.

Walsh, J.E., X. Zhou, D. Portis, and M.C. Serreze, 1994. "Atmospheric contribution to hydrologic variations in the Arctic." *Atmosphere-Ocean*, 32(4): 733–755.

Wang, J., and M. Ikeda, 2001. "Arctic sea-ice oscillation: regional and seasonal perspectives." *Annals of Glaciology*, 33: 481–492.

Wania, F., 1997. "Modelling the behaviour of non-polar organic chemicals in an ageing snowpack." *Chemosphere*, 35: 2345–2363.

Wania, F., 1998a. "The potential of organic chemicals for long-range transport and deposition in polar regions based on calculations with a global distribution model." WECC–Report 2/98, August 1998, 30 pp.

Wania, F., 1998b. "An integrated criterion for the persistence of organic chemicals based on model calculations." WECC-Report 1/98, July 1998, 31 pp.

Wania, F., 1999a. "On the origin of elevated levels of persistent chemicals in the environment." *Environmental Science and Pollution Research,* 6: 11–19.

Wania, F., 1999b. "Global modelling of polychlorinated biphenyls." WECC–Report 1/1999, June 1999, 23 pp.

Wania, F., 2001. "Quantifying and reducing uncertainty in model calculations of global pollutant fate." pp. 138–143. In: *Synopsis of Research Conducted under the 1998/99 Northern Contaminants Program.* S. Kalhok, (ed.), Indian and Northern Affairs Canada, Ottawa, ON.

Wania, F., and G.L. Daly, 2002. "Estimating the contribution of degradation in air and deposition to the deep sea to the global loss of PCBs." *Atmospheric Environment,* 36: 5581–5593.

Wania, F., and C.B. Dugani, 2003. "Assessing the Long-Range Transport Potential of Polybrominated Diphenyl Ethers: A Comparison of Four Multimedia Models." *Environmental Toxicology and Chemistry,* in press.

Wania, F., and D. Mackay, 1993a. "An approach to modelling the global distribution of toxaphene: A discussion of feasibility and desirability." *Chemosphere,* 27: 2079–2094.

Wania, F., and D. Mackay, 1993b. "Global fractionation and cold condensation of low volatility organochlorine compounds in polar regions." *Ambio,* 22:10–18.

Wania, F., and D. Mackay, 1995. "A global distribution model for persistent organic chemicals." *The Science of the Total Environment,* 160/161: 211–232.

Wania F., and D. Mackay, 1996. "Tracking the distribution of persistent organic pollutants." *Environmental Science and Technology,* 30: A390–A396.

Wania F., and D. Mackay, 1999a. "Global chemical fate of α-hexachlorocyclohexane. 2. Use of a global distribution model for mass balancing, source apportionment, and trend prediction." *Environmental Toxicology and Chemistry,* 18: 1400–1407.

Wania, F., and D. Mackay, 1999b. "The evolution of mass balance models of persistent organic pollutant fate in the environment." *Environmental Pollution,* 100: 223–240.

Wania, F., and D. Mackay, 2000a. "The global distribution model. A non-steady state multi-compartmental mass balance model of the fate of persistent organic pollutants in the global environment." *Technical Report and Computer Program,* 21 p. (www.utsc.utoronto.ca/~wania).

Wania, F., and D. Mackay, 2000b. "A comparison of overall persistence values and atmospheric travel distances calculated by various multi-media fate models." *WECC-Report 2/2000.* July 2000, 42 p.

Wania, F., and M.S. McLachlan, 2001. "Estimating the influence of forests on the overall fate of semi-volatile organic compounds using a multimedia fate model." *Environmental Science and Technology,* 35: 582–590.

Wania, F., J.T. Hoff, C.Q. Jia, and D. Mackay, 1998. "The effects of snow and ice on the environmental behaviour of hydrophobic organic chemicals." *Environmental Pollution.* 102: 25–41.

Wania, F., D. Mackay, and J.T. Hoff, 1999a. "The importance of snow scavenging of polychlorinated biphenyl and polycyclic aromatic hydrocarbon vapors." *Environmental Science and Technology,* 33: 195–197.

Wania, F., D. Mackay, M.S. McLachlan, A. Sweetman, and K.C. Jones, 1999b. "Global modelling of polychlorinated biphenyls." pp. 61–70. In: *Synopsis of Research Conducted under the 1998/99 Northern Contaminants Program.* S. Kalhok (ed.). Indian and Northern Affairs Canada, Ottawa, ON.

Wania, F., D. Mackay, Y–F. Li, T.F. Bidleman, and A. Strand, 1999c. "Global chemical fate of -hexachlorocyclohexane. 1. Evaluation of a global distribution model." *Environmental Toxicology and Chem*istry, 18: 1390–1399.

Wania, F., R. Semkin, J.T. Hoff, and D. Mackay, 1999d. "Modelling the fate of non-polar organic chemicals during the melting of an Arctic snowpack." *Hydrological Processes,* 13: 2245–2256.

Wania, F., Y.D. Lei, G. Daly, and M.S. McLachlan, 2000. "Quantifying and reducing uncertainty in model calculations of global pollutant fate." pp. 142–152. In: *Synopsis of Research Conducted under the 1999–2000 Northern Contaminants Program*. S. Kalhok, (ed.). Indian and Northern Affairs Canada, Ottawa, ON.

Wania, F., Y.D. Lei, and A. Wong, 2001. "Quantifying and reducing uncertainty in model calculations of global pollutant fate." pp. 138. In: *Synopsis of research conducted under the 2000–2001 Northern Contaminants Program.* S. Kalhok, (ed.). Indian and Northern Affairs Canada, Ottawa, ON.

Wassmann, P., M. Vernet, B.G. Mitchell, and F. Reyh, 1990. "Mass sedimentation of *Phaeocystis pouchettii* in the Barents Sea." *Marine Ecology Progress Series,* 66: 183–195.

Wassmann, P., 2001. "Vernal export and retention of biogenic matter in the north-eastern North Atlantic and adjacent Arctic Ocean: the role of the Norwegian Atlantic Current and toopography." *Memoirs of the National Institute of Polar Research, Special Issue*, 54: 377–392.

Watanabe, I., M. Kawano, Y. Wang, and R. Tatsukawa, 1992. "Polybrominated Dibenzo-*p*-dioxins (PBDD) and -dibenzofurans (PBDFs) in atmospheric air in Taiwan and Japan." *Organohalogen Cmpounds*, 9: 309–312.

Watanabe, O., and R. Tatsukawa, 1990. "Anthropogenic brominated aromatics in Japanese environment." In: (Kemi 1990) *Proceedings of Workshop on brominated aromatic flameretardants.* Swedish Chemical Inspectorate, Solna, Sweden.

Watanabe, O., T. Kashimoto, and R. Tatsukawa,1987. "Polybrominated biphenyl ethers in marine fish, shellfish and river and marine sediments in Japan." *Chemosphere,* 16: 2389–2396.

Weatherhead, E.C., and C.M. Morseth, 1998. "Climate change, ozone, and ultraviolet radiation." *AMAP Assessment Report: Arctic Pollution Issues. Arctic Monitoring and Assessment Programme (AMAP), Oslo, Norway, xii +859 pp.*

Webster, E., D. Mackay, and F. Wania, 1998. "Evaluating environmental persistence." *Environmental Toxicology and Chemistry*, 17: 2148–2158.

Wegmann, F., 1999. Untersuchung von α–HCH und DDT bezüglich Ferntransport, "Grass-Hopper-Effect" und Depositions- und Revolatilisierungs-Gleichgewicht. Diploma thesis, Department of Chemistry, ETH Zürich.

Welch, H.E., 1995. "Marine conservation in the Canadian Arctic: A regional overview." *Northern Perspectives*, 23: 5–17.

Welch, H.E., 1998. "A year on the ice: The SHEBA/JOIS Project." *Arctic*, 51: 293–300.

Welch, H.E., D.C.G. Muir, B.N. Billeck, W.L. Lockhart, G.J. Brunskill, H.J. Kling, M.P. Olson, and R.M. Lemoine, 1991. "Brown snow: A long-range transport event in the Canadian Arctic." *Environmental Science and Technology*, 25: 280–286.

Welch, H.E., M.A. Bergmann, T.D. Siferd, K.A. Martin, M.F. Curtis, R.E. Crawford, R. J. Conover, and H. Hop, 1992. "Energy flow through the marine ecosystem of the Lancaster Sound region, Arctic Canada." *Arctic,* 45: 343–357.

Weller, G., and M. Lange, 1999. "Impacts of Global Climate Change in the Arctic Regions." *International Arctic Science Committee*, Center for Global Change and Arctic System Research, University of Alaska, University of Alaska, Fairbanks. pp. 1–59.

Whittle, D.M., R.M. Kiriluk, A.A. Carswell, M.J. Keir, and D.C. MacEachen, 2000. "Toxaphene congeners in the Canadian Great Lakes basin: temporal and spatial food web dynamics." *Chemosphere,* 40: 1221–1226.

Wiberg, K., R.J. Letcher, C.D. Sandau, R.J. Norstrom, M. Tysklind and T.F. Bidleman. 2000. "The enantioselective bioaccumulation of chiral chlordane and α-HCH contaminants in the polar bear food chain." *Environmental Science and Technology*, 34: 2668–2674.

Wiberg, K., T. Harner, J.L. Wideman and T.F. Bidleman, 2001. "Chiral analysis of organochlorine pesticides in Alabama soils." *Chemosphere*, 45: 843–848.

Wideqvist, U., 1999. "Analysis of Complex Mixtures of some Polychlorinated Organic Compounds in the Environment." *Ph.D. dissertation*, Stockholm University, 379 p.

Wiig, Ø., A. Derocher, and S. Belikov, 1999. "Ringed seal (*Phoca hispida*) breeding in the drifting pack ice of the Barents Sea." *Marine Mammal Science,* 15: 595–598.

Wilkening, K.E., L.A. Barrie, and M. Engle, 2000. "Trans-Pacific air pollution." *Science,* 290: 65–66.

Willett, K, E. Ulrich, and R.A. Hites, 1998. "Differential toxicity and environmental fate of hexachlorocyclohexane isomers." *Environmental Science and Technology*, 32: 2197–2207.

Williams, D.T., B. Kennedy, and G.L. LeBel, 1993. "Chlorinated naphthalenes in human adipose tissue from Ontario municipalities." *Chemosphere*, 27: 795–806.

Willis, B., M.J. Crookes, J. Diment, and S.D. Dobson, 1994. "Environmental hazard assessment: chlorinated paraffins." *Toxic Substances Division*. UK Dept. of the Environment.

Witt, K., and K.H. Niessen, 2000. "Toxaphenes and chlorinated naphthalenes in adipose tissue of children." *Journal of Pediatric Gastroenterology and Nutrition*, 30: 164–169.

Wolfe, D.A., M.J. Hameedi, J.A. Galt, G. Watabayashi, J. Short, C. O'Claire, S. Rice, J. Michel, J.R. Payne, J. Braddock, S. Hanna, and D. Sale, 1994. "The fate of the oil spilled from the Exxon Valdez." *Environmental Science and Technolgy*, 28: 561A–568A.

Wong, A., Y.D. Lei, M. Alaee, and F. Wania, 2001. "Vapor pressures of the polybrominated diphenylethers." *Journal of Chemical and Engineering Data*, 46: 239–242.

Woodgate, R.A., K. Aagaard, R.D. Muench, J. Gunn, G. Björk, B. Rudels, A.T. Roach, and U. Schauer, 2001. "The Arctic Ocean boundary current along the Eurasian slope and adjacent Lomonosov Ridge: Water mass properties, transports and transformations from moored instruments." *Deep-Sea Research I*, 48: 1757–1792.

Woollett, J.M., A.S. Henshaw, and C.P. Wake, 2000. "Palaeoecological implications of archaeological seal bone assemblages: case studies from Labrador and Baffin Island." *Arctic*, 53: 395–413.

Worthing, C.R., 1983. "The Pesticide Manual." A World Compendium, 7th ed. The British Crop Protection Council, Lavenham Press Ltd., Lavenham, Suffolk.

Worthy, D.E.J., N.B.A. Trivett, and C. Blanchette, 1999. "Observatory." pp. 1–1–1–18. In: *Canadian Baseline Program Summary of Progress to 1998. An overview of the measurements underway in the Canadian Baseline Program, and results to date.* M. Racki, (ed.). January, 1999, Toronto, Ontario, Canada.

Wu, W.Z., K.W. Schramm, B. Henkelmann, Y. Xu, A. Yediler, and A. Kettrup, 1997. "PCDD/Fs, PCBs, HCHs and HCB in sediments and soils of Ya–Er lake area in China: Results on residual levels and correlation to the organic carbon and the particle size." *Chemosphere*, 34: 191–202.

Xiao, Z.F., J. Munthe, W.H. Schroeder, and O. Lindqvist, 1991. "Vertical fluxes of volatile mercury over forest soil and lake surfaces in Sweden." *Tellus*, 43(B): 267–279.

Yablokov, A.V., V.K. Karasev, V.M. Rumyantsev, M. Kokeyev, O.I. Petrov, V.N. Lystov, and A.F. Melyanenkov, 1993. "Russian Federation state report on sea burial of radioactive wastes." *Admin. Pres. Russ. Fed.*, Moscow, Russia, 108 pp.

Yamashita, N., K. Kannan, T. Imagawa, A. Miyazaki, and J.P. Giesy, 2000. "Concentrations and profiles of polychloronaphthalene congeners in eighteen technical polychlorinated biphenyl preparations." *Environmental Science and Technology*, 34: 4236–4241

Yunker, M.B., R.W. Macdonald, W.J. Cretney, B.R. Fowler, and F.A. McLaughlin, 1993. "Alkane, terpene, and polycyclic aromatic hydrocarbon geochemistry of the Mackenzie River and Mackenzie Shelf: Riverine contributions to Beaufort Sea coastal sediment." *Geochimica et Cosmochimica Acta*, 57: 3041–3061.

Yunker, M.B., R.W. Macdonald, D.J. Veltkamp, and W.J. Cretney, 1995. "Terrestrial and marine biomarkers in a seasonally ice-covered Arctic estuary — Integration of multivariate and biomarker approaches." *Marine Chemistry,* 49: 1–50.

Yunker, M.B., S.M. Backus, E. Graf Pannatier, D.S. Jeffries, and R.W. Macdonald, 2002. "Sources and significance of alkane and PAH hydrocarbons in Canadian Arctic rivers." *Estuarine and Coastal Shelf Science,* 55, 1–31.

Zedeck, M.S., 1980. "Polycyclic aromatic hydrocarbons. A review." *Journal of Environmental Pathology and Toxicology*, 3: 537–567.

Zell. M., and K. Ballschmiter,1980. "Baseline studies of the global pollution. II. Global occurrence of hexachlorocyclohexane (HCB) and polychlorocamphenes (toxaphene) (PCC) in biological samples." *Fresenius' Journal of Analytical Chemistry,* 300: 387–402.

Zhang, J., D. Rothrock, and M. Steele, 2000. "Recent changes in the Arctic sea ice: The interplay between ice dynamics and thermodynamics." *Journal of Climate,* 13: 3099–3114.

Zhang, T., R.G. Barry, K. Knowles, J.A. Heginbottom, and J. Brown. 1999a. "Statistics and characteristics of permafrost and ground-ice distribution in the Northern Hemisphere." *Polar Geography*, 2: 132–154.

Zhang, X., K.W. Schramm, B. Hinkelmann, C. Klimm, A. Kaune, A. Kettrup, and P. Lu, 1999b. "A method to estimate the octanol-air partition coefficient of semi-volatile organic compounds." *Analytical Chemistry,* 71: 3834–3838.

Zhang, X., S.A. Naidu, J.J. Kelley, S.C. Jewett, D. Dasher, and L.K. Duffy, 2001. "Baseline concentrations of total mercury and methylmercury in salmon returning via the Bering Strait." *Marine Pollution Bulletin,* 42: 993–997.

Zhulidov, A.V., J.V. Headley, D.F. Pavlov, R.D. Robarts, L.G. Korotova, V.F.O.V. Fadeev, Y. Volovik, and V. Khlobystov,1998. "Distribution of organochlorine insecticides in rivers of the Russian Federation." *Journal of Environmental Quality,* 27: 1356–1366.

Zhulidov, A.V., R.D. Robarts, J.V. Headley, K. Liber, D.A. Zhulidov, A.V Zhulidova, and D.F. Pavlov, 2002. "Levels of DDT and hexachlorocyclohexanes in burbot (Lota lota) from Russian Arctic rivers." *The Science of the Total Environment*, 293: 231–246.

Zwiers, F.W., 2002. "The 20-year forecast." *Nature,* 416: 690–691.